The Organic Chemistry of Nickel

VOLUME II

Organic Synthesis

ORGANOMETALLIC CHEMISTRY
A Series of Monographs

EDITORS

P. M. MAITLIS
THE UNIVERSITY
SHEFFIELD, ENGLAND

F. G. A. STONE
UNIVERSITY OF BRISTOL
BRISTOL, ENGLAND

ROBERT WEST
UNIVERSITY OF WISCONSIN
MADISON, WISCONSIN

BRIAN G. RAMSEY: Electronic Transitions in Organometalloids, 1969.

R. C. POLLER: The Chemistry of Organotin Compounds, 1970.

RUSSELL N. GRIMES: Carboranes, 1970.

PETER M. MAITLIS: The Organic Chemistry of Palladium, Volumes I and II, 1971.

DONALD S. MATTESON: Organometallic Reaction Mechanisms of the Nontransition Elements, 1974.

RICHARD F. HECK: Organotransition Metal Chemistry: A Mechanistic Approach, 1974.

P. W. JOLLY AND G. WILKE: The Organic Chemistry of Nickel, Volume I, Organonickel Complexes, 1974. Volume II, Organic Synthesis, 1975.

P. C. WAILES, R. S. P. COUTTS, AND H. WEIGOLD: Organometallic Chemistry of Titanium, Zirconium, and Hafnium, 1974.

U. BELLUCO: Organometallic and Coordination Chemistry of Platinum, 1974.

P. S. BRATERMAN: Metal Carbonyl Spectra, 1974.

L. MALATESTA AND S. CENINI: Zerovalent Compounds of Metals, 1974.

THOMAS ONAK: Organoborane Chemistry, 1975.

R. P. A. SNEEDEN: Organochromium Compounds, 1975.

The Organic Chemistry of Nickel

P. W. Jolly and G. Wilke

Max-Planck-Institut für Kohlenforschung
Mülheim-Ruhr, Germany

VOLUME II
Organic Synthesis

ACADEMIC PRESS New York San Francisco London 1975
A Subsidiary of Harcourt Brace Jovanovich, Publishers

ACADEMIC PRESS, INC.
111 Fifth Avenue, New York, New York 10003

United Kingdom Edition published by
ACADEMIC PRESS, INC. (LONDON) LTD.
24/28 Oval Road, London NW1

Library of Congress Cataloging in Publication Data

Jolly, P W
    The organic chemistry of nickel.

    (Organometallic chemistry)
    Includes bibliographical references.
    CONTENTS: v. 1.   Organonickel complexes.–
v. 2.   Organic synthesis.
    1.   Organonickel compounds   I.   Wilke, Günther,
(date)   joint author.   II.   Title.
QD412.N6J64   547'.05'6252   73-19711
ISBN 0–12–388402–0 (v. 2)

IN MEMORIAM

Karl Ziegler

(1898–1973)

Contents

Chapter I. **The Oligomerization of Olefins and Related Reactions**

Chapter II. **The Oligomerization of Alkynes and Related Reactions**

Chapter VI. **Carbonylation and Related Reactions**

Preface

Organonickel chemistry began in 1890 with the isolation of nickel tetra-carbonyl, gained momentum with the discovery of the Reppe catalysts in 1940 and nickelocene in 1953, and obtained its present character in 1960 with the discovery of cyclododecatriene nickel. The research devoted to this field has found an outlet in some 3500 primary publications, and this number is increasing at the rate of approximately twelve each week. These figures are modest in comparison with those for other transition metals, but do emphasize the need which exists for a series of volumes devoted to the organic chemistry of the transition metals.

There is more than one system that could be used to organize the subject matter, and, in adopting that based on the individual elements, the editors have chosen a system the success of which can only be judged as the series progresses. Our task has been made easier by the publication of the volumes devoted to palladium, and the usefulness of both books will increase still further with the appearance of the work on platinum which is in preparation. The extent of the relevant literature necessitates division of the topic into two volumes devoted to a description of the chemistry of the organonickel complexes and to the use of nickel in organic synthesis.

We feel that these volumes appear at a timely moment in the development of the organic chemistry of nickel. Until recently, chemists have been mainly concerned with the synthesis and reactions of organometal complexes, and the contribution of quantitative theoretical work has been disappointingly small. However, in the last few years the first sophisticated attempts have been made to describe the bonding in some of these systems, and it can be anticipated that the situation will change rapidly. Organic synthesis using organonickel complexes has also reached a transition state. The first generation of organonickel catalysts, the mononuclear catalysts, has been well explored, and chemists are now turning their attention to what may well be the second generation of catalysts, the multinuclear catalysts.

It is doubtful whether it would be possible today for practicing chemists to write any book of this nature quickly enough for it to be useful without considerable peripheral help, and we wish to acknowledge the assistance of Frau M. Pauling, Frl. R. Pupka, Herrn H. Nussbicker, and Herrn H. Schmitz for collecting and photocopying the relevant literature; Herrn K. H. Boll, Dr. E. Illmeier, and Dr. E. Ziegler for developing and programming the computer-controlled system used to process the literature; our colleagues at this Institute for many constructive suggestions; and, in particular, Dr. H. D. Empsall for reading and criticizing the final manuscript.

P. W. JOLLY

G. WILKE

Introduction

This volume of "The Organic Chemistry of Nickel" is concerned with the use of nickel in organic synthesis. In general, we have limited the discussion to reactions involving homogeneous catalysts. Heterogeneous systems have been included only where necessary to give balance, and stoichiometric reactions only where they have found general synthetic application. Of the available methods for organizing the material, that based upon the nature of the active catalyst would be most illuminating; unfortunately, the gaps in our knowledge prevent this, and instead we have adopted a more conventional layout, devoting separate chapters to the reactions of olefins, alkynes, and 1,3-dienes as well as to coupling and carbonylation reactions. The only important topic which falls outside both the scope of this treatment and the experience of the authors is the use of nickel in "template" syntheses, and the reader is referred to references 1–3 for authoritative accounts.

Transition-metal catalyzed reactions do not lend themselves to conventional mechanistic investigations, and, as a result, this aspect has been neglected. The formulation of a convincing mechanism requires the use of product analysis and kinetic data combined with the isolation of intermediate organometallic complexes and the preparation and study of model compounds, preferably such in which the metal atom is the same as that used in the catalysis. Although the reactions involving nickel have been more thoroughly investigated than most, many of the proposed mechanisms are largely speculative. Nevertheless, evidence has accumulated which indicates that these reactions proceed by a series of oxidative addition–reductive elimination steps which are instigated by changes in the coordination sphere around the nickel atom. Most attention has been given, however, to the fate of the organic ligands, and a deeper understanding requires a more precise knowledge of the effect of coordinating ligands both on the electronic state of the metal atom and on the geometry of the intermediates formed.

The literature is based on *Chemical Abstracts*. A short list of the more important general review articles has been included at the end of each chapter; a more comprehensive list, essentially complete through 1972, has been compiled by M. I. Bruce (4). We would be grateful if important errors or omissions were brought to our attention so that in one center, at least, a complete account of the organic chemistry of nickel is available.

P. W. JOLLY
G. WILKE

References

1. N. F. Curtis. *Coord. Chem. Rev.* **3,** 3 (1968).
2. L. F. Lindoy and D. H. Busch. *Prep. Inorg. React.* **6,** 1 (1971).
3. D. St. C. Black and A. J. Hartshorn. *Coord. Chem. Rev.* **9,** 219 (1972/1973).
4. M. I. Bruce. *Advan. Organometal. Chem.* **12,** 379 (1974); *Advan. Organometal. Chem.* **11,** 447 (1973); *Advan. Organometal. Chem.* **10,** 273 (1972).

Contents of Volume I

CHAPTER I

The Oligomerization of Olefins and Related Reactions

I. Introduction

The discovery in 1952 by K. Ziegler and his co-workers that ethylene reacts with triethylaluminum in the presence of nickel to produce 1-butene has had such important consequences that the historical development is worth recalling (46, 195–201, P148*).

In 1949 it had been discovered that long-chain aluminum-alkyls could be produced by the repeated insertion of ethylene molecules into the Al–C bond. This growth reaction terminated, however, after the statistical insertion of

$$\text{al}-CH_2CH_3 + CH_2{=}CH_2 \longrightarrow \text{al}-CH_2CH_2CH_2CH_3 \xrightarrow{nCH_2:CH_2}$$

$$\text{al}-CH_2CH_2(CH_2CH_2)_nCH_2CH_3 \longrightarrow \text{al}-H + CH_2{=}CH(CH_2CH_2)_nCH_2CH_3$$

$$\text{al}-H + CH_2{=}CH_2 \longrightarrow \text{al}-CH_2CH_3$$

approximately 100 olefin molecules. At the end of a long series of experiments a new and initially unwelcome effect suddenly made its appearance: termination occurred after one insertion step and 1-butene was produced practically quantitatively. A search for the origin of this phenomenon showed that it was associated with traces of nickel salts present in the autoclave,

$$\text{al}-CH_2CH_2CH_2CH_3 + CH_2{=}CH_2 \xrightarrow{[Ni]} \text{al}-CH_2CH_3 + CH_2{=}CHCH_2CH_3$$

* A number preceded by the letter "P" refers to the Patent literature at the end of the chapter.

which was made of chrome–nickel steel and, in contrast to normal practice, had been cleaned with nitric acid. This effect became known as the "nickel effect." As it was suspected that metallic cocatalysts may have caused premature chain termination in the original insertion reactions, a systematic study of the effect of other transition metals was undertaken and resulted, in the autumn of 1953, in the discovery of the Ziegler catalysts and the low-pressure polymerization of ethylene.

The discovery of the nickel effect was a turning point in organonickel chemistry. Not only was the combination nickel salt–aluminum-alkyl to become a standard catalyst in organic synthesis but also the recognition that traces of acetylene were necessary to prolong the life of the catalyst was an early indication of the important role that would be played in these reactions by molecules capable of acting as ligands.

The nickel effect subsequently attracted surprisingly little attention. However, the commercial interest in low molecular weight olefins and their oligomerization and polymerization products has led to an enormous activity in the field of Ziegler catalysis, and a variety of nickel-based Ziegler catalysts are now known. Although of no significance in olefin polymerization, these systems are particularly versatile for the dimerization of mono-olefins, in particular of propylene. A recent aspect that may become of more significance is the synthesis of optically active codimers using catalysts modified by optically active phosphines.

In the discussion that follows, we have devoted separate sections to the nickel effect, to the oligomerization and co-oligomerization of olefins, and to their polymerization. These are followed by short accounts of the mechanistically related isomerization and hydrogenation of olefins and the hydrosilylation and hydrocyanation reactions.

A list of important review articles concerned with the transition metal-catalyzed oligomerization of olefins in general is to be found at the end of the chapter.

II. The Nickel Effect (Table I-1)

The nature and discovery of the nickel effect have been mentioned in the Introduction to this chapter. A convenient method for preparing the catalyst consists of adding diethylaluminum ethoxide to a mixture of nickel acetylacetonate and phenylacetylene (1:30) in xylene and adding the resulting orange-red solution to triethylaluminum. When this mixture is used it is possible to convert ethylene into butene in over 90% yield under relatively mild conditions.

In spite of its historical importance, the reaction has been relatively neglected (25, 33, 46, 47, 63, 107, 119, 125, P5, P21, P254) and until recently its origin has been obscure. The rate of dimerization in the presence of nickel is the same as the rate of the multiple insertion of ethylene molecules into the Al–C bond, and it is therefore assumed that this reaction also takes place at the aluminum atom, and not at the nickel atom (198, see however 202), in contrast to the reactions described in Section III. The intermediacy of an aluminum-hydride can be excluded because such compounds act as effective catalyst poisons (23, 46). Insight into the mechanism has been obtained from studying the interaction of zerovalent nickel-olefin complexes with aluminum-alkyls. Particularly relevant is the quantitative liberation of butene from the

$$(CH_2{=}CH_2)_3Ni + Al(C_4H_9)_3 \longrightarrow 3C_4H_8 + Al(CH_2CH_3)_3 + Ni$$

stoichiometric reaction of tris(ethylene)nickel with tributylaluminum. Based on these results the following mechanism has been proposed.

The nickel(2+) compound introduced as the catalytic component is re-duced by the trialkylaluminum, generating nickel(0) which then reacts with the olefins present in the reaction mixture to form an olefin complex, e.g., tris(ethylene)nickel. A multicenter bond is formed between the nickel(0) com-plex and the trialkylaluminum in which the α-C atom of the trialkylaluminum bridges the nickel and aluminum atoms. It can be seen from models that in an

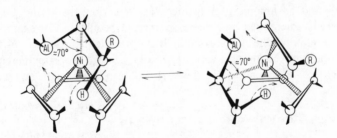

arrangement of this sort the C atoms of the complexed olefinic double bond, as well as the β-H atom of the alkyl group, closely approach the Al atom, thereby permitting an electrocyclic reorganization process. (46)

The postulated interaction between the aluminum-alkyl and the nickel should be contrasted to that found in complexes **1** and **2**, which are formed by reacting trimethylaluminum with the appropriate nickel-methyl complexes (46, 203, 204).

The dimerization of ethylene using triethylaluminum is a special case of a general reaction that is of potential interest for the preparation of other aluminum-trialkyls (Table I-1). The equilibrium can be displaced in favor of

$$Al(CH_2CH_2R)_3 + 3CH_2{=}CHCH_3 \rightleftharpoons Al(CH_2CH_2CH_3)_3 + 3CH_2{=}CHR$$

one of the components by altering the concentration of either the reacting olefin or the liberated olefin. The success of this reaction rests on the preferred addition of the aluminum atom to the C_1 carbon atom of the reacting olefin (i.e., Markownikoff addition), the reverse addition ($Al \rightarrow C_2$) being insignificant.

Isomerization of the olefin components has been reported occasionally (9, 62, 198, 199, 207) and is said to be suppressed in the presence of alkynes or olefins (P253).

Similar behavior to that observed with aluminum-alkyls has been reported briefly for reactions involving the alkyls of beryllium, zinc, and boron (62, 86) and in more detail for those of magnesium (208–213, 326). The reaction of Grignard reagents with olefins was first reported as early as 1924 by Job and Reich, who observed that ethylene reacted with phenylmagnesium bromide in the presence of nickel chloride to give styrene as one of the products (213). The reaction has subsequently been extended to include a variety of olefins and alkyl- and arylmagnesium halides. The reactions involving Grignard reagents are characterized by a decrease in regiospecificity. For example, the reaction of ethylmagnesium bromide with propylene produces 74.5% n-propylmagnesium bromide ($Mg \rightarrow C_1$) and 25.5% isopropylmagnesium bromide ($Mg \rightarrow C_2$), whereas in the case of the reaction with styrene only

$$CH_3CH_2MgBr + CH_3CH{=}CH_2 \xrightarrow{\text{[Ni]}} CH_2{=}CH_2 + \begin{array}{l} CH_3CH_2CH_2MgBr\ (74.5\%) \\ + (CH_3)_2CHMgBr\ (25.5\%) \end{array}$$

$Mg \rightarrow C_2$ addition, to give 2-phenylethylmagnesium bromide, is observed. Nickel-catalyzed isomerization of the resulting Grignard reagent has been observed for 2-phenylethylmagnesium but is only of minor importance for alkyl-substituted alkylmagnesium derivatives.

$$CH_3CH_2MgBr + C_6H_5CH{=}CH_2 \xrightarrow{\text{[Ni]}} CH_2{=}CH_2 + C_6H_5C(CH_3)HMgBr$$

The proposed intermediacy of a nickel-alkyl species in the reactions involving organomagnesium compounds must be viewed with some skepticism in the light of the results discussed above for the corresponding aluminum-alkyls.

TABLE I-1

THE NICKEL-CATALYZED REACTION OF ALUMINUM-ALKYLS WITH OLEFINS

$$al - CH_2CH_2R + CH_2{=}CHR' \rightleftharpoons al - CH_2CH_2R' + CH_2{=}CHR$$

Aluminum-alkyl	Olefin	Nickel catalyst	Ref.
$Al(CH_2CH_2CH_2CH_3)_3$	$CH_2{:}CH_2$	$Ni(acac)_2$	25, 33, 46, 47, 63, 107, 119, 125, 195–201, P5, P21, P148, P240, P253, P254
$Al[CH_2CH(CH_3)_2]_3$	$CH_2{:}C(CH_2CH_3)C_4H_9$	$Ni(busal)_2{}^a$	61
	$CH_2{:}C(C_6H_5)CH_2CH_3$	$Ni(busal)_2$	61
	$CH_2{:}C(C_6H_5)iso\text{-}C_3H_7$	$Ni(busal)_2$	61
	$CH_2{:}CHCH(CH_3)(CH_2)_3iso\text{-}C_3H_7$	$Ni(busal)_2$	62
	$CH_2{:}CH(CH_2)_5CH_3$	$Ni(acac)_2$	9, P241
	$CH_2{:}CH(CH_2)_5CH_3$	$[(C_6H_5)_3P]_4Ni$	P241
$Al[CH_2CH(CH_3)CH_2CH_3]_3$	$CH_2{:}CH_2$	$Ni(acac)_2$	3, 88, P157
	$CH_2{:}CHCH_2CH(CH_3)CH_2CH_3$	$Ni(mesal)_2{}^b$	62, 86, 205
$Al(CH_2CH_2R)_3{}^c$	$CH_2{:}CH_2$	$Ni(acac)_2$	199
	$CH_2{:}CHCHCH_3$	$Ni(acac)_2$	207
$\begin{array}{l} CH_2CH_2al{}^d \\ | \\ CH_2al \end{array}$	$CH_2{:}CH_2$	$Ni(acac)_2$	206

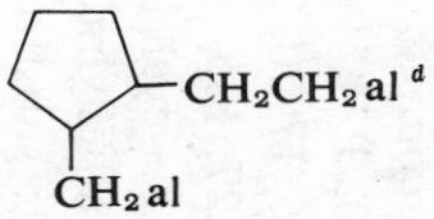

^{a} busal $\equiv$

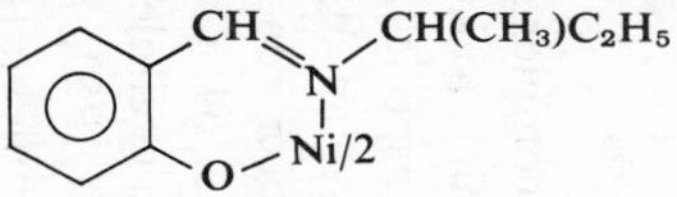

$CH{=}N{-}CH(CH_3)C_2H_5$ ... $Ni/2$ ... O

^{b} mesal $\equiv$

$CH{=}N{-}CH_3$ ... $Ni/2$ ... O

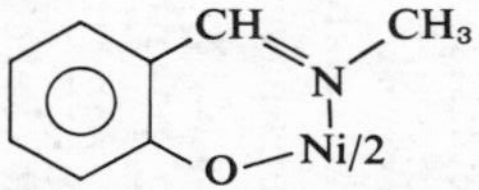

^{c} Higher aluminum-alkyl produced by reacting $Al(C_2H_5)_3$ with ethylene.

^{d} Formed from 1,5-COD + $2al\text{-}CH_2CH(CH_3)_2$.

A reaction perhaps related to these is that of isoprene with magnesium in THF, which is catalyzed by $[(C_6H_5)_3P]_2NiBr_2$–C_4H_9MgBr and which is reported to give diisoprene(1,4)magnesium (P255; see also refs. 298 and 349).

III. The Oligomerization and Co-oligomerization of Olefins
(Tables I-5, I-7–I-10)

A. *Preparation of the Catalyst*

Catalysts active for the oligomerization of olefins can be prepared from practically every type of nickel compound. In general, these must be activated by a Lewis acid and in some cases a reducing agent is necessary. In practice, the nickel can be in the form of a homogeneous catalyst or a supported catalyst.

The nickel component in the homogeneous systems is most frequently introduced as a nickel $(2+)$ salt, such as nickel acetylacetonate or a nickel halide. The activity depends to some extent on the nature of the nickel salt and, in addition to the acetylacetonate, particularly active systems are claimed to be formed from the nickel salts of diisobutyrylmethane and dibenzoylmethane (55, P14). Other systems have been produced from bis(N,N-dialkylcarbyldithiocarbamato)nickel **(3)**, bis(picolinato)nickel, bis(β-mercaptoethylamine)nickel **(4)**, and bis(alkylbenzenesulfonato)nickel (P80, P85, P86, P119). In addition, catalysts have been prepared from ionic nickel

        3   **4**

coordination complexes, e.g., $[R_4P]^+[R_3PNiCl_3]^-$ (30, 32, P97, P100, P105, P108, P111), nickel-nitrosyl complexes (P45, P80, P85, P89, P102, P111), and the nickel(1+) species tristriphenylphosphinenickel chloride (50, P35).

The catalyst can also be formed from organonickel complexes and reactions have been reported involving a nickel-hydride (98), nickel-alkyls or -aryls (2, 8, 57, 64, 65, 67, 68, 99, 108, 115, 137, 350, P72, P81, P88, P92, P102, P259), nickel-olefin complexes, e.g., $(COD)_2Ni$ (7, 69, 95, 97, 338, 341, P15, P32, P35, P50, P70, P74, P111, P114, P115, P123, P125, P153); π-allylnickel complexes (4, 6, 17–20, 28, 44, 48, 78, 87, 90, 96, 97, 102, 104, 106, 108, 110–112, 114, 116, 127, 135, 344, P36, P48, P59, P88, P93, P109, P113, P115, P130, P135–P137); π-cyclobutadienenickel complexes (79–83, P110, P112, P113); π-cyclopentadienylnickel complexes (44, 78, 85, 118, 127, 133, P102, P111, P130, P135); substituted nickel-carbonyl complexes (50, 131,

357, P31, P32, P50, P91, P111, P119, P142, P149); tetrakisligand nickel complexes (29, 50, P35, P60, P108, P111); and Raney nickel (P20, P54, P57).

In almost every case the effect of the addition and variation of ligands on the product distribution has been investigated. Particularly detailed studies have been reported for catalysts based on inorganic nickel salts (89, 91, 100), π-allylnickel complexes (17–19, 111, 112, 344, P137), and π-cyclobutadiene-nickel complexes (79). Phosphines and phosphites are the standard ligands; more exotic ligands include chelating phosphines (53, 54, 112, P106, P141), phosphorus pentoxide (P71), phosphine oxides (P33, P88), phosphorus ylids (108, P153), diphenylphosphinoacetate $[(C_6H_5)_2PCH_2CO_2K]$ and related systems (P95, P115, P122–P124, P153, P155, P248, P260), the aminoamide $cyclo\text{-}C_6H_{11}NHCH(CONH cyclo\text{-}C_6H_{11})_2$ (P79, P250), as well as optically

active phosphines, e.g., $P\left(CH_2-\left\langle\overline{\diagup\diagup}\right\rangle\right)_3$ (see Section III, C).

The full activity of the catalyst is only normally observed in the presence of a Lewis acid (see, for example, P118). In the majority of cases the Lewis acid is an alkylaluminum sesquichloride [e.g., $(C_2H_5)_3Al_2Cl_3$] or an aluminum trihalide. Other metal halides may also be used and these include $TiCl_4$ (4, 6, 44, 78, 96, 104, 127, P36, P104) and BX_3 (50, 57, 64, 65, 67, 68, 131, 338, 341, 350, P15, P28, P61, P92, P140). It has been claimed that transition metal fluorides are more effective than the chlorides (P50, P149). Systematic studies of the effect of varying the concentration of the Lewis acid are rare. Apparently, the optimum value depends on the nature of both the Lewis acid and the nickel component. Typical results are shown in Table I-2.

In cases involving inorganic nickel(2+) salts the generation of an active catalyst requires the addition of a reducing agent, which may be present as an alkylaluminum sesquichloride or dialkylaluminum alkoxide (55, 357, P6, P7, P9, P13, P14, P21, P32, P35, P70, P134, P151), trialkylaluminum (1, 5, 53–56, 66, P15, P17, P56, P61–P64, P93, P98, P102, P136, P140, P141), butyllithium (11, 29, 55, P60), sodium borohydride (103, P95, P124, P155, P260), trialkylboron (55, P24), ethyl Grignard (55, P31), diethylberyllium (P102, P110, P111, P113), diethylzinc (P31), or potassium (P18).

The effect of additives on the activity of the catalyst is a favorite area in the patent literature and those investigated include tetraethyltin (P94), triethylamine (P34, P35), *tert*-butylhypochlorite (P39), water (P43, P49, P51–P55, P57, P105), sulfides (P102, P110, P111), alkyl halides (5, P17, P20), and olefins (P7, P25, P29, P30, P66).

Although the majority of the catalysts consist of the combination nickel complex–Lewis acid, the Lewis acid is not always essential. For example,

TABLE I-2

OPTIMUM LEWIS ACID:NICKEL RATIOS

Lewis acid	Nickel component	Lewis acid/nickel comp.	Olefin	Ref.
$AlBr_3$	$(\pi\text{-}C_3H_5NiBr)_2$	1:1	C_3H_6	116
$AlCl_3$	$[(C_6H_5)_3P]_2Ni(CO)_2$	5:1	C_3H_6	50
$(C_2H_5)_3Al_2Cl_3$	$(\pi\text{-}C_3H_5NiBr)_2$	3–4:1	C_3H_6	112
	$(COD)_2Ni$	5:1	$C_2H_4\text{-}C_4H_6$	95
	$[(iso\text{-}C_3H_7)_4P][(iso\text{-}C_3H_7)_3PNiCl_3]$	$\sim 50{:}1$	C_3H_6	32
$(iso\text{-}C_4H_9)_2AlCl$	Ni oleate	50–100:1	C_3H_6	78, 91
$(C_2H_5)_2AlOC_2H_5$	$Ni(acac)_2$	1:1	C_3H_6	55
$BF_3 \cdot O(C_2H_5)_2$	$[(C_6H_5)_3P]_2NiX(R)$	$> 20{:}1$	C_2H_4	64
$TiCl_4$	$(\pi\text{-}C_3H_5NiBr)_2$	2:1	C_3H_6	6

ethylene is slowly dimerized at moderate temperatures by $[(C_6H_5)_3P]_2$-NiBr(aryl) complexes (57) and π-allylNiX systems (P137) or by nickelocene at 200° (85, 118), whereas more active catalysts have been obtained by treating $(COD)_2Ni$ with CF_3CO_2H (P114) or diphenylphosphinoacetate and related systems (P115, P122, P123, P155). Propylene is slowly dimerized by π-allylNiX catalysts (78, P59), as is 1,5-hexadiene (58, 104), whereas the same catalysts readily dimerize styrene at room temperature (28, 48, 87, P48). Of considerable mechanistic interest is the observation that the nickel-hydride $HNi[P(OR)_3]_4{}^+$ catalyzes the codimerization of butadiene and ethylene (98).

The solvent plays rather an ambiguous role in the reaction. In general aromatic or halohydrocarbons (e.g., C_6H_5Cl, CH_2Cl_2) are preferred but individual cases are known in which the reaction proceeds satisfactorily in saturated hydrocarbons (e.g., 2, 29, P9) and alcohols (e.g., 11, P153), or even in the absence of solvent. A study has been made of the effect of various halohydrocarbons on the codimerization of butadiene and ethylene (5) while the dimerization of propylene, using $[\pi\text{-}(CH_3)_4C_4NiCl_2]_2\text{-}(C_2H_5)_3Al_2Cl_3\text{-}Lig$ as catalyst, has been shown to proceed five times faster in chlorobenzene than in benzene (80).

The nickel component in the supported catalysts is frequently nickel oxide. However, more recently catalysts have been prepared by reacting the support with an organonickel complex, e.g., $(COD)_2Ni$ (P23, P114, P115, P126), π-allylnickel complexes (132, P120, P127), π-cyclopentadienylnickel complexes (P90, P129, P131, P132) and substituted nickel-carbonyl complexes (P4). The support or the nickel component may be pretreated with diethyl-aluminum sesquichloride (P1–P3, P23, P38, P42, P57, P82, P116, P128, P150, P263) or triethylaluminum (P11, P26, P120). The most frequently implemented support is silica–alumina, which may be activated by treatment with a sulfur-containing ligand (P116, P129, P252). Other supports include activated charcoal (P11), polystyrene (332), polyphosphinostyrene (P23), and polyvinylpyridine (P42, P82).

B. The Oligomerization Reaction

The catalysts described in the preceding section convert monoolefins into a mixture of dimers, trimers, and higher oligomers in which the dimer predominates. The higher oligomers are probably formed as the result of co-oligomerization with the monomer. In practically all cases the reaction is accompanied by isomerization of the products; this aspect is discussed in Section VI.

The kinetics of the dimerization and codimerization of simple olefins have been investigated for several catalysts (Table I-3). In general, the reaction is

TABLE 1-3

THE KINETICS OF NICKEL-CATALYZED OLEFIN DIMERIZATION

Olefin	Catalyst	Rate law	Activation energy (kcal/mole)	Ref.
$CH_2:CH_2$	$\pi\text{-}(CH_3)_4C_4NiCl_2\text{-}(C_2H_5)_3Al_2Cl_3\text{-}P(C_4H_9)_3$	$[Cat][Olef]^2$	7.1 ± 0.2	81
$CH_3CH:CH_2$	$\pi\text{-}(CH_3)_4C_4NiCl_2\text{-}(C_2H_5)_3Al_2Cl_3\text{-}P(C_4H_9)_3$	$[Cat][Olef]^2$	9.6 ± 0.2	80
	$(\pi\text{-}CH_3CHCHCH_2NiCl)_2\text{-}TiCl_4$	$[Cat][Olef]^2$	15.2	6
	$Ni(acac)_2\text{-}(C_2H_5)_3Al_2Cl_3\text{-}P(C_6H_5)_3$	$[Cat]^2[Olef]^2$	9.55	52
$CH_2:CH_2\text{-}CH_3CH:CH_2$	$\pi\text{-}(CH_3)_4C_4NiCl_2\text{-}(C_2H_5)_3Al_2Cl_3\text{-}P(C_4H_9)_3$	$[Cat][Olef_1][Olef_2]$	8.0 ± 0.5	82
$CH_2:CH_2\text{-}butene$	$\pi\text{-}(CH_3)_4C_4NiCl_2\text{-}(C_2H_5)_3Al_2Cl_3\text{-}P(C_4H_9)_3$	$[Cat][Olef_1][Olef_2]$	10.6 ± 0.5	81
$C_6H_5CH:CH_2$	$(\pi\text{-}C_3H_5NiI)_2$	$[Cat][Olef]$	—	48

first order in catalyst concentration and second order in olefin concentration. Quantitative studies of the effect of varying the Lewis acid concentration have not been made. The activation energy needed to catalytically dimerize ethylene or propylene is ca. 30 kcal/mole lower than that needed for the thermal gas-phase dimerization.

The rate of reaction decreases in the order

$$CH_2:CH_2 > CH_3CH:CH_2 > \text{cycloolefin} > CH_3CH:CHCH_3$$

The rate of dimerization of ethylene, using $\pi\text{-}(CH_3)_4C_4NiCl_2\text{-}(C_2H_5)_3Al_2Cl_3\text{-}P(C_4H_9)_3$ as catalyst, has been shown to be approximately 40 times faster than that of propylene (83). The high activities that can be obtained are illustrated by the dimerization of propylene using $(\pi\text{-allylNiX})_2\text{-AlX}_3\text{-Lig}$ as catalyst: the rate of conversion at 35–40° and 15 atm is 15 kg/gm nickel per hour with a catalyst consumption of 50 mg of nickel per kilogram of product, which corresponds to the conversion of 30,000 propylene molecules per molecule of catalyst (19). Other terminal olefins (e.g., 1-pentene and 1-hexene) can be oligomerized at room temperature or slightly above. The reactions of nonterminal olefins (e.g., 2-butene and 2-pentene) probably involve an initial isomerization to the terminal olefin.

The oligomerization of propylene has been studied in some detail (the C_6 olefins produced are of considerable interest as fuel additives and as basic chemicals for the polymer industry). The main product of the nickel-catalyzed oligomerization is a mixture of hexenes, methylpentenes, and 2,3-dimethylbutenes, as well as lesser amounts of C_9 olefins. The composition of the product can be varied over a wide range by introducing ligands, normally phosphines. Particularly detailed studies of the effect of varying the phosphine are to be found in references 19 and 112 and the mechanistic implications are discussed in Section III,D. The results using a $(\pi\text{-allylNiX})_2\text{-}(C_2H_5)_3Al_2Cl_3\text{-}$ Lig catalyst are shown in Table I-4: the highest yields of methylpentene (80%) are obtained in the presence of $P(CH_3)_3$ and those of 2,3-dimethylbutene (87.8%) in the presence of $(iso\text{-}C_3H_7)_2Ptert\text{-}C_4H_9$. Linear hexenes predominate in the reaction catalyzed by the ligand-free catalyst $Ni(acac)_2\text{-}(C_2H_5)_2AlOC_2H_5$ (55, P7, P9, P13, P14, P134, P151) and by $\{[(CH_3)_2N]_3PO\}_2\text{-}NiBr_2\text{-}(C_2H_5)_3Al_2Cl_3$ (P33).

The codimerization of ethylene and propylene is accompanied by considerable dimerization of the monomers and only ca. 50% codimer is obtained. The product consists of pentenes and 2-methylbutene and its composition is influenced by the presence of a ligand: $P(cyclo\text{-}C_6H_{11})_3$ produces mainly methylbutene (ca. 90%), whereas in the absence of a ligand mainly pentenes (ca. 60%) are obtained (19, 39, 82, 112).

The codimerization of ethylene with a cyclic olefin occurs readily to give a vinyl derivative or, as a result of isomerization, the corresponding ethylidene

TABLE I-4

THE INFLUENCE OF PHOSPHINES ON THE NICKEL-CATALYZED DIMERIZATION OF PROPYLENE^{a,b}

Phosphine	Σ n-Hexene	Σ Methylpentene	Σ 2,3-Dimethylbutene	Dimer formation (gm/hr)^c
P(C$_6$H$_5$)$_3$	19.8	76.0	4.2	ca. 300
(C$_6$H$_5$)$_2$PCH$_2$P(C$_6$H$_5$)$_2$	21.6	73.9	4.5	126
(C$_6$H$_5$)$_2$PCH$_2$C$_6$H$_5$	12.2	83.0	4.7	58
(C$_6$H$_5$)$_2$P(CH$_2$)$_3$P(C$_6$H$_5$)$_2$	19.2	75.4	5.1	291
P(CH$_3$)$_3$	20.1	73.3	6.6	7
(C$_6$H$_5$)$_2$Piso-C$_3$H$_7$	9.9	80.3	9.8	ca. 400
P(C$_2$H$_5$)$_3$	14.4	73.0	12.6	388
P(C$_4$H$_9$)$_3$	9.2	69.7	21.1	ca. 350
P(CH$_2$C$_6$H$_5$)$_3$	7.1	69.6	23.3	87
($cyclo$-C$_6$H$_{11}$)$_2$P-P($cyclo$-C$_6$H$_{11}$)$_2$	6.7	63.6	29.2	318
P($cyclo$-C$_6$H$_{11}$)$_3$	4.4	46.5	49.2	132
P(iso-C$_3$H$_7$)$_3$	3.3	37.9	58.8	250
($tert$-C$_4$H$_9$)$_2$PCH$_3$	1.8	30.3	67.9	300
($tert$-C$_4$H$_9$)$_2$PCH$_2$CH$_3$	1.2	24.5	74.0	189
(iso-C$_3$H$_7$)$_2$P$tert$-C$_4$H$_9$	0.6	22.3	77.0	292
($tert$-C$_4$H$_9$)$_2$Piso-C$_3$H$_7$	0.1	11.9	87.8	332
	0.6	70.1	29.1	347

^{a} Catalyst: (π-C$_3$H$_5$NiX)$_2$-(C$_2$H$_5$)$_3$Al$_2$Cl$_3$-Lig; −20°C; 1 atm.
^{b} From refs. 19, 112.
^{c} For uniform catalytic conditions.

derivative, e.g., **5** and **6**, derived from bicycloheptene. Incorporation of a second molecule of ethylene has been observed in the reaction involving bicycloheptene, and the C_{11} olefins **7** and **8** have been identified (110).

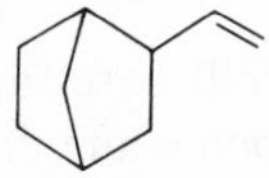

    5 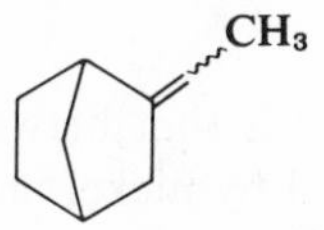   **6** 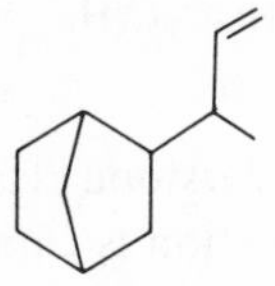   **7** 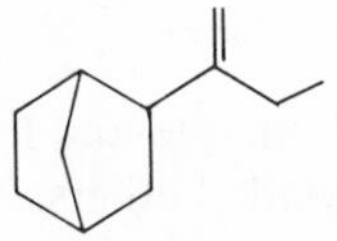   **8**

The same catalysts that are effective for the oligomerization and co-oligomerization of monoolefins are also able to codimerize ethylene or propylene with conjugated dienes. The product is a 1,4-diene. The reaction of ethylene with butadiene produces a mixture of *cis-* and *trans*-1,4-hexadiene with lesser amounts of 3-methyl-1,4-pentadiene. The ratio of *trans-* to *cis*-1,4-hexadiene depends on the phosphine present, being 1.5:1 for P(*cyclo*-C_6H_{11})₃ and 6:1 for (C_6H_5)₂PC_6F_5 (95). Cyclic 1,3-dienes react similarly. The reaction of 1,3-cyclooctadiene with propylene produces, in addition to the expected codimers **9–11**, the bicyclic compounds **12** and **13** (109). The distribution of the products in this reaction is also phosphine dependent: P(*cyclo*-C_6H_{11})₃ produces mainly **11** and P(C_6H_5)₃ produces mainly a mixture of **9** and **10**.

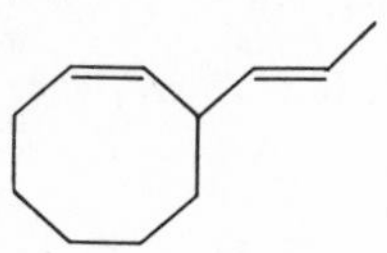

    9 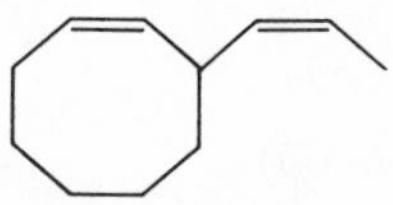   **10** 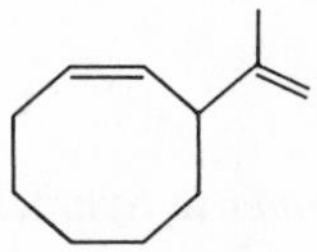   **11**

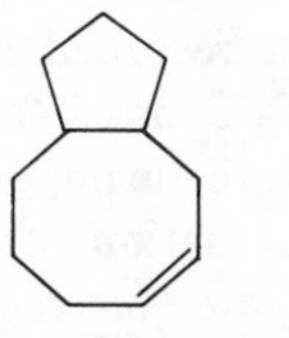

    12 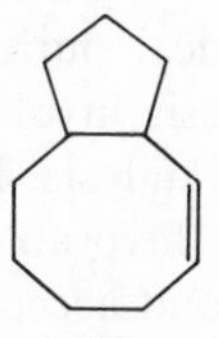   **13**

For completeness we should mention two reactions that may be related to those discussed above, although no mechanistic information is available. The reaction of the triene, cycloheptatriene, with acrylic acid ester is catalyzed by triphenylphosphinenickel tricarbonyl and is unusual in that the codimerization is accompanied by hydrogen migration to give an α-cycloheptatrienyl propionic acid ester, which subsequently isomerizes by a series of 1,5-hydrogen shifts (21).

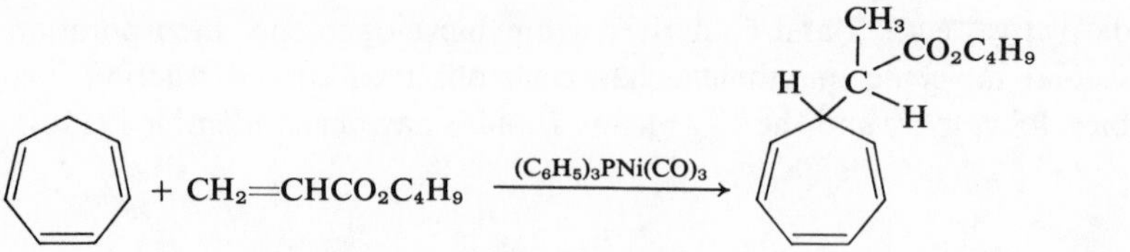

A unique and little understood reaction is that between Schiff bases and vinylalkyl ethers. The reaction is catalyzed by nickel tetracarbonyl and produces 2-substituted 4-alkoxy-1,2,3,4-tetrahydroquinoline derivatives in good yield. N-Propylideneaniline does not react with vinyl ethers; instead dimerization of the Schiff base occurs (134).

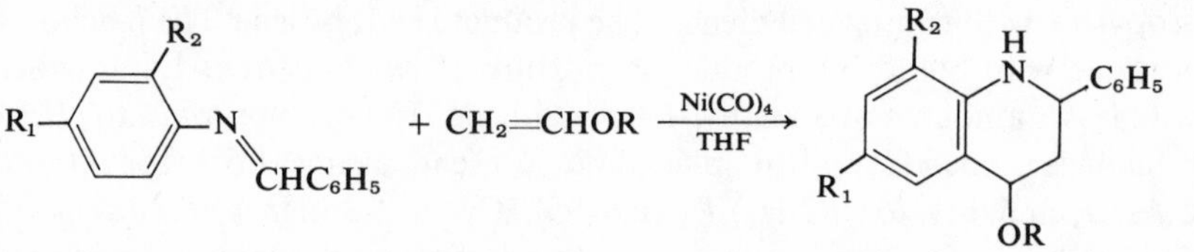

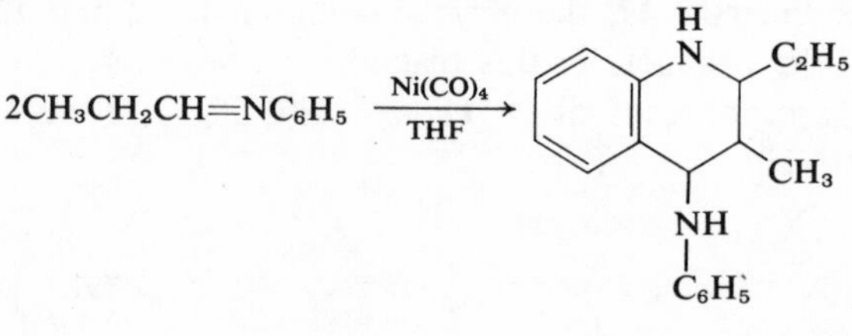

C. Asymmetric Syntheses (Table I-5)

An exciting recent development is the discovery that optically active codimers can be synthesized using a catalyst modified by an optically active phosphine. The highest optical purities yet recorded in transition metal-catalyzed asymmetric synthesis involving C–C bond formation have been obtained using nickel-based catalysts. The synthetic importance of the nickel-catalyzed reactions lies in a favorable combination of high optical purity with high yield of codimer and acceptable rates of reaction, and it can be safely predicted that this area will attract considerable attention. The high optical purity is, in part, a result of the high activity of the catalyst, which enables the codimerization reactions to be carried out under very mild conditions (e.g., −65°), thereby allowing full implementation of the small differences in the free activation enthalpy for formation of the diastereomeric intermediates. Only one research group has been active in this area and their results are to be found in references 17, 19, 106, 109–111, and P159. Asymmetric synthesis in general is reviewed in reference 106.

The catalyst commonly used is of the type $(\pi\text{-allylNiX})_2\text{-}(C_2H_5)_3Al_2Cl_3\text{-}PR_3$, although it has also been shown that Ni(acac)$_2$ can be used equally effectively as the nickel component. Two types of optically active phosphine have been studied: the so-called Horner phosphines, in which the asymmetric center is at the phosphorus, e.g., $C_6H_5P(CH_3)tert\text{-}C_4H_9$, and phosphines in which the optically active center (or centers) is in the alkyl substituent, e.g., P(menthyl)$_2$CH$_3$. Surprisingly, initial results with the Horner phosphines indicate that the optical induction is inferior to that using the phosphines containing optically active substituents and they have not been studied in detail (111). The most effective phosphines have been found to be those containing a menthyl group, e.g., $(-)$-(menthyl)$_2$P*iso*-C$_3$H$_7$ (**14**) or a myrtanyl group, e.g., $(-)$-*trans*-(myrtanyl)$_3$P (**15**).

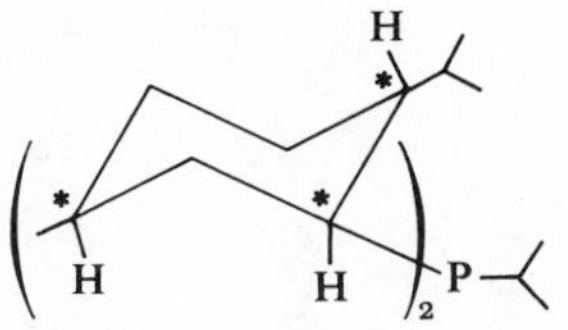

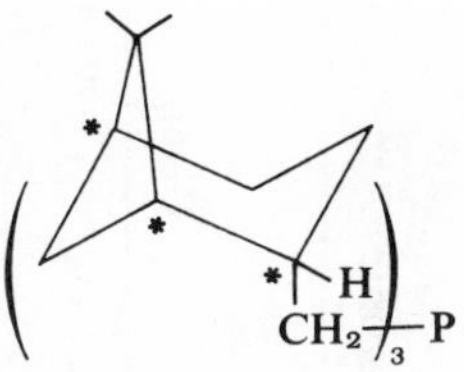

The reactions that have been investigated are shown in Table I-5. The most successful syntheses involve the codimerization of ethylene with a cyclic olefin. The increased reactivity of strained olefins makes them particularly suitable for asymmetric syntheses at low temperatures. The primary product of the reaction between bicycloheptene and ethylene is *exo*-2-vinylbicycloheptane (**16**); i.e., the ethylene is incorporated exclusively at the sterically less hindered side of the molecule. The optical purity of **16** increases on lowering

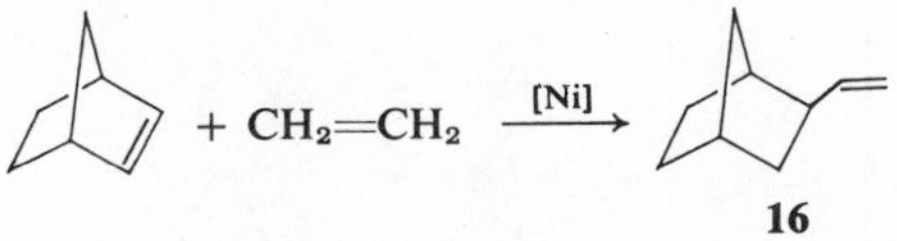

the reaction temperature and a value of 80.6% is obtained at $-97°$. The rate of reaction at this temperature is rather low (17.2% conversion of bicycloheptene in 18 hr) but at slightly higher temperatures ($-83°$) the rate is considerably faster (100% conversion in 9 hr, other parameters being constant) and the optical purity is still as high as 75%. The linear relationship between the optical purity and the reaction temperature has been used to show that the difference in the activation energy ($\Delta\Delta E\ddagger$) between $(+)$- and $(-)$-*exo*-2-vinylbicycloheptane (**16**) is ca. 1.5 kcal/mole.

A similar increase in optical purity on lowering the reaction temperature

TABLE I-5

ASYMMETRIC CODIMERIZATION[a,b]

Codimer	Optical purity (%)	React. temp. (°C)	Olefin	Ligand[c]
(+)-(S)-CH₂:CHCH(CH₃)CH₂CH₃	46 64	0 -40	$CH_2:CH_2$ 	Butene	(−)-(menthyl)₂PCH₃
(+)-(S)-CH₂:CHCH(CH₃)C₆H₅ (−)-(R)-CH₂:CHCH(CH₃)C₆H₅	12.3 21.8	0 -60	$CH_2:CH_2$ $CH_2:CH_2$	C₆H₅CH:CH₂	(−)-(menthyl)₂PCH₃ (−)-(menthyl)₂*Piso*-C₃H₇
(−)-3S-	27 53	24 -75	$CH_2:CH_2$		(−)-(menthyl)₂PCH₃
(+)-1S,2S,4R-	30.0 80.6	10 -97	$CH_2:CH_2$		(−)-(menthyl)₂*Piso*-C₃H₇
(+)-1R,4S,5S-	77.5	-65	$CH_2:CH_2$		(−)-(menthyl)₂*Piso*-C₃H₇
(+)-1R,4S-	20.8	0	$CH_2:CH_2$		(−)-(menthyl)₂*Piso*-C₃H₇

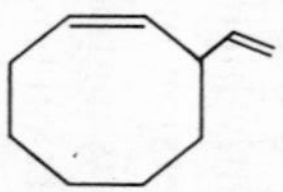

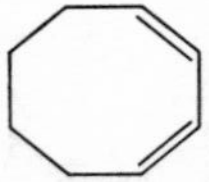

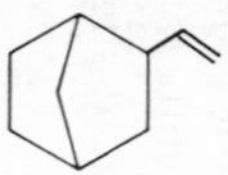

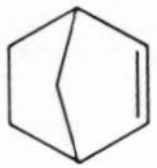

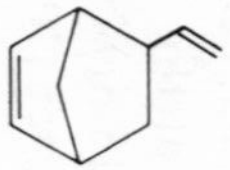

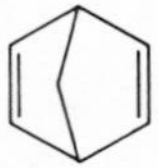

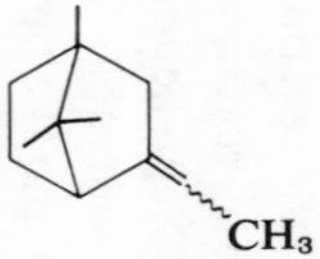

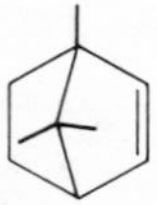

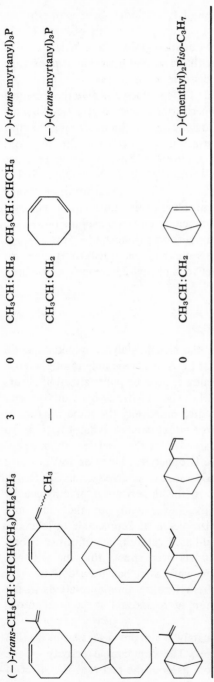

^{a} From refs. 17, 19, 106, 109–111.

^{b} Catalyst $(\pi\text{-}C_3H_5NiX)_2\text{-}(C_2H_5)_3Al_2Cl_3\text{-Lig.}$

^{c} P-Menthyl = P—⟨...⟩ ; P-*trans*-myrtanyl = P—CH₂—⟨...⟩

has been observed in the codimerization of 1,3-cyclooctadiene with ethylene to give 3-vinylcyclooctene. The highest optical yield in this reaction (70%) however, is obtained at $0°$ using $(-)$-(menthyl)$_2$PCH$_3$ in a Ni:P ratio of $1:3.8$. It has been suggested here either that a second molecule of phosphine interacts with the nickel atom or that dissociation is suppressed.

The reaction between bornylene and ethylene does not give the expected codimer (3-vinylbornane). Instead, isomerization occurs to give 3-ethylidene-bornane, which is readily converted into epicamphor. An interesting example of an enantioelective catalytic codimerization is observed in this reaction: $(+)$-bornylene reacts preferentially out of the racemic mixture in the presence of a catalyst modified by $(-)$-(menthyl)$_2$P*iso*-C$_3$H$_7$ or $(-)$-(menthyl)$_2$PCH$_3$.

An unusual effect has also been observed in the codimerization of styrene and ethylene to 3-phenyl-1-butene: the absolute configuration of the codimer depends on the modifying phosphine; dimenthylmethylphosphine producing $(+)$-*S*-3-phenyl-1-butene and dimenthylisopropylphosphine the $(-)$-*R* isomer. However, the possibility that a secondary enantioelective isomerization occurs to give the optically inactive 2-phenyl-2-butene has not been excluded.

D. Mechanistic Considerations

The postulation of a mechanism for the nickel-catalyzed oligomerization of olefins is complicated by the fact that Lewis acids are able to oligomerize olefins in the absence of nickel. For example, cationic polymerization occurs in the presence of both BF$_3$ and TiCl$_4$, whereas it has been shown that $(C_2H_5)_3Al_2Cl_3$ dimerizes propylene to give practically the same mixture of C$_6$ olefins obtained using the ligand-free nickel catalyst Ni(acac)$_2$–$(C_2H_5)_3$-Al$_2$Cl$_3$ (56). Aluminum-alkyls act as catalysts for the insertion oligomerization of olefins; for example, propylene is dimerized, albeit at $180°$, to give almost exclusively 2-methylpentene (139) and, as we have seen in Section II, nickel plays the role of a cocatalyst in the dimerization of ethylene by aluminum-alkyls. However, neither the Lewis acid nor the aluminum component alone is able to oligomerize olefins at appreciable rates under the mild conditions at which the nickel-based catalysts are effective. Moreover, in the presence of phosphines it has been shown that the adducts, e.g., AlBr$_3$–P(C$_6$H$_5$)$_3$, are inactive (116) whereas it has been convincingly demonstrated that the dimerization of ethylene by aluminum-alkyls in the presence of nickel occurs at the aluminum atom and not at the nickel atom (see Section II).

It is probable that the multitude of catalytic systems discussed in Section III,A involve a common active species. The comparable activity of the various systems supports this suggestion, as does the fact that the effect of added phosphine on the distribution of the products depends mainly on the nature of the phosphine (other components being constant) and not on the nature of the π-bonded group (if any) originally bonded to the nickel component. Evidence has accumulated that indicates the active species is a nickel-hydride.

The reaction of cyclooctene with the phosphine-free catalysts $(\pi$-$C_3H_5NiX)_2$–$(C_2H_5)_3Al_2Cl_3$ (X = Br, acac) and $Ni(acac)_2$–$(C_2H_5)_3Al_2Cl_3$ have been investigated in some detail (19): in this reaction cyclooctene is dimerized to a mixture of 1-cyclooctylcyclooctene and dicyclooctylidene and the nickel originally present as π-C_3H_5Niacac (**17**) or $Ni(acac)_2$ is recovered as π-dicyclooctynylNiacac (**18**). The fate of the π-allyl group attached originally

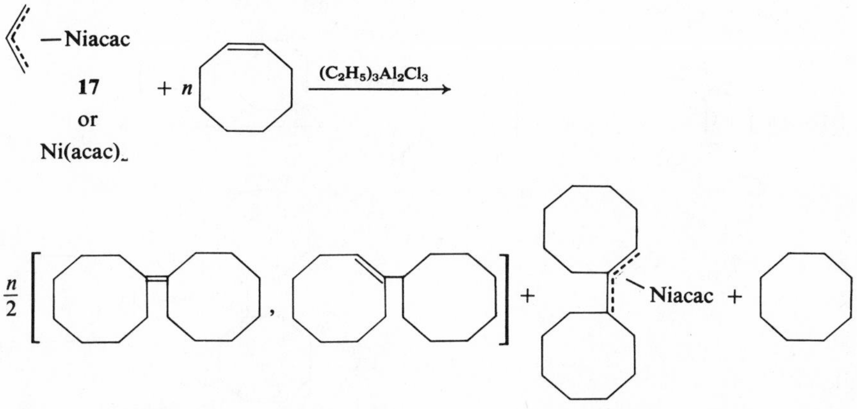

—Niacac

17

or

$Ni(acac)_2$ $+ n$ [octene] $\xrightarrow{(C_2H_5)_3Al_2Cl_3}$

$\dfrac{n}{2}$ [...] $+$ ... Niacac $+$

18

to the nickel atom has been studied by removing samples during the reaction and determining the concentration of the π-allylnickel species by adding ammonia to cause disproportionation into bis(π-allyl)nickel. It could be shown that the highest rate of reaction is observed at the start of the reaction, at which point none of the nickel is present as the original π-allylnickel complex **17**. As the reaction proceeds the rate decreases and the concentration of the π-dicyclooctynylnickel species **18** increases and finally accounts for 70–80% of the nickel initially present. By carrying the reaction out at $-40°$ (at which temperature the rate is very low) it has been shown that 80–85% of the original π-C_3H_5 groups add to the cyclooctene to form $C_{11}H_{18}$ olefins (mainly bicyclo[6.3.0]undecene-3). Hydrogen is not evolved in this reaction which indicates that the transfer is accompanied by the formation of a nickel-hydride species. The final conversion of this HNiX species into the π-dicyclooctynyl complex **18** requires the abstraction of a hydrogen atom from the cyclooctene dimer and is accompanied by hydrogenation of the monomer to give cyclooctane.

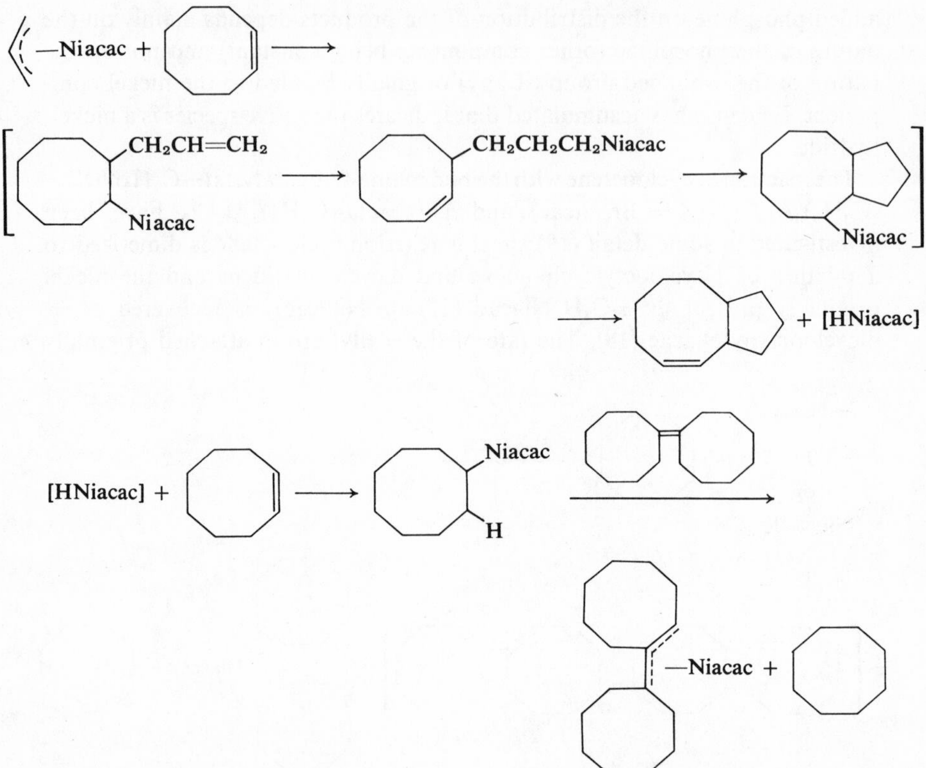

18

In other cases the nickel-hydride is probably formed directly (e.g., $NiCl_2$–$NaBH_4$) or as a result of β-elimination of an olefin from an intermediate nickel-alkyl (e.g., NiX_2–$(C_2H_5)_3Al_2Cl_3$). It has also been shown that a hydride can be formed indirectly by olefin insertion even in those systems in which direct formation is not possible (109), e.g.,

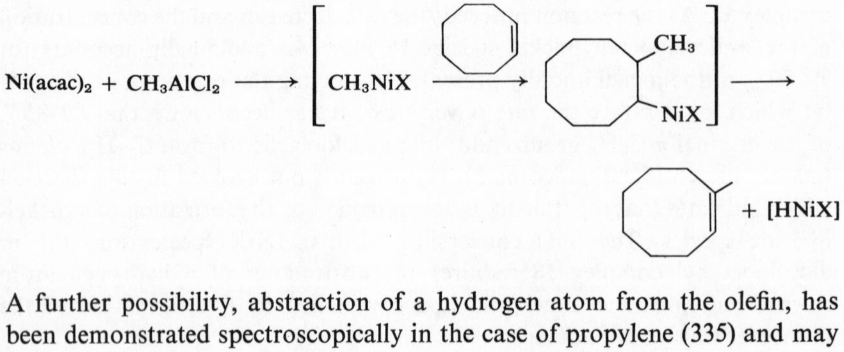

A further possibility, abstraction of a hydrogen atom from the olefin, has been demonstrated spectroscopically in the case of propylene (335) and may

account for the formation of the HNiX system in reactions involving such zerovalent nickel systems as $[(C_6H_5)_3P]_2Ni(CO)_2$.

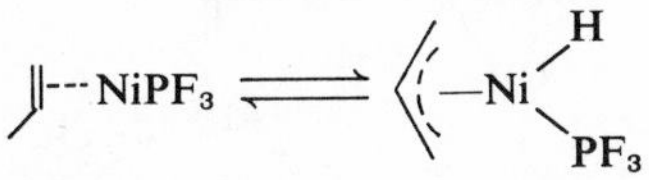

$$\parallel\!\cdots\!NiPF_3 \rightleftharpoons \left\langle\!\!\left(\!-Ni\!\begin{array}{c}H\\ \\PF_3\end{array}\right.\right.$$

It is unlikely that the nickel-hydride discussed above as the active species is ever formed in significant concentrations because it is expected to react immediately with the olefin molecules also bonded to the nickel atom. The catalytically active species can be visualized as consisting of a square planar hybridized nickel atom interacting with a hydride (or alkyl), a phosphine, an electronegative group X, and olefin molecules.

It is probable that the Lewis acid interacts with the nickel through the group X. An indication of the nature of this interaction has been obtained from an x-ray structural study of the adduct formed by π-C_3H_5NiCl-$[P(cyclo$-$C_6H_{11})_3]$ with CH_3AlCl_2: the nickel and aluminum are bridged by a chlorine atom (106; see Volume I, p. 356). It is generally assumed that the

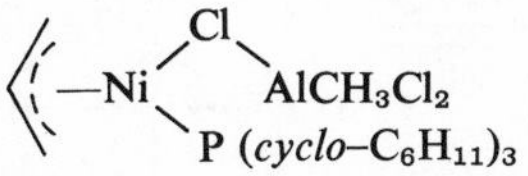

$$\left\langle\!\!\left(\!-Ni\!\begin{array}{c}Cl\\ \\P\ (cyclo\text{–}C_6H_{11})_3\end{array}\!\!AlCH_3Cl_2\right.\right.$$

Lewis acid in these complexes decreases the charge on the metal. A second important function of the Lewis acid is probably to react with redundant phosphine molecules, thereby creating free coordination sites at the nickel. This is perhaps particularly important for catalysts based on $(R_3P)_2NiX_2$ compounds.

The effectiveness of halohydrocarbons as solvents for the oligomerization reaction is probably associated with their polar, weakly basic nature, which enables them to act as effective solvents for polar species. A further possibility, that the halohydrocarbon functions as a source of halide, has never been substantiated but may be of importance in those reactions involving trialkylaluminum.

A simplified mechanism for the oligomerization of ethylene is shown below and consists of three distinct steps: (a) insertion into the Ni–H bond, (b) insertion into a Ni–C bond, and (c) olefin elimination. It is probable that the fifth coordination position at the nickel is involved in the olefin elimination step.

The situation for propylene is complicated by the asymmetry of the olefin. Recent results suggest that the first step is kinetically controlled and produces principally a Ni–n-propyl species (Ni $\rightarrow$ C$_1$) that rearranges to the thermodynamically more stable Ni–isopropyl species (Ni $\rightarrow$ C$_2$) (141).

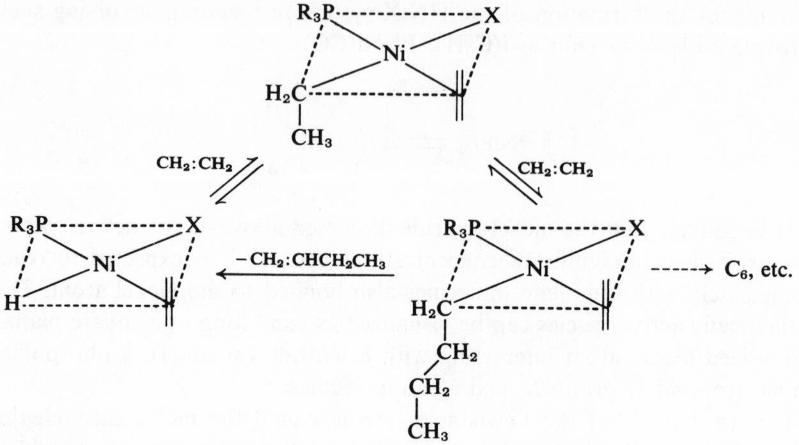

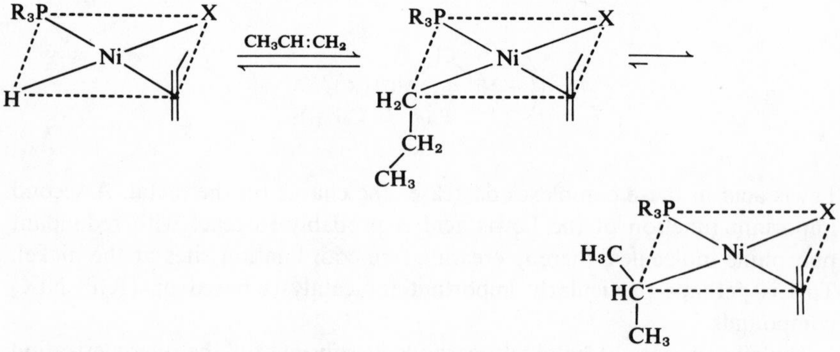

The same choice exists in the second step and leads to the situation shown on p. 23. It is apparent from this scheme that the distribution of the products enables an estimate to be made of the direction of addition (Ni $\rightarrow$ C$_1$ or Ni $\rightarrow$ C$_2$). For the first step this is given by

$$\frac{\% \text{ Ni} \rightarrow \text{C}_1}{\% \text{ Ni} \rightarrow \text{C}_2} = \frac{\% \text{ (4-methyl-1-pentene + 4-methyl-2-pentene + 2,3-dimethyl-1-butene)}}{\% \text{ (2-methyl-1-pentene + 1-hexene + 2-hexene)}}$$

and for the second step by

$$\frac{\% \text{ Ni} \rightarrow \text{C}_1}{\% \text{ Ni} \rightarrow \text{C}_2} = \frac{\% \text{ (4-methyl-1-pentene + 4-methyl-2-pentene + 1-hexene + 2-hexene)}}{\% \text{ (2,3-dimethyl-1-butene + 2-methyl-1-pentene)}}$$

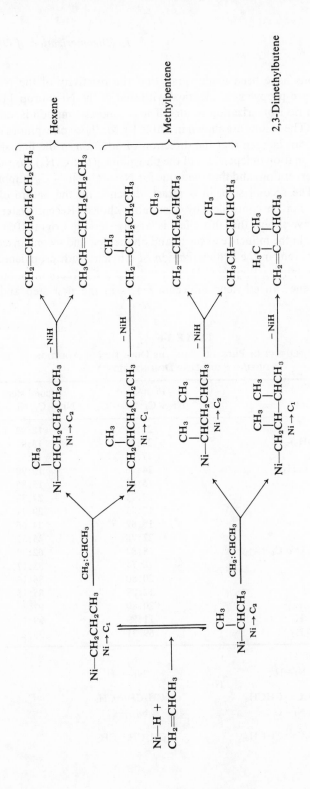

Two assumptions have been made: first, that the reactivity of the Ni–*iso*-C_3H_7 bond toward propylene is identical with that of the Ni–*n*-propyl bond, and second that no isomerization occurs. The second assumption is valid at low conversion. The results are shown in Table I-6 for those phosphines listed in Table I-4. It can be seen that the direction of addition in the first step is to a first approximation independent of the phosphine [$(tert$-$C_4H_9)_2Piso$-C_3H_7 is, however, an exception] and that the value for Ni $\rightarrow$ C_1/Ni $\rightarrow$ C_2 is approximately 20/80. The second step, in contrast, depends on the nature of the phosphine and can be controlled by a suitable choice. Recent systematic investigations have shown that this effect is largely steric in origin (105).

The difference in the influence of the ligand on the first and second insertion steps is also apparent in the codimerization of ethylene with propylene: use of a catalyst containing $P(cyclo$-$C_6H_{11})_3$ produces a C_5 fraction containing 90% methylbutene; an ethylene molecule reacts in the first step and the

TABLE I-6

THE INFLUENCE OF PHOSPHINES ON THE DIRECTION OF ADDITION
DURING PROPYLENE DIMERIZATION[a]

Phosphine	First step[b] Ni $\rightarrow$ C_1 : Ni $\rightarrow$ C_2	Second step[c] Ni $\rightarrow$ C_1 : Ni $\rightarrow$ C_2
$P(C_6H_5)_3$	25:75	18:82
$(C_6H_5)_2PCH_2P(C_6H_5)_2$	21:79	13:87
$(C_6H_5)_2PCH_2C_6H_5$	27:73	13:87
$(C_6H_5)_2P(CH_2)_3P(C_6H_5)_2$	24:76	10:90
$P(CH_3)_3$	15:85	15:85
$(C_6H_5)_2Piso$-C_3H_7	29:71	27:73
$P(C_2H_5)_3$	17:83	29:71
$P(C_4H_9)_3$	18:82	34:66
$P(CH_2C_6H_5)_3$	22:78	45:55
$(cyclo$-$C_6H_{11})_2PP(cyclo$-$C_6H_{11})_2$	18:82	62:38
$P(cyclo$-$C_6H_{11})_3$	26:74	83:17
$P(iso$-$C_3H_7)_3$	20:80	86:14
$(tert$-$C_4H_9)_2PCH_3$	12:88	85:15
$(tert$-$C_4H_9)_2PC_2H_5$	20:80	97:3
$(iso$-$C_3H_7)_2Ptert$-C_4H_9	11:89	99:1
$(tert$-$C_4H_9)_2Piso$-C_3H_7	69:31	98:2

[a] Taken from ref. 112.

[b] $Ni \rightarrow C_1 \equiv \overset{Ni-H}{\underset{H_2C=CHCH_3}{\vdots\ \vdots}}$; $Ni \rightarrow C_2 \equiv \overset{Ni-H}{\underset{CH_3CH=CH_2}{\vdots\ \vdots}}$.

[c] $Ni \rightarrow C_1 \equiv \overset{Ni-C}{\underset{H_2C=CHCH_3}{\vdots\ \vdots}}$; $Ni \rightarrow C_2 \equiv \overset{Ni-C}{\underset{CH_3CH=CH_2}{\vdots\ \vdots}}$.

following insertion of propylene into the Ni–C_2H_5 bond occurs preferentially in an Ni $\rightarrow$ C_1 sense, as predicted from Table I-6.

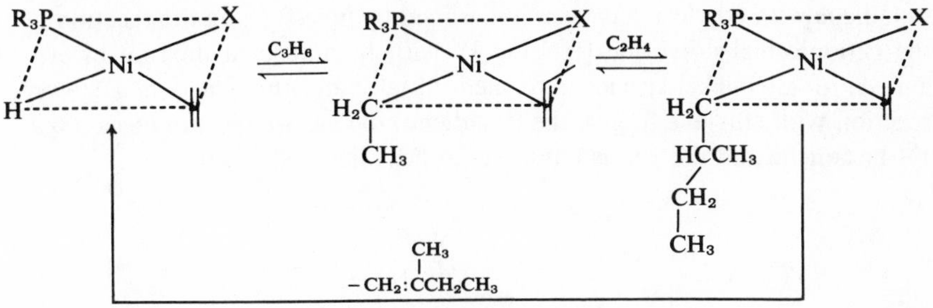

Partial support for the mechanism outlined above has been obtained by studying model complexes. The formation of a nickel-alkyl by the addition of a nickel-hydride to an olefin has been observed in the reactions of the π-allylnickel hydride **19** with ethylene (143; see Volume I, p. 147), and in the reactions of the $HNiX(PR_3)$ complexes with propylene (142). In the latter

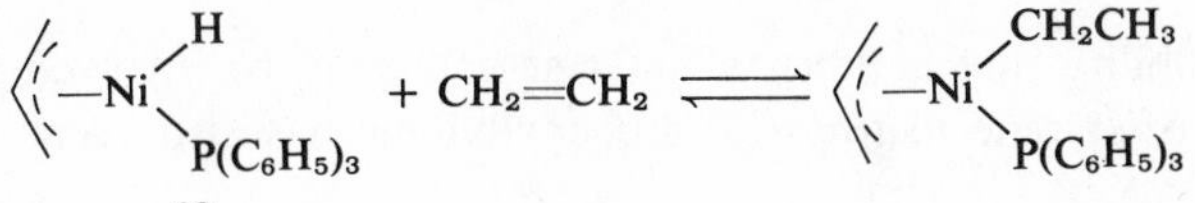

case the ratio of the nickel-propyl species **20** and **21** has been determined by adding CCl_4 and estimating the ratio of *n*-propylchloride to isopropyl-chloride. The results mirror those found in the catalytic process (see Volume

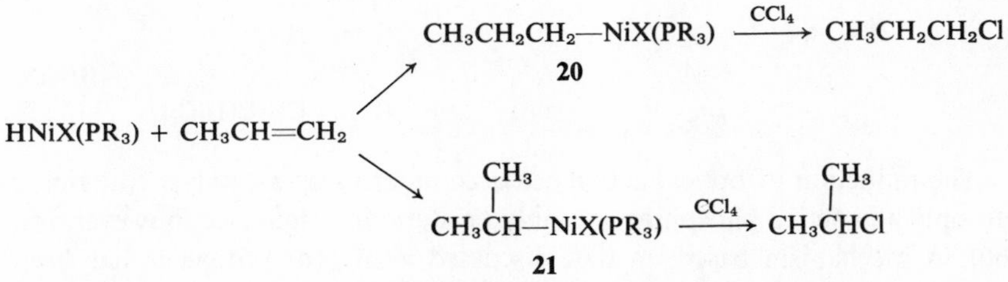

I, p. 142). The second step in the catalytic reaction (olefin insertion into a Ni–C bond) has not been observed in a model system.

The role of the Lewis acid in the catalytic reaction has received little attention. A recent study indicates that variation of the Lewis acid affects the first step in the dimerization of propylene but has no significant effect on the second, and it is suggested that the nature of the anion influences the rate at which the Ni–*n*-propyl species isomerizes to the thermodynamically more stable Ni–isopropyl species (141).

The relevance of a nickel-hydride intermediate in the codimerization of butadiene with ethylene has been underlined by investigating the reaction of the cationic nickel-hydride $[HNiLig_4]^+$ with butadiene: a mixture of *syn*- and *anti*-π-allylnickel species is formed initially and this reacts in a slower reaction with ethylene to give the hexadiene; the *syn* isomer produces *trans*-1,4-hexadiene and the *anti* isomer, *cis*-1,4-hexadiene (98, 140).

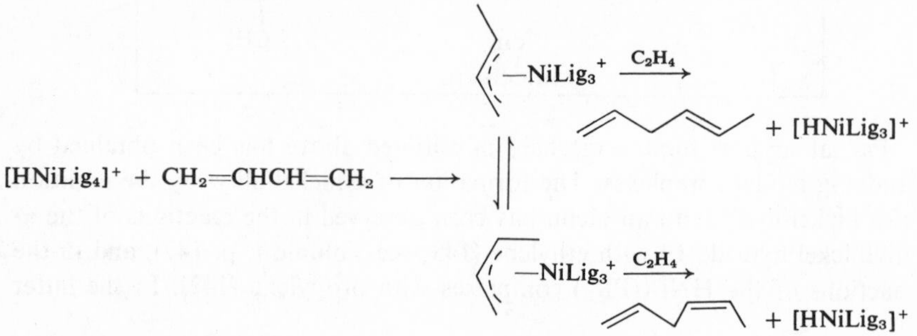

—NiLig3+
C2H4
+ [HNiLig3]+
[HNiLig4]+ + CH2=CHCH=CH2
—NiLig3+
C2H4
+ [HNiLig3]+

The possibility that a π-benzyl intermediate may be involved in the dimerization of styrene to *trans*-1,3-diphenylbutene has also been considered (48).

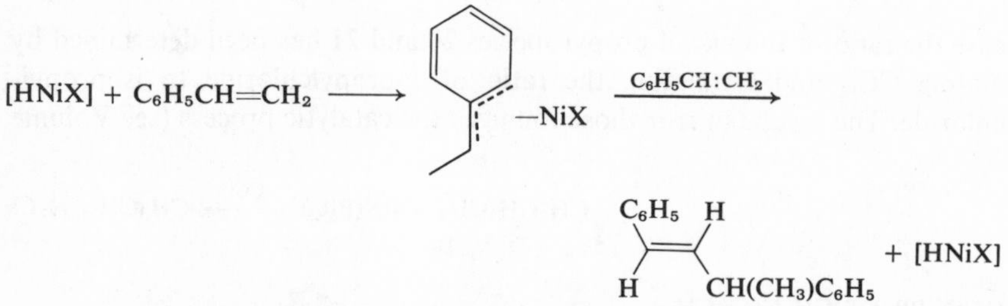

[HNiX] + C6H5CH=CH2
—NiX
C6H5CH:CH2
C6H5 H
H CH(CH3)C6H5
+ [HNiX]

The induction of optical activity in a codimer using a catalyst containing an optically active phosphine can only be steric in origin (see, however, ref. 60). A mechanism based on that discussed above for propylene has been proposed for the codimerization of bicycloheptene and ethylene (Fig. I-1) (106). It is assumed that the bicycloheptene molecule is initially complexed *exo* to the nickel atom, with the double bond perpendicular to the plane formed by the remaining ligands (**22**). The first assumption is supported by the geometry found crystallographically for (bicycloheptene)$_3$nickel (138; see 46), whereas the second is the typical Zeise's salt arrangement. Insertion of the olefin molecule into the Ni–H bond to give a nickel-alkyl is accompanied by

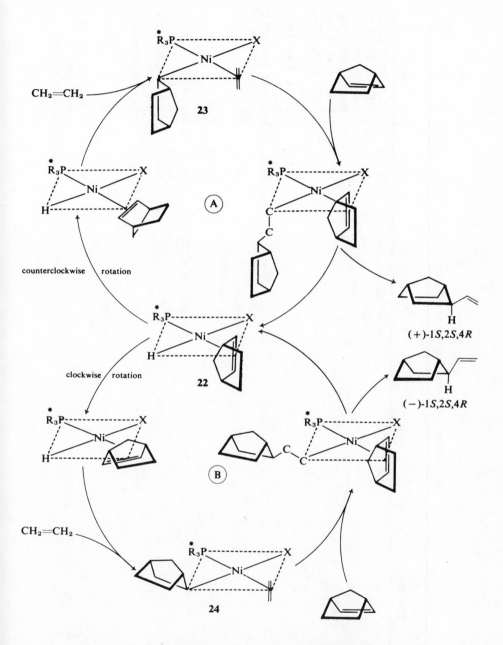

Fig. I-1. The formation of 2-vinylbicycloheptane (106).

TABLE I-7

THE OLIGOMERIZATION OF OLEFINS: HOMOGENEOUS CATALYSTS

Product	Olefin	Nickel comp.^{a}	Lewis acid	Ref.^{b}
C_4, C_6 Olefin	$CH_2:CH_2$	Nickel salt, e.g., $Ni(acac)_2$, Lig_2NiBr_2	R_2AlX	5, 34, 35, 40, 41, 43, 55, 357, P33, P37, P41, P68, P75, P80, P85, P88, P98, P99, P100, P102, P104, P108– P110, P118, P134, P144
			MX_n	64, P34, P96, P102, P140
			Misc.	P95, P102, P110, P124, P141, P153, P260
			—	137
		Nickel-alkyl, e.g., $Lig_2NiBr(R)$	R_2AlX	P102
			MX_n	8, 64, 65, 67
			—	P114, P153
		Ni-olefin, e.g., $(COD)_2Ni$	R_2AlX	P70, P111
			MX_n	7, 338, 341, P35, P70, P74
			Misc.	P114, P115, P122, P123, P153
			—	P137
		π-Allylnickel, e.g., $(\pi\text{-}C_3H_5NiBr)_2$	R_2AlX	4, 90, 111, 112, 114, P113, P137
			MX_n	4, 18, 19, 36, P36

π-Cyclobutadienenickel, e.g., $[\pi\text{-}(CH_3)_4C_4NiCl_2]_2$	R_2AlX	79, 81, 83, P110, P113
	Misc.	P113
	—	85, 118, P135
π-Cyclopentadienylnickel, e.g., $(\pi\text{-}C_5H_5)_2Ni$	R_2AlX	P102, P110, P111
$Lig_nNi(CO)_{4-n}$	R_2AlX	357, P91, P111
Lig_4Ni	R_2AlX	P111
	MX_n	29, P35, P60
	Misc.	P111
$Lig_2NiBr(NO)$	R_2AlX	P102
$(LigNiCS_2)_2$	R_2AlX	P111
Nickel salt, e.g., $Ni(acac)_2$, Lig_2NiBr_2	R_2AlX	26, 32, 34, 35, 40, 43–45, 49, 51, 52, 55, 56, 89, 91, 100, 127, 351, P7–P9, P12–P14, P22, P25, P27–P30, P33, P37, P38, P40, P41, P43, P45, P53, P56–P58, P66, P71, P73, P75–P80, P85, P86, P88, P97–P105, P107–P109, P118, P119, P133, P134, P139, P143, P147, P151, P250, P256
$CH_3CH:CH_2$	MX_n	50, P10, P22, P28, P34, P35, P61–P64, P102
C_6, (C_9) Olefin	Misc.	56, P17, P18, P61–P64, P136

(continued)

TABLE I-7 (*continued*)

Product	Olefin	Nickel comp.^{a}	Lewis acid	Ref.^{b}
		Nickel-alkyl, e.g., Lig$_2$NiX(R)	R$_2$AlX	2, 8, P81, P88, P259
		Nickel-olefin, e.g., (COD)$_2$Ni	MX$_n$ R$_2$AlX	8, 115, P72 P74, P111
		π-Allylnickel, e.g., (π-C$_3$H$_5$NiBr)$_2$	MX$_n$ R$_2$AlX	7, 341, P35, P50, P74 78, 90, 102, 108, 112, 114, 127, P88, P93, P109, P113, P136, P137
			MX$_n$	6, 18–20, 44, 78, 102, 116, 127, 135, 344, P36
		π-Cyclobutadienenickel, e.g., [π-(CH$_3$)$_4$C$_4$NiCl$_2$]$_2$	—	78, P59, P137 79, 80, 83, P113
		π-Cyclopentadienylnickel, e.g., (π-C$_5$H$_5$)$_2$Ni	R$_2$AlX	44, 78, 127, P111
		Lig$_n$Ni(CO)$_{4-n}$	R$_2$AlX MX$_n$	P111 50, 131, P50, P111, P149
		Lig$_4$Ni	R$_2$AlX MX$_n$	P108, P111 50, P35
		Lig$_2$NiX(NO)	R$_2$AlX	P45, P80, P85, P89
		Raney nickel	R$_2$AlCl	P20, P54, P57
		Nickel salt, e.g., Ni(acac)$_2$	R$_2$AlX	30, 42, 55, P42, P51, P53, P80, P85, P87
		π-AllylNiX	Misc. AlBr$_3$	P18, P24 P137
C$_8$ Olefin	CH$_2$:CHCH$_2$CH$_3$ and CH$_3$CH:CHCH$_3$			

C_{10} Olefin	$CH_2:CH(CH_2)_2CH_3$ and $CH_3CH:CHCH_2CH_3$	Nickel salt, e.g., $Ni(acac)_2$	R_2AlX	P80, P85, P88, P133, P134
C_{12} Olefin	$CH_2:CH(CH_2)_3CH_3$	Nickel salt	R_2AlX Misc.	55 P24
C_{16} Olefin	$CH_2:CH(CH_2)_5CH_3$	Nickel salt	R_2AlX	P99, P100
C_{12} Diolefin	$CH_2:CHCH_2CH:CHCH_3$	Nickel salt π-AllylNiX	R_2AlX —	58 58, 104
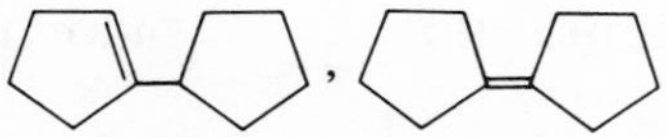		Nickel salt	R_2AlX	P39, P258
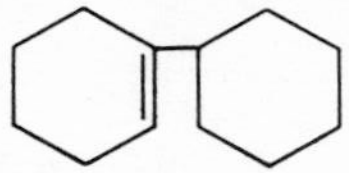	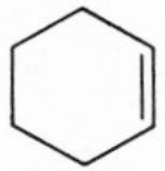	Nickel salt	R_2AlX	P67
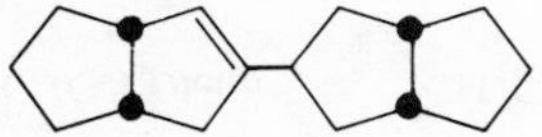	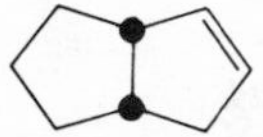	Nickel salt	R_2AlX	113
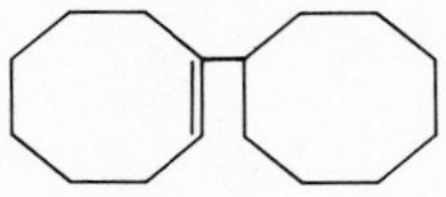 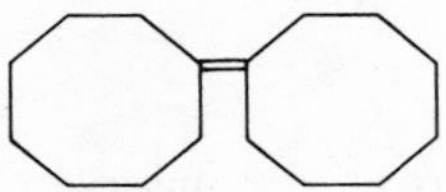	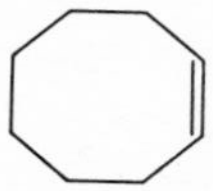	Nickel salt	R_2AlX	19, 20, 109
$C_6H_5CH:CHCH(CH_3)C_6H_5$ Dicyanobutene	$C_6H_5CH:CH_2$ $CH_2:CHCN$	π-AllylNiX π-AllylNiX $Ni(acac)_2$	R_2AlX — R_3Al	19 28, 48, 87, P48 P47

[a] In the majority of cases the nickel component is associated with a ligand.

[b] The references refer not only to the catalyst shown in the table but also to related systems.

TABLE I-8

THE OLIGOMERIZATION OF OLEFINS: SUPPORTED CATALYSTS

Product	Olefin	Nickel comp.	Support	Ref.
C_4, C_6 Olefin	$CH_2:CH_2$	Nickel salt, e.g., $NiCl_2$, NiO	$SiO_2-Al_2O_3$	37, 120, 124, 126, 128–130, P128, P156
			Al_2O_3	P3, P26, P116
			Misc.	66, 122, 123, P38, P83
		$(COD)_2Ni-Lig$	$SiO_2-Al_2O_3$	P114, P126
		π-Cyclopentadienylnickel, e.g., $(\pi-C_5H_5)_2Ni$	$SiO_2-Al_2O_3$	P90, P129, P131, P132
		$Lig_2Ni(CO)_2$	$SiO_2-Al_2O_3$	P4
		$[(C_6H_5)_3P]_4Ni$	Halopolystyrene	332
$C_6(C_9)$ Olefin	$CH_3CH:CH_2$	Nickel salt, e.g., $NiCl_2$	$SiO_2-Al_2O_3$	37, 38, 56, 126, P2, P150, P156
			Al_2O_3	P3, P26, P263
			Misc.	84, P11, P42, P44, P82
		$(COD)_2Ni$	Polyphosphinostyrene	P23
		$Lig_2Ni(CO)_2$	$SiO_2-Al_2O_3$	P4
		π-AllylNiCl	$SiO_2-Al_2O_3$	132, P120, P127, P128
		π-Cyclopentadienylnickel	$SiO_2-Al_2O_3$	P132
C_8 Olefin	$CH_2:CHCH_2CH_3$ and $CH_3CH:CHCH_3$	Nickel salt	$SiO_2-Al_2O_3$	P128
			Al_2O_3	P26
			Misc.	P42, P262
		π-Cyclopentadienylnickel	$SiO_2-Al_2O_3$	P132
C_{12} Olefin	$(CH_3)_2C:CH_2$	Ni-S	$SiO_2-Al_2O_3$	P252
	Hexene	NiO_2	$SiO_2-Al_2O_3$	P152

TABLE I-9

The Co-oligomerization of Olefins: Homogeneous Catalysts

Product	Olefin	Nickel comp.^{a}	Lewis acid	Ref.^{b}
C_5 olefin	$CH_2:CH_2$	$CH_3CH:CH_2$	Nickel salt, e.g., $Ni(acac)_2$, Lig_2NiX_2	R_2AlX	34, 39, 43, 55, 96, P6, P52, P57, P75, P84, P88, P102, P104, P112, P118, P261
			Nickel-alkyl	MX_n	8, 68
			Nickel-olefin	MX_n	69
			π-AllylNiX	R_2AlX	96, 112, P137
				MX_n	96
			π-Cyclobutadiene-nickel, e.g., $[\pi\text{-}(CH_3)_4C_4NiCl_2]_2$	R_2AlX	79, 82, 83, P112
			Lig_4Ni		P111
C_6, C_8 Olefin	$CH_2:CH_2$	$CH_2:CHCH_2CH_3$ and $CH_3CH:CHCH_3$	Nickel salt, e.g., $Ni(acac)_2$	R_2AlX	30, 55, P6, P69, P80, P85, P110, P112
			π-AllylNiX	MX_n	106, P137
			π-Cyclobutadiene-nickel	R_2AlX	83
			$Lig_2Ni(CO)_2$	R_2AlX	P111
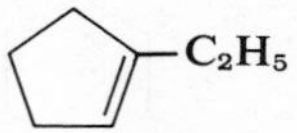 (cyclopentene with C_2H_5)	$CH_2:CH_2$		Nickel salt	R_2AlX	P88
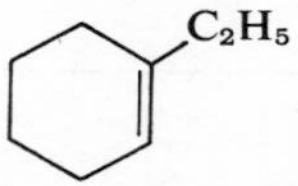 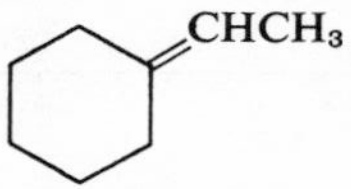 (cyclohexene with C_2H_5, $=CHCH_3$)	$CH_2:CH_2$	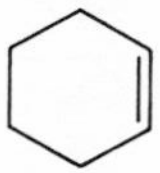	π-AllylNiX	MX_n	P137

33

(continued)

TABLE I-9 (*continued*)

Product	Olefin	Nickel comp.^{a}	Lewis acid	Ref.^{b}
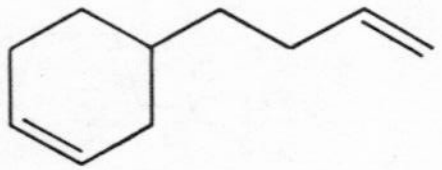	$CH_2:CH_2$	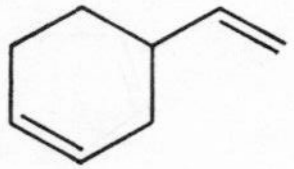 Nickel salt	R_2AlX	P16
Phenylbutene, phenylhexene, diphenylhexene	$CH_2:CH_2$	Nickel-olefin	R_2AlX	P15
	$C_6H_5CH:CH_2$	Nickel-alkyl	MX_n	P15
		π-AllylNiX	MX_n	57, 350, P92
			R_2AlX	110
$CH_2:CHC(CH_3)_2C_6H_5$	$CH_2:CH_2$	Lig_4Ni	MX_n	29, P60
Ethylphenylbenzene	$CH_2:CH_2$	Lig_4Ni	MX_n	29, P60
Dibutenylbenzene	$CH_2:CH_2$	Lig_4Ni	MX_n	29, P60
$C_6H_5C(CH_3):CH_2$		Lig_4Ni	MX_n	29, P60
$C_2H_5C_6H_4CH:CH_2$				
$C_6H_4(CH:CH_2)_2$				
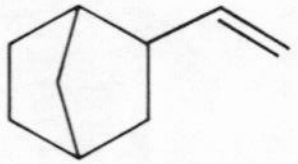 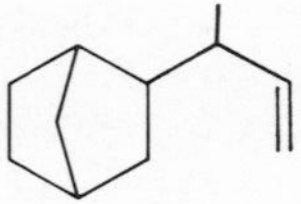 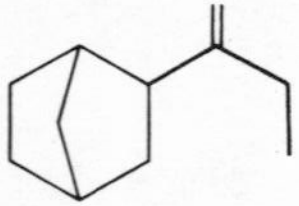	$CH_2:CH_2$ 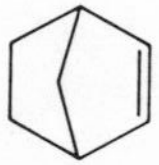	π-AllylNiX	MX_n	104, 106, 110, 111
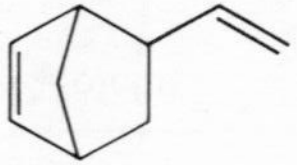	$CH_2:CH_2$	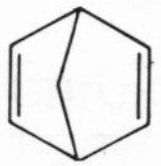 Nickel salt	R_2AlCl	P106
		π-AllylNiX	R_2AlCl	106, 110
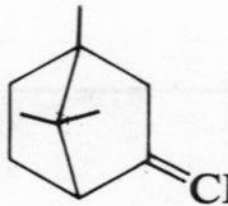	$CH_2:CH_2$ 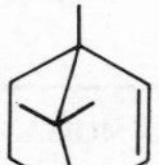	π-AllylNiX	R_2AlX	106, 110

34

Product		Diene	Catalyst	Cocatalyst	References
CH₂:CHCH₂CH:CHCH₃, CH₂:CHCH(CH₃)CH:CH₂	CH₂:CH₂	CH₂:CHCH:CH₂	Nickel salt	R_2AlX	5, 70, P31, P33, P138, P154, P158
			[HNiLig₄]⁺	R_3Al	1, 5, 53, 54, 70, P234
			(COD)₂Ni	—	98
			π-Cyclopentadienyl-nickel	R_2AlX	95, 97
			Lig₂Ni(CO)₂	R_2AlX	P130
				R_2AlX	P31, P32
				MX_n	P50
CH₂:CHCH₂C(CH₃):CHCH₃, CH₂:C(CH₃)CHCH₃CH:CH₂	CH₂:CH₂	CH₂:C(CH₃)CH:CH₂	Nickel salt	R_2AlX	70, P31, P158
CH₂:CHCH₂CCl:CHCH₃	CH₂:CH₂	CH₂:CClCH:CH₂	Nickel salt	R_2AlX	70, P31, P158
CH₂:CHCH(CH₃)CH:CHCH₃	CH₂:CH₂	CH₂:CHCH:CHCH₃	Nickel salt	R_2AlX	70, P31, P158
CH₂:CHCH(C₂H₅)CH:CHCH₃, CH₂:CHCH(CH₃)CH:CHC₂H₅	CH₂:CH₂	CH₃CH:CHCH:CHCH₃	Nickel salt	R_2AlX	P31, P158
CH₂:CHCH₂CCl:CClCH₃	CH₂:CH₂	CH₂:CClCCl:CH₂	Nickel salt	R_2AlX	P31, P158
[vinylcyclohexene structure]	CH₂:CH₂	[cyclohexadiene structure]	Nickel salt	R_2AlX	70, 359, P31, P158
[vinylcyclooctene structure]	CH₂:CH₂	[cyclooctadiene structure]	Nickel salt	R_2AlX	70, 109, 359, P31, P158
			π-AllylNiX	R_2AlX	17, 19, 106, 111, P159
[vinylcycloheptene structure]	CH₂:CH₂	[cycloheptadiene structure]	Nickel salt	R_2AlX	359

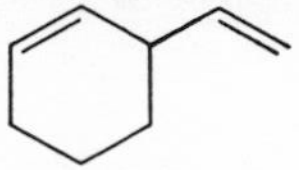

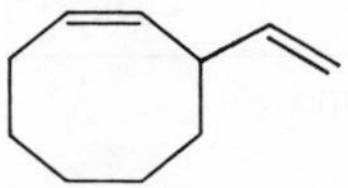

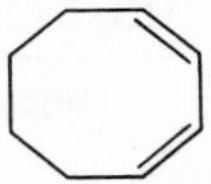

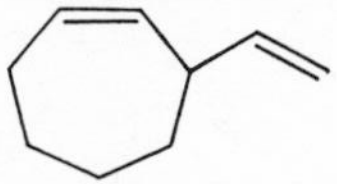

(continued)

35

TABLE I-9 (*continued*)

Product	Olefin		Nickel comp.[a]	Lewis acid	Ref.[b]
C_7 Olefin	$CH_3CH{:}CH_2$	$CH_2{:}CHCH_2CH_3$ and $CH_3CH{:}CHCH_3$	Nickel salt	R_2AlX	55, P6, P28, P45, P46, P49, P51, P52, P55, P57, P108
				MX_n	P28
			π-AllylNiX	MX_n	106, P137
			$Lig_2Ni(CO)_2$	MX_n	P50
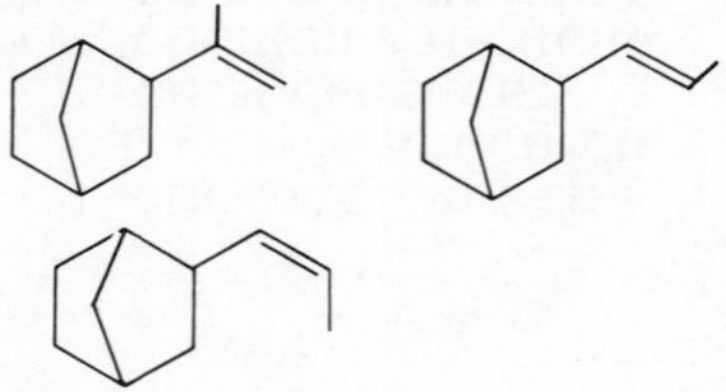	$CH_3CH{:}CH_2$	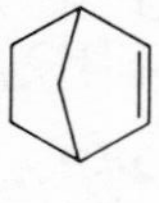	π-AllylNiX	R_2AlX	110
C_8 Olefin	$CH_3CH{:}CH_2$	$CH_2{:}CH(CH_2)_2CH_3$	Nickel salt	R_2AlX	P88
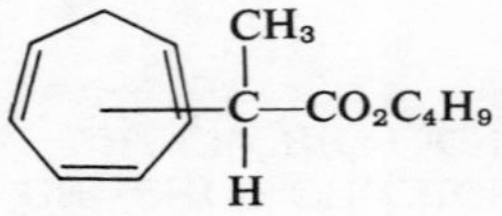	$CH_2{:}CHCO_2$-C_4H_9	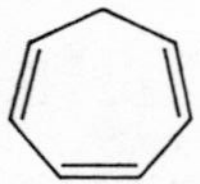	$(C_6H_5)_3PNi(CO)_3$	—	21

Reactant	Product	Catalyst	Co-catalyst	References
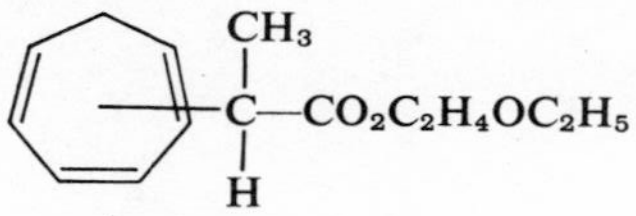 cycloheptatriene–CH_3–C(H)–$CO_2C_2H_4OC_2H_5$	$CH_2:CHCO_2$-$C_2H_4OC_2H_5$ + cycloheptatriene	 $(C_6H_5)_3PNi(CO)_3$	—	21
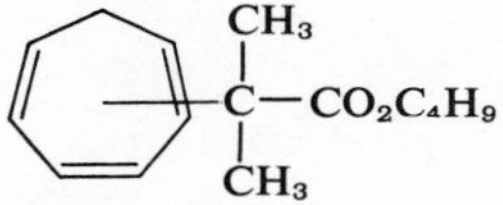 cycloheptatriene–CH_3–C(CH_3)–$CO_2C_4H_9$	$CH_2:C(CH_3)$-$CO_2C_4H_9$ + cycloheptatriene	 $(C_6H_5)_3PNi(CO)_3$	—	21
$CH_2:C(CH_3)CH_2CH:CHCH_3$	$CH_3CH:CH_2$	Nickel salt	R_2AlX	70, P31, P158
$CH_2:C(CH_3)CH_2C(CH_3):CHCH_3$	$CH_3CH:CH_2$	Nickel salt	R_2AlX	P31, P158
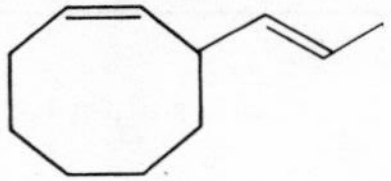 cyclooctadiene structure	$CH_3CH:CH_2$ + cyclooctadiene	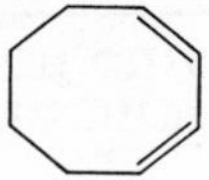 Nickel salt	R_2AlX	109, P159
		π-AllylNiX	R_2AlX	19
			MX_n	10
$CH_2:C(C_4H_9)CH_2CH:CHCH_3$	$CH_2:CH$-$(CH_2)_3CH_3$ + cyclooctadiene	Nickel salt	R_2AlX	70, P31, P158
$CH_2:CHCH(OCH_3)(CH_2)_3CH:CH_2$, $CH_3OCH_2CH:CH(CH_2)_3CH:CH_2$	CH_3OCH_2-$CH:CHCH_3$ + cyclooctadiene	Nickel salt	C_4H_9Li	11

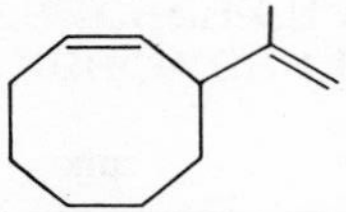

[a] In the majority of cases the nickel component is associated with a donor ligand.
[b] The references refer not only to the catalyst shown in the table but also to related systems.

37

TABLE I-10

THE CO-OLIGOMERIZATION OF OLEFINS: SUPPORTED CATALYSTS

Product	Olefin	Nickel comp.	Support	Ref.
C_5 Olefins	$CH_3CH:CH_2$	Nickel salt	$SiO_2-Al_2O_3$ Misc.	37, 126, P156 31, 59, P42
$CH_2:CHCH_2CH:CHCH_3$, $CH_2:CHCH(CH_3)CH:CHC_2H_5$	$CH_2:CH_2$	Nickel salt	$SiO_2-Al_2O_3$	P1, P150
C_7 Olefins	$CH_3CH:CH_2$	$Lig_2Ni(CO)_2$ Nickel salt	$SiO_2-Al_2O_3$ $SiO_2-Al_2O_3$	P4 P26, P156
	$CH_2:CHCH_2CH_3$ or $CH_3CH:CHCH_3$			

rotation of the bicycloheptene molecule, which can occur either in a clockwise manner (cycle B) or in a counterclockwise manner (cycle A). In the presence of a nickel-bonded optically active phosphine, the barrier to rotation is different in each case and the intermediate nickel-alkyl complexes **23** and **24** are formed in different concentrations. As a result, the ensuing ethylene insertion and codimer elimination leads to the preferential formation of one of the enantiomers.

Simulation of this process, using a model of the intermediate complex **22** in which the optically active phosphine is $(-)$-(menthyl)$_2$PCH$_3$, shows that clockwise rotation is associated with an unfavorable interaction of the bridging methylene group of the bicycloheptene with one of the menthyl groups; such an interaction is absent in the case of the counterclockwise rotation and hence $(+)$-1S,2S,4R-2-vinylbicycloheptane is expected to be formed preferentially. This is indeed observed. The suggested interference between the menthyl group and the organic ligand is supported by an x-ray structural determination of π-CH$_3$CHCHCHCH$_3$NiCH$_3$[$(-)$-(menthyl)$_2$-PCH$_3$] in which one of the isopropyl groups associated with the menthyl substituent is seen to closely approach the nickel atom (106, 121) and this arrangement may be supposed to be statistically preferred in solution.

IV. The Oligomerization of Strained Olefins and Alkanes
(Tables I-11 and I-12)

The unusual behavior exhibited by strained olefins and alkanes in oligomerization reactions justifies a separate treatment. These reactions, in contrast to those discussed in Section III, are in general catalyzed by zerovalent nickel complexes and as such have much in common with the oligomerization of butadiene and other 1,3-dienes discussed in Chapter III. The frequency with which two basic reactions are observed is characteristic: (1) $2_\pi + 2_\pi$ addition to give cyclobutane derivatives, e.g., **25** and **26**, formed by dimerization of methylenecyclopropane and 3,3-dimethylcyclopropene, respectively, and (2)

      25   **26**

$2_\sigma + 2_\pi$ addition to give cyclopentane derivatives, e.g., **27** and **28**, derived from the dimerization of methylenecyclopropane and the dimerization of bicycloheptadiene.

Methylenecyclopropane undergoes both types of reaction in the presence of zerovalent nickel complexes; i.e., both **25** and **27** are formed. In addition, rearrangement to butadiene followed by dimerization to methylenevinyl-cyclopentane (29) has been observed in the presence of $[(C_4H_9)_3P]_2NiBr_2$–C_4H_9Li (216). An interesting example of asymmetric addition is observed

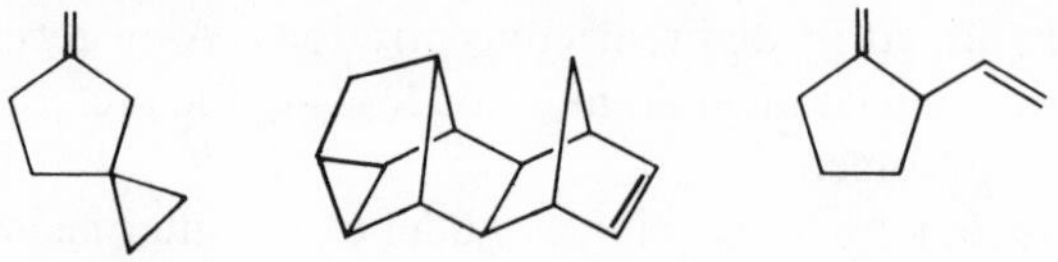

27   28   29

in the reaction between methylenecyclopropane and bicycloheptadiene catalyzed by $(COD)_2Ni$ modified with the optically active phosphine (−)-benzylmethylphenylphosphine: the cyclobutane derivative 30 is produced in an optically active form (but of unknown optical purity) (76).

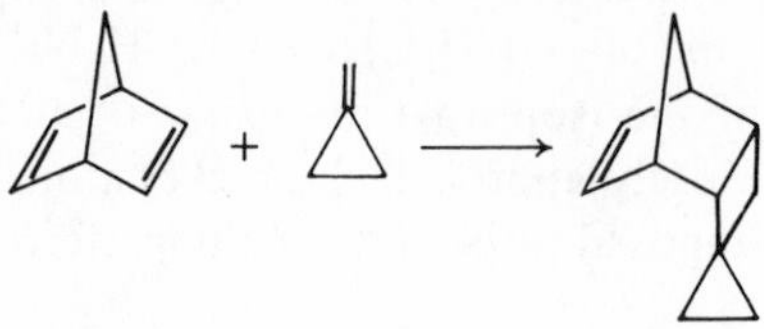

30

Although a detailed mechanism for the reactions of methylenecyclopropane cannot be given, the suggested intermediacy of bis-σ-alkylnickel species, e.g., 31 and 32 as precursors of 25 and 27, is supported by the isolation of the bipyridyl adduct corresponding to 31 (222). The possibility that a trimethyl-

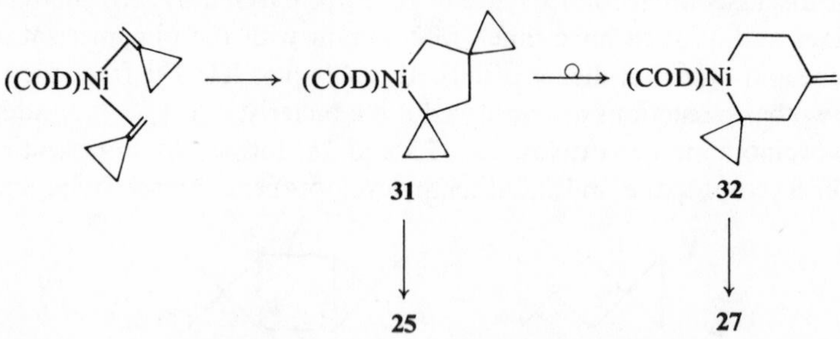

31   32

25   27

enemethane–nickel intermediate is involved is considered unlikely because 33 and 34 are isolated from the reaction of methylacrylate with 2,2-dimethyl-methylenecyclopropane and isopropylidenecyclopropane, respectively. These two reactions can be expected to give the same codimer or mixture of co-dimers if an intermediate dimethyltrimethylenemethane derivative (i.e., 35) is involved (74). The rearrangement of methylenecyclopropane to butadiene

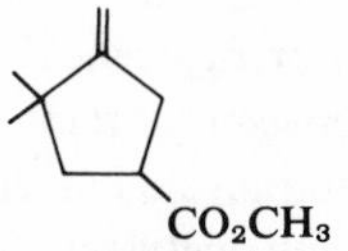

$$\underset{\textbf{33}}{\text{CO}_2\text{CH}_3} \qquad \underset{\textbf{34}}{\text{CO}_2\text{CH}_3} \qquad \underset{\textbf{35}}{\text{—Ni}}$$

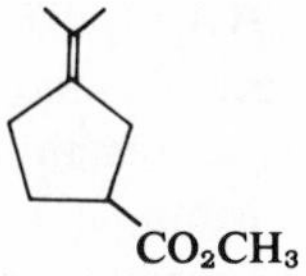

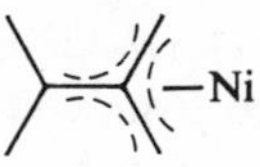

mentioned above has a precedent in the stoichiometric thermal rearrangement of the bisphosphite nickel-methylenecyclopropane complex **36** (221). This reaction necessarily involves a hydrogen migration but its mechanism is not understood.

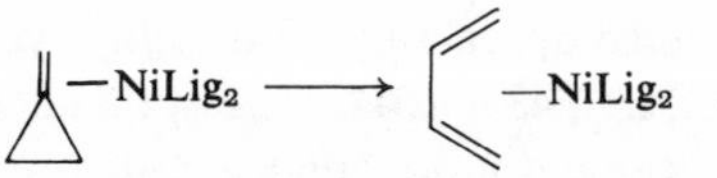

$$\underset{\textbf{36}}{\text{—NiLig}_2} \longrightarrow \text{—NiLig}_2 \qquad \text{Lig} = \text{P(OC}_6\text{H}_4\text{-}o\text{-C}_6\text{H}_5)_3$$

Bicycloheptadiene also takes part in both $2_\pi + 2_\pi$ and $2_\sigma + 2_\pi$ addition reactions in the presence of zerovalent nickel catalysts as well as of $(\text{R}_3\text{P})_2\text{Ni(CN)}_2$. The structure of most of the oligomers has been well established (10, 24, 101; a review is to be found in ref. 218) and from the nickel-catalyzed reactions five dimers and a trimer have been isolated. The dimers have been identified as the exo-trans-exo, endo-trans-endo and exo-trans-endo isomers of pentacyclo[8.2.1.1.^{4,7}.0^{2,9}.0^{3,8}]tetradecadiene (**37–39**) and the nortricyclic derivatives **28** and **40**. The structure of the trimer is unknown but it is probably an isomer of **41** (P117, P146). A novel dimerization reaction

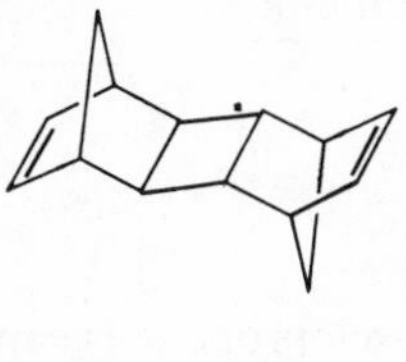

37 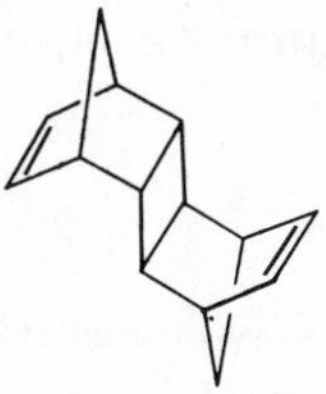   38 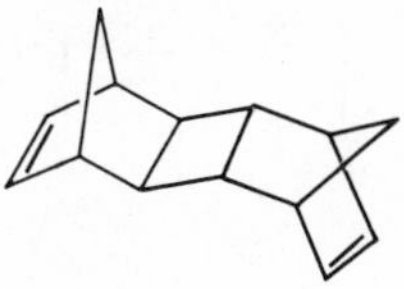   39

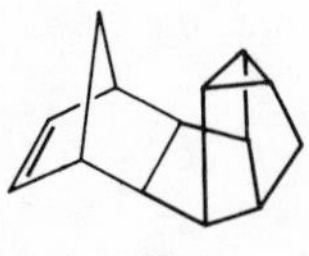

40 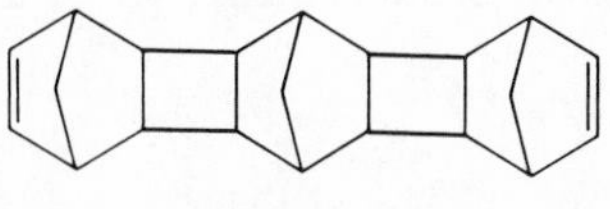   41

has been observed in the presence of $[(\text{C}_4\text{H}_9)_3\text{P}]_2\text{NiX}_2\text{–NaBH}_4$ using amines as the solvent: *exo*-5-*o*-tolyl-2-bicycloheptene is formed in addition to **37** and **39** (337). In addition, extensive polymer formation has been reported to occur in the presence of $[(C_6H_5)_3P]_2NiX_2$ (X = Cl, Br, I) and $Ni(CO)_4$–$BF_3 \cdot O(C_2H_5)_2$ (94, 220). Both cyclobutane and nortricyclene derivatives have been isolated from codimerization reactions between bicycloheptadiene and olefins or alkynes. An unusual compound, the tetracycloundecadiene **42**, has been isolated in traces from the reaction involving acetylene (92).

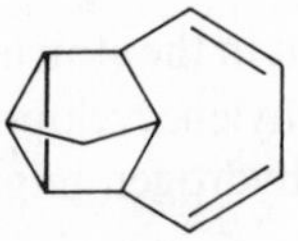

42

The structures of the products formed by reacting 3,3-dimethylcyclopropene or bicyclo[1.1.0]butane with olefins, e.g., **43** and **44**, suggest that here carbene intermediates, which add to the olefinic double bond, are involved (72, 77, 214, 347).

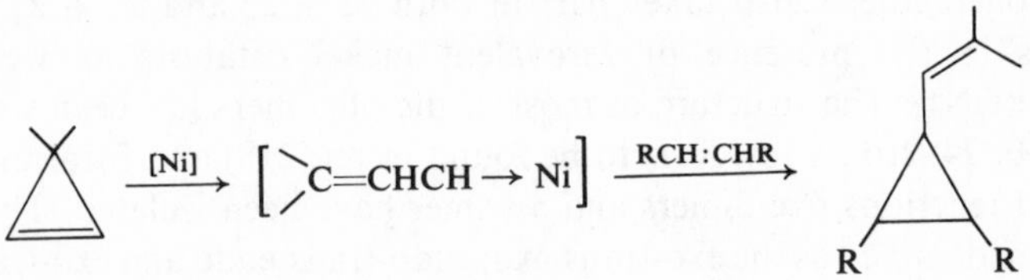

43

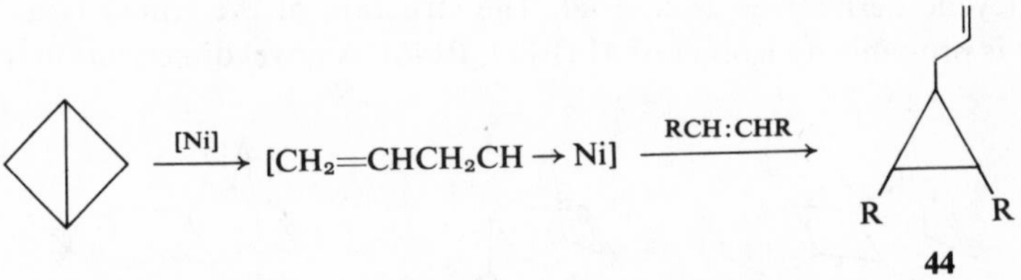

44

The dimerization reaction involving methylenecyclopropane is suppressed in the presence of a phosphine-modified zerovalent nickel catalyst. Instead, trimerization is observed to give, depending on the nature of the phosphine, either mainly linear products, e.g., **45**, or a cyclic trimer, e.g., **46** (13),

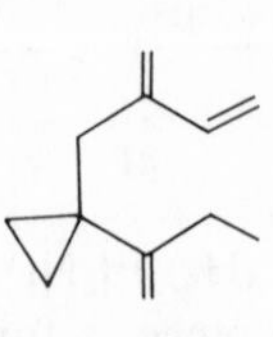

45 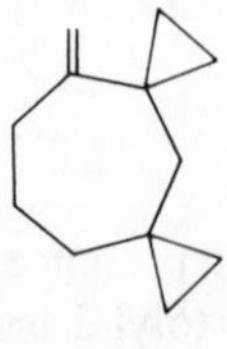   **46**

THE OLIGOMERIZATION OF STRAINED OLEFINS AND RELATED REACTIONS

Product	Olefin	Catalyst	Ref.
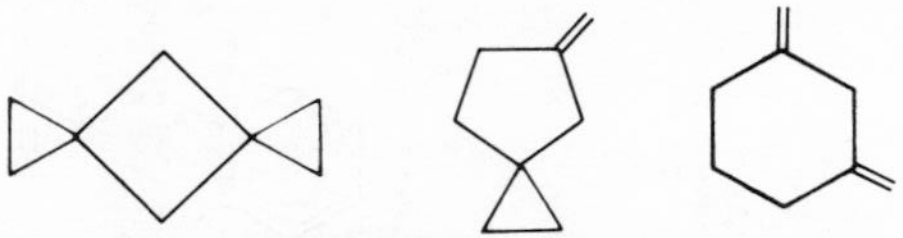		$(COD)_2Ni$	14, 15
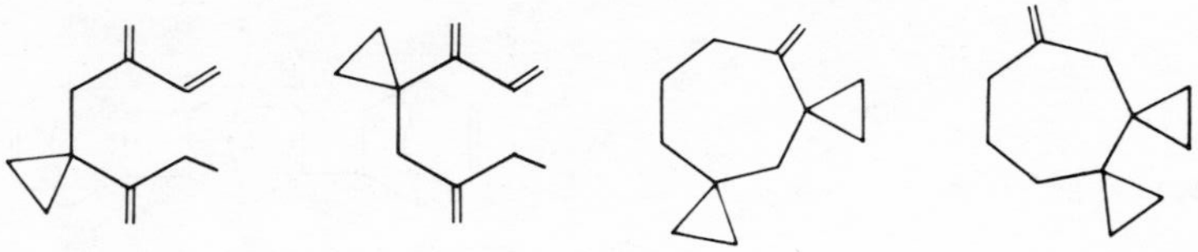		$(COD)_2Ni-PR_3$	13
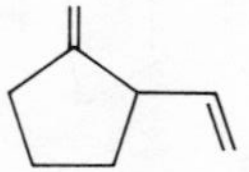		$(R_3P)_2NiBr_2-C_4H_9Li$	216
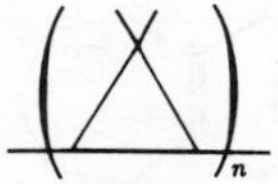		$(COD)_2Ni$	214
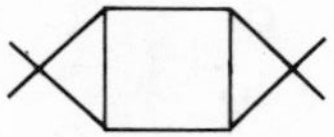		$[(C_6H_5)_3P]_4Ni$	215
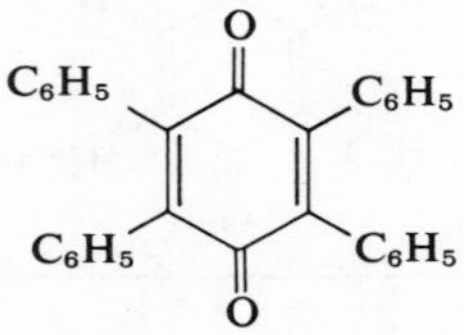	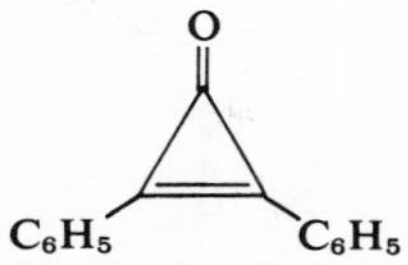	$(COD)_2Ni$	75

(continued)

43

TABLE I-11 *(continued)*

Product	Olefin	Catalyst	Ref.
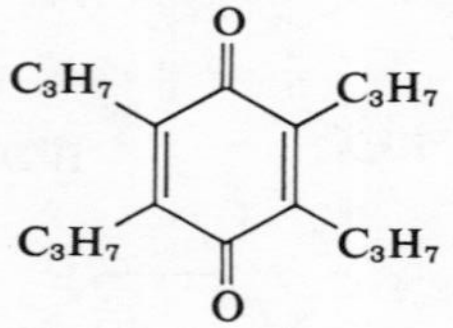	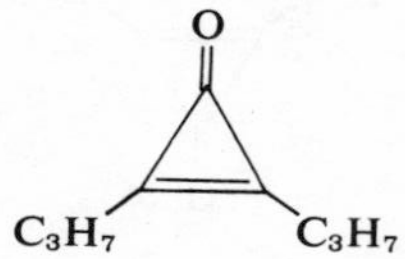	$(COD)_2Ni$	75
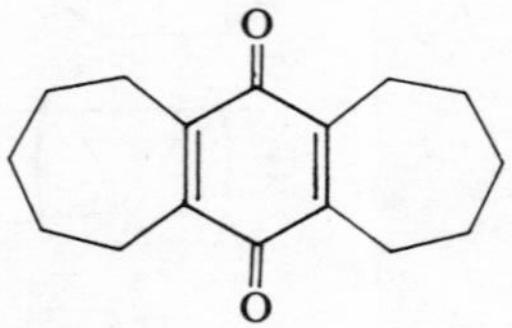	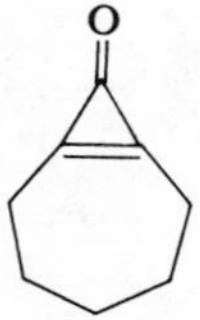	$(COD)_2Ni$	75
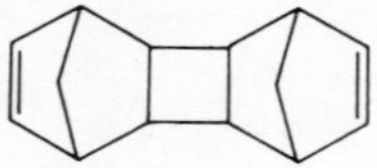 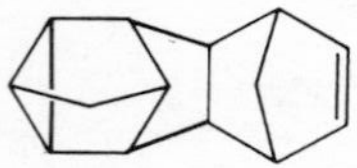	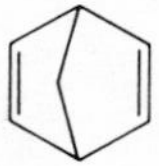	$Ni(CO)_4$	12, 16, 92–94
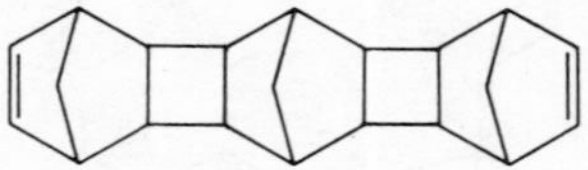		$(COD)_2Ni$	101, 219, 323, P117, P121, P145, P146
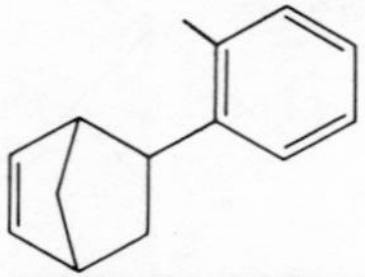		$[(C_4H_9)_3P]_2NiX_2-$ $NaBH_4$	337

44

TABLE I-12

THE CO-OLIGOMERIZATION OF STRAINED OLEFINS AND RELATED REACTIONS

Product	Olefin	Catalyst	Ref.
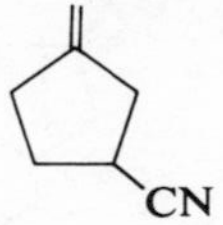 (cyclopentane with =CH₂ and CN)	 $CH_2:CHCN$	$(CH_2:CHCN)_2Ni$	74
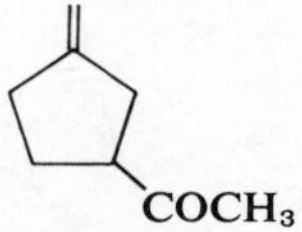 (cyclopentane with =CH₂ and $COCH_3$)	$CH_2:CHCOCH_3$	$(CH_2:CHCN)_2Ni$	74
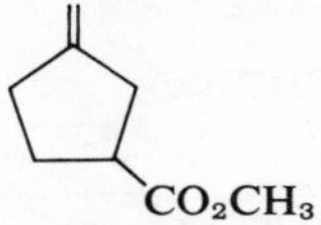 (cyclopentane with =CH₂ and CO_2CH_3)	$CH_2:CHCO_2CH_3$	$(CH_2:CHCN)_2Ni$	14, 71, 74
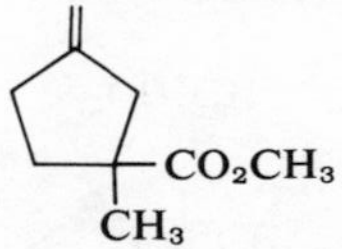 (cyclopentane with =CH₂ and CO_2CH_3, CH_3)	$CH_2:C(CH_3)CO_2CH_3$	$(COD)_2Ni$	14
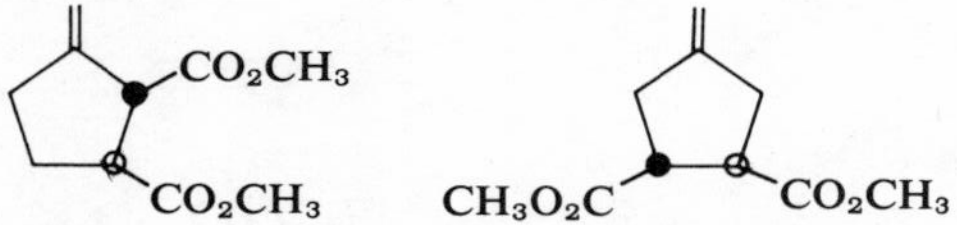 (cyclopentane with =CH₂ and CO_2CH_3, CO_2CH_3)	$trans\text{-}CH_3CO_2CH:CHCO_2CH_3$	$(CH_2:CHCN)_2Ni$	71
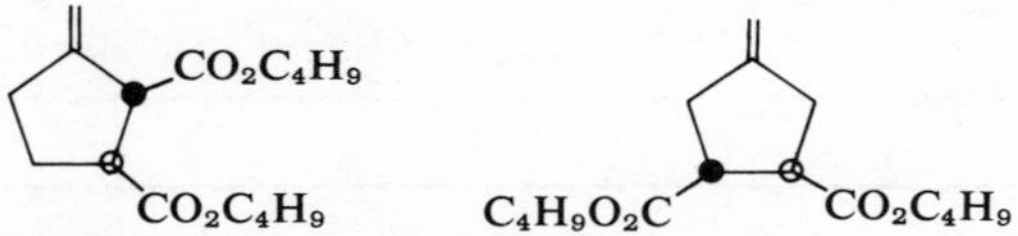 (cyclopentane with =CH₂ and $CO_2C_4H_9$, $CO_2C_4H_9$)	$trans\text{-}C_4H_9CO_2CH:CHCO_2C_4H_9$	$(COD)_2Ni$	14

(continued)

45

TABLE I-12 (*continued*)

Product	Olefin	Catalyst	Ref.
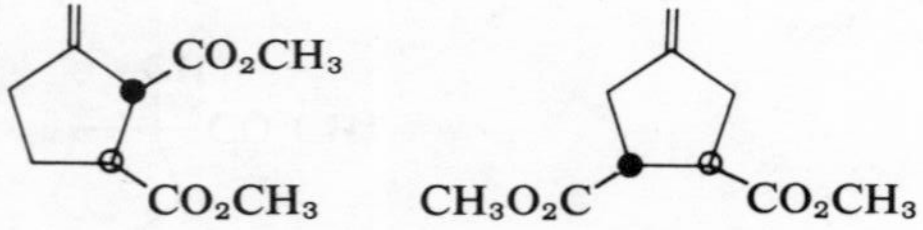	cis-CH_3CO_2CH:$CHCO_2CH_3$	$(CH_2$:$CHCN)_2Ni$	71
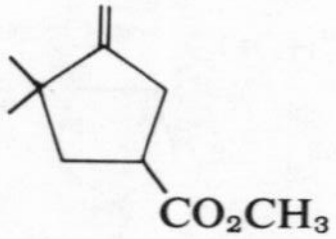	 CH_2:$CHCO_2CH_3$	$(CH_2$:$CHCN)_2Ni$	74
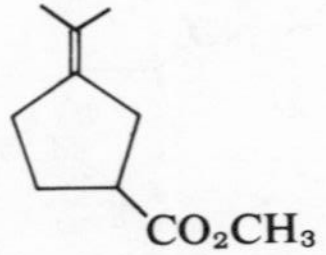	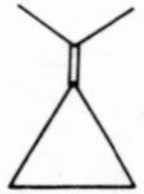 CH_2:$CHCO_2CH_3$	$(CH_2$:$CHCN)_2Ni$	74
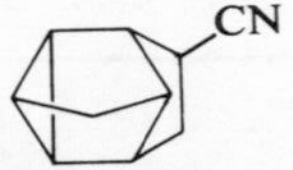	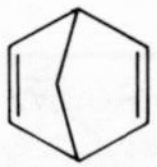 CH_2:$CHCN$	$(CH_2$:$CHCN)_2Ni$	92, 93, P19

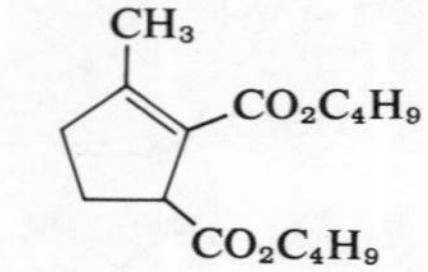

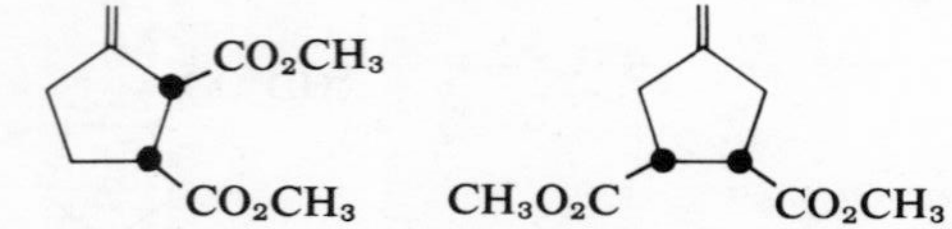

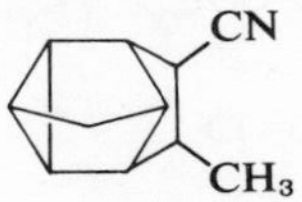	$CH_3CH:CHCN$	$[(C_6H_5)_3P]_2Ni(CO)_2$	92
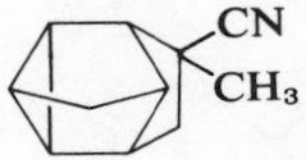	$CH_2:C(CH_3)CN$	$[(C_6H_5)_3P]_2Ni(CO)_2$	92
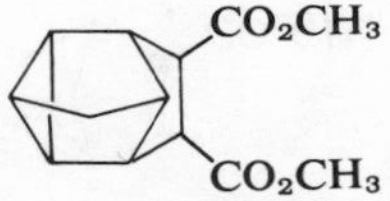	$cis\text{-}CH_3CO_2CH:CHCO_2CH_3$	$[(C_6H_5)_3P]_2Ni(CO)_2$	92
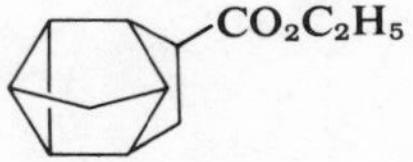	$CH_2:CHCO_2C_2H_5$	$(C_6H_5)_3PNi(CO)_3$	93, P19, P65
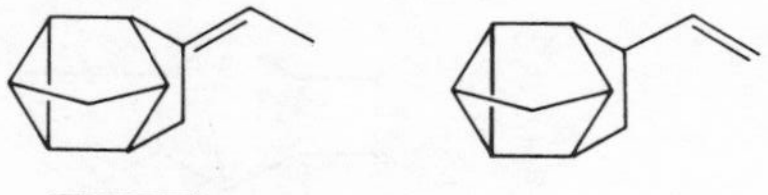	$CH_2:CHCH:CH_2$	$[(C_6H_5)_3P]_2NiCl_2-(C_2H_5)_2AlCl$	22
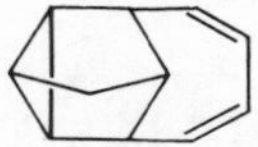	$HC:CH$	$[(C_6H_5)_3P]_2Ni(CO)_2$	92
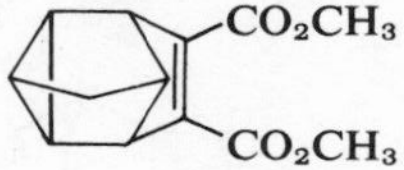	$CH_3CO_2C:CCO_2CH_3$	$[(C_6H_5)_3P]_2Ni(CO)_2$	92
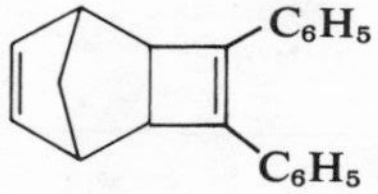	$C_6H_5C:CC_6H_5$	$[(C_6H_5)_3P]_2Ni(CO)_2$	92

(*continued*)

47

TABLE I-12 (*continued*)

Product	Olefin	Catalyst	Ref.
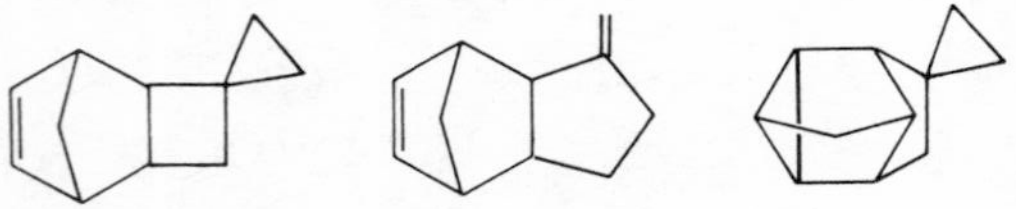		$(COD)_2Ni$	76
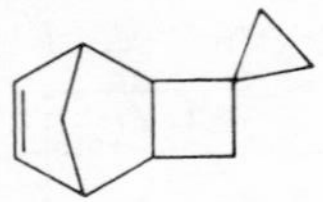		$(COD)_2Ni-P(C_6H_5)_3$	76
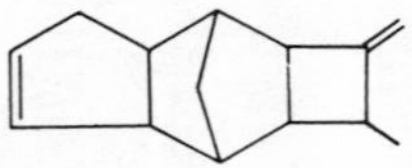	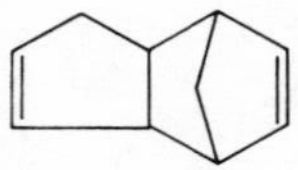 $CH_2:CHCH:CH_2$	$[(C_4H_9)_3P]_2NiBr_2-NaBH_4$	103, 358
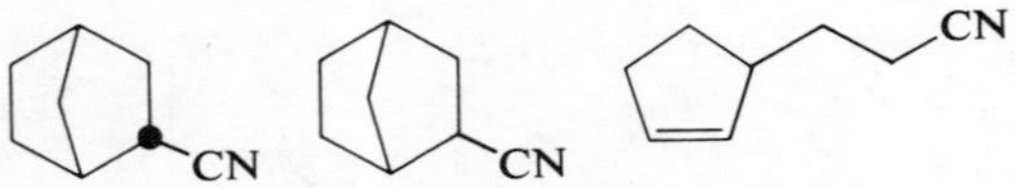	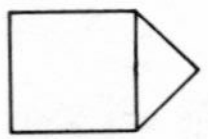 $CH_2:CHCN$	$(CH_2:CHCN)_2Ni$	73
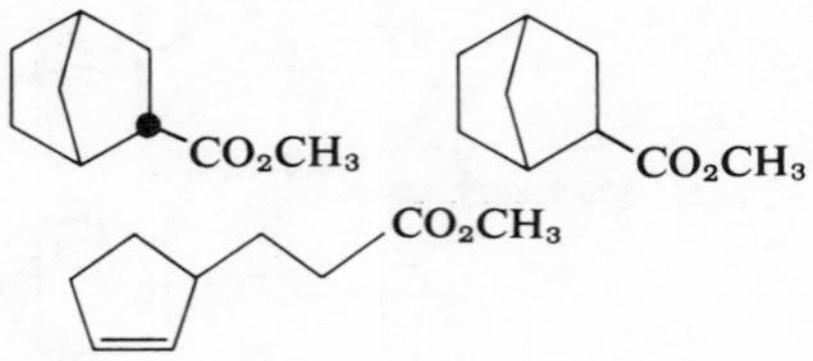	$CH_2:CHCO_2CH_3$	$(CH_2:CHCN)_2Ni$	73, 339
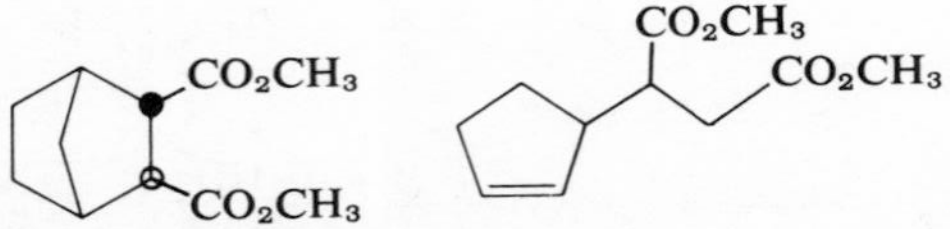	*trans*-$CH_3CO_2CH:CHCO_2CH_3$	$(CH_2:CHCN)_2Ni$	73, 339

48

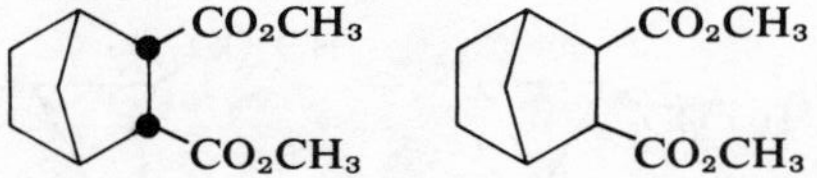

cis-$CH_3CO_2CH{:}CHCO_2CH_3$ $(CH_2{:}CHCN)_2Ni$ 73, 339

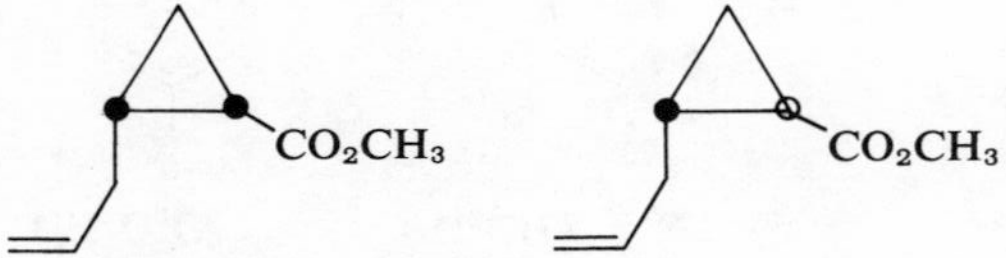

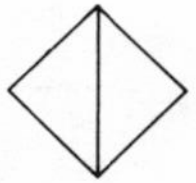

$CH_2{:}CHCO_2CH_3$ $(CH_2{:}CHCN)_2Ni$ 72, 347

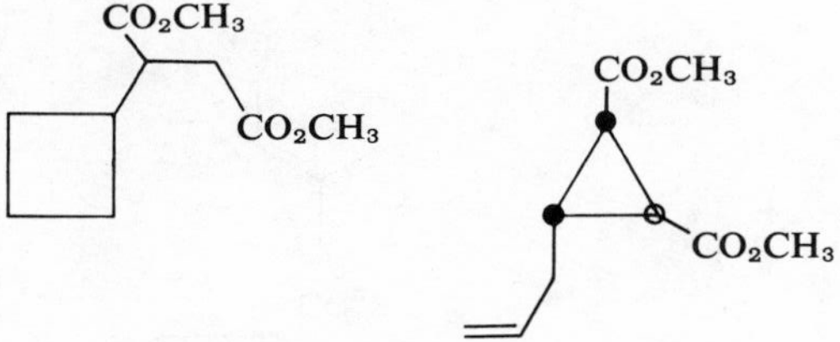

$trans$-$CH_3CO_2CH{:}CHCO_2CH_3$ $(CH_2{:}CHCN)_2Ni$ 72

cis-$CH_3CO_2CH{:}CHCO_2CH_3$ $(CH_2{:}CHCN)_2Ni$ 72

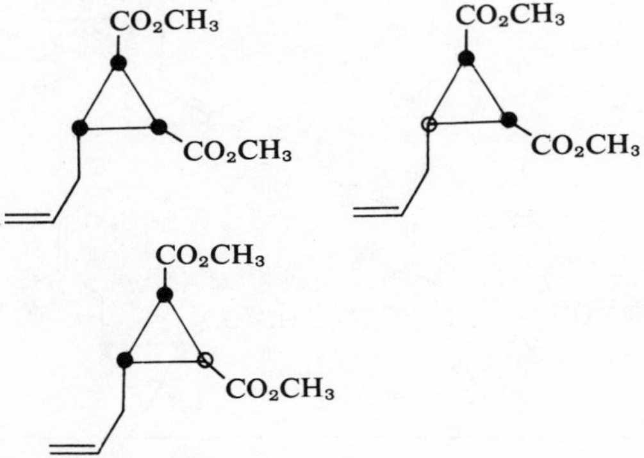

49

(continued)

TABLE 1-12 (*continued*)

Product	Olefin	Catalyst	Ref.
	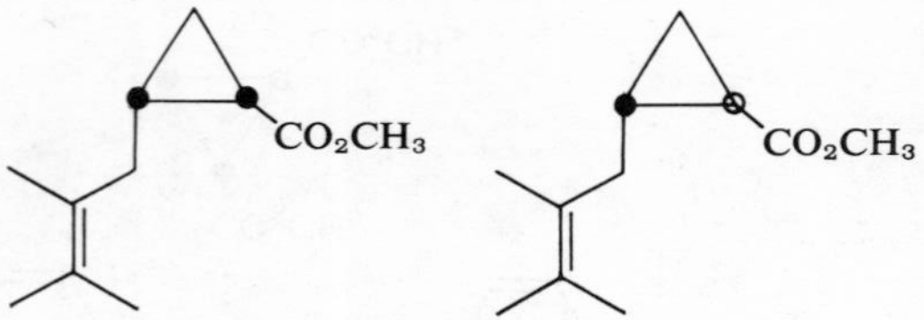 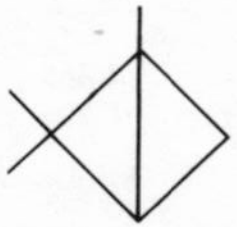 $CH_2:CHCO_2CH_3$	$(CH_2:CHCN)_2Ni$	72, 77
	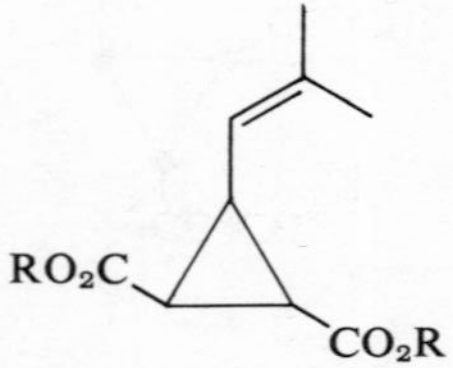  *trans*-$RO_2CCH:CHCO_2R$	$(COD)_2Ni$	214
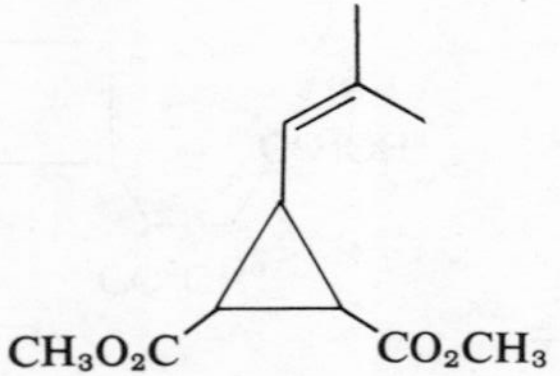	*cis*-$CH_3O_2CCH:CHCO_2CH_3$	$(COD)_2Ni$	214
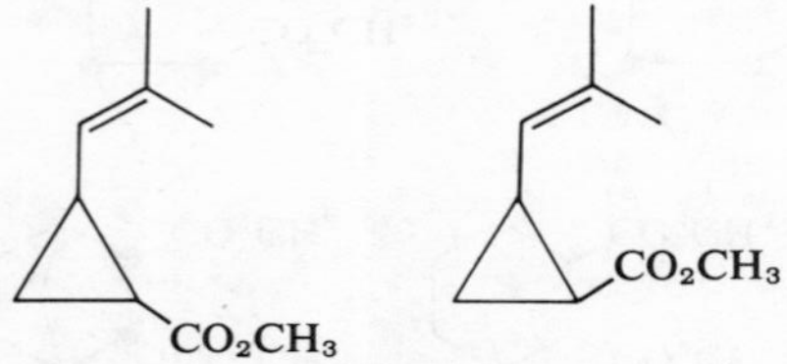	$CH_2:CHCO_2CH_3$	$(COD)_2Ni$	214

50

V. The Polymerization of Olefins (Table I-13)

Although nickel is not normally associated with the catalytic polymerization of monoolefins, a number of active nickel-containing systems have been reported. Nickel complexes are also able to initiate free radical polymerization processes (recent review articles include refs. 156 and 187).

The polymerization of ethylene or propylene has been reported to occur in the presence of activated nickel salts (163, P165) and nickel-containing Ziegler catalysts (see, for example, 158, 180, P10, P160, P163, P169, P177–P179). However, of greater interest are the reactions catalyzed by systems based on bis(cyclooctadiene)nickel that do not contain an aluminum component: the nickel is modified by acids, e.g., thiolactic acid or o-diphenylphosphinobenzoic acid, or ylids, e.g., $(C_6H_5)_3P:CHCO_2C_2H_5$, and is capable of polymerizing ethylene at relatively low temperatures and pressures (e.g., 65° and 40 atm). It has been suggested in the first case that a nickel-hydride may be the active species (P182).

The oligomerization of olefins using the nickel salt–Lewis acid combination discussed in Section III is generally accompanied by the production of minor amounts of polymeric material. The degree of oligomerization using phosphine-modified catalysts has been found to depend on both the basicity and steric nature of the phosphine and in the case of ethylene a quantitative formation of polyethylene at $-20°$ and atmospheric pressure has been observed using as catalyst $(\pi\text{-}C_3H_5NiCl)_2\text{–}(C_2H_5)_3Al_2Cl_3\text{–}P(tert\text{-}C_4H_9)_3$ (Ni:P = 1:4) (19, 112).

Cyclic olefins are readily polymerized by a variety of transition metal compounds. Catalysts containing nickel produce polymer formed exclusively by reaction of the double bond (in contrast to ring opening) and in the case of the reaction of cyclobutene catalyzed by $(\pi\text{-}C_3H_5NiBr)_2$ it has been shown that the product is the *erythro* diisotactic isomer **47** (177, 178, 188, 329). It is

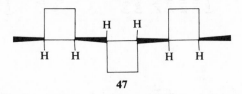

47

convenient to mention here that silacyclobutanes, e.g., $R_2Si(CH_2)_2CH_2$, also readily polymerize in the presence of π-allylnickel complexes to give linear polymers, e.g., $+R_2Si(CH_2)_3+_n$, as the result of cleavage of an Si–C bond (324, 325, P239).

The free radical polymerization of vinyl monomers (e.g., $CH_2:C(CH_3)\text{-}CO_2CH_3$ and $C_6H_5CH:CH_2$) is initiated by most transition metals and a

TABLE I-13

THE POLYMERIZATION OF OLEFINS

Olefin	Nickel comp.	Cocatalyst	Ref.^{a}
$CH_2:CH_2$	$(COD)_2Ni$	$CH_3CHSHCO_2H$	P122, P180, P182–P185,
		$o\text{-}HO_2CC_6H_4P(C_6H_5)_2$	P122, P180, P182–P185 P248
		$(C_6H_5)_3P:CHCO_2C_2H_5$	P153
		$(C_2H_5)_3Al_2Cl_3$	19, 112
$CH_2:CHCH_3$	$(\pi\text{-}C_3H_5NiCl)_2\text{–}P(tert\text{-}C_4H_9)_3$	$Ti\text{-}Al(C_2H_5)_3$	180, P169; see also 158
$CH_2:CHCH_2CH_3$	$NiCl_2$	$TiCl_3\text{-}Al(C_2H_5)_3$	P178
$CH_2:CHCH_2iso\text{-}C_3H_7$	$NiCl_2 \cdot 4py$	$TiCl_3\text{-}(C_2H_5)_2AlCl$	190, P174
$CH_2:CHCH_2tert\text{-}C_4H_9$	$Ni(acac)_2$	$TiCl_3\text{-}(C_2H_5)_2AlCl$	P174
(cyclobutane)	$Ni(acac)_2$ $(\pi\text{-}C_4H_7NiX)_2$	—	165, 166, 177, 178, 188
	$(\pi\text{-}C_4H_7NiX)_2$	$AlBr_3$	165, 166
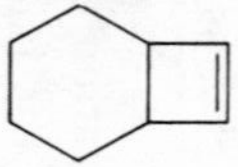 (bicyclo structure)	$(\pi\text{-}C_3H_5NiX)_2$	—	174
 (cyclopentene)	$(\pi\text{-}C_4H_7NiCl)_2$	$AlBr_3$	165, 166
	$Ni(CO)_4$	WCl_6	P176
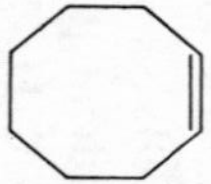 (cyclooctene)	$(\pi\text{-}C_3H_5NiCl)_2$	$AlBr_3$	165
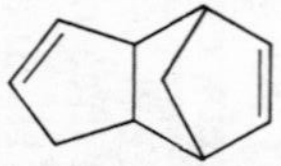 (dicyclopentadiene)	$(\pi\text{-}C_3H_5NiBr)_2$	—	329
$C_6H_5CH:CH_2$	$Ni(CO)_4$	CCl_4	327, P168; see also 328
	$[(C_6H_5O)_3P]_4Ni$	Maleic anhydride	145

Monomer	Catalyst	Additive	References[a]
CH₂:C(CH₃)CO₂CH₃	(COD)₂Ni	CCl₄	147, P167
	(π-C₃H₅NiX)₂	RCO₂H	160
	(π-C₃H₅NiX)₂	—	28, 167, 176, 331, P48
	(π-C₅H₅)₂Ni	(C₆H₅COO)₂	168–170, 172; see also 194
	NiX₂	(C₆H₅COO)₂	159
	NiX₂		96, 163, 164, 175, 330
	Ni(CO)₄	AlCl₃–Al	186, P98
	[(C₆H₅O)₃P]₄Ni	CCl₄	148–151, 153, 327
	[(OC)₃NiP(C₆H₅)₂]₂	CCl₄	144–147, 153–155, P164, P167
	[(π-C₅H₅NiCO)₂	CCl₄	171, P170–P172
	[(C₆H₅)₃P]₂Ni(C₂H₄)	CCl₄	149, 152, 340
	(π-C₃H₅)₂Ni		P166
	BipyNi(C₂H₅)₂	CCl₄	157, 159, P173
CH₂:CHCHO	Ni(acac)₂2py	Poly-*p*-chlorostyrene	161
CH₂:CHCOCH₃	BipyNi(C₂H₅)₂		162
	NiCl₂	Poly-*p*-chlorostyrene	161
CH₂:CHiso-C₄H₉	(π-C₄H₇NiX)₂	Al(C₄H₉)₃	P244
	Ni(CO)₄	(C₆H₅COO)₂	168–170, 172, 186; see also 194
CH₂:CHCl	Ni(CO)₄	CCl₄	328
	NiCl₂	CCl₄	183
CH₂:CHCN	(Duroquinone)₂Ni	Al(C₂H₅)₃–Cl₂	173, P181
	BipyNi(C₆H₅CN)	CH₂Cl₂	182
	BipyNi(C₂H₅)₂	—	184
	BipyNi(C₆H₅CN)	—	185
CH₂:C(CH₃)CN	Ni(CO)₄	—	184
CH₂:C(CN)₂	Ni(CO)₄	—	191
(CN)₂C:C(CN)₂	[(C₆H₅)₂P]₂Ni(acac)₂	CHCl₃	192, 193; see also 179
π-C₅H₅Fe-π-C₅H₄CH:CH₂			181

[a] The references refer not only to the catalyst shown in the table but also to related systems.

variety of active catalysts based on nickel have been reported. The reactions of methyl methacrylate and styrene have been most thoroughly investigated, and evidence for the formation of free radicals has been obtained from reactions catalyzed by nickel tetracarbonyl (150, 327) and tetrakistriphenylphosphite nickel (144–147, 153–155, 189, 336) in carbon tetrachloride, as well as by bis(cyclooctadiene)nickel–RCO_2H (160). Kinetic studies indicate that in the first case the initial reaction is the substitution of a ligand molecule by the olefin and this is followed by reaction with CCl_4 to generate CCl_3

$$[(C_6H_5O)_3P]_4Ni \; \underset{P(OC_6H_5)_3}{\overset{\text{olefin}}{\rightleftharpoons}} \; [(C_6H_5O)_3P]_3Ni(\text{olefin}) \; \underset{-\text{olefin}}{\overset{CCl_4}{\longrightarrow}} \; [(C_6H_5O)_3P]_3NiCl + CCl_3^{\cdot}$$

radicals. In the second case it is suggested that radicals are generated by cleavage of a nickel-alkyl bond. The activity of $[(CO)_3NiP(C_6H_5)_2]_2$

$$(COD)_2Ni + RCO_2H \longrightarrow (COD)Ni\underset{H}{\overset{OCOR}{<}} \overset{C_6H_5CH:CH_2}{\longrightarrow}$$

$$(COD)Ni\underset{\underset{C_6H_5}{\overset{|}{CHCH_3}}}{\overset{OCOR}{<}} \longrightarrow C_6H_5\overset{\cdot}{C}HCH_3 + [Ni(1+)]$$

in CCl_4 for the polymerization of methyl methacrylate is thought to be associated with the facile cleavage of the P–P bond and generation of the radical $[(OC)_3NiP(C_6H_5)_2]^{\cdot}$ (171). Copolymerization of vinyl monomers involving nickel catalysts has received relatively little attention. Copolymerization of maleic anhydride with methyl methacrylate or styrene has been investigated using a tetrakisligand nickel catalyst: the presence of CCl_4 is not essential, radicals being formed presumably by direct transfer of an electron from the nickel to the complexed electron-deficient olefin (145). The formation of a 1:1 copolymer from equimolar quantities of methyl methacrylate and styrene using as catalyst $(COD)_2Ni–CH_3CHSHCO_2H$ is additional evidence for the intermediacy of radicals in this process (160; see also 328). It has been reported that graft polymers between PVC and methyl methacrylate are formed in the reaction catalyzed by the nickel-olefin complexes $Lig_2Ni-(CH_2:CH_2)$ and $Lig_2Ni(COD)$ (P161, P162).

VI. The Isomerization of Olefins

In this section we are concerned with investigations devoted specifically to the isomerization of olefins. These can be divided into two types, double bond migration and skeletal rearrangement, and these are treated separately below.

A. Double Bond Migration (Table I-15)

Bearing in mind that similar nickel-hydride intermediates are involved, it is not surprising that isomerization is observed during most of the oligomerization and hydrogenation reactions discussed in Sections III and VII.

Two basic mechanisms have been discussed for the isomerization of olefins using transition metals catalysts: (a) addition and elimination of a metalhydride,

$$RCH_2CH{=}CH_2 + M{-}H \;\rightleftharpoons\; RCH_2{-}\underset{\underset{M}{\mid}}{CH}{-}CH_3 \;\rightleftharpoons\; RCH{=}CHCH_3 + M{-}H$$

and (b) rearrangement through a π-allylmetal hydride (226, 227),

$$RCH_2CH{=}CH_2 \;\rightleftharpoons\; \quad \;\rightleftharpoons\; RCH{=}CHCH_3$$

The nickel-catalyzed isomerization reactions are believed to proceed only through the addition and elimination of a nickel-hydride, although examples of both types of reaction have been observed in the chemistry of organonickel complexes.

The evidence for the intermediate formation of a nickel-hydride in the nickel-catalyzed oligomerization reactions has been discussed in Section III,D. A particularly detailed study of the isomerization of olefins has been made for the system $\pi\text{-}C_3H_5NiCl(PR_3)\text{-}R_nAlX_{3-n}$: the activity of the catalyst depends on the nature of both the phosphine and the Lewis acid. The most effective Lewis acid is $C_2H_5AlCl_2$, with a nickel–aluminum ratio of $1:3$–4, and it is suggested that the excess Lewis acid interacts indirectly with the nickel-bonded halogen atom in the active species to form halogen-bridged Lewis acid aggregates. The rate of isomerization is also dependent on the phosphine. Phosphines having low steric requirements, e.g., $P(CH_3)_3$, are apparently particularly effective, and in some cases isomerization occurs under very mild conditions (e.g., $-40°$) without significant oligomerization. The choice of phosphine for optimal isomerization depends, however, on the reaction conditions and the nature of the olefin. Isomerization occurs in a stepwise manner and the final isomer concentrations approach the thermodynamic equilibrium values. Terminal olefins are more readily isomerized than disubstituted olefins and these more readily than trisubstituted olefins. There is some evidence that the isomerization of a terminal olefin produces the *cis*-2-olefin preferentially and that this then isomerizes to the *trans*-2-olefin.

The isomerization of 4-methyl-1-pentene in the presence of π-C_3H_5NiCl-$[P(CH_3)_3]$–$C_2H_5AlCl_2$ is described to illustrate these points (112). If the reaction is carried out at 0° the thermodynamic equilibrium mixture, consisting mainly of 2-methyl-2-pentene, is obtained (Table I-14). If the reaction

TABLE I-14

ISOMERIZATION OF 4-METHYL-1-PENTENE AT 0° [a,b]

Olefin	Concentration (%)
	0.2
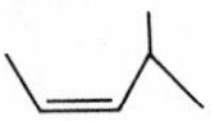	0.9
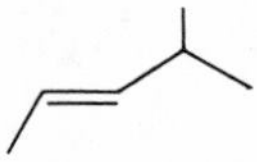	7.1–7.3
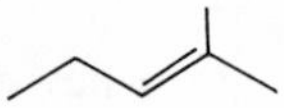	83.6–84.0
	7.8–8.6

[a] Catalyst π-C_3H_5NiCl$[P(CH_3)_3]$–$C_2H_5AlCl_2$.
[b] From ref. 112.

is carried out at low temperature, the individual isomerization steps can be followed. At −58° isomerization to *cis*-4-methyl-2-pentene occurs and is followed by isomerization to the trans isomer (Fig. I-2), which is then converted into 2-methyl-2-pentene at −40°. If the reaction is carried out at −58° with a catalyst modified with tricyclohexylphosphine, it is possible to produce *cis*-4-methyl-2-pentene in 78% conversion, whereas the use of a phosphine-free catalyst at the same temperature gives *trans*-4-methyl-2-pentene in 91% conversion. The isomerization of the methylpentenes is of practical importance because these are produced by the nickel-catalyzed dimerization of propylene: by using an isomerization catalyst it is possible to convert the propylene dimer into a mixture consisting almost exclusively of 2-methyl-2-pentene and 2,3-dimethyl-2-butene (18–20, 112, 116, P197).

The most dramatic example of the preferential formation of a *cis*-2-olefin is to be found in the isomerization of 1-butene using the exotic catalyst $[(C_6H_5)_3P]_2NiX_2$–$P(C_6H_5)_3$–Zn–$SnCl_2$. A ratio of *cis*-2-butene:*trans*-2-butene as high as 98:2 has been observed. Isomerization to the thermodynamically more stable *trans*-olefin occurs only after all the 1-butene has been converted (233). The origin of this effect has not been satisfactorily explained.

The isomerization is not limited to monoolefins and what is probably the most convenient synthesis of 1,3-cyclooctadiene is based on the isomerization of 1,5-cyclooctadiene with the catalyst $Ni(acac)_2-(C_2H_5)_3Al_2Cl_3-P(CH_3)_3$. The reaction occurs in a stepwise manner and, if the reaction is terminated before completion, it is possible to isolate 1,4-cyclooctadiene in up to 40% yield. *cis,trans*-1,5-Cyclodecadiene and 1,7-octadiene react anomalously; *cis,cis*-1,6-cyclodecadiene and 1,6- and 2,6-octadiene, respectively, are produced (112, 113). This behavior is perhaps associated with the favorable chelating properties of these diolefins.

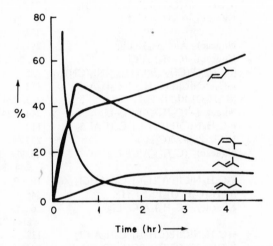

Fig. I-2. The isomerization of 4-methyl-1-pentene at $-58°$ (112).

Vinylcycloalkenes and -cycloalkanes are readily isomerized to the corresponding ethylidene derivatives. 1,2-Divinylcyclohexane is isomerized by a phosphine-free catalyst to give a high yield of 1-vinyl-2-ethylidenecyclohexane (**48**), whereas the use of a trimethylphosphine-modified catalyst produces *cis,trans*-1,2-diethylidenecyclohexane (**49**) as the main product (113).

48   **49**

TABLE I-15
NICKEL-CATALYZED DOUBLE BOND ISOMERIZATION

Olefin	Catalyst	Ref.
Butene	$Ni(acac)_2-Al(C_2H_5)_3-Lig$	26, 247, 253, P94
	$Ni(acac)_2-(C_2H_5)_3Al_2Cl_3-Lig$	26, 51, 242, 252, P193, P194
	$[(C_6H_5)_3P]_2NiX_2-P(C_6H_5)_3-SnCl_2-Zn$	233
	$Na_2[Ni(CN)_4]-NaBH_4$	241, 245
	$\pi-C_3H_5NiCl[P(CH_3)_3]-C_2H_5AlCl_2$	19, 20, 112, 116, 235, 250
	$(C_6H_5)_3PNi(CO)_3-AlBr_3$	26
	$[(C_2H_5O)_3P]_4Ni-HX$	226, 228, 248
	$(COD)_2Ni-(C_6H_5)_2PCH_2CO_2CH_3$	P196
Methylbutene	$Ni(acac)_2-(C_2H_5)_2AlCl$	26, 242
	$NiCl_2 \cdot 2py-Al(C_2H_5)_3$	253
Pentene	$Ni(acac)_2-Al(C_2H_5)_3-Lig$	247, 253
	$Ni(acac)_2-(C_2H_5)_2AlCl$	26, 242, 252
	$[(C_6H_5)_2P(CH_2)_4P(C_6H_5)_2]_3Ni_2(CN)_2$	225
	$\pi-C_3H_5NiCl[P(C_6H_5)_3]-C_2H_5AlCl_2$	112
	$[(C_2H_5O)_3P]_4Ni-HX$	223, 224
Dimethylbutene	$Ni(acac)_2-(C_2H_5)_3Al_2Cl_3-Lig$	244, P247
	$\pi-C_3H_5NiCl[P(CH_3)_3]-C_2H_5AlCl_2$	112
Methylpentene	$NiCl_2 \cdot 2py-Al(C_2H_5)_3$	253
	$Ni(acac)_2-(C_2H_5)_2AlCl-Lig$	26, 43, 127, 190, 230, 244, 247, 252, P193
	$\pi-C_3H_5NiCl[P(CH_3)_3]-C_2H_5AlCl_2$	19, 20, 112, P198
	$[C_6H_5(C_2H_5O)_2P]_4Ni-HX$	346
Hexene	$Ni(acac)_2-(C_2H_5)_2AlCl-Lig$	26, 43, 242, 244; see also P243
	$\pi-C_3H_5NiCl[P(CH_3)_3]-C_2H_5AlCl_2$	112
	$(COD)_2Ni-(C_6H_5)_2PCH_2CO_2CH_3$	P196
Dimethylpentene	$Ni(acac)_2-(C_2H_5)_2AlCl$	26, 252
Octene	$NiX_2-(C_2H_5)_2Al_2Cl_3$	242
	$Ni(CO)_4-HCl$	229
Decene	$NiBr_2-RMgBr$	254
Dodecene	$NiBr_2-RMgBr$	254
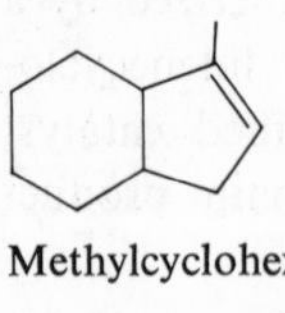	$Ni(acac)_2-(C_2H_5)_3Al_2Cl_3-P(cyclo-C_6H_{11})_3$	113
Methylcyclohexene	$(\pi-C_5H_5)_2Ni$	P199
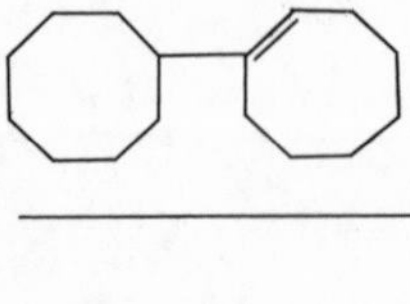	$Ni(acac)_2-(C_2H_5)_3Al_2Cl_3$	112

(continued)

TABLE I-15 (*continued*)

Olefin	Catalyst	Ref.
3-Allylcyclooctene	$(\pi\text{-}C_3H_5NiCl)_2\text{-}(C_2H_5)_3Al_2Cl_3\text{-}$ $P(cyclo\text{-}C_6H_{11})_3$	19
Pentadiene	$[(C_2H_5O)_3P]_4Ni\text{-}HX$	231, 249
3-Methylpentadiene	$[(C_4H_9)_3P]_2NiCl_2\text{-}(iso\text{-}C_4H_9)_2AlCl$	239
Hexadiene	$NiX_2\text{-}(iso\text{-}C_4H_9)_2AlCl\text{-}Lig$	58, 113, 234, 239
	$(\pi\text{-}C_4H_7NiX)_2$	58, 104
	$[(RO)_3P]_3Ni\text{-}HX$	231, 238
Octadiene	$Ni(acac)_2\text{-}(C_2H_5)_3Al_2Cl_3\text{-}P(cyclo\text{-}$ $C_6H_{11})_3$	113
	$\pi\text{-}C_3H_5NiCl[P(C_6H_5)_3]\text{-}C_2H_5AlCl_2$	112
Vinylcyclohexene	$[(C_4H_9)_3P]_2NiCl_2\text{-}(C_2H_5)_2AlCl$	P191, P192, P195, P242
	$(CH_2:CHCN)_2Ni\text{-}(C_4H_9)_2AlCl$	P16
	$(\pi\text{-}C_5H_5)_2Ni$	P199
Divinylcyclohexane	$Ni(acac)_2\text{-}(C_2H_5)_3Al_2Cl_3\text{-}Lig$	113
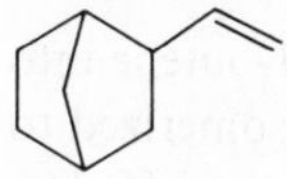	$(\pi\text{-}C_3H_5NiCl)_2\text{-}(C_2H_5)_3Al_2Cl_3\text{-}Lig$	109
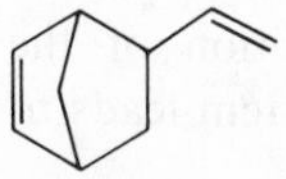	$(\pi\text{-}C_3H_5NiCl)_2\text{-}(C_2H_5)_3Al_2Cl_3\text{-}Lig$	109
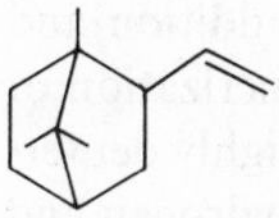	$(\pi\text{-}C_3H_5NiCl)_2\text{-}(C_2H_5)_3Al_2Cl_3\text{-}Lig$	110
1,5-Cyclooctadiene	$Ni(acac)_2\text{-}(C_2H_5)_3Al_2Cl_3\text{-}P(CH_3)_3$	113; see also 117
	$Ni(acac)_2\text{-}Al(C_2H_5)_3$	P175
	$(\pi\text{-}C_5H_5)_2Ni$	P199
	$[(C_6H_5)_3P]_2Ni(CN)_2\text{-}NaBH_4$	246
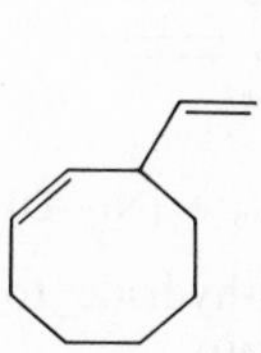	$(\pi\text{-}C_3H_5NiCl)_2\text{-}(C_2H_5)_3Al_2Cl_3\text{-}Lig$	109, 111
	$Ni(acac)_2\text{-}Al(C_2H_5)_3$ $Ni(acac)_2\text{-}(C_2H_5)_3Al_2Cl_3\text{-}P(cyclo\text{-}$ $C_6H_{11})_3$	232 113

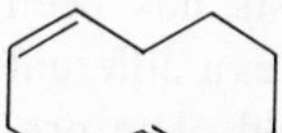

No double bond migration is observed on reacting the triene 1,5,9-cyclo-dodecatriene with $Ni(acac)_2-(C_2H_5)_3Al_2Cl_3-P(CH_3)_3$. However, in the case of the trans,cis,cis isomer a rapid isomerization to the trans,trans,cis isomer occurs (113).

A second important class of isomerization catalysts is prepared by treating tetrakisligand nickel complexes with acids (223–225, 228, 248, 346). Stable nickel-hydride species have been isolated from this system (see Volume I, p. 148) and a kinetic study of the isomerization of 1-butene indicates that the active species is $[HNiLig_4]^+$. The relative rate constants for the isomerization of 1-butene are shown below.

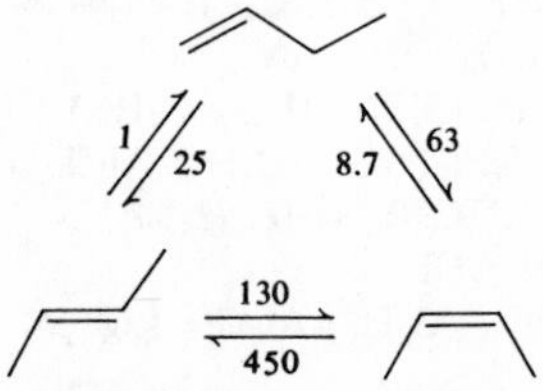

Here, also, it is found that initially a preferential conversion of 1-butene into *cis*-2-butene occurs (226). 3-Methyl-1-pentene is quantitatively isomerized to a mixture of *cis*- and *trans*-3-methyl-2-pentene in the presence of $[C_6H_5-(C_2H_5O)_2P]_4Ni$–*p*-toluenesulfonic acid. Interestingly, the addition of the optically active phosphine $(-)R\text{-}CH_3*P(C_6H_5)C_3H_7$ to this system leads to the preferential isomerization of the $(-)R$ olefin isomer (346).

Evidence that the mechanism of isomerization involves the addition and elimination of a nickel-hydride has been obtained from the isomerization of $1,2\text{-}d_2\text{-}1$-pentene: the product includes C_5 olefins that are more highly deuterated than the reactant and, in addition, a redistribution of the hydrogen and deuterium bonded to C_1 and C_2 occurs (223). Both of these processes are most easily explained assuming the addition and elimination of a nickel-hydride.

$$CHD{=}CDCH_2CH_2CH_3 + [Ni{-}H] \rightleftharpoons CH_2D{-}\underset{\underset{Ni}{|}}{CD}{-}CH_2CH_2CH_3 \rightleftharpoons$$

$$CH_2{=}CDCH_2CH_2CH_3 + [Ni{-}D]$$

Butadiene, which is known to react with the cationic nickel-hydride to form a π-allylnickel system, hinders the isomerization process (249).

The isomerization of 1,5-cyclooctadiene with nickel catalysts has been discussed above. In the absence of donor ligands, the reaction takes a different course and bicyclo[3.3.0]octene-2 (**50**) is produced in high yield. The preferred catalyst for this reaction is $Ni(acac)_2-(C_2H_5)_3Al_2Cl_3$ (113, 236, P198) but active catalysts have also been reported involving π-allylnickel halides

(113, P186, P192, P198) or π-cyclopentadienylnickel complexes (P195) as well as $[(C_6H_5)_3P]_2NiI_2-CH_3CO_2H$ (117). The reaction is accompanied by the formation of traces ($<5\%$) of 1,3-cyclooctadiene and bicyclo[3.3.0]-octene-$\Delta^{1,5}$ (**51**). A mechanism, involving a π,σ-cyclooctenylnickel intermediate has been suggested. 3-Methyl-1,5-cyclooctadiene reacts similarly to

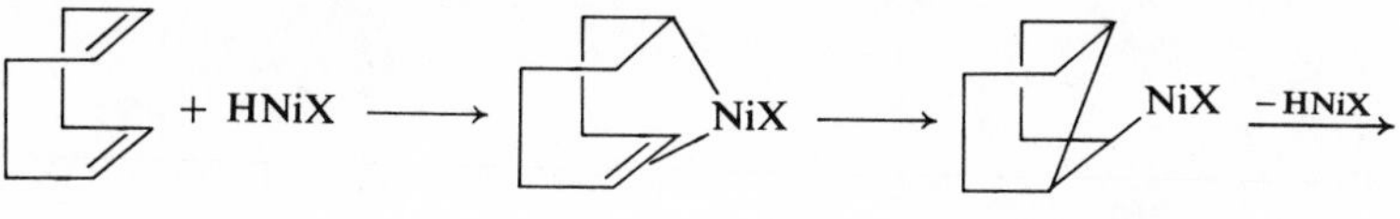

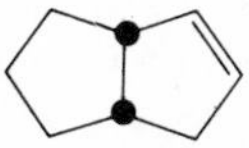

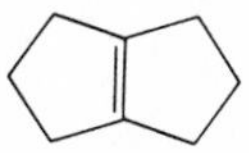

give a mixture of the four possible isomers of methylbicyclooctene (113). Other 1,5-dienes show a less pronounced tendency to form five-membered rings: 2-methylbicyclo[3.4.0]nonane (**52**) and a mixture of bicyclo[6.3.0]-undecene-2 and -3 (**53** and **54**) have been isolated from the reactions of *cis*-1,2-divinylcyclohexane and 3-allylcyclooctene, whereas a bicyclic compound, provisionally assigned structure **55**, has been isolated from the reaction with 1,1-biscyclooctene-2 (19, 113).

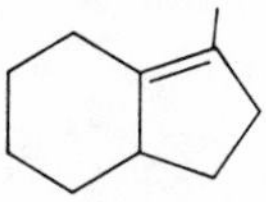

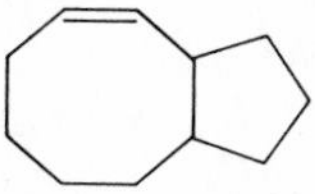

B. Skeletal Rearrangements *(Table I-16)*

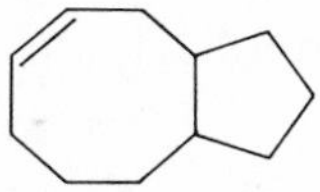

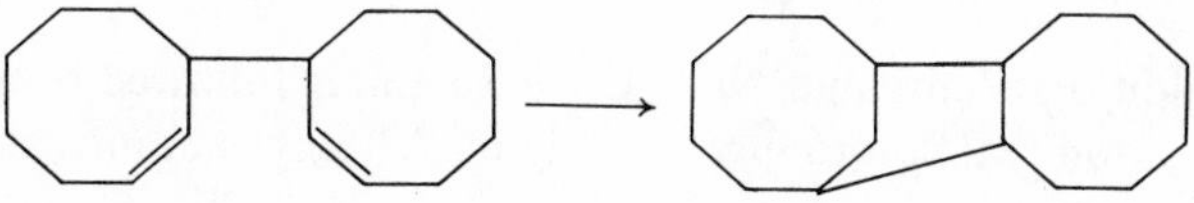

Two types of nickel-catalyzed skeletal rearrangement of dienes have been observed: (a) conversion of a 1,4-diene into a branched-chain 1,3-diene, e.g., 1,3-pentadiene into 2-methylbutadiene (isoprene); and (b) conversion of a branched-chain 1,4-diene into a linear 1,4-diene, e.g., 3-methyl-1,4-pentadiene into 1,4-hexadiene. The first process is catalyzed by both $\text{Lig}_n\text{Ni-HX}$ systems as well as by $[(C_4H_9)_3P]_2NiCl_2-(iso-C_4H_9)_2AlCl$, whereas the second has been reported to occur only with the last mentioned catalyst. The nature of these rearrangements and their probable mechanisms have been elucidated with the aid of deuterated and methyl-substituted 1,4-dienes. These results indicate that the skeletal rearrangements involved are those shown below.

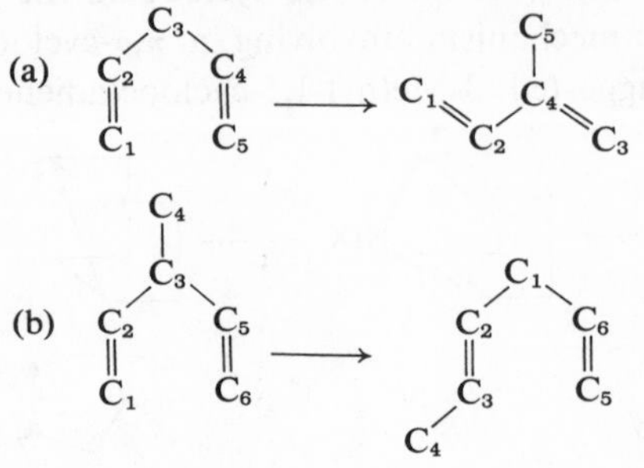

(a)

(b)

It is suggested that both processes involve the initial addition of a nickel-hydride: in case (a) addition occurs in a $Ni \rightarrow C_2$ sense and is followed by rearrangement through a cyclopropylcarbinylnickel intermediate.

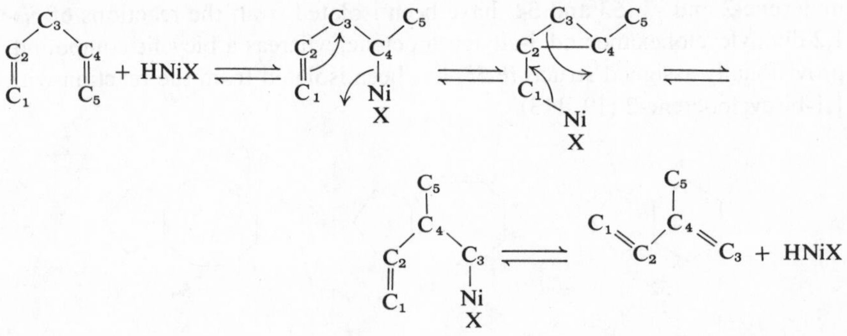

In case (b) addition occurs in a $Ni \rightarrow C_1$ sense and is followed by a novel β-carbon elimination reaction to give a π-allylnickel(ethylene) intermediate.

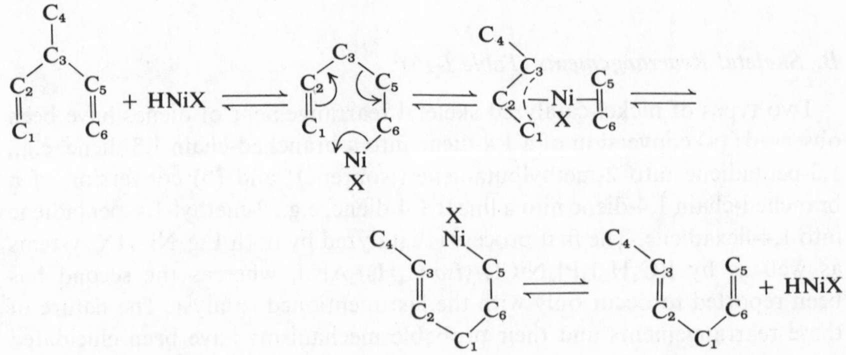

The postulated intermediacy of a cyclopropylcarbinyl species in the first reaction is supported by the observed nickel-catalyzed isomerization of *cis*- and *trans*-2-methylvinylcyclopropane into 2-methyl-1,3-pentadiene and 2,4-hexadiene (240, 354, 355).

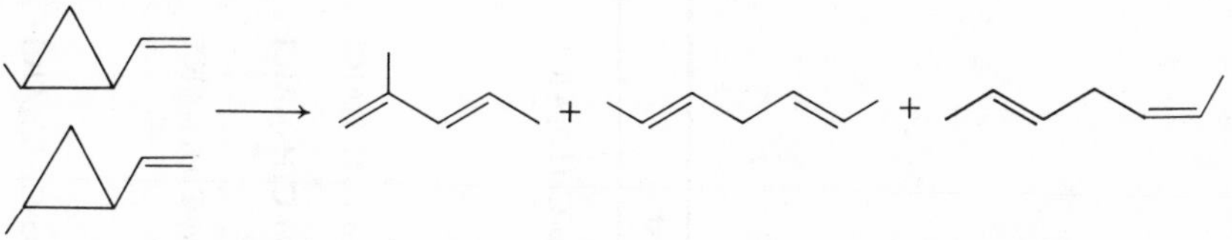

The β-carbon elimination proposed in the second mechanism is the reverse of the addition step observed in the codimerization of olefins with 1,3-dienes and its relevance is indicated by the isolation of appreciable quantities of a C_7 olefin (2-methyl-1,4-hexadiene) from the reaction of 3-methyl-1,4-pentadiene (C_6) with propylene. This result is most easily explained by assuming that the C_6 diolefin undergoes β-carbon elimination to give a π-crotylnickel(ethylene) species from which the ethylene is displaced by propylene (237, 239, 353, 356).

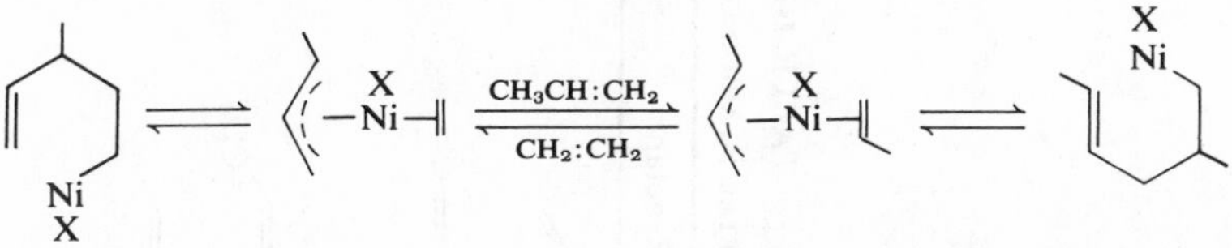

A reaction possibly related to these is the conversion of methylbutene-nitrile to linear pentenenitriles in the presence of tetrakisligand nickel complexes. For example, 2-methyl-3-butenenitrile is converted in the presence of $[(C_2H_5O)_3P]_4Ni$ into a mixture of 3- and 4-pentenenitrile (P187–P190).

Although the isomerization of strained cycloalkanes on transition metal catalysts has attracted considerable attention in recent years, the utilization of nickel has been largely neglected and reports are limited to the isomerization of the 1,1'-bishomocubane system 56 by bis(cyclooctadiene)nickel in the presence of methylacrylate, the isomerization of benzotricyclooctatriene (57) to benzocyclooctatetraene by bis(cyclooctadiene)nickel (243, 251), and the

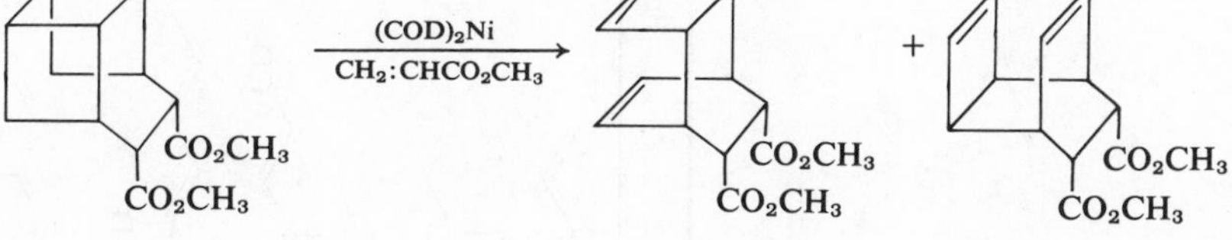

TABLE I-16

THE SKELETAL REARRANGEMENT OF DIENES

Product	Reactant	Catalyst	Ref.
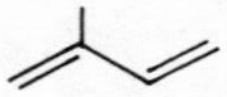		$[(C_4H_9)_3P]_2NiCl_2-(iso\text{-}C_4H_9)_2AlCl$	240, 353
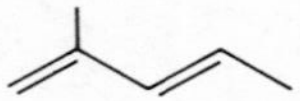		$(R_3P)_nNi\text{-}HCl$	231, 238
		$(R_3P)_2NiBr_2\text{-}NaBH_4$	231
		$[(C_4H_9)_3P]_2NiCl_2-(iso\text{-}C_4H_9)_2AlCl$	239, 353
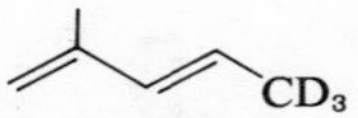	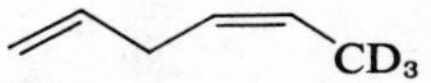	$[(C_4H_9)_3P]_2NiCl_2-(iso\text{-}C_4H_9)_2AlCl$	255, 354
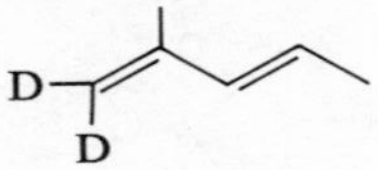	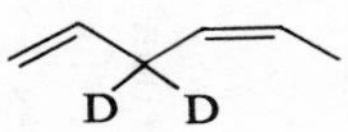	$[(C_4H_9)_3P]_2NiCl_2-(iso\text{-}C_4H_9)_2AlCl$	255, 354
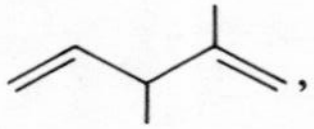 		$[(C_4H_9)_3P]_2NiCl_2-(iso\text{-}C_4H_9)_2AlCl$	255, 353

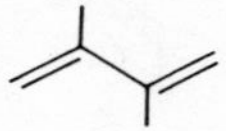  $[(C_4H_9)_3P]_2NiCl_2-(iso-C_4H_9)_2AlCl$	255, 353
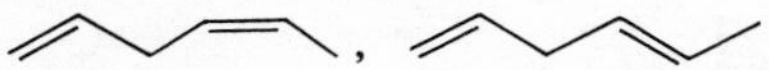  $[(C_4H_9)_3P]_2NiCl_2-(iso-C_4H_9)_2AlCl$	239, 353
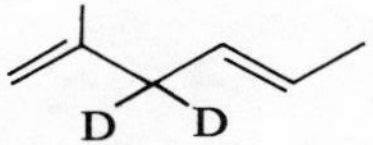 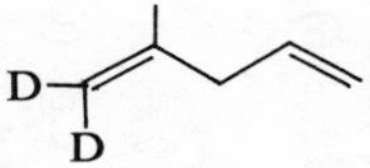 $[(C_4H_9)_3P]_2NiCl_2-(iso-C_4H_9)_2AlCl$	237, 356
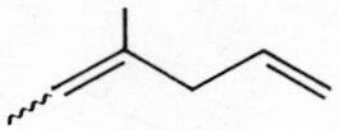 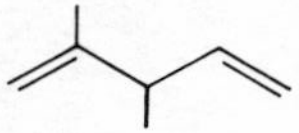 $[(C_4H_9)_3P]_2NiCl_2-(iso-C_4H_9)_2AlCl$	237, 353
$[(C_4H_9)_3P]_2NiCl_2-(iso-C_4H_9)_2AlCl$	240, 355
$Lig_2Ni(CH_2:CH_2)-HCl$	238, 355

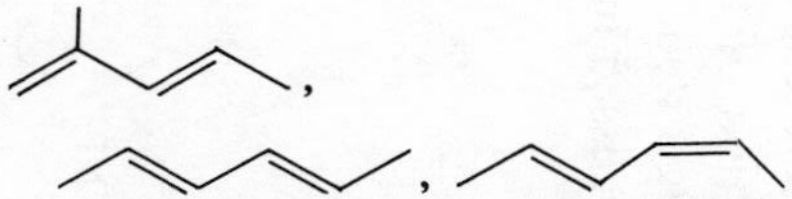

conversion of hexamethyl-Dewarbenzene into hexamethylbenzene (321) and of quadricyclene into tricycloheptadiene (322). The activity of these catalysts apparently depends on the ability of the organic system to displace the ligands attached to the nickel: neither nickel tetracarbonyl nor bis(cyclooctadiene)-nickel alone causes the bishomocubane system to isomerize, and dibenzotri-cyclooctadiene (58) is not isomerized in the presence of bis(cyclooctadiene)-nickel.

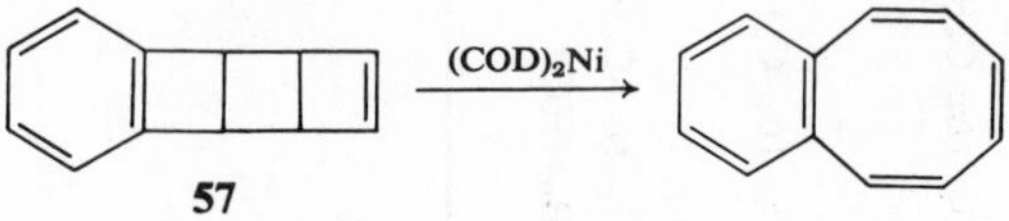

$$\text{57} \xrightarrow{(COD)_2Ni}$$

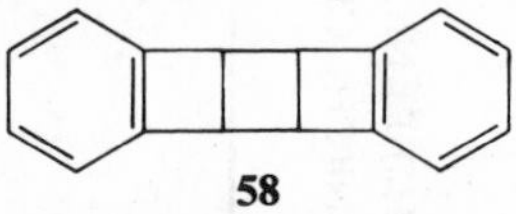

$$\text{58}$$

The related reactions of bicyclo[1.1.0]butane and bicyclo[2.1.0]pentane with olefins in the presence of bis(acrylonitrile)nickel (72, 73, 339, 347) have been discussed in Section IV.

C. Miscellaneous Rearrangements

It is convenient to include in this section the nickel-catalyzed rearrangement of triarylphosphorus ylids to diarylbenzylphosphines (108, 316). The reaction

$$Ar_3P{=}CH_2 \xrightarrow{[Ni]} Ar_2PCH_2Ar$$

is catalyzed by a variety of nickel-olefin complexes, e.g. $(COD)_2Ni$ and $[(C_6H_5)_3P]_2Ni(CH_2{:}CH_2)$, as well as tetrakisphosphine nickel complexes and π-allylnickel chloride. The rearrangement is confined to triarylphosphorus ylids and is reminiscent of the Stevens rearrangement of ammonium ylids. A mechanism involving interaction of the nickel atom with an o-H atom of one of the rings has been suggested. Complexes formed by phosphorus ylids with nickel have been discussed in Volume I, p. 39.

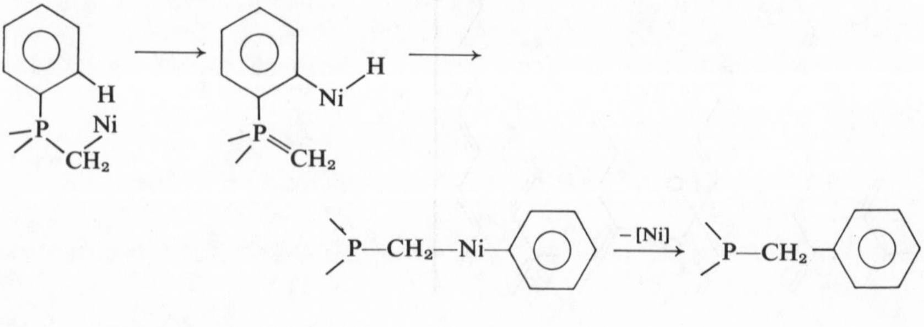

VII. The Hydrogenation of Olefins

A detailed discussion of the nickel-catalyzed hydrogenation of olefins falls outside the scope of this book as mainly heterogeneous systems are involved. However, a number of the catalysts described in previous sections as being active for the oligomerization and isomerization of olefins are also active hydrogenation catalysts and, because common intermediates are believed to be involved, these are briefly mentioned.

Nickel salts, with or without additional ligands, can be used directly as hydrogenation catalysts, although their activity is low and elevated temperatures are necessary (117, 258, 262, 274–278, P218, P219, P238; see also 320, 342). More active catalysts may be prepared by treating nickel salts, in the presence of hydrogen, with Grignard reagents (271, 272), aluminum-alkyls (264, 265, 268–270, 273, 279, 289, 317, P216, P217, P225, P226, P245, P246, P257), butyllithium (318, P235), or sodium borohydride (246, 257, 263, P222). Highly active catalysts can also be obtained by treating zerovalent nickel-olefin complexes with hydrogen. For example, bis(cyclooctadiene)nickel reacts with hydrogen, after an induction period, to give a catalyst which is even able to hydrogenate benzene at 20°C and atmospheric pressure; this behavior is attributed to the formation of finely divided nickel (345). A series of publications is concerned with what is evidently a heterogeneous catalyst formed by reacting sodium borohydride with aqueous or alcoholic solutions of nickel salts (280–287).

In addition, a number of examples have been reported in which nickel-catalyzed transfer hydrogenation occurs in the absence of hydrogen. Particularly active are the systems $[Ni_2(CN)_6]^{4-}-H_2O$ (258–261; see also 343) and $[(C_6H_5)_3P]_2NiX_2-o-HOC_6H_4OH$ (267). Less active systems include $[(C_6H_5)_3P]_2NiX_2$ in THF or benzene (278) and $Ni(acac)_2-iso-C_3H_7OH$ (266).

Detailed attention has been given to the hydrogenation of polyunsaturated fatty acids (257, 258, 262, 263, 266, 277, 278; see also 348) and the course of reaction is found to depend on the nature of the catalyst. With some catalysts (e.g., Lig_2NiX_2) isomerization to the conjugated system precedes hydrogenation and a *trans*-monoene results (258, 277, 278), whereas in other cases (e.g., $NiCl_2-NaBH_4-DMF$) hydrogenation occurs without initial isomerization (257, 263). Isomerization followed by hydrogenation enables 1,5-cyclooctadiene to be selectively converted into cyclooctene (117, P226).

The rate of hydrogenation of substituted olefins catalyzed by the system $Ni(2-ethylhexanoate)_2-Al(C_2H_5)_3$ has been shown to decrease in the order monosubstituted > unsymmetrically disubstituted > cyclic > symmetrically disubstituted and is similar to that obtained with Raney nickel (265).

There is considerable interest in modifying the properties of a polymer by hydrogenation and a variety of nickel-containing Ziegler catalysts have been used for this purpose (see, for example, 265, 289, 290, P200–P215, P236, P237).

It should also be mentioned that H–D exchange between C_2H_4 and C_2D_4 in the presence of hydrogen is catalyzed at room temperature by $[(C_6H_5)_3P]_2$-NiX_2, even though the rate of exchange between H_2 and D_2 is very slow (256). Nickelocene also promotes exchange between H_2 and D_2 (296, 297).

It is generally assumed that the hydrogenation is the result of the addition of a nickel-hydride intermediate to the olefin molecule to give a nickel-alkyl species that then reacts further with hydrogen. The plausibility of the first

$$RCH{=}CH_2 + HNiX \rightleftharpoons RCH_2CH_2NiX \xrightarrow{H_2} RCH_2CH_3 + HNiX$$

step is underlined by the reaction of the model system $HNiX(PR_3)$ with propylene mentioned on page 25. The second step, in which cleavage of the hydrogen molecule occurs, is less well founded; it is possible that intermolecular cleavage occurs to give an hydridonickel-alkyl species, which then couples. A model reaction possibly of relevance to this process is that of the coordinatively unsaturated nickel complex **59** with hydrogen to give the bridging binuclear hydride **60** (46).

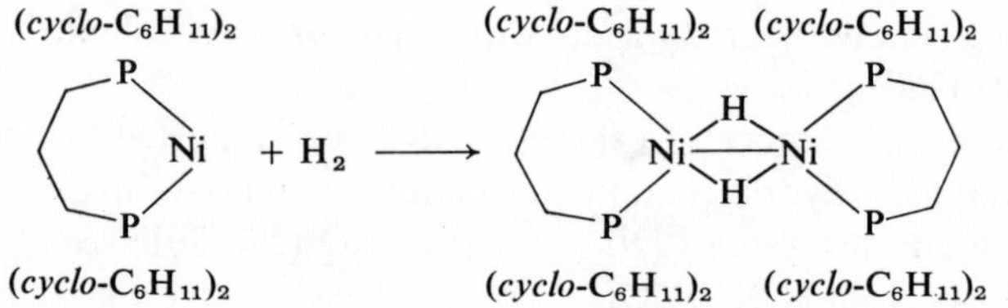

It has been spectroscopically demonstrated that an intermediate π-crotyl-nickel complex is produced during the transfer hydrogenation of butadiene catalyzed by $[Ni_2(CN)_6]^{4-}$ in water. Kinetic evidence suggests that a binuclear butadiene complex is formed initially (259). Related π-allyl systems have been isolated from the reaction of 1,3-dienes with cationic tetrakisligand nickel-hydride complexes (249).

Miscellaneous Hydrogenation

Several of the systems described above are active for the hydrogenation or reduction of organic compounds other than olefins.

Alkynes are readily hydrogenated by the system NiX_2–AlR_3 (265, 268, P217, P225). In the case of hexyne it has been shown that the reaction occurs in a stepwise manner, with preferred cis addition in the first step. Acetylene is selectively converted into ethylene in the presence of an aqueous solution of $[Ni_2(CN)_6]^{4-}$ (P224). Aromatic systems may also be hydrogenated by using nickel salts and aluminum-alkyls as catalysts (265, 291–293, see also 345, P217).

The reduction of aromatic nitro compounds leads to azoxy compounds in the presence of sodium borohydride and nickel cyanide (295), and to amines in the presence of sodium borohydride and $[(C_6H_5)_3P]_2NiX_2$ (294; see also 319), as well as of nickel tetracarbonyl (P220). The nickel-catalyzed conversion of nitriles into amines has also been reported (261, 265, 288, P217, P221, P223).

VIII. The Hydrosilylation and Hydrocyanation of Olefins

The hydrosilylation reaction (Table I-17) is catalyzed by both divalent and zerovalent nickel complexes. The divalent nickel systems include bisphosphine nickel dihalides (299, 301–304, 306, 308, 309, P227), nickel-alkyls (352) as well as nickelocene (352), whereas the zerovalent nickel complexes include nickel tetracarbonyl (300, 310; see also 311), $(\pi\text{-}C_5H_5NiCO)_2$ (305), nickel-olefin complexes (300, 301), $[(C_6H_5O)_3P]_4Ni$ (299), and Raney nickel (307). The activity of the catalyst is increased on addition of CuCl (299).

The reactivity of the silanes with 1-octene decreases in the order $HSiCl_3 > HSiCl_2CH_3 > HSiCl(CH_3)_2 \gg HSi(CH_3)_3$ and the reaction is confined to the hydrosilylation of terminal or cyclic olefins (302). The direction of addition depends to some extent on the nature of the modifying ligand. In general, hydrosilylation proceeds in an $Si \rightarrow C_1$ sense to give a linear product but in the presence of 1,2-bis(dimethylphosphino)-1,2-dicarbo-*closo*-dodeca-carborane (**61**) comparable amounts of $Si \rightarrow C_1$ and $Si \rightarrow C_2$ addition are observed (304). The direction of addition to styrene also depends on the

$$2RCH\!=\!CH_2 + 2HSiCl_2CH_3 \longrightarrow RCH_2CH_2SiCl_2CH_3 + \overset{\overset{\textstyle CH_3}{|}}{R}CHSiCl_2CH_3$$

nature of the ligand present (299).

It has proved possible, using a catalyst modified by the optically active ligand benzylmethylphenylphosphine, to effect an asymmetric hydrosilylation

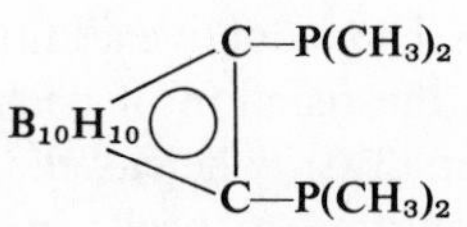

61 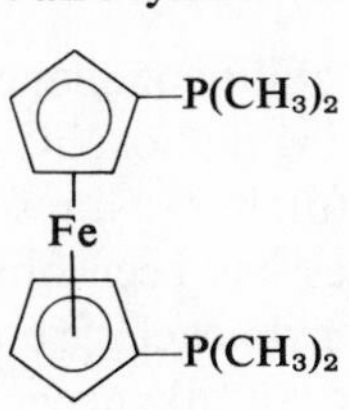   **62**

and, from the reaction between α-methylstyrene and methyldichlorosilane, 2-phenylpropylmethyldichlorosilane with an optical purity of 17.6% has been isolated (306).

$$\underset{\displaystyle C_6H_5\overset{\displaystyle CH_3}{\underset{|}{C}}=CH_2}{} + HSiCl_2CH_3 \xrightarrow{[C_6H_5CH_2\overset{*}{P}(CH_3)C_6H_5]_2NiCl_2} C_6H_5\overset{*}{\underset{H}{\overset{CH_3}{\underset{|}{C}}}}-CH_2SiCl_2CH_3$$

Hydrosilylation is accompanied by olefin isomerization and exchange of a silicon-bonded chlorine atom for a hydrogen atom. The reactions of 2-pentene lead to n-pentylsilanes and, regardless of whether 1- or 2-pentene is reacted, isomerization of the recovered pentene is observed. Isomerization is also observed in the reaction involving 1,4-cyclohexadiene or 1,5-cyclo-octadiene. In the first case a mixture of 3- and 4-silylcyclohexene is obtained and in the second, the three possible substituted cyclooctenes (302–304, 310). The extent of Cl–H exchange depends on the basicity of the ligand present in the catalyst and in the presence of strongly basic phosphines, such as 1,1'-bis(dimethylphosphino)ferrocene (62), is the principal reaction (301–303, 306, P227).

$$RCH{=}CH_2 + 2HSiCl_2CH_3 \longrightarrow RCH_2CH_2SiCl(CH_3)H + CH_3SiCl_3$$

The mechanism of the hydrosilylation reaction has been suggested to involve the oxidative addition of the Si–H group to a zerovalent nickel-olefin complex [which, in the case of the reactions involving nickel(2+) salts, is produced by reducing the salt with the silane] and is followed by insertion

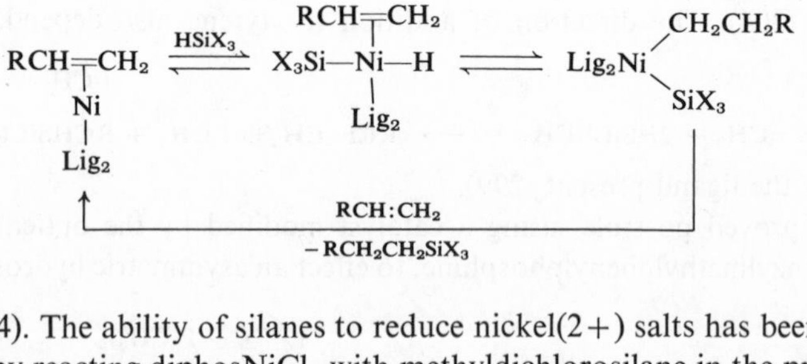

(301, 304). The ability of silanes to reduce nickel(2+) salts has been demonstrated by reacting diphosNiCl$_2$ with methyldichlorosilane in the presence of excess diphos: (diphos)$_2$Ni results (301). The oxidative addition of silanes to nickel complexes has been observed in the reaction of trichlorosilane with π-cyclopentadienylnickel carbonyl dimer (312), whereas the addition of an SiCl$_3$ moiety to an alkyne has been reported to occur in the reaction of bipyNi(SiCl$_3$)$_2$ with diphenylacetylene (313).

$$(\pi\text{-}C_5H_5NiCO)_2 + HSiCl_3 \longrightarrow \pi\text{-}C_5H_5Ni\text{---}SiCl_3(CO) + [\pi\text{-}C_5H_5NiH(CO)]$$

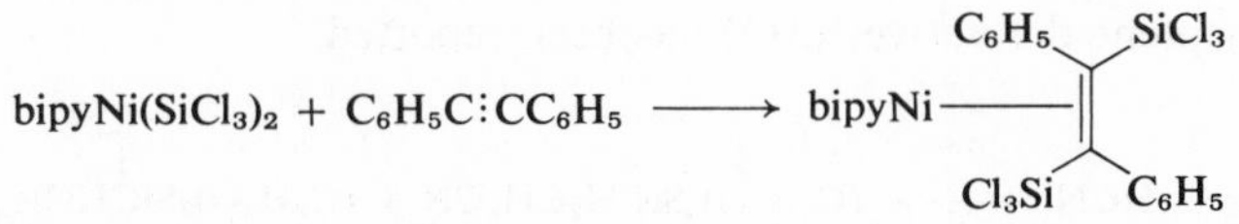

$$bipyNi(SiCl_3)_2 + C_6H_5C\!:\!CC_6H_5 \longrightarrow bipyNi\underset{Cl_3Si}{\overset{C_6H_5}{\underset{\diagdown}{\diagup}}}\!\!\!\!\!\!\!\!=\!\!\!\!\!\!\!\!\underset{C_6H_5}{\overset{SiCl_3}{\underset{\diagup}{\diagdown}}}$$

A discussion of the hydrosilylation of alkynes and 1,3-dienes is to be found in Chapter II, Section IV, and Chapter III, Section VI.

The hydrocyanation of olefins has received attention mainly from industrial chemists, who have been principally concerned with the addition of HCN to nonconjugated pentenenitriles to form adiponitrile. Catalysts for this reaction include nickel tetracarbonyl (P233) and tetrakisligand nickel complexes (314, 315, P228–P232, P249, P251) in the presence of various promoters, e.g., $AlCl_3$, $ZnCl_2$, or $(C_6H_5)_3B$. In addition to pentenenitrile, the

TABLE I-17

THE HYDROSILYLATION OF OLEFINS
$RCH\!=\!CHR + HSiX_3 \rightarrow RCH_2CHRSiX_3$

Olefin	Silane	Ref.
$CH_2\!:\!CH_2$	$HSiCl_2CH_3$	302
$CH_2\!:\!CHCH_3$	$HSiCl_2CH_3$	302, 304
$CH_2\!:\!CHCH_2CH_3$	$HSiCl_2CH_3$	302
$CH_2\!:\!C(CH_3)_2$	$HSiCl_2CH_3$	304
$CH_2\!:\!CH(CH_2)_2CH_3$	$HSiCl_2CH_3$	302, 303
$CH_2\!:\!CH(CH_2)_3CH_3$	$HSiCl_2CH_3$	302–304, P227
$CH_2\!:\!CH(CH_2)_4CH_3$	$HSiCl_2CH_3$	302, 303
$CH_2\!:\!CH(CH_2)_5CH_3$	$HSiCl_3$	302–304, 308, 352
	$HSiCl_2CH_3$	300–304, 352
	$HSiCl(CH_3)_2$	302
	$HSiCl_2C_6H_5$	304
Cyclohexene	$HSiCl_2CH_3$	302
1,4-Cyclohexadiene	$HSiCl_2CH_3$	302
1,5-Cyclooctadiene	$HSiCl_2CH_3$	302, 303
$(CH_3)_3SiCH\!:\!CH_2$	$HSiCl_2CH_3$	304
$CH_3Cl_2SiCH\!:\!CH_2$	$HSiCl_2CH_3$	304
$C_6H_5CH\!:\!CH_2$	$HSiCl_3$	299, 304, 305, 308, 310
	$HSiCl_2CH_3$	302, 310, 352
$C_6H_5C(CH_3)\!:\!CH_2$	$HSiCl_2CH_3$	306
$CH_2\!:\!CHCH_2CN$	$HSiCl_3$	308
$CH_2\!:\!CHCN$	$HSiCl_3$	307–309
	$HSiCl_2CH_3$	300, 302; see also 333
	$HSi(C_2H_5)_3$	300, 311, see also 333
$CH_2\!:\!CHCHO$	$HSi(C_2H_5)_3$	300
$CH_2\!:\!CHCO_2CH_3$	$HSiCl_2CH_3$	302
$CH_2\!:\!CHOCOCH_3$	$HSiCl_2CH_3$	302

hydrocyanation of hexene (334, P249), triethoxyvinylsilane (315), and a variety of bicycloheptene derivatives (315) has been reported.

$$(C_2H_5O)_3SiCH{=}CH_2 + HCN \longrightarrow (C_2H_5O)_3SiCH_2CH_2CN + (C_2H_5O)_3Si\overset{\overset{\displaystyle CN}{\displaystyle |}}{C}HCH_3$$

A reaction perhaps related to these is the alkylation by olefins of aromatic hydrocarbons which is observed in the presence of the catalyst $C_2H_5AlCl_2$–$NiCl_2$ (136, 217; see also 27).

References

1. V. M. Akhmedov, M. A. Mardanov, A. A. Khanmetov, and L. I. Zakharkin, *J. Org. Chem. USSR* **7**, 2610 (1971).
2. M. Abedini, *Bull. Fac. Sci., Tehran Univ.* **2**, 1 (1971).
3. H. Ando, O. Inomata, M. Enomoto, Y. Shimizu, H. Nishikata, and T. Amemiya, *Sekiyu Gakkai Shi* **11**, 528 (1968); *Chem. Abstr.* **70**, 3160 (1969).
4. S. G. Abasova, A. I. Leshcheva, E. A. Mushina, V. S. Feldblyum, and B. A. Krentsel, *Bull. Acad. Sci. USSR* p. 608 (1972).
5. V. M. Akhmedov, A. A. Khanmetov, M. A. Mardanov, and L. I. Zakharkin, *J. Org. Chem. USSR* **9**, 442 (1973).
6. S. G. Abasova, E. I. Mushina, L. S. Muraveva, B. A. Krentsel, and V. S. Feldblyum, *Neftekhimiya* **13**, 46 (1973); *Chem. Abstr.* **78**, 135221 (1973).
7. T. Arakawa and S. Saeki, *Kogyo Kagaku Zasshi* **71**, 1028 (1968).
8. T. Arakawa, *Kogyo Kagaku Zasshi* **70**, 1738 (1967).
9. F. Asinger, B. Fell, and R. Janssen, *Chem. Ber.* **97**, 2515 (1964).
10. D. R. Arnold, D. J. Trecker, and E. B. Whipple, *J. Amer. Chem. Soc.* **87**, 2596 (1965).
11. J. Beger, C. Duschek, and H. Füllbier, *Z. Chem.* **13**, 59 (1973).
12. C. W. Bird, R. C. Cookson, and J. Hudec, *Chem. Ind. (London)* p. 20 (1960).
13. P. Binger and J. McMeeking, *Angew. Chem.* **85**, 1053 (1973).
14. P. Binger, *Synthesis* p. 427 (1973).
15. P. Binger, *Angew. Chem.* **84**, 352 (1972).
16. C. W. Bird, D. L. Colinese, R. C. Cookson, J. Hudec, and R. O. Williams, *Tetrahedron Lett.* p. 373 (1961).
17. B. Bogdanović, B. Henc, B. Meister, H. Pauling, and G. Wilke, *Angew. Chem.* **84**, 1070 (1972).
18. B. Bogdanović and G. Wilke, *World Petrol Congr., Proc., 7th, 1967* Vol. 5, p. 351 (1968).
19. B. Bogdanović, B. Henc, H.-G. Karmann, H.-G. Nüssel, D. Walter, and G. Wilke, *Ind. Eng. Chem.* **62**, 34 (1970).
20. B. Bogdanović and G. Wilke, *Brennst. Chem.* **49**, 323 (1968).
21. K. von Bredow, G. Helferich, and C. D. Weis, *Helv. Chim. Acta* **55**, 553 (1972).
22. A. Carbonaro, F. Cambisi, and G. Dall'Asta, *J. Org. Chem.* **36**, 1443 (1971).
23. S. Cesca, W. Marconi, and M. L. Santostasi, *J. Polym. Sci., Part B* **7**, 547 (1969).
24. L. G. Cannell, *Tetrahedron Lett.* p. 5967 (1966).
25. I. M. Chernikova, I. I. Pisman, and M. A. Dalin, *Azerb. Khim. Zh.* p. 35 (1965); *Chem. Abstr.* **64**, 19385 (1966).

26. Y. Chauvin, H.-H. Phung, N. Guichard-Loudet, and G. Lefebvre, *Bull. Soc. Chim. Fr.* p. 3223 (1966).
27. K. M. Minachev, Y. I. Isakov, Y. T. Eidus, A. L. Lapidus, and V. P. Kalinin, *Izv. Akad. Nauk SSSR, Ser. Khim.* p. 1920 (1973); *Chem. Abstr.* 79, 136636 (1973).
28. F. Dawans, *Tetrahedron Lett.* p. 1943 (1971).
29. C. Dixon, E. W. Duck, and D. K. Jenkins, *Organometal. Chem. Syn.* 1, 77 (1970/1971).
30. G. G. Eberhardt and H. K. Myers, *J. Catal.* 26, 459 (1972).
31. E. Echigoya and A. Kobayashi, *Chem. Lett.* p. 277 (1972).
32. G. G. Eberhardt and W. P. Griffin, *J. Catal.* 16, 245 (1970).
33. J. J. Eisch and M. W. Foxton, *J. Organometal. Chem.* 12, P33 (1968).
34. J. Ewers, *Erdoel Kohle* 21, 763 (1968).
35. J. Ewers, *Angew. Chem.* 78, 593 (1966).
36. V. S. Feldblyum and A. I. Leshcheva, *J. Org. Chem. USSR* 8, 1115 (1972).
37. V. S. Feldblyum, T. I. Baranova, R. B. Svitych, B. A. Dolgoplosk, E. I. Tinyakova, and K. L. Makovetskii, *J. Org. Chem. USSR* 8, 657 (1972).
38. V. S. Feldblyum and T. I. Baranova, *J. Org. Chem. USSR* 7, 2348 (1971).
39. V. S. Feldblyum, A. I. Leshcheva, and N. V. Petrushanskaya, *J. Org. Chem. USSR* 6, 1114 (1970).
40. V. S. Feldblyum, A. I. Leshcheva, and N. V. Petrushanskaya, *J. Org. Chem. USSR* 6, 2419 (1970).
41. V. S. Feldblyum, A. I. Leshcheva, and N. V. Obeshchalova, *J. Org. Chem. USSR* 6, 205 (1970).
42. V. S. Feldblyum, A. I. Leshcheva, N. V. Obeshchalova, O. P. Yablonskii, and N. M. Pashchenko, *Neftekhimiya* 8, 533 (1968); *Chem. Abstr.* 70, 37061 (1964).
43. V. S. Feldblyum, N. V. Obeshchalova, and A. I. Leshcheva, *Proc. Acad. Sci. USSR* 172, 19 (1966).
44. V. S. Feldblyum and N. V. Obeshchalova, *Proc. Acad. Sci. USSR* 172, 56 (1967).
45. V. S. Feldblyum, N. V. Obeshchalova, A. I. Leshcheva, and T. I. Baranova, *Neftekhimiya* 7, 379 (1967); *Chem. Abstr.* 67, 90333 (1967).
46. K. Fischer, K. Jonas, P. Misbach, R. Stabba, and G. Wilke, *Angew. Chem.* 85, 1002 (1973).
47. G. Ghymes, A. Grobler, A. Simon, I. Kada, and I. Andor, *Magy. Kem. Lapja* 20, 570 (1965); *Chem. Abstr.* 64, 6471 (1966).
48. G. Henrici-Olivé, S. Olivé, and E. Schmidt, *J. Organometal. Chem.* 39, 201 (1972).
49. E. Hayashi, S. Narui, M. Ota, M. Sakai, Y. Sakakibara, and N. Uchino, *Kogyo Kagaku Zasshi* 74, 1834 (1971).
50. G. Hata and A. Miyake, *Chem. Ind. (London)* p. 921 (1967).
51. F. Hojabri, *J. Appl. Chem. Biotechnol.* 21, 90 (1971).
52. F. Hojabri, *Erdoel Kohle* 23, 724 (1970).
53. Y. Inoue, T. Kagawa, Y. Uchida, and H. Hashimoto, *Bull. Chem. Soc. Jap.* 45, 1996 (1972).
54. Y. Inoue, T. Kagawa, and H. Hashimoto, *Tetrahedron Lett.* p. 1099 (1970).
55. J. R. Jones and T. J. Symes, *J. Chem. Soc., C* p. 1124 (1971).
56. J. R. Jones, *J. Chem. Soc., C* p. 1117 (1971).
57. N. Kawata, K.-I. Maruya, T. Mizoroki, and A. Ozaki, *Bull. Chem. Soc. Jap.* 44 3217 (1971).
58. E. N. Kropacheva, B. A. Dolgoplosk, I. I. Ermkova, I. G. Zhuchikhina, and I. Y. Tsereteli, *Proc. Acad. Sci. USSR* 187, 677 (1969).
59. A. Kobayashi and E. Echigoya, *Nippon Kagaku Kaishi*, p. 668 (1973).

60. N. T. Anh, O. Eisenstein, J. M. Lefour, and M. E. T. H. Dau, *J. Amer. Chem. Soc.* **95**, 6146 (1973).
61. L. Lardicci, G. P. Giacomelli, R. Menicagli, and P. Pino, *Organometal. Chem. Syn.* **1**, 447 (1972).
62. L. Lardicci, G. P. Giacomelli, P. Salvadori, and P. Pino, *J. Amer. Chem. Soc.* **93**, 5794 (1971).
63. G. Lucas, *Abstr. 142nd Meet.*, *Amer. Chem. Soc.* Paper 68Q (1962).
64. K. Maruya, T. Mizoroki, and A. Ozaki, *Bull. Chem. Soc. Jap.* **45**, 2255 (1972).
65. K. Maruyama, T. Kuroki, T. Mizoroki, and A. Ozaki, *Bull. Chem. Soc. Jap.* **44**, 2002 (1971).
66. K. Maruya, N. Ando, and A. Ozaki, *Nippon Kagaku Zasshi* **91**, 1125 (1970).
67. K. Maruya, T. Mizoroki, and A. Ozaki, *Bull. Chem. Soc. Jap.* **43**, 3630 (1970).
68. K. Maruya, T. Mizoroki, and A. Ozaki, *Bull. Chem. Soc. Jap.* **46**, 993 (1973).
69. S. Miyamoto and T. Arakawa, *Kogyo Kagaku Zasshi* **74**, 1394 (1971).
70. R. G. Miller, T. J. Kealy, and A. L. Barney, *J. Amer. Chem. Soc.* **89**, 3756 (1967).
71. R. Noyori, Y. Kumagai, I. Umeda, and H. Takaya, *J. Amer. Chem. Soc.* **94**, 4018 (1972).
72. R. Noyori, T. Suzuki, Y. Kumagai, and H. Takaya, *J. Amer. Chem. Soc.* **93**, 5894 (1971).
73. R. Noyori, T. Suzuki, and H. Takaya, *J. Amer. Chem. Soc.* **93**, 5896 (1971).
74. R. Noyori, T. Odagi, and H. Takaya, *J. Amer. Chem. Soc.* **92**, 5780 (1970).
75. R. Noyori, I. Umeda, and H. Takaya, *Chem. Lett.* p. 1189 (1972).
76. R. Noyori, T. Ishigami, N. Hayashi, and H. Takaya, *J. Amer. Chem. Soc.* **95**, 1674 (1973).
77. R. Noyori, *Tetrahedron Lett.* p. 1691 (1973).
78. N. V. Obeshchalova, V. S. Feldblyum, and N. M. Pashchenko, *J. Org. Chem. USSR* **4** 982 (1968).
79. O.-T. Onsager, H. Wang, and U. Blindheim, *Helv. Chim. Acta* **52**, 187 (1969).
80. O.-T. Onsager, H. Wang, and U. Blindheim, *Helv. Chim. Acta* **52**, 196 (1969).
81. O.-T. Onsager, H. Wang, and U. Blindheim, *Helv. Chim. Acta* **52**, 215 (1969).
82. O.-T. Onsager, H. Wang, and U. Blindheim, *Helv. Chim. Acta* **52**, 224 (1969).
83. O.-T. Onsager, H. Wang, and U. Blindheim, *Helv. Chim. Acta* **52**, 230 (1969).
84. G. M. Panchenkov, A. I. Kuznetsov, and M. K. Shauki, *Neftekhimiya* **11**, 18 (1971); *Chem. Abstr.* **74**, 111488 (1971).
85. I. Pavlik and J. Klikorka, *Wiss. Z. Techn. Hochsch. Chem. "Carl Schorlemmer" Leuna-Merseburg* **15**, 73 (1973).
86. P. Pino, L. Lardicci, P. Palagi, and G. P. Giacomelli, *Chim. Ind. (Milan)* **50**, 355 (1968).
87. L. I. Redkina, K. L. Makovetskii, E. I. Tinyakova, and B. A. Dolgoplosk, *Proc. Acad. Sci. USSR* **186**, 105 (1969).
88. E. Y. Saridzhalinskaya, O. N. Babaev, Y. G. Kambarov, and M. A. Lalin, *Prom. Sin. Kauch Nauch Tekh. Sb.* p. 3 (1970); *Chem. Abstr.* **76**, 33714 (1972).
89. Y. Sakakibara, T. Tagano, M. Sakai, and N. Uchino, *Bull. Inst. Chem. Res., Kyoto Univ.* **50**, 375 (1972).
90. A. Schott, H. Schott, G. Wilke, J. Brandt, H. Hoberg, and E. G. Hoffmann, *Justus Liebigs Ann. Chem.*, p. 508 (1973).
91. F. K. Schmidt, V. S. Tkach, and A. V. Kalabina, *Neftekhimiya* **12**, 76 (1972).
92. G. N. Schrauzer and P. Glockner, *Chem. Ber.* **97**, 2451 (1964).
93. G. N. Schrauzer and S. Eichler, *Chem. Ber.* **95**, 2764 (1962).
94. G. N. Schrauzer, R. K. Y. Ho., and G. Schlesinger, *Tetrahedron Lett.* p. 543 (1970).

95. A. C. L. Su and J. W. Collette, *J. Organometal. Chem.* **36**, 177 (1972).
96. A. A. Surovtsev, N. A. Borisova, A. I. Leshcheva, E. A. Mushina, V. S. Feldblyum, and B. A. Krentsel, *Neftekhimiya* **12**, 687 (1972); *Chem. Abstr.* **78**, 98668 (1973).
97. A. C. L. Su and J. W. Collette, *Polym. Prepr., Amer. Chem. Soc., Div. Polym. Chem.* **12**, 415 (1971).
98. C. A. Tolman, *J. Amer. Chem. Soc.* **92**, 6777 (1970).
99. M. Uchino, A. Yamamoto, and S. Ikeda, *J. Organometal. Chem.* **24**, C63 (1970).
100. M. Uchino, Y. Chauvin, and G. Lefebvre, *C.R. Acad. Sci., Ser. C* **265**, 103 (1967).
101. G. E. Voecks, P. W. Jennings, G. D. Smith, and C. N. Caughlan, *J. Org. Chem.* **37**, 1460 (1972).
102. G. Wilke *et al., Angew Chem.* **78**, 157 (1966).
103. S. Yoshikawa, S. Nishimura, J. Kiji, and J. Furukawa, *Tetrahedron Lett.* p. 3071 (1973).
104. I. G. Zhuchikhina, I. I. Ermakova, E. N. Kropacheva, I. A. Zevakin, and B. A. Dolgoplosk, *Proc. Acad. Sci. USSR* **200**, 856 (1971).
105. B. Bogdanović, H.-G. Karmann, B. Meister, H.-G. Nüssel, and G. Wilke, *Abstr. Int. Conf. Organometal. Chem., 5th, 1971* Vol. I, p. 201 (1971).
106. B. Bogdanović, B. Henc, A. Lösler, B. Meister, H. Pauling, and G. Wilke, *Angew. Chem.* **85**, 1013 (1973).
107. K. Fischer, Ph.D. Thesis, University of Bochum, 1973.
108. F. Heydenreich, Ph.D. Thesis, University of Bochum, 1971.
109. B. Henc, Ph.D. Thesis, University of Bochum, 1971.
110. A. Lösler, Ph.D. Thesis, University of Bochum, 1973.
111. B. Meister, Ph.D. Thesis, University of Bochum, 1971.
112. H.-G. Karmann, Ph.D. Thesis, University of Bochum, 1970.
113. H.-G. Nüssel, Ph.D. Thesis, University of Bochum, 1970.
114. A. Schott, Ph.D. Thesis, University of Bochum, 1970.
115. H. Schott, Ph.D. Thesis, Technische Hochschule Aachen, 1965.
116. D. Walter, Ph.D. Thesis, Technische Hochschule Aachen, 1965.
117. H. Itatani and J. C. Bailar, *Ind. Eng. Chem., Prod. Res. Develop.* **11**, 146 (1972).
118. M. Tsutsui and T. Koyano, *J. Polym. Sci., Part A-1* **5**, 681 (1967).
119. I. I. Pisman, I. M. Chernikova, M. A. Dalin, P. K. Taktarov, and Y. A. Agadzhanov, *Khim. Prom. (Moscow)* **43**, 328 (1967); *Chem. Abstr.* **67**, 108104 (1967).
120. I. I. Pisman, M. A. Dalin, and G. V. Vasil'kovskaya, *Azerb. Khim. Zh.*, p. 69 (1964); *Chem. Abstr.* **62**, 14473 (1965).
121. B. L. Barnett and C. Krüger, unpublished results (1973).
122. S. Y. Pshezhetskii and A. T. Gladyshev, *J. Phys. Chem. USSR* **15**, 333 (1941); *Chem. Abstr.* **36**, 6400 (1942).
123. S. Y. Pshezhetskii, *J. Phys. Chem. USSR* **14**, 1376 (1940); *Chem. Abstr.* **35**, 3884, (1941).
124. Y. T. Eidus and B. K. Nefedov, *Neftekhimiya* **1**, 786 (1961); *Chem. Abstr.* **57**, 8808 (1962).
125. I. I. Pisman, I. M. Chernikova, and M. A. Dalin, *Azerb. Khim. Zh.* p. 73 (1968); *Chem. Abstr.* **69**, 105782 (1968).
126. J. P. Hogan, R. L. Banks, W. C. Lanning, and A. Clerk, *Ind. Eng. Chem.* **47**, 752 (1955).
127. V. S. Feldblyum, *Proc. Int. Congr. Catal., 4th, 1968* p. 222 (1971).
128. H. Uchida and H. Imai, *Bull. Chem. Soc. Jap.* **35**, 989 (1962).
129. A. Takahashi, N. Mogi, and H. Takahama, *Kogyo Kagaku Zasshi* **66**, 1472 (1963).
130. V. C. F. Holm, G. C. Bailey, and A. Clark, *Ind. Eng. Chem.* **49**, 250 (1957).

131. M. Born, Y. Chauvin, G. Lefebvre, and N.-H. Phung, *C.R. Acad. Sci., Ser. C* **268**, 1600 (1969).
132. S. Malinowski and W. Skupinski, *Rocz. Chem.* **48**, 359 (1974).
133. M. Tsutsui, J. Aryoshi, T. Koyano, and M. N. Levy, *Advan. Chem. Ser.* **70**, 266 (1968).
134, T. Joh and N. Hagihara, *Nippon Kagaku Zasshi* **91**, 378 (1970).
135. U. Birkenstock, H. Bönnemann, B. Bogdanović, D. Walter, and G. Wilke, *Advan. Chem. Ser.* **70**, 250 (1968).
136. N. N. Korneev, A. F. Popov, and G. S. Soloveva, *Khim. Prom. (Moscow)* **49**, 7 (1973).
137. K. Maruya, K. Nishio, N. Kawata, Y. Nakamura, T. Mizoroki, and A. Ozaki, *Nippon Kagaku Kaishi*, p. 1385 (1973).
138. C. Krüger and Y.-H. Tsay, unpublished results (1973).
139. See H. Lehmkuhl and K. Ziegler, *Houben-Weyl* **13**/4, 195 (1970).
140. C. A. Tolman, *J. Amer. Chem. Soc.* **92**, 6785 (1970).
141. B. Bogdanović and J. E. Galle, unpublished results (1973).
142. H. Bönnemann, C. Grard, W. Kopp, and G. Wilke, *Plenary Lect., Spec. Lect. Int. Congr. Pure Appl. Chem., 23rd, 1971* Vol. 6, p. 265 (1971).
143. H. Bönnemann, Ph.D. Thesis, Technische Hochschule Aachen, 1967.
144. J. Ashworth and C. H. Bamford, *J. Chem. Soc., Faraday Trans. 1*, p. 314 (1973).
145. C. H. Bamford and E. O. Hughes, *J. Chem. Soc., Faraday Trans. 1*, p. 1474 (1972).
146. C. H. Bamford and E. O. Hughes, *Proc. Roy. Soc., Ser. A* **326**, 469 (1972).
147. C. H. Bamford and E. O. Hughes, *Proc. Roy. Soc., Ser. A* **326**, 489 (1972).
148. C. H. Bamford, M. S. Blackie, and C. A. Finch, *Chem. Ind. (London)* p. 1763 (1962).
149. C. H. Bamford, *J. Polym. Sci., Part C* **4**, 1571 (1963).
150. C. H. Bamford and C. A. Finch, *Trans. Faraday Soc.* **59**, 548 (1963).
151. C. H. Bamford and C. A. Finch, *Z. Naturforsch. B* **17**, 804 (1962).
152. C. H. Bamford and C. A. Finch, *Z. Naturforsch. B* **17**, 500 (1962).
153. C. H. Bamford, G. C. Eastmond, and D. Whittle, *J. Organometal. Chem.* **17**, P33 (1969).
154. C. H. Bamford and K. Hargreaves, *Nature (London)* **209**, 292 (1966).
155. C. H. Bamford and K. Hargreaves, *Proc. Roy. Soc., Ser. A* **297**, 425 (1967).
156. C. H. Bamford, *Pure Appl. Chem.* **34**, 173 (1973).
157. D. G. H. Ballard, W. H. Janes, and T. Medinger, *J. Chem. Soc., B* p. 1168 (1968).
158. J. Boor, *J. Polym. Sci., Part A* **3**, 995 (1965).
159. J. Furukawa and T. Tsuruta, *Kogyo Kagaku Zasshi* **60**, 802 (1957).
160. G. Henrici-Olivé and S. Olivé, *J. Polym. Sci., Polym. Chem. Ed.* **11**, 1953 (1973).
161. S. Ikeda and T. Harimoto, *J. Organometal Chem.* **60**, C67 (1973).
162. Y. Kitahama and S. Ishida, *Makromol. Chem.* **119**, 64 (1968).
163. E. G. Kastning, H. Naarmann, H. Reis, and C. Berding, *Angew. Chem.* **77**, 313 (1965).
164. K. Kaeriyama and Y. Shimura, *Makromol. Chem.* **167**, 129 (1973).
165. V. A. Kormer, I. A. Poletayeva, and T. L. Yufa, *J. Polym. Sci., Part A-1* **10**, 251 (1972).
166. V. A. Kormer, T. L. Yufa, I. A. Poletaeva, B. D. Babitsku, and Z. D. Stepanova, *Dokl. Akad. Nauk SSSR* **185**, 873 (1969).
167. K. L. Makovetskii, I. Y. Ostrovskaya, L. I. Redykina, and I. L. Kershenbaum, *Kinet. Mech. Polyreact., Int. Symp. Macromol. Chem. Prepr.* **2**, 415 (1969).
168. T. Matsumoto, J. Furukawa, and H. Morimura, *J. Polym. Sci., Part A-1* **9**, 875 (1971).

169. T. Matsumoto, J. Furukawa, and H. Morimura, *J. Polym. Sci.*, *Part B* **6**, 869 (1968).
170. T. Matsumoto, J. Furukawa, and H. Morimura, *J. Polym. Sci.*, *Part B* **7**, 541 (1969).
171. R. H. B. Mais, P. G. Owston, D. T. Thompson, and A. M. Wood, *J. Chem. Soc.*, *A* p. 1744 (1967).
172. T. Matsumoto and J. Furukawa, *J. Macromol. Sci.*, *Chem.* **6**, 281 (1972).
173. K. S. Minsker, Y. A. Sangalov, and G. A. Razuwayev, *J. Polym. Sci.*, *Part C* **16**, 1489 (1967).
174. G. Dall'Asta and G. Motroni, *J. Polym. Sci.*, *Part A-1* **6**, 2405 (1968).
175. I. Y. Ostrovskaya, K. L. Makovetskii, B. A. Dolgoplosk, and E. I. Tinyakova, *Izv. Akad. Nauk SSSR, Ser. Khim.* p. 1632 (1967).
176. I. Y. Ostrovskaya, K. L. Makovetskii, E. I. Tinyakova, and B. A. Dolgoplosk, *Proc. Acad. Sci. USSR* **181**, 701 (1968).
177. L. Porri, G. Natta, and M. C. Gallazzi, *J. Polym. Sci.*, *Part C* **16**, 2525 (1967).
178. L. Porri, G. Natta, and M. C. Gallazzi, *Chim. Ind.* (*Milan*) **46**, 428 (1964).
179. M. Rougée and C. André, *J. Polym. Sci.*, *Part C* **16**, 3167 (1967).
180. G. A. Short and E. C. Shokal, *J. Polym. Sci.*, *Part B* **3**, 859 (1965).
181. C. Simionescu, T. Lixandru, I. Negulescu, I. Mazilu, and L. Tataru, *Makromol. Chem.* **163**, 59 (1973).
182. G. N. Schrauzer and H. Thyret, *Z. Naturforsch. B* **17** 73 (1962).
183. W. Strohmeier and H. Grübel, *Z. Naturforsch. B* **22**, 553 (1967).
184. M. Uchino and S. Ikeda, *J. Organometal. Chem.* **33**, C41 (1971).
185. A. Yamamoto and S. Ikeda, *J. Amer. Chem. Soc.* **89**, 5989 (1967).
186. J. Zachoval, J. Kalal, B. Veruovic, and L. Stefka, *Collect. Czech. Chem. Commun.* **30**, 1326 (1965).
187. D. G. H. Ballard, *Advan. Catal.* **23**, 263 (1973).
188. G. Dall'Asta, *J. Polym. Sci.*, *Part A-1* **6**, 2397 (1968).
189. J. Ashworth and C. H. Bamford, *J. Chem. Soc.*, *Faraday Trans. 1*, p. 302 (1973).
190. J. Lefebvre and I. Chovin, *Nov. Neftekhim. Protsessy Perspekt. Razv. Neftekhim.* p. 314 (1970); *Chem. Abstr.* **76**, 72847 (1972).
191. G. N. Schrauzer, S. Eichler, and D. A. Brown, *Chem. Ber.* **95**, 2755 (1962).
192. O. W. Webster, W. Mahler, and R. E. Benson, *J. Org. Chem.* **25**, 1470 (1960).
193. G. N. Schrauzer and S. Eichler, *Chem. Ind.* (*London*) p. 1270 (1961).
194. L. R. Wallace and J. F. Harrod, *Macromolecules* **4**, 656 (1971).
195. G. Wilke, *Pure Appl. Chem.* **17**, 179 (1968).
196. K. Ziegler, *Advan. Organometal. Chem.* **6**, 1 (1968).
197. K. Ziegler, *Angew. Chem.* **76**, 545 (1964).
198. K. Ziegler, H.-G. Gellert, E. Holzkamp, G. Wilke, E. W. Duck, and W.-R. Kroll. *Justus Liebigs Ann. Chem.* **629**, 172 (1960).
199. K. Ziegler, H.-G. Gellert, K. Zosel, E. Holzkamp, J. Schneider, M. Söll, and W.-R. Kroll. *Justus Liebigs Ann. Chem.* **629**, 121 (1960).
200. K. Ziegler, *Brennst.-Chem.* **35**, 321 (1954).
201. K. Ziegler, E. Holzkamp, H. Breil, and H. Martin, *Angew. Chem.* **67**, 543 (1955).
202. L. F. Albright and C. S. Smith, *Amer. Inst. Chem. Eng. J.* **14**, 325 (1968).
203. P. Misbach, Ph.D. Thesis, University of Bochum, 1969.
204. H. Schenkluhn, Ph.D. Thesis, University of Bochum, 1971.
205. G. P. Giacomelli, L. Lardicci, and P. Pino, *J. Organometal. Chem.* **33**, 105 (1971).
206. E. Marcus, D. L. MacPeek, and S. W. Tinsley, *J. Org. Chem.* **36**, 381 (1971).
207. K. Ziegler, W.-R. Kroll, W. Larbig, and O. W. Steudel, *Justus Liebigs Ann. Chem.* **629**, 53 (1960).
208. L. Farady, L. Bencze, and L. Marko, *J. Organometal. Chem.* **10**, 505 (1967).

209. L. Farady, L. Bencze, and L. Marko, *J. Organometal. Chem.* **17**, 107 (1969).
210. L. Farady and L. Marko, *Magy. Kem. Foly.* **75**, 121 (1969).
211. L. Farady and L. Marko, *J. Organometal. Chem.* **28**, 159 (1971).
212. L. Farady, *Kem. Kozlem.* **38**, 37 (1972).
213. A. Job and R. Reich, *C.R. Acad. Sci.* **179**, 330 (1924).
214. P. Binger and J. McMeeking, *Angew. Chem.* **86**, 518 (1974).
215. P. Binger, G. Schroth, and J. McMeeking, *Angew. Chem.* **86**, 518 (1974).
216. H. Takaya, N. Hayashi, T. Ishigami, and R. Noyori, *Chem. Lett.* p. 813 (1973).
217. A. F. Popov, N. N. Korneev, and G. S. Soloveva, *Khim. Prom. (Moscow)* **48**, 90 (1972).
218. G. N. Schrauzer, *Advan. Catal.* **18**, 373 (1968).
219. R. V. Meyer, Ph.D. Thesis, University of Bochum, 1973.
220. G. N. Schrauzer, B. N. Bastian, and G. J. Fosselius, *J. Amer. Chem. Soc.* **88**, 4890 (1966).
221. M. Englert, P. W. Jolly, and G. Wilke, *Angew. Chem.* **83**, 84 (1971).
222. P. Binger and J. McMeeking, unpublished results (1973).
223. D. Bingham, D. W. Webster, and P. B. Wells, *J. Chem. Soc., Dalton* p. 1928 (1972).
224. B. Corain, *Chem. Ind. (London)* p. 1465 (1971).
225. B. Corain, *Gazz. Chim. Ital.* **102**, 687 (1972).
226. R. Cramer and R. V. Lindsey, *J. Amer. Chem. Soc.* **88**, 3534 (1966).
227. R. Cramer, *J. Amer. Chem. Soc.* **88**, 2272 (1966).
228. B. Corain and G. Puosi, *J. Catal.* **30**, 403 (1973).
229. B. Fell and J. M. J. Tetteroo, *Angew. Chem.* **77**, 813 (1965).
230. V. S. Feldblyum and N. V. Obeshchalova, *Kinet. Katal.* **11**, 893 (1970).
231. L. W. Gosser and G. W. Parshall, *Tetrahedron Lett.* p. 2555 (1971).
232. P. Heimbach, *Angew. Chem.* **78**, 604 (1966).
233. H. Kanai, *Chem. Commun.* p. 203 (1972).
234. E. N. Kropacheva, B. A. Dolgoplosk, I. I. Ermakova, I. G. Zhuchikhina, and I. Y. Tsereteli, *Kinet. Mech. Polyreact., Int. Symp. Macromol. Chem., Prepr.* **2**, 329 (1969); *Chem. Abstr.* **75**, 77872 (1971).
235. I. A. Kopeva, E. I. Tinyakova, and B. A. Dolgoplosk, *Bull. Acad. Sci. USSR* p. 1336 (1970).
236. N. A. Maly, H. Menapace, and M. F. Farona, *J. Catal.* **29**, 182 (1973).
237. R. G. Miller, H. J. Golden, D. J. Baker, and R. D. Stauffer, *J. Amer. Chem. Soc.* **93**, 6308 (1971).
238. R. G. Miller, P. A. Pinke, R. D. Stauffer, and H. J. Golden, *J. Organometal. Chem.* **29**, C42 (1971).
239. R. G. Miller, *J. Amer. Chem. Soc.* **89**, 2785 (1967).
240. R. G. Miller and P. A. Pinke, *J. Amer. Chem. Soc.* **90**, 4500 (1968).
241. T. Mizuta, H. Samejima, and T. Kwan, *Bull. Chem. Soc. Jap.* **41**, 727 (1968).
242. N. V. Obeshchalova, V. S. Feldblyum, M. E. Basner, and V. S. Dzyuba, *Zh. Org. Khim.* **4**, 574 (1968).
243. R. Pettit, H. Sugahara, J. Wristers, and W. Merk, *Discuss. Faraday Soc.* **47**, 71 (1969).
244. Y. Sakakibara, M. Mukai, M. Sakai, and N. Uchino, *Nippon Kagaku Kasshi*, p. 1457 (1972).
245. H. Samejima, T. Mizuta, and T. Kwan, *Nippon Kagaku Zasshi* **89**, 1028 (1968).
246. H. Samajima, T. Mizuta, H. Yamamoto, and T. Kwan, *Bull. Chem. Soc. Jap.* **42**, 2722 (1969).
247. T. Suzuki and Y. Takegami, *Kogyo Kagaku Zasshi* **74**, 1371 (1971).

248. C. A. Tolman, *J. Amer. Chem. Soc.* **94**, 2994 (1972).
249. C. A. Tolman, *J. Amer. Chem. Soc.* **92**, 6785 (1970).
250. J. Thomson and M. C. Baird, *Can. J. Chem.* **48**, 3443 (1970).
251. H. Takaya, M. Yamakawa, and R. Noyori, *Chem. Lett.* p. 781 (1973).
252. G. Lefebvre and Y. Chauvin, *World Petrol. Congr., Proc., 7th, 1967* Vol. 5, p. 343 (1968).
253. Y. Chauvin and G. Lefebvre, *C.R. Acad. Sci. Ser. C* **258**, 2105 (1964).
254. R. P. A. Sneeden and H. H. Zeiss, *J. Organometal. Chem.* **22**, 713 (1970).
255. R. G. Miller, P. A. Pinke, and D. J. Baker, *J. Amer. Chem. Soc.* **92**, 4490 (1970).
256. N. Ando, K. I. Maruya, T. Mizoroki, and A. Ozaki, *J. Catal.* **20**, 299 (1971).
257. P. Abley and F. J. McQuillin, *J. Catal.* **24**, 536 (1972).
258. J. C. Bailar and H. Itatani, *J. Amer. Chem. Soc.* **89**, 1592 (1967).
259. D. Bingham and M. G. Burnett, *J. Chem. Soc., A,* p. 1782 (1971).
260. M. G. Burnett, *Chem. Commun.* p. 507 (1965).
261. W. H. Dennis, D. H. Rosenblatt, R. R. Richmond, G. A. Finseth, and G. T. Davis, *Tetrahedron Lett.* p. 1831 (1968).
262. E. A. Emken, E. N. Frankel, and R. O. Butterfield, *J. Amer. Oil Chem. Soc.* **43**, 14 (1966).
263. A. G. Hinze and D. J. Frost, *J. Catal.* **24**, 541 (1972).
264. I. V. Kalechits, V. G. Lipovich, and F. K. Shmidt, *Neftekhimiya* **6**, 813 (1966); *Chem. Abstr.* **66**, 94632 (1967).
265. S. J. Lapporte, *Ann. N.Y. Acad. Sci.* **158**, 510 (1969).
266. S. Nanya, M. Hanai, and K. Fukuzumi, *Kogyo Kagaku Zasshi* **72**, 2005 (1969).
267. T. Nishiguchi and K. Fukuzumi, *Chem. Commun.* p. 139 (1971).
268. N. F. Noskova, N. I. Marusich, and D. V. Sokol'skii, *Tr. Inst. Khim. Nauk, Akad. Nauk Kaz. SSR* **30**, 3 (1970); *Chem. Abstr.* **74**, 6837 (1971).
269. M. F. Sloan, A. S. Matlack, and D. S. Breslow, *J. Amer. Chem. Soc.* **85**, 4014 (1963).
270. Y. Takegami, T. Ueno, and T. Fujii, *Bull. Chem. Soc. Jap.* **42**, 1663 (1969).
271. Y. Takegami, T. Ueno, and T. Sakala, *Kogya Kagaku Zasshi* **68**, 2373 (1965).
272. Y. Takegami, T. Ueno, and K. Kawajiri, *Kogyo Kagaku Zasshi* **66**, 1068 (1963).
273. Y. Tajima and E. Kunioka, *J. Org. Chem.* **33**, 1689 (1968).
274. V. A. Tulupov, *Russ. J. Phys. Chem.* **32**, 727 (1958).
275. V. A. Tulupov, *Russ. J. Phys. Chem.* **37**, 365 (1963).
276. P. Abley and F. J. McQuillin, *Discuss. Faraday Soc.* **46**, 31 (1968).
277. E. N. Frankel, H. Itatani, and J. C. Bailar, *J. Amer. Oil Chem. Soc.* **49**, 132 (1972).
278. H. Itatani and J. C. Bailar, *J. Amer. Chem. Soc.* **89**, 1600 (1967).
279. I. V. Kalechits and F. K. Shmidt, *Kinet. Katal.* **7**, 614 (1966); *Chem. Abstr.* **65**, 16817 (1966).
280. R. Paul, P. Buisson, and N. Joseph, *C.R. Acad. Sci., Ser. C* **232**, 627 (1951).
281. R. Paul, R. Buisson, and N. Joseph, *Ind. Eng. Chem.* **44**, 1006 (1952).
282. C. A. Brown and H. C. Brown, *J. Amer. Chem. Soc.* **85**, 1003 (1963).
283. H. C. Brown and C. A. Brown, *J. Amer. Chem. Soc.* **85**, 1005 (1963).
284. C. A. Brown, *Chem. Commun.* p. 952 (1969).
285. C. A. Brown, *J. Org. Chem.* **35**, 1900 (1970).
286. H. I. Schlesinger, H. C. Brown, A. E. Finholt, J. R. Gilbreath, H. R. Hoekstra, and E. K. Hyde, *J. Amer. Chem. Soc.* **75** 215 (1953).
287. C. A. Brown and V. K. Ahuja, *J. Org. Chem.* **38**, 2226 (1973).
288. T. Satoh, S. Suzuki, Y. Suzuki, Y. Miyaji, and Z. Imai, *Tetrahedron Lett.* p. 4555 (1969).
289. D. R. Witt and J. P. Hogan, *J. Polym. Sci., Part A-1* **8**, 2689 (1970).

290. E. W. Duck, J. M. Locke, and C. J. Mallinson, *Justus Liebigs Ann. Chem.* **719**, 69 (1968).
291. S. J. Lapporte and W. R. Schuett, *J. Org. Chem.* **28**, 1947 (1964).
292. V. G. Lipovich, F. K. Shmidt, and I. V. Kalechits, *Kinet. Katal.* **8**, 1300 (1967); *Chem. Abstr.* **69**, 35147 (1968).
293. V. G. Lipovich, F. K. Shmidt, and I. V. Kalechits, *Kinet. Katal.* **8**, 939 (1967); *Chem. Abstr.* **68**, 59185 (1968).
294. K. Hanaya, N. Fujita, and H. Kudo, *Chem. Ind. (London)* p. 794 (1973).
295. K. Hanaya and T. Ono, *Nippon Kagaku Zasshi* **92**, 1225 (1971).
296. P. Jiru and H. Kuchynka, *Z. Phys. Chem. (Frankfurt am Main)* [*N.S.*] **25**, 286 (1960).
297. M. Ichikawa, M. Soma, T. Onishi, and K. Tamaru, *Trans. Faraday Soc.* **63**, 2528 (1967).
298. M. Yang, K. Yamamoto, N. Otake, M. Ando, and K. Takase, *Tetrahedron Lett.* p. 3843 (1970).
299. E. W. Bennett and P. J. Orenski, *J. Organometal. Chem.* **28**, 137 (1971).
300. E. T. Chukovskaya and R. K. Freidlina, *Izv. Akad. Nauk SSSR, Otd. Khim. Nauk* p. 761 (1963); *Chem. Abstr.* **59**, 7551 (1963).
301. Y. Kiso, M. Kumada, K. Maeda, K. Sumitani, and K. Tamao, *J. Organometal. Chem.* **50**, 311 (1973).
302. Y. Kiso, M. Kumada, K. Tamao, and M. Umeno, *J. Organometal. Chem.* **50**, 297 (1973).
303. M. Kumada, Y. Kiso, and M. Umeno, *Chem. Commun.* p. 611 (1970).
304. M. Kumada, K. Sumitani, Y. Kiso, and K. Tamao, *J. Organometal. Chem.* **50**, 319 (1973).
305. P. Svoboda, P. Sedlmayer, and J. Hetflejs, *Collect. Czech. Chem. Commun.* **38**, 1783 (1973).
306. K. Yamamoto, Y. Uramoto, and M. Kumada, *J. Organometal. Chem.* **31**, C9 (1971).
307. A. D. Petrov, V. F. Mironov, V. M. Vdovin, and S. I. Sadykh-Zade, *Izv. Akad. Nauk SSSR, Otd. Khim. Nauk* p. 256 (1956); *Chem. Abstr.* **50**, 13726 (1956).
308. S. Nozakura, *Bull. Chem. Soc. Jap.* **29**, 784 (1956).
309. S. Nozakura and S. Konotsune, *Bull. Chem. Soc. Jap.* **29**, 322 (1956).
310. M. Capka, P. Svoboda, and J. Hetflejs, *Collect. Czech. Chem. Commun.* **38**, 3830 (1973).
311. A. N. Nesmeynanov, R. K. Freidlina, E. C. Chukovskaya, R. G. Petrova, and A. B. Belyavsky, *Tetrahedron* **17**, 61 (1962).
312. W. Jetz and W. A. G. Graham, *J. Amer. Chem. Soc.* **89**, 2773 (1967).
313. Y. Kiso, K. Tameo, and M. Kumada, *Chem. Commun.* p. 1208 (1972).
314. E. S. Brown and E. A. Rick, *Chem. Commun.* p. 112 (1969).
315. E. S. Brown, E. A. Rick, and F. D. Mendicino, *J. Organometal. Chem.* **38**, 37 (1972).
316. P. Heydenreich, A. Mollbach, G. Wilke, H. Dreeskamp, E. G. Hoffmann, G. Schroth, K. Seevogel, and W. Stempfle, *Isr. J. Chem.* **10**, 293 (1972).
317. R. Giezynski and S. Pasynkiewicz, *Przem. Chem.* **52**, 746 (1973); *Organometal. Compounds* **24** (4), i–v (1974).
318. J. C. Falk, *J. Org. Chem.* **36**, 1445 (1971).
319. B. C. S. Rao and G. P. Thakar, *J. Sci. Ind. Res., Sect. B* **20**, 317 (1961).
320. V. M. Frolov, O. P. Parenago, L. P. Shuikina, and B. A. Dolgoplosk, *Katal. Reakts. Zhidk. Faze*, p. 304 (1972); *Chem. Abstr.* **79**, 114841 (1973).
321. C. J. Attridge and S. J. Maddock, *J. Organometal. Chem.* **26**, C65 (1971).
322. J. Manassen, *J. Catal.* **18**, 38 (1970).

323. P. S. Skell, J. J. Havel, D. L. Williams-Smith, and M. J. McGlinchey, *Chem. Commun.* p. 1098 (1972).
324. V. A. Poletaev, V. M. Vdovin, and N. S. Nametkin, *Proc. Acad. Sci. USSR* **208**, 128 (1973).
325. V. S. Poletaev, V. M. Vdovin, and N. S. Nanetkin, *Proc. Acad. Sci. USSR* **203**, 379 (1972).
326. M. Tamura and J. K. Kochi, *Bull. Chem. Soc. Jap.* **44**, 3063 (1971).
327. C. H. Bamford, G. C. Eastmond, and P. Murphy, *Trans. Faraday Soc.* **66**, 2598 (1970).
328. S. Aoki, C. Shirafuji, and T. Otsu, *Polym. J.* **2**, 257 (1971).
329. G. Dall'Asta, G. Motroni, R. Manetti, and C. Tosi, *Makromol. Chem.* **130**, 153 (1969).
330. I. Y. Ostrovskaya, K. L. Makovetskii, G. P. Karpacheva, E. I. Tinyakova, and B. A. Dolgoplosk, *Proc. Acad. Sci. USSR* **197**, 350 (1971).
331. I. Y. Ostrovskaya, K. L. Makovetskii, E. I. Tinyakova, and B. A. Dolgoplosk, *Proc. Acad. Sci. USSR* **181**, 701 (1968).
332. N. Kawata, T. Mizoroki, A. Ozaki, and M. Ohkawara, *Chem. Lett.* p. 1165 (1973).
333. R. K. Freidlina, Ts'ao-I. and E. T. Chukovskaya, *Proc. Acad. Sci. USSR* **132**, 483 (1960).
334. B. W. Taylor and H. E. Swift, *J. Catal.* **26**, 254 (1972).
335. H. Bönnemann, *Angew. Chem.* **82**, 699 (1970).
336. C. H. Bamford and I. Sakamoto, *J. Chem. Soc., Faraday Trans. 1* p. 330 (1974).
337. S. Yoshikawa, K. Aoki, J. Kiji, and J. Furukawa, *Tetrahedron* **30**, 405 (1974).
338. N. V. Petrushanskaya, A. I. Kurapova, and V. S. Feldblyum, *Proc. Acad. Sci. USSR* **211**, 593 (1973).
339. R. Noyori, Y. Kumagai, and H. Takaya, *J. Amer. Chem. Soc.* **96**, 634 (1974).
340. E. V. Bykova and G. A. Berezhok, *Vysokomol. Soedin., Ser. B* **15**, 266 (1973); *Chem. Abstr.* **79**, 66876 (1973).
341. N. V. Petrushanskaya, A. I. Kurapova, and V. S. Feldblyum, *Zh. Org. Khim.* **9**, 2620 (1973); *Chem. Abstr.* **80**, 59357 (1974).
342. V. M. Frolov, O. P. Parenago, and B. A. Dolgoplosk, *Probl. Kinet. Katal.* **15**, 125 (1973); *Chem. Abstr.* **79**, 136390 (1973).
343. L. K. Friedlin, L. I. Gvinter, L. N. Suvorova, and S. S. Danielova, *Izv. Akad. Nauk SSSR, Ser. Khim.* p. 2260 (1973).
344. G. Wilke, *Proc. Robert A. Welch Found. Conf. Chem. Res.* **9**, 165 (1965).
345. B. Bogdanović, M. Kröner, and G. Wilke, *Justus Liebigs Ann. Chem.* **699**, 1 (1966); M. Kröner, Ph.D. Thesis, Technische Hochschule Aachen, 1961.
346. G. Strukul, M. Bonivento, R. Ros, and M. Graziana, *Tetrahedron Lett.* p. 1791 (1974).
347. R. Noyori, H. Kawauchi, and H. Takaya, *Tetrahedron Lett.* p. 1749 (1974).
348. I. Heertje, G. K. Koch, and W. J. Wösten, *J. Catal.* **32**, 337 (1974).
349. Y. Nakano, K. Natsukawa, H. Yasuda, and H. Tani, *Tetrahedron Lett.* p. 2833 (1972).
350. N. Kawata, K. Maruya, T. Mizoroki, and A. Ozaki, *Bull. Chem. Soc. Jap.* **47**, 413 (1974).
351. F. K. Shmidt, A. V. Kalabina, V. S. Tkach, and V. G. Lipovich, *Khim. Aromat. Nepredel'n Soedin.* p. 127 (1971); *Chem. Abstr.* **80**, 132717 (1974).
352. Y. Kiso, K. Tamao, and M. Kumada, *J. Organometal. Chem.* **76**, 95 (1974).
353. R. G. Miller, P. A. Pinke, R. D. Stauffer, H. J. Golden, and D. J. Baker, *J. Amer. Chem. Soc.* **96**, 4211 (1974).

354. P. A. Pinke and R. G. Miller, *J. Amer. Chem. Soc.* **96**, 4221 (1974).
355. P. A. Pinke, R. D. Stauffer, and R. G. Miller, *J. Amer. Chem. Soc.* **96**, 4229 (1974).
356. H. J. Golden, D. J. Baker, and R. G. Miller, *J. Amer. Chem. Soc.* **96**, 4235 (1974).
357. Y. Sakakibara, T. Yamasaki, M. Hikita, M. Sakai, and N. Uchino, *Nippon Kagaku Kaishi*, p. 910 (1974).
358. J. Kiji, S. Yoshikawa, E. Sasakawa, S. Nishimura, and J. Furukawa, *J. Organometal Chem.* **80**, 267 (1974).
359. B. Adler, J. Beger, C. Duschek, C. Gericke, W. Pritzkow, and H. Schmidt, *J. Prakt. Chem.* **316**, 449 (1974).

Patents

P1. J. S. Yoo (Atlantic Richfield Co.), U.S. Patent 3,679,772 (1972); *Chem. Abstr.* **77**, 125938 (1972).
P2. J. S. Yoo and R. L. Milam (Atlantic Richfield Co.), U.S. Patent 3,697,617 (1972); *Chem. Abstr.* **78**, 3652 (1973).
P3. W. Schulz, H. Mix, E. Kurras, F.-W. Wilcke, J. Reihsig, H. Fuhrmann, and I. Grassert (Akademie der Wissenschaften der DDR), Ger. Offen, 2,252,856 (1973).
P4. J. S. Yoo (Atlantic Richfield Co.), Ger. Offen. 2,131,814 (1971); *Chem. Abstr.* **76**, 85362 (1972).
P5. I. I. Pisman, I. M. Chernikova, and M. A. Dalin (All Union Scientific Research Technol. Inst.), U.S.S.R. Patent 174,620 (1965); *Chem. Abstr.* **64**, 1954 (1966).
P6. British Petroleum Co., Fr. Patent 1,575,732 (1969); *Chem. Abstr.* **76**, 60315 (1972).
P7. J. R. Jones and L. Priestley (British Petroleum Co.), U.S. Patent 3,505,425 (1970); see *Chem. Abstr.* **72**, 12040 (1970).
P8. J. K. Hambling and J. R. Jones (British Petroleum Co.), U.S. Patent 3,417,160 (1968); see *Chem. Abstr.* **69**, 43388 (1968).
P9. J. K. Hambling and J. R. Jones (British Petroleum Co.), Brit. Patent 1,123,474 (1969); *Chem. Abstr.* **69**, 105858 (1968).
P10. British Petroleum Co., Fr. Patent 1,507,007 (1967); *Chem. Abstr.* **70**, 4792 (1969).
P11. J. K. Hambling and J. R. Jones (British Petroleum Co.), Brit. Patent 1,110,974 (1968); Conc. Neth. Appl. 6,609,513 (1967); *Chem. Abstr.* **67**, 53611 (1967).
P12. J. K. Hambling and J. R. Jones (British Petroleum Co.), Brit. Patent 1,101,657 (1968); Conc. Neth. Appl. 6,609,512 (1967); *Chem. Abstr.* **67**, 53609 (1967).
P13. J. K. Hambling and J. R. Jones (British Petroleum Co.), Brit. Patent 1,101,498 (1968); Conc. Neth. Appl. 6,608,574 (1967); *Chem. Abstr.* **67**, 2739 (1967).
P14. J. R. Jones and T. J. Symes (British Petroleum Co.), Ger. Offen. 1,813,115 (1969); *Chem. Abstr.* **71**, 123489 (1969).
P15. W. Schneider (B. F. Goodrich Co.), U.S. Patent 3,408,416 (1968); *Chem. Abstr.* **70**, 11211 (1969).
P16. W. Schneider (B. F. Goodrich Co.), U.S. Patent 3,412,164 (1968); *Chem. Abstr.* **70**, 87141 (1969).
P17. W. Schneider (B. F. Goodrich Co.), U.S. Patent 3,452,115 (1969); see *Chem. Abstr.* **72**, 3108 (1970).
P18. K. Schloemer, H. Kroeper, and H.-M. Weitz (B.A.S.F.), U.S. Patent 3,622,648 (1972); see *Chem. Abstr.* **71**, 123494 (1969).
P19. G. N. Schrauzer and S. Eichler (B.A.S.F.), Ger. Patent 1,186,052 (1969); *Chem. Abstr.* **62**, 16086 (1965).

P20. H. Müller and D. Mangold (B.A.S.F.), Ger. Offen. 1,643,655 (1971).

P21. H. Kröper and K. Schlömer (B.A.S.F.), Ger. Patent 1,178,419 (1965); *Chem. Abstr.* **61**, 15973 (1964).

P22. H. Morimura, A. Kachi, and S. Konotsune (Chisso Corp.), Jap. Kokai 73 17,588 (1973); *Chem. Abstr.* **79**, 5922 (1973).

P23. H. Pracejus and H. Jahr (Deutsche Akademie der Wissenschaften), Ger. Offen. 2,230,739 (1973).

P24. M. Dubeck and D. R. Brackenridge (Ethyl Corp.), U.S. Patent 3,647,883 (1972); *Chem. Abstr.* **76**, 112649 (1972).

P25. E. H. Drew and A. H. Neal (Esso Research Eng. Co), U.S. Patent 3,564,070 (1971); see *Chem. Abstr.* **72**, 2992 (1970).

P26. J. K. Mertzweiller and H. M. Tenney (Esso Research Eng. Co.), U.S. Patent 3,562,351 (1971); see *Chem. Abstr.* **71**, 49214 (1969).

P27. Y. P. L. M. Castille and P. T. Parker (Esso Research Eng. Co.), Ger. Offen. 2,021,524 (1971); *Chem. Abstr.* **74**, 140889 (1971).

P28. Y. P. L. M. Castille and P. T. Parker (Esso Research Eng. Co.), Ger. Offen. 2,021,523 (1971); *Chem. Abstr.* **74**, 124765 (1971).

P29. E. H. Drew (Esso Research Eng. Co.), U.S. Patent 3,390,201 (1969); *Chem. Abstr.* **69**, 51513 (1968).

P30. E. H. Drew and A. H. Neal (Esso Research Eng. Co.), Brit. Patent 1,167,289 (1969); *Chem. Abstr.* **72**, 2992 (1970).

P31. F. J. Kealy (E.I. du Pont), U.S. Patent 3,306,948 (1967); Conc. Neth. Appl. 6,402,444 (1964); *Chem. Abstr.* **62**, 10334 (1965).

P32. J. W. Collette and A. C. L. Su (E.I. du Pont), U.S. Patent 3,565,967 (1971); *Chem. Abstr.* **74**, 140892 (1971).

P33. W. Herwig (Farbwerke Hoechst AG.), U.S. Patent 3,507,930 (1970); see *Chem. Abstr.* **72**, 21310 (1930).

P34. F. Röhrscheid (Farbwerke Hoechst AG.), Ger. Offen. 1,643,976 (1971); Conc. Fr. Patent 1,581,030 (1969); *Chem. Abstr.* **73**, 35938 (1970).

P35. Farbwerke Hoechst AG, Fr. Patent 1,576,134 (1969); *Chem. Abstr.* **72**, 78358 (1970).

P36. Farbwerke Hoechst AG, Fr. Patent 1,491,963 (1967); Conc. Neth. Appl. 6,612,735 (1967); *Chem. Abstr.* **67**, 53608 (1967).

P37. V. S. Feldblyum *et al.*, U.S.S.R. Patent 290,706 (1972); *Chem. Abstr.* **78**, 147319 (1973).

P38. V. S. Feldblyum *et al.*, U.S.S.R. Patent 382,598 (1973); *Chem. Abstr.* **79**, 65770 (1973).

P39. P. Günther and W. Oberkirch (Farbenfabriken Bayer AG.), Ger. Offen. 1,793,059 (1972); see *Chem. Abstr.* **74**, 32167 (1971).

P40. H. E. Swift and C.-Y. Wu (Gulf Research), U.S. Patent 3,622,649 (1971); *Chem. Abstr.* **76**, 33758 (1972).

P41. Gelsenberg Benzin AG, Fr. Patent 1,547,921 (1968); *Chem. Abstr.* **71**, 50646 (1969).

P42. P. L. Ragg (I.C.I.), Ger. Offen. 1,937,232 (1970); *Chem. Abstr.* **72**, 79631 (1970).

P43. E. D. Smith (I.C.I.), Brit. Patent 1,140,821 (1969); *Chem. Abstr.* **70**, 96143 (1969).

P44. R. J. Sampson, D. Jackson, and J. M. Thomas (I.C.I.), U.S. Patent 3,565,971 (1971); see *Chem. Abstr.* **71**, 49212 (1969).

P45. K. A. Taylor, W. Hewertson, and J. A. Leonard (I.C.I.), Brit. Patent 1,131,146 (1966); *Chem. Abstr.* **70**, 37139 (1969).

P46. R. W. Dunning, K. A. Taylor, and J. Walker (I.C.I.), Brit. Patent 1,164,855 (1969); *Chem. Abstr.* **71**, 123486 (1969).

P47. D. A. Cornforth, D. Y. Waddan, and D. Williams (I.C.I.), Brit. Patent 1,123,097 (1968); *Chem. Abstr.* **69**, 67597 (1968).

P48. F. Dawans (Institut Français du Pétrole), Fr. Patent 2,079,592 (1971); *Chem. Abstr.* **77**, 61505 (1972).

P49. J. Gaillard (Institut Français du Pétrole), Fr. Patent 2,114,114 (1972); *Chem. Abstr.* **78**, 62876 (1973).

P50. Y. Chauvin (Institut Français du Pétrole), Fr. Patent 2,051,973 (1971); *Chem. Abstr.* **76**, 141520 (1972).

P51. Y. Chauvin and J. Gaillard (Institut Français du Pétrole), Ger. Offen. 2,024,093 (1970); *Chem. Abstr.* **74**, 41866 (1971).

P52. Y. Chauvin, G. Lefebvre, and M. Uchino (Institut Français du Pétrole), Fr. Patent 1,583,731 (1969); *Chem. Abstr.* **73**, 76634 (1970).

P53. Y. Chauvin, G. Lefebvre, and M. Uchino (Institut Français du Pétrole), Fr. Patent 1,540,270 (1968); *Chem. Abstr.* **71**, 80633 (1969).

P54. Y. Chauvin, G. Lefebvre, and M. Uchino (Institut Français du Pétrole), Fr. Patent 1,540,269 (1968); *Chem. Abstr.* **71**, 80634 (1969).

P55. Y. Chauvin, J. Gaillard, G. Lefebvre, and J. P. Wauquier (Institut Français du Pétrole), Ger. Offen. 1,926,771 (1969); *Chem. Abstr.* **72**, 78357 (1970).

P56. N.-H. Phung and G. Lefebvre (Institut Français du Pétrole), Fr. Patent 1,549,177 (1968); *Chem. Abstr.* **72**, 89736 (1970).

P57. Institut Français du Pétrole, Brit. Patent 1,160,399 (1969); see *Chem. Abstr.* **71**, 123487 (1969).

P58. F. K. Shmidt, V. S. Tkach, and A. V. Kalabina (Irkutsk State Univ.), U.S.S.R. Patent 379,554 (1973); *Chem. Abstr.* **79**, 52761 (1973).

P59. S. G. Abasova *et al.* (Topchiev, A. V., Institute Petrochem. Syn.), U.S.S.R. Patent 289,102 (1970); *Chem. Abstr.* **74**, 142621 (1971).

P60. D. K. Jenkins and C. G. P. Dixon (International Syn. Rubber Co.), U.S. Patent 3,641,176 (1972); Conc. Fr. Patent 1,599,774 (1970); *Chem. Abstr.* **75**, 35368 (1971).

P61. H. Mori, K. Ikeda, I. Nagaoha, S. Hirayanagi, A. Kogure, and S. Shimizu (Japan Syn. Rubber Co), Jap. Patent 72 22,207 (1972); *Chem. Abstr.* **77**, 100695 (1972).

P62. H. Mori, I. Nagaoka, S. Hirayanagi, and A. Kichinoki (Japan Syn. Rubber Co.), Jap. Patent 72 49,563 (1972); *Chem. Abstr.* **78**, 110508 (1973).

P63. H. Mori, K. Ikeda, I. Nagaoha, S. Hirayanagi, I. Shimizu, and A. Kogure (Japan Syn. Rubber Co.), Jap. Patent 72 22,206 (1972); *Chem. Abstr.* **77**, 87808 (1972).

P64. H. Mori, K. Ikeda, I. Nagaoka, S. Hirayanagi, I. Shimizu, and A. Kogure (Japan Syn. Rubber Co.), Jap. Patent 72 24,523 (1972); *Chem. Abstr.* **77**, 87812 (1972).

P65. C. D. Weis and A. C. Rochat (J. R. Geigy AG), Ger. Offen. 1,931,152 (1970); *Chem. Abstr.* **72**, 66495 (1970).

P66. E. Takagi, M. Matsui, K. Otsu, and Y. Iwama (Mitsubishi Chem. Ind. Co.), Jap. Patent 72 38,402 (1972); *Chem. Abstr.* **78**, 15478 (1973).

P67. M. Yamaguchi, Y. Tsunoda, and K. Kakihara (Mitsubishi Chem. Ind. Co.), Jap. Patent 72 45,737 (1972); *Chem. Abstr.* **78**, 42902 (1973).

P68. E. Takagi, M. Matsui, K. Otsu, and Y. Iwama (Mitsubishi Chem. Ind. Co.), Jap. Patent 72 24,524 (1972); *Chem. Abstr.* **77**, 100699 (1972).

P69. E. Takagi, M. Matsui, O. Kunimiki, and Y. Iwama (Mitsubishi Chem. Ind. Co.), Jap. Patent 72 24,525 (1972); *Chem. Abstr.* **77**, 100698 (1972).

P70. T. Arakawa and K. Saheki (Mitsui Petrochem. Ind. Ltd.), Jap. Patent 72 22,208 (1972); *Chem. Abstr.* **77**, 100694 (1972).

P71. E. Takagi, M. Matsui, K. Otsu, and Y. Iwama (Mitsubishi Chem. Ind. Co.), Jap. Patent 72 23,286 (1972); *Chem. Abstr.* **77**, 87819 (1972).

P72. T. Arakawa (Mitsui Petrochem. Ind. Ltd.), Jap. Patent 72 22,204 (1972); *Chem. Abstr.* **77**, 87807 (1972).

P73. E. Takagi, M. Matsui, K. Otsu, and Y. Iwama (Mitsubishi Chem. Ind. Co.), Jap. Patent 71 38,766 (1971); *Chem. Abstr.* **76**, 24668 (1972).

P74. Mitsui Petrochem. Ind. Ltd., Fr. Patent 1,567,630 (1969); *Chem. Abstr.* **72**, 111618 (1970).

P75. V. S. Feldblyum *et al.* (Nauchno-Issledovatelsky Inst.), Brit. Patent 1,164,882 (1969); *Chem. Abstr.* **72**, 2994 (1970).

P76. A. H. Neal and P. T. Parker, U.S. Patent 3,686,352 (1972); *Chem. Abstr.* **77**, 151423 (1972).

P77. Nauchno-Issledovatelsky Inst., Fr. Patent 1,420,952 (1965); *Chem. Abstr.* **65**, 13539 (1966).

P78. I. J. Tjurjaev *et al.* (Nauchno-Issledovatelsky Inst.), Brit. Patent 1,051,564 (1966); *Chem. Abstr.* **66**, 65044 (1967).

P79. H. E. Dunn (Phillips Petroleum Co.), U.S. Patent 3,651,111 (1972); *Chem. Abstr.* **76**, 13991 (1972).

P80. H. E. Dunn (Phillips Petroleum Co.), U.S. Patent 3,644,218 (1972); *Chem. Abstr.* **77**, 61249 (1972).

P81. H. E. Dunn (Phillips Petroleum Co), U.S. Patent 3,684,588 (1972); *Chem. Abstr.* **78**, 29197 (1973).

P82. H. E. Dunn (Phillips Petroleum Co.), U.S. Patent 3,737,474 (1973); *Chem. Abstr.* **79**, 31449 (1973).

P83. T. Hutson and C. O. Carter (Phillips Petroleum Co.), U.S. Patent 3,631,121 (1971); *Chem. Abstr.* **76**, 58937 (1972).

P84. E. A. Zuech (Phillips Petroleum Co.), U.S. Patent 3,590,095 (1971); *Chem. Abstr.* **75**, 76133 (1971).

P85. H. E. Dunn (Phillips Petroleum Co.), U.S. Patent 3,558,738 (1971); *Chem. Abstr.* **74**, 87308 (1971).

P86. H. E. Dunn (Phillips Petroleum Co.), U.S. Patent 3,592,870 (1971); *Chem. Abstr.* **75**, 76135 (1971).

P87. R. E. Dixon and J. F. Hutto (Phillips Petroleum Co.), U.S. Patent 3,544,649 (1970); *Chem. Abstr.* **72**, 99989 (1970).

P88. E. A. Zuech (Phillips Petroleum Co.), U.S. Patent 3,485,881 (1969); *Chem. Abstr.* **72**, 99989 (1970).

P89. P. L. Maxfield (Phillips Petroleum Co.), U.S. Patent 3,427,365 (1969); *Chem. Abstr.* **70**, 87002 (1969).

P90. D. W. Walker and E. L. Czenkusch (Phillips Petroleum Co.), U.S. Patent 3,134,824 (1964); *Chem. Abstr.* **61**, 5810 (1964).

P91. G. Nowlin and H. D. Lyons (Phillips Petroleum Co.), U.S. Patent 2,969,408 (1961); *Chem. Abstr.* **55**, 16009 (1961).

P92. A. Ozaki, T. Mizoroki, and M. Kanagawa (President Tokyo Inst. Technol.), Ger. Offen. 2,211,745 (1973); *Chem. Abstr.* **78**, 110835 (1973).

P93. N. V. Petrushanskaya and V. S. Feldblyum, U.S.S.R. Patent 290,764 (1971); *Chem. Abstr.* **74**, 140893 (1971).

P94. E. W. Duck (Petrochemicals Ltd.), Brit. Patent 837,350 (1960); *Chem. Z.* **4**, 2368 (1966).

P95. R. F. Mason (Shell Oil Co.), U.S. Patent 3,686,351 (1972); *Chem. Abstr.* **77**, 151422 (1972).

P96. W. A. Butte (Sun Oil Co.), U.S. Patent 3,564,072 (1971).

P97. G. G. Eberhardt and W. P. Griffin (Sun Oil Co.), U.S. Patent 3,472,911 (1969); *Chem. Abstr.* **72**, 57475 (1970).

P98. W. P. Griffin and W. A. Butte (Sun Oil Co.), Ger. Offen. 1,810,027 (1969); *Chem. Abstr.* **71**, 92057 (1969).

P99. G. G. Eberhardt (Sun Oil Co.), U.S. Patent 3,482,001 (1969); *Chem. Abstr.* **72**, 42720 (1970).

P100. G. G. Eberhardt and W. P. Griffin (Sun Oil Co.), U.S. Patent 3,459,825 (1969); *Chem. Abstr.* **71**, 112383 (1969).

P101. W. P. Griffin (Sun Oil Co.), U.S. Patent 3,467,726 (1969); *Chem. Abstr.* **71**, 123492 (1969).

P102. N. Bergem, U. Blindheim, U.-T. Onsager, and H. Wang (Sentralinstitut Ind. Forskning), U.S. Patent 3,709,953 (1973); see *Chem. Abstr.* **71**, 49204 (1969).

P103. Scholven-Chemie A. G., Fr. Addn. 95,635 (1971); *Chem. Abstr.* **77**, 113781 (1972).

P104. V. S. Feldblyum, N. V. Obeschalova, A. I. Lescheva, L. D. Kononova, and L. S. Krotova (Nauchno-Issledovatelsky Institut), Fr. Patent 1,588,162 (1970); *Chem. Abstr.* **76**, 153182 (1972).

P105. G. Desgrandchamps, H. Hemmer, and M. Haurie (Société Nat. d'Aquitaine), Fr. Patent 2,070,554 (1971); *Chem. Abstr.* **77**, 4840 (1972).

P106. H. M. J. C. Creemers (Stamicarbon N.V.), Ger. Offen. 2,152,827 (1972); *Chem. Abstr.* **77**, 19259 (1972).

P107. G. Desgrandchamps, H. Hemmer, and M. Haurie (Société Nat. d'Aquitaine), Ger. Offen. 2,001,923 (1970); *Chem. Abstr.* **74**, 3302 (1971).

P108. G. Desgrandchamps, H. Hemmer, and M. Haurie (Société Nat. d'Aquitaine). Ger. Offen. 1,964,701 (1970); *Chem. Abstr.* **74**, 3303 (1971).

P109. Scholven-Chemie A. G., Brit. Patent 1,151,550 (1969); Conc. Neth. Appl. 6,612,339 (1967); *Chem. Abstr.* **67**, 63689 (1967).

P110. N. Bergem, U. Blindheim, O.-T. Onsager, and H. Wang (Sentralinstitut Ind. Forskning), S. Afr. Patent 67 04,671 (1967); *Chem. Abstr.* **70**, 68530 (1969).

P111. N. Bergem, U. Blindheim, O.-T. Onsager, and H. Wang (Sentralinstitut Ind. Forskning), S. Afr. Patent 67 05,408 (1967); *Chem. Abstr.* **70**, 37132 (1969).

P112. N. Bergem, U. Blindheim, O.-T. Onsager, and H. Wang (Sentralinstitut Ind. Forskning), Ger. Offen. 1,768,887 (1972); see *Chem. Abstr.* **71**, 12502 (1969).

P113. Sentralinstitut Ind. Forskning, Brit. Patent 1,124,123 (1968); Conc. Neth. Appl, 6,601,770 (1966); *Chem. Abstr.* **67**, 2738 (1967).

P114. H. van Zwet, R. S. Bauer, and W. Keim (Shell Int. Res.), Ger. Offen. 2,062,293 (1971); *Chem. Abstr.* **75**, 98942 (1971).

P115. P. W. Glockner, W. Keim, R. F. Mason, and R. S. Bauer (Shell Int. Res.), Ger. Offen. 2,053,758 (1971); *Chem. Abstr.* **75**, 88072 (1971).

P116. P. W. Glockner (Shell. Int. Res.), Ger. Offen. 1,931,060 (1970); *Chem. Abstr.* **72**, 66352 (1970).

P117. Shell International Research, Brit. Patent 979,778 (1965); Conc. Belg. Patent 626,407 (1963); *Chem. Abstr.* **60**, 13164 (1964).

P118. B. C. Roest and E. L. T. M. Spitzer (Shell Oil Co.), U.S. Patent 3,321,546 (1967); Conc. Fr. Patent 1,385,503 (1965); *Chem. Abstr.* **62**, 14496 (1965).

P119. E. L. T. M. Spitzer (Shell Oil Co.), U.S. Patent 3,327,015 (1967); Conc. Fr. Patent 1,385,503 (1965); *Chem. Abstr.* **62**, 14496 (1965).

P120. Shell International Research, Neth. Appl. 68 13,667 (1969); *Chem. Abstr.* **71**, 80626 (1969).

P121. L. G. Cannell (Shell Oil Co.), U.S. Patent 3,258,502 (1960); Conc. Neth. Appl. 6,506,276 (1965); *Chem. Abstr.* 64, 11104 (1966).

P122. Shell International Research, Neth. Appl. 70 16,039 (1971); *Chem. Abstr.* 75, 110729 (1971).

P123. Shell International Research, Neth. Appl. 70 16,037 (1971); *Chem. Abstr.* 75, 110727 (1971).

P124. R. F. Mason (Shell Oil Co.), U.S. Patent 3,676,523 (1972); *Chem. Abstr.* 77, 100710 (1972).

P125. F. F. Farley (Shell Oil Co.), U.S. Patent 3,647,906 (1972); *Chem. Abstr.* 76, 112643 (1972).

P126. R. S. Bauer, H. Chung, P. W. Glockner, W. Keim, and H. van Zwet (Shell Oil Co.), U.S. Patent 3,644,563 (1972); see *Chem. Abstr.* 75, 110729 (1971).

P127. E. F. Magoon, L. G. Cannell, and J. H. Raley (Shell Oil Co.), U.S. Patent 3,620,981 (1971); see *Chem. Abstr.* 71, 80626 (1969).

P128. L. G. Cannell (Shell Oil Co.), U.S. Patent 3,592,869 (1971); *Chem. Abstr.* 75, 98144 (1971).

P129. K. W. Barnett and P. W. Glockner (Shell Oil Co.), U.S. Patent 3,527,838 (1970); see *Chem. Abstr.* 73, 77852 (1970).

P130. J. D. McClure (Shell Oil Co.), U.S. Patent 3,530,197 (1970); *Chem. Abstr.* 73, 130600 (1970).

P131. K. W. Barnett and J. H. Raley (Shell Oil Co.), U.S. Patent 3,532,765 (1970); *Chem. Abstr.* 73, 120047 (1970).

P132. K. W. Barnett and J. H. Raley (Shell Oil Co.), U.S. Patent 3,459,826 (1969); *Chem. Abstr.* 71, 112384 (1969).

P133. L. G. Cannell and E. F. Magoon (Shell Oil Co.), U.S. Patent 3,355,510 (1967); *Chem. Abstr.* 68, 49017 (1968).

P134. L. G. Cannell and E. F. Magoon (Shell Oil Co.), U.S. Patent 3,424,815 (1969); *Chem. Abstr.* 70, 77279 (1969).

P135. J. D. McClure and K. W. Barnett (Shell Oil Co.), U.S. Patent 3,424,816 (1969); *Chem. Abstr.* 70, 68871 (1969).

P136. R. Rienäcker and G. Wilke (Studiengesellschaft Kohle mbH), Ger. Offen. 1,493,217 (1969); see *Chem. Abstr.* 68, 49439 (1968).

P137. G. Wilke (Studiengesellschaft Kohle mbH), U.S. Patent 3,379,706 (1968); Conc. Neth. Appl. 64 09,179 (1965); *Chem. Abstr.* 63, 5770 (1965).

P138. G. Hata and A. Miyake (Toyo Rayon Co. Ltd.), Jap. Patent 70 07,285 (1970); *Chem. Abstr.* 73, 26394 (1970).

P139. T. Hata, K. Takahashi, and A. Miyake (Toray Industries Inc.), Jap. Patent 72 22,807 (1972); *Chem. Abstr.* 77, 87818 (1972).

P140. S. Izawa, S. Yamada, and N. Kunimoto (Toyo Soda Manufg. Co. Ltd.), Jap. Patent 72 22,205 (1972); *Chem. Abstr.* 77, 87809 (1972).

P141. M. Iwamoto and S. Sodeguchi (Toray Industries Inc.), Jap. Patent 71 38,765 (1971); *Chem. Abstr.* 76, 154446 (1972).

P142. H. Ito and K. Kimura (Toa Gosei Chem. Ind. Co.), Ger. Offen. 1,917,884 (1969); *Chem. Abstr.* 72, 21329 (1970).

P143. T. Hata and A. Miyake (Toray Industries Inc.), Jap. Patent 71 03,161 (1971); *Chem. Abstr.* 74, 124757 (1971).

P144. T. Hata and A. Miyake (Toray Industries Inc.), Jap. Patent 71 03,162 (1971); *Chem. Abstr.* 74, 124758 (1971).

P145. E. A. Rick and R. L. Pruett (Union Carbide Corp.), U.S. Patent 3,458,550 (1969); *Chem. Abstr.* 72, 25572 (1970),

P146. R. L. Pruett and E. A. Rick (Union Carbide Corp.), U.S. Patent 3,440,294 (1969); *Chem. Abstr.* **71**, 3056 (1969).

P147. J. Ewers and H.-W. Voges (VEBA-Chemie AG.), Ger. Offen. 1,643,045 (1971); see *Chem. Abstr.* **75**, 117955 (1971).

P148. K. Ziegler and E. Holzkamp, Ger. Patent 964,642 (1957); *Chem. Abstr.* **54**, 967 (1960).

P149. Y. Chauvin (Institut Français du Pétrole), Ger. Offen. 2,032,140 (1971); *Chem. Abstr.* **74**, 87306 (1971).

P150. T. Hill (B.P. Chemicals Ltd.), Ger. Offen. 2,004,361 (1970); *Chem. Abstr.* **74**, 54350 (1971).

P151. J. R. Jones and I. Priestley (British Petroleum Co. Ltd.), Ger. Offen. 1,643,716 (1971); see *Chem. Abstr.* **70**, 87001 (1969).

P152. F. Parker and R. W. Ral (British Hydrocarbon Chemicals Ltd.), Ger. Offen. 1,568,071 (1970); see *Chem. Abstr.* **66**, 4696 (1967).

P153. R. Bauer, H. Chung, K. W. Barnett, P. W. Glockner, and W. Keim (Shell Oil Co.), U.S. Patent 3,686,159 (1972); see *Chem. Abstr.* **76**, 15196 (1972).

P154. T. Yoshida and S. Yuguchi (Toyo Rayon Co. Ltd.), Jap. Patent 70 07,522 (1970); *Chem. Abstr.* **73**, 24887 (1970).

P155. R. F. Mason (Shell Oil Co.), U.S. Patent 3,737,475 (1973).

P156. J. S. Yoo and H. Erickson (Atlantic Richfield Co.), U.S. Patent 3,755,490 (1973).

P157. T. Amemiya, A. Tsuneya, M. Suzuki, and Y. Baba (Bureau of Ind. Technics), Jap. Patent 2662('59) (1959); *Chem. Abstr.* **53**, 18862 (1959).

P158. T. J. Kealy (E. I. du Pont), Fr. Patent 1,388,305 (1965); *Chem. Abstr.* **63**, 6858 (1965).

P159. G. Wilke, B. Bogdanović, and H. Pauling (Studiengesellschaft Kohle mbG.), Ger. Offen. 2,039,125 (1972); *Chem. Abstr.* **76**, 126478 (1972).

P160. I. Aishima, Y. Takahashi, H. Morita, and T. Ikegami (Asaki Chem. Ind. Co. Ltd.), Jap. Kokai 72 34,879 (1972); *Chem. Abstr.* **78**, 137085 (1973).

P161. A. Nishihara, S. Miyaji, H. Kokura, and H. Kawasaki (Asahi Electro-Chem. Co. Ltd.), Japan Kokai 72 29,480 (1972); *Chem. Abstr.* **79**, 92860 (1973).

P162. A. Nishihara, S. Miyaji, and M. Kokura (Asahi Electro-Chem. Co. Ltd.), Jap. Kokai 72 34,474 (1972); *Chem. Abstr.* **78**, 137113 (1973).

P163. K. Ziegler, Brit. Patent 826,638 (1960); *Chem. Abstr.* **54**, 16019 (1960).

P164. C. H. Bamford and K. Hargreaves, Brit. Patent 1,181,062 (1970); *Chem. Abstr.* **72**, 112033 (1970).

P165. O. Ozaki (Tokyo Institute of Technol.), Jap. Kokai 73 28,087 (1973); *Chem. Abstr.* **79**, 79532 (1973).

P166. H. Naarmann, H. P. Hofmann, and E. G. Kastning (B.A.S.F.), Brit. Patent 1,089,465 (1967); see *Chem. Abstr.* **64**, 14308 (1966).

P167. H. Kröper and H. M. Weitz (B.A.S.F.), Ger. Patent 1,026,959 (1958); *Chem. Abstr.* **55**, 1078 (1961).

P168. H. Kröper and H. M. Weitz (B.A.S.F.), Ger. Patent 1,051,003 (1959); *Chem. Abstr.* **55**, 2180 (1961).

P169. S. Konotsune, A. Kachi, and Y. Iwahashi (Chisso Corp.), Ger. Offen. 1,928,136 (1969); *Chem. Abstr.* **72**, 44305 (1970).

P170. I.C.I., Fr. Patent 1,500,264 (1967); *Chem. Abstr.* **69**, 60094 (1968).

P171. C. H. Bamford, F. J. Duncan, and R. J. W. Reynolds (I.C.I.), U.S. Patent 3,433,774 (1969); see *Chem. Abstr.* **67**, 65318 (1967).

P172. C. H. Bamford and D. T. Thompson (I.C.I.), Brit. Patent 1,086,066 (1967); *Chem. Abstr.* **67**, 117527 (1967).

P173. C. H. Ballard, W. H. Janes, and J. D. Seddon (I.C.I.), Brit Patent 1,099,116 (1968); *Chem. Abstr.* **68**, 60056 (1968).

P174. Y. Chauvin and G. Lefebvre (Institut Français du Pétrole), U.S. Patent 3,563,967 (1971); see *Chem. Abstr.* **70**, 48016 (1969).

P175. D. Wittenberg and H. Seibt (B.A.S.F.), Ger. Patent 1,136,329 (1962); *Chem. Abstr.* **58**, 4442 (1963).

P176. V. A. Kormer, B. D. Babitsky, T. L. Jufa, and I. A. Poletaeva, Ger. Offen. 1,944,753 (1970); *Chem. Abstr.* **73**, 121392 (1970).

P177. A. S. Vandi, F. Valeretto, and M. Ragazzini (Montecatini Edison S.p.A.), Ger. Offen. 1,645,060 (1970); see *Chem. Abstr.* **68**, 30408 (1968).

P178. Montecatini Soc. Gen. and K. Ziegler, Brit. Patent 891,646 (1962); *Chem. Abstr.* **57**, 2424 (1962).

P179. R. H. Gaeth (Phillips Petroleum Co.), U.S. Patent 3,725,379 (1973); *Chem. Abstr.* **79**, 6530 (1973).

P180. R. S. Bauer, H. Chung, W. Keim, and H. van Zwet (Shell Int. Res. Maatsch. N.V.), Ger. Offen. 2,101,391 (1971); *Chem. Abstr.* **75**, 130323 (1971).

P181. Solvic Soc. Anon., Belg. Patent 570,027 (1959); *Chem. Abstr.* **53**, 16594 (1959).

P182. D. M. Singleton, P. W. Glockner, and W. Keim (Shell Int. Res. Maatsch. N.V.), Ger. Offen. 2,159,370 (1972); *Chem. Abstr.* **77**, 89124 (1972).

P183. R. C. Morris, R. S. Bauer, H. Chung, W. Keim, and H. van Zwet (Shell Int. Res. Maatsch. N.V.), Ger. Offen. 2,062,335 (1971); *Chem. Abstr.* **75**, 89076 (1971).

P184. R. S. Bauer, H. Chung, P. W. Glockner, W. Keim, and H. van Zwet (Shell Oil Co.), U.S. Patent 3,635,937 (1972); see *Chem. Abstr.* **75**, 110729 (1971).

P185. R. S. Bauer, H. Chung, L. G. Cannell, W. Keim, and H. van Zwet (Shell Oil Co.), U.S. Patent 3,637,636 (1972); *Chem. Abstr.* **75**, 130322 (1971).

P186. D. Wittenberg and H. Müller (B.A.S.F.), Ger. Patent 1,240,852 (1967); *Chem. Abstr.* **67**, 108331 (1967).

P187. W. C. Drinkard (E.I. du Pont), Ger. Offen. 2,221,113 (1972); *Chem. Abstr.* **78**, 42868 (1973).

P188. Y. T. Chia (E.I. du Pont), U.S. Patent 3,676,481 (1972); see *Chem. Abstr.* **71**, 90876 (1969).

P189. W. C. Drinkard and R. V. Lindsey (E.I. du Pont), Ger. Offen. 2,149,175 (1972); *Chem. Abstr.* **77**, 4982 (1972).

P190. Y. T. Chia (E.I. du Pont), Ger. Offen. 1,805,278 (1969); *Chem. Abstr.* **71**, 90876 (1969).

P191. L. C. Kreider (Goodrich-Gulf Chem. Co.), U.S. Patent 3,390,193 (1968); *Chem. Abstr.* **69**, 43523 (1968).

P192. Montecatini Soc. Gen., Ital. Patent 792,187 (1967); *Chem. Abstr.* **76**, 59064 (1972).

P193. E. Takagi, M. Matsui, K. Otsu, and Y. Iwama (Mitsubishi Chem. Ind.), Jap. Patent 72 28,761 (1972); *Chem. Abstr.* **77**, 125931 (1972).

P194. C. E. Smith and B. J. White (Phillips Petroleum Co.), U.S. Patent 3,641,184 (1972); *Chem. Abstr.* **76**, 99070 (1972).

P195. P. L. Maxfield (Phillips Petroleum Co.), U.S. Patent 3,471,581 (1969); *Chem. Abstr.* **71**, 123730 (1969).

P196. M. K. Carter, P. W. Glockner, and J. L. van Winkle (Shell Int. Res.), Ger. Offen. 2,120,977 (1971); *Chem. Abstr.* **76**, 33766 (1972).

P197. G. Wilke and B. Bogdanović (Studiengesellschaft Kohle mbH.), Ger. Offen. 1,924,628 (1970); *Chem. Abstr.* **74**, 87309 (1971).

P198. G. Wilke, B. Bogdanović, and H. G. Nüssel (Studiengesellschaft Kohle mbH), Ger. Offen 2,063,149 (1972); *Chem. Abstr.* **77**, 100933 (1972).

P199. S. W. Tinsley, E. A. Rick, and J. E. McKeon (Union Carbide Corp.), U.S. Patent 3,375,287 (1968); *Chem. Abstr.* **69**, 35573 (1968).

P200. T. Yoshimoto, S. Kaneko, H. Yoshii, and T. Sasaki (Bridgestone Tire Co.), Jap. Patent 71 17,126 (1971); *Chem. Abstr.* **75**, 152761 (1971).

P201. T. Yoshimoto, S. Kaneko, and H. Okado (Bridgestone Tire Co.), Jap. Patent 71 17,130 (1971); *Chem. Abstr.* **75**, 152762 (1971).

P202. T. Yoshimoto, S. Kaneko, and H. Okado (Bridgestone Tire Co.), Jap. Patent 71 02,832 (1971); *Chem. Abstr.* **75**, 7137 (1971).

P203. T. Yoshimoto, T. Narumiya, S. Kaneko, H. Yoshii, and T. Takamatsu (Bridgestone Tire Co.), Jap. Patent 70 39,556 (1970); *Chem. Abstr.* **74**, 127210 (1971).

P204. T. Yoshimoto, S. Kaneko, T. Narumiya, and H. Yoshii (Bridgestone Tire Co.), U.S. Patent 3,541,064 (1970); *Chem. Abstr.* **74**, 23500 (1971).

P205. Bridgestone Tire Co., Fr. Patent 1,581,146 (1969); *Chem. Abstr.* **73**, 4738 (1970).

P206. Bridgestone Tire Co., Fr. Patent 1,572,717 (1969); *Chem. Abstr.* **72**, 44810 (1970).

P207. T. Yoshimoto, T. Narumiya, S. Kaneko, H. Yoshii, and T. Takamatsu (Bridgestone Tire Co.), S. Afr. Patent 68 07,486 (1969); *Chem. Abstr.* **71**, 92454 (1969).

P208. T. Yoshimoto, H. Yoshii, S. Kaneko, and T. Sasaki (Bridgestone Tire Co.), Ger. Offen. 1,920,403 (1969); *Chem. Abstr.* **72**, 13609 (1970).

P209. E. W. Duck, J. M. Locke, and C. J. Mallinson (International Synthetic Rubber Co.), Brit. Patent 1,229,573 (1971); *Chem. Abstr.* **75**, 50244 (1971).

P210. A. N. de Vault (Phillips Petroleum Co.), U.S. Patent 3,696,088 (1972); *Chem. Abstr.* **78**, 44503 (1973).

P211. C. W. Strobel (Phillips Petroleum Co.), U.S. Patent 3,646,142 (1972); *Chem. Abstr.* **76**, 154487 (1972).

P212. H. L. Hassell (Shell Int. Res.), Ger. Offen. 2,242,190 (1973); *Chem. Abstr.* **79**, 5881 (1973).

P213. M. M. Wald and M. G. Quam (Shell Oil Co.), U.S. Patent 3,700,633 (1972); *Chem. Abstr.* **78**, 17353 (1973).

P214. R. J. A. Eckert and J. Heemskerk (Shell Int. Res.), Ger. Offen. 2,132,336 (1972); *Chem. Abstr.* **76**, 128357 (1972).

P215. L. E. de Winkler (Shell Int. Res.), Ger. Offen. 2,125,413 (1971); *Chem. Abstr.* **76**, 86893 (1972).

P216. J. S. Yoo (Atlantic Richfield Co.), U.S. Patent 3,671,565 (1972); *Chem. Abstr.* **77**, 100865 (1972).

P217. S. J. Lapporte (California Research Corp.), U.S. Patent 3,205,278 (1965); see *Chem. Abstr.* **64**, 600 (1966).

P218. L. W. Gosser (E.I. du Pont), Ger. Offen. 1,940,303 (1970); *Chem. Abstr.* **72**, 100161 (1970).

P219. M. T. Musser (E.I. du Pont), U.S. Patent 3,631,210 (1971); *Chem. Abstr.* **76**, 72116 (1972).

P220. W. W. Prichard (E.I. du Pont), U.S. Patent 2,671,807 (1954); *Chem. Abstr.* **49**, 4017 (1955).

P221. D. R. Levering (Hercules Powder Co.), U.S. Patent 3,152,184 (1964); *Chem. Abstr.* **62**, 427 (1965).

P222. R. F. Heck (Hercules Powder Co.), U.S. Patent 3,270,087 (1966); *Chem. Abstr.* **65**, 16857 (1966).

P223. D. Y. Waddan and D. Williams (I.C.I.), Ger. Offen. 1,904,613 (1969); *Chem. Abstr.* **71**, 123525 (1970).

P224. M. S. Spencer (I.C.I.), U.S. Patent 2,966,534 (1960); *Chem. Abstr.* **55**, 8288 (1961).

P225. C. Lassau, R. Stern, and L. Sajus (Institut Français du Pétrole), Ger. Offen. 2,116,313 (1971); *Chem. Abstr.* **76**, 24664 (1972).

P226. D. C. Tabler (Phillips Petroleum Co.), U.S. Patent 3,692,852 (1972); *Chem. Abstr.* **77**, 139467 (1972).

P227. G. Chandra (Dow Corning Ltd.), Ger. Offen. 2,302,231 (1973); *Chem. Abstr.* **79**, 105400 (1973).

P228. C. M. King, W. C. Seidel, and C. A. Tolman (E.I. du Pont), Ger. Offen. 2,237,703 (1973); *Chem. Abstr.* **78**, 135700 (1973).

P229. W. C. Drinkard (E.I. du Pont), U.S. Patent 3,655,723 (1972); *Chem. Abstr.* **77**, 4986 (1972).

P230. W. C. Drinkard (E.I. du Pont), Fr. Patent 2,069,411 (1971); *Chem. Abstr.* **77**, 4399 (1972).

P231. R. G. Downing and R. A. Fouty (E. I. du Pont), Ger. Offen. 1,930,267 (1969); *Chem. Abstr.* **72**, 78489 (1970).

P232. E. I. du Pont, Brit. Patent 1,178,950 (1970); *Chem. Abstr.* **72**, 89831 (1970).

P233. P. Arthur and B. C. Pratt (E.I. du Pont), U.S. Patent 2,571,099 (1951); *Chem. Abstr.* **46**, 3068 (1952).

P234. S. Hashimoto and Y. Inoue, Jap. Patent 73 30,602 (1973); *Chem. Abstr.* **80**, 26724 (1974).

P235. E. W. Duck and J. R. Hawkins (Int. Syn. Rubber Co.), Ger. Offen. 2,207,782 (1972); *Chem. Abstr.* **77**, 165261 (1972).

P236. W. C. Kray (Shell Int. Res.), Ger. Offen. 2,051,251 (1971); *Chem. Abstr.* **75**, 37623 (1971).

P237. P. M. Duinker, E. L. T. M. Spitzer, and J. A. Waterman (Shell Int. Res.), Ger. Patent 1,215,372 (1970); see *Chem. Abstr.* **63**, 9878 (1965).

P238. J. T. Anrigo (Universal Oil Products Co.), U.S. Patent 3,294,853 (1966); *Chem. Abstr.* **66**, 46145 (1967).

P239. N. S. Nametkin *et al.* (Topchiev A.V. Institute Petrochem. Syn.), U.S.S.R. Patent 322,344 (1971); *Chem. Abstr.* **76**, 127741 (1972).

P240. E. Ichiki, Y. Inoue, Y. Kondo, and T. Yako (Sumitomo Chem. Co. Ltd.), Ger. Offen. 2,022, 278 (1970); *Chem. Abstr.* **74**, 13257 (1971).

P241. G. Zoche, E. W. Muller, and F. W. A. G. K. Korte (Shell Intern. Research), Brit. Patent 1,019,968 (1966); *Chem. Abstr.* **64**, 15742 (1966).

P242. J. E. Lyons and G. L. Johnson (Sun. Res. Develop. Co.), Ger. Offen. 2,214,928 (1972); *Chem. Abstr.* **78**, 17338 (1973).

P243. G. Wilkinson (Johnson, Matthey and Co. Ltd.), Brit. Patent 1,130,749 (1968); *Chem. Abstr.* **70**, 11094 (1969).

P244. F. P. Gay (E.I. du Pont), U.S. Patent 3,047,554 (1962); *Chem. Abstr.* **57**, 13999 (1962).

P245. M. Born, D. Durand, and C. Lasau (Institut Français du Pétrole), Ger. Offen. 2,327,230 (1974); *Chem. Abstr.* **80**, 82221 (1974).

P246. M. Born, C. Lasau, Dinh Chan Trinh, and Vu Quang Dang (Institut Français du Pétrole), Ger. Offen. 2,310,468 (1973); *Chem. Abstr.* **79**, 136449 (1973).

P247. V. S. Feldblyum *et al.*, U.S.S.R. Patent 405,849 (1973); *Chem. Abstr.* **80**, 82043 (1974).

P248. R. F. Mason and G. R. Wicker (Shell Int. Res.), Ger. Offen. 2,264,088 (1973); *Chem. Abstr.* **79**, 136457 (1973).

P249. B. W. Taylor and H. E. Swift (Gulf Research and Develop. Co.), U.S. Patent 3,778,462 (1973); *Chem. Abstr.* **80**, 59520 (1974).

P250. H. E. Dunn (Phillips Petroleum Co.), U.S. Patent 3,760,027 (1973); *Chem. Abstr.* **80**, 70271 (1974).

P251. Y-T. Chia, W. C. Drinkard, and E. N. Squire (E.I. du Pont), U.S. Patent 3,766,237 (1973); *Chem. Abstr.* **80**, 70373 (1974).

P252. P. G. Bercik and A. M. Henke (Gulf Research Develop. Co.), Ger. Offen. 2,321,907 (1973); *Chem. Abstr.* **80**, 39047 (1974).

P253. K. L. Motz (Continental Oil Co.), U.S. Patent 3,784,623 (1974); *Chem. Abstr.* **80**, 70286 (1974).

P254. D. M. Coyne, H. L. Hackett, and R. L. Poe (Continental Oil Co.), U.S. Patent 2,978,523 (1961); *Chem. Abstr.* **55**, 16419 (1961).

P255. T. Sakaguchi, S. Akutagawa, and A. Komatsu (Takasago Perfumery Co., Ltd.), Jap. Patent 72 20,005 (1972); *Chem. Abstr.* **77**, 101866 (1972).

P256. H. Mix, E. Kurras, F. W. Wilcke, J. Reihsig, W. Schulz, H. Fuhrmann, I. Grassert, W. Fuchs, and J. Meissner, Ger. (East) Patent 99,556 (1973); *Chem. Abstr.* **80**, 132760 (1974).

P257. M. Born, T. D. Chan, D. V. Quang, and C. Lassau (Institut Français du Pétrole), Fr. Demande 2,177,478 (1973); *Chem. Abstr.* **80**, 120211 (1974).

P258. W. Schneider (B.F. Goodrich) U.S. Patent 3,808,283 (1974); *Chem. Abstr.* **81**, 13174 (1974).

P259. D. R. Fahey (Phillips Petroleum Co.) U.S. Patent 3,800,000 (1974); *Chem. Abstr.* **81**, 42059 (1974).

P260. E. F. Lutz (Shell Oil Co.) U.S. Patent 3,825,615 (1974).

P261. S. Isawa, Y. Yamada, and N. Kunimoto (Toyo Soda) Jap. Patent 73 39,923 (1973); *Chem. Abstr.* **81**, 26171 (1974).

P262. L. Forni and R. Invernizzi (Soc. Ital. Resine) Ger. Offen. 2,347,235 (1974); *Chem. Abstr.* **81**, 3322 (1974).

Reviews

A selection of general review articles relevant to the material discussed in this chapter is listed below.

A. Andreetta, F. Conti, and G. F. Ferrari, Selective homogeneous hydrogenation of dienes and polyenes to monoenes. *Aspects Homogen Catal.* **1**, 204 (1970).

D. G. H. Ballard, Pi and sigma transition metal carbon compounds as catalysts for the polymerization of vinyl monomers and olefins. *Advan. Catal.* **23**, 263 (1973).

C. W. Bird, "Transition Metal Intermediates in Organic Synthesis." Academic Press, New York, 1967.

B. Bogdanović, Asymmetrische Synthesen mit homogen Übergansmetallkatalysatoren. *Angew. Chem.* **85**, 1013 (1973).

R. S. Coffey, Recent advances in homogeneous hydrogenation of carbon-carbon multiple bonds. *Aspects Homogen. Catal.* **1**, 5 (1970).

N. R. Davies, The isomerization of olefins catalyzed by palladium and other transition-metal complexes. *Rev. Pure Appl. Chem.* **17**, 83 (1967).

G. Dolcetti and N. W. Hoffmann, Homogeneous hydrogenation of organic compounds catalyzed by transition metal complexes and salts. *Inorg. Chim. Acta* **9**, 269 (1974).

V. S. Feldblyum and N. V. Obeschalova, Dimerization of alkenes. *Russ. Chem. Rev.* **37**, 789 (1968).

E. N. Frankel and H. J. Dutton, Hydrogenation with homogeneous and heterogeneous catalysts. *Top. Lipid Chem.* **1**, 161 (1970).

J. Habeshaw, Oligomers and co-oligomers of propylene. *In* "Propylene and its Industrial Derivatives" (E. G. Hancock, ed.), p. 115, Ernest Benn Ltd., London, 1973.

R. E. Harmon, S. K. Gupta, and D. J. Brown, Hydrogenation of organic compounds using homogeneous catalysts. *Chem. Rev.* **73**, 21 (1973).

J. Hetflejs and J. Langova, The homogeneous catalytic dimerization of olefins. (In Czech.) *Chem. Listy* **67**, 590 (1973).

A. J. Hubert and H. Reimlinger, The isomerization of olefins. Part II. *Synthesis* p. 405 (1970).

Y. Izumi, Methoden der asymmetrischen Synthese—enantioselektive katalytische Hydrierung. *Angew. Chem.* **83**, 956 (1971).

N. Kohler and F. Dawans, La catalyse par des dérivés de métaux de transition déposés sur des supports polymériques organiques. *Rev. Inst. Fr. Petrole Ann. Combust. Liquids* **27**, 105 (1972).

J. Kwiatek, Hydrogenation and dehydrogenation. In "Transition Metals in Homogeneous Catalysis" (G. N. Schrauzer, ed.), p. 14. Dekker, New York, 1971.

G. Lefebvre and Y. Chauvin, Dimerization and codimerization of olefinic compounds by co-ordination catalysis. *Aspects Homogen. Catal.* **1**, 108 (1970).

F. J. McQuillin, Homogeneously catalyzed hydrogenation. *Progr. Org. Chem.* **8**, 314 (1973).

L. Marko and B. Heil, Asymmetric homogeneous hydrogenation and related reactions. *Catal. Rev.* **8**, 269 (1973).

J. W. Scott and D. Valentine, Asymmetric synthesis. *Science* **184**, 943 (1974).

W. Strohmeier, Homogene katalytische Hydrierung. *Fortschr. Chem. Forsch.* **25**, 71 (1972).

C. A. Tolman, Role of transition metal hydrides in homogeneous catalysis. *In* "Transition Metal Hydrides" (E. L. Muetterties, ed.), p. 271. Dekker, New York, 1971.

M. E. Volpin and I. S. Kolomnikov, Homogeneous hydrogenation. *Russ. Chem. Rev.* **38**, 273 (1969).

CHAPTER II

The Oligomerization of Alkynes and Related Reactions

I. Introduction

The cyclotetramerization of acetylene to cyclooctatetraene (COT), discovered in 1940 by W. Reppe and T. Toepel, is one of the most intriguing of the reactions catalyzed by nickel. Bearing in mind that COT had been previously synthesized (by R. Willstätter and E. Waser in 1911) in a yield of only 1–2% by a classical 13-stage organic synthesis starting from pseudo-pelletierine (1) (an alkaloid isolated from the bark of the pomegranate tree) (108, 109), it seems incredible that COT can be prepared catalytically in around 70% yield simply by allowing acetylene to interact with a nickel salt (39, P19). More than 30 years later this reaction still retains some of its

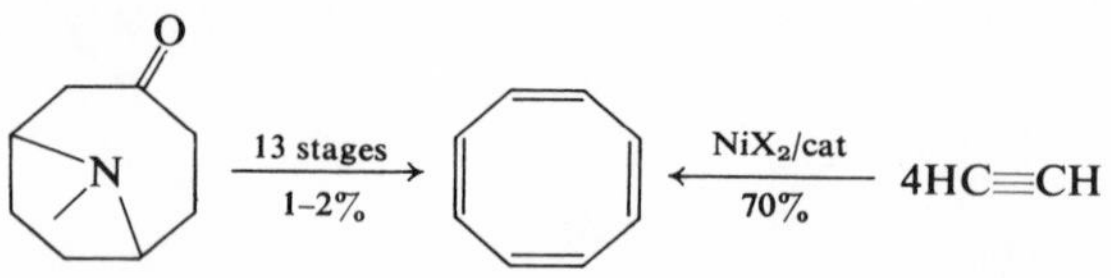

mystery and the reputation of being closer to the arts than the sciences, partly because of the danger of working with acetylene under pressure and partly because of the continued absence of a satisfactory mechanism for the process. It is perhaps a little disappointing that no technical use has been found for cyclooctatetraene, which, with the prospect of cheaper acetylene (40), can be expected to become a relatively inexpensive starting material.

The discovery that the cyclotetramerization can be directed into a cyclotrimerization by using phosphine-modified catalysts is also due to Reppe (38, P20). Although of no significance for preparing benzene, it is a convenient method for synthesizing substituted benzene derivatives from substituted alkynes.

A discussion of the oligomerization of acetylene and substituted alkynes is followed by sections devoted to the co-oligomerization of alkynes with olefins and to the related oligomerization of allene. A number of telomerization reactions involving alkynes or allenes has also been included.

Important review articles in which the transition metal-catalyzed oligomerization of alkynes or allene is discussed are included at the end of the chapter.

II. The Oligomerization of Alkynes

A. *The Oligomerization of Acetylene* (*Table II-3*)

The cyclo-oligomerization of acetylene to COT is catalyzed by a variety of nickel salts of which the cyanide, acetylacetonate, and ethylacetoacetate have received the most attention. The reaction is normally carried out at 80–120° with an acetylene pressure of 10–25 atm (an account of the necessary technology is contained in ref. 165). The most commonly used solvents are benzene, THF, or dioxan and anhydrous conditions are essential; these were achieved in early work by adding calcium carbide or ethylene oxide. It is claimed that mixtures of two different nickel salts [e.g., $Ni(acac)_2$–Ni(ethylacetoacetate)$_2$] or the use of a variety of additives (e.g., barium oxide, iron butoxide, and organotin and organolead compounds) increase the efficiency of the process and also suppress the formation of the black polymer, cuprene (P14, P15, P19, P23, P41).

The yield of COT is approximately 70%. The remaining 30% is mainly benzene, resins, and cuprene. A number of other compounds are formed in trace quantities and these include vinylacetylene (5, P40), styrene (38), vinylcyclooctatetraene (15, 16, 22, 39), *cis*-1-phenyl-1,3-butadiene (15, 16, P24), naphthalene (15, 39), and azulene (15, 39). In addition, a yellow oil having the composition $C_{12}H_{12}$ was reported in the original publication (39) but it has apparently received no further attention. The yield of the higher oligomers increases on raising the reaction temperature and at 140° represents 20% of the product.

The formation of the black polymer cuprene is to be avoided because it removes the catalyst and introduces technical problems for the large-scale production of COT. A suggestion has been made that it is formed by cross-linking of crystalline linear *trans*-polyacetylene (64), which is, moreover, the only product in a number of reactions involving two component catalysts (e.g., $NiCl_2$–$NaBH_4$). Experiments using a mixture of acetylene and vinylacetylene indicate that styrene and vinylcyclooctatetraene are co-oligomerization products (21, 22). Presumably, 1-phenylbutadiene is the product of the

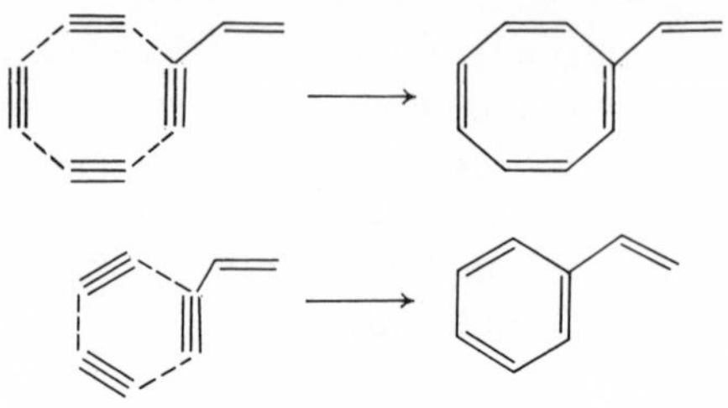

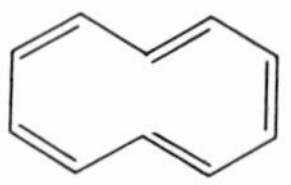

2

co-oligomerization of acetylene with 1,3-hexadiene-5-yne, although this latter compound has never been identified in the reaction mixture. Azulene and naphthalene are perhaps the rearrangement products of cyclodecapentaene (**2**).

As can be seen from Table II-3, the cyclotetramerization reaction can be directed into a cyclotrimerization reaction by introducing a phosphine or phosphite, whereas the presence of a bidentate ligand suppresses the cyclo-oligomerization reaction completely (43). The only reaction observed in co-ordinating solvents (e.g., pyridine or quinoline) is polymerization (145, 146) or dimerization to vinylacetylene (5, P40).

B. *The Oligomerization of Substituted Alkynes*

For many years it was believed that the cyclotetramerization reaction necessarily involved acetylene. However, with appropriate catalysts high yields of tetrasubstituted cyclooctatetraene derivatives have been obtained by oligomerizing $HOCH_2C\vdots CH$, $HOC(CH_3)_2C\vdots CH$ and $RO_2CC\vdots CH$. The reaction is, however, apparently limited to monosubstituted alkynes and cannot be used to prepare octasubstituted COT derivatives. The catalysts active for the cyclotetramerization are either ligand free [e.g., $Ni(acac)_2$ and NiX_2–$NaBH_4$] or contain ligands that are readily displaced by the alkyne [e.g., $(Cl_3P)_4Ni$, $(COD)_2Ni$, or $(\pi\text{-}C_3H_5)_2Ni$]. The introduction of ligands that are less easily displaced by the alkyne, e.g., $P(C_6H_5)_3$, changes the direction of reaction from cyclotetramerization to cyclotrimerization and, in the case of monosubstituted alkynes, mixtures of 1,3,5- and 1,2,4-trisubstituted benzene derivatives are obtained. Running parallel to the cyclo-oligomerization reaction is a linear oligomerization reaction, and in many cases linear oligomers are the principal products (13, 31, 33, 105). Linear oligomerization has not been observed for disubstituted alkynes.

The reaction of 3-methyl-3-butyne-2-ol [HOC(CH$_3$)$_2$C$\vdots$CH] has been extensively studied and, as it clearly demonstrates the various effects that may be observed, can be discussed in detail (see Tables II-1 and II-2). A linear dimer (3),* cyclic trimers (4 and 5), cyclic tetramers (6, 7, and 8) and polymeric material have been isolated.

RCH=CH—C≡CR

3   4 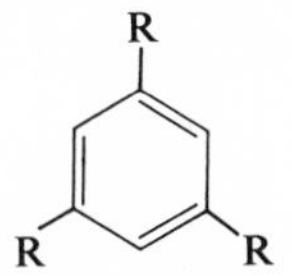   5 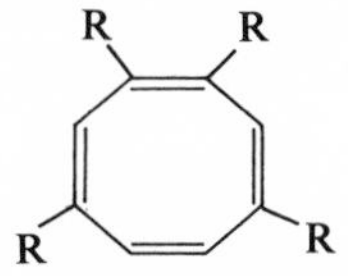   6

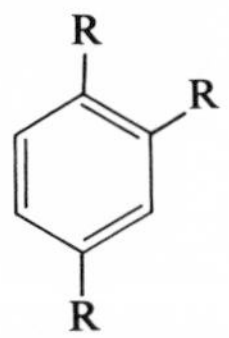

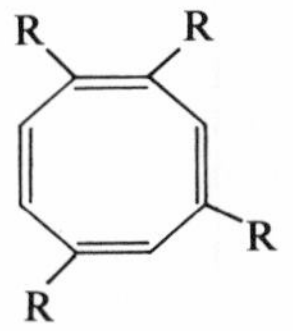

7 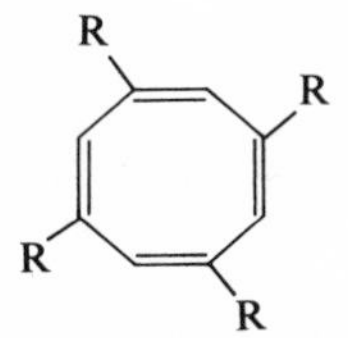   8

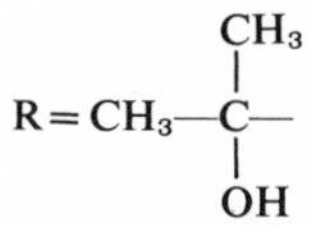

$$R = CH_3 - \overset{\displaystyle CH_3}{\underset{\displaystyle OH}{\overset{|}{\underset{|}{C}}}} -$$

Although there are a number of discrepancies in the results and one suspects that in some cases a complete mass balance has not been obtained, it can nevertheless be seen from Table II-1 that in the absence of added ligands the main products are the dimer 3 and the 1,2,4,7-tetrasubstituted COT derivative 6 as well as polymer. The addition of a donor ligand suppresses the formation of tetramer and polymer and a mixture of dimer and trimers results. The cyclotrimerization can in turn be suppressed by carrying the reaction out in a coordinating solvent (e.g., DMF or pyridine) and the only product is the linear dimer 3.

The effect of varying the ligand on the composition of the product is shown in Table II-2. Any consistent mechanism must explain such divergent observations as those that not only bipyridyl but also triphenylphosphite (Ni:Lig = 1:1) completely suppress the formation of trimer and that triphenylphosphine and tricyclohexylphosphine cause predominant formation of the 1,3,5-substituted benzene derivative, whereas tributylphosphine yields the 1,2,4 isomer. A further remarkable effect is observed in the cyclotrimerization of 1-hexyne with π-crotylnickel halide dimer as catalyst: the chloride produces almost exclusively 1,3,5-tributylbenzene, whereas the iodide gives the 1,2,4 isomer (36).

---

* A second dimer viz. RC$\vdots$CC(R):CH$_2$ has been recently identified (167).

TABLE II-1

THE OLIGOMERIZATION OF $HOC(CH_3)_2C{:}CH$

Catalyst	Conversion (%)	Dimer (%) 3	Trimer (%) 4	5	6	Tetramer (%) 7	8	Polymer (%)	Ref.
$Ni(acac)_2$	64	65	—	—	35	—	—	—	89
$Ni(acac)_2{-}P(C_6H_5)_3$	91	25	3	27	—	—	—	45	10
$Ni(CO)_4$	68	—	0.3	16.4	69.9	—	0.9	13	11
$[(C_6H_5)_3P]_2Ni(CO)_2$	90	~31.5	~5.7	~25.8	—	—	—	37	10
$NiCl_2{\cdot}6H_2O{-}NaBH_4$	55	—	0.7	2.1	55.2	—	2.4	39.5	10, 11
$NiCl_2{\cdot}6H_2O{-}NaBH_4{-}P(C_6H_5)_3$	70	~25	—	~25	—	—	—	54	10, 11
$(\pi{-}C_3H_5)_2Ni$	64	66.3	0.9	0.2	7.2	—	1.8	23.6	4, 9
$(\pi{-}C_3H_5)_2Ni{-}P(C_6H_5)_3$	82	14.8	3.2	80	—	—	—	2.0	4, 9

TABLE II-2

THE EFFECT OF ADDED LIGANDS ON THE OLIGOMERIZATION OF $HOC(CH_3)_2C\vdots CH^a$

Nickel component	Ligand	Conver-sion (%)	Dimer (%) 3	Trimer (%) 4	5	Polymer (%)
$Ni(N\text{-}C_6H_5\text{salicylaldimine})_2$	$P(OC_6H_5)_3$	25	100	—	—	—
$Ni(N\text{-}C_6H_5\text{salicylaldimine})_2$	$P(C_6H_5)_3$	92	62	—	38	—
$Ni(N\text{-}C_6H_5\text{salicylaldimine})_2$	$P(C_4H_9)_3$	80	72	—	18	—
$Ni(N\text{-}C_6H_5\text{salicylaldimine})_2$	Bipy	62	100	—	—	—
$(\pi\text{-}C_3H_5)_2Ni$	$P(C_6H_5)_3$	82	14.8	3.2	80	2.0
$(\pi\text{-}C_3H_5)_2Ni$	$P(C_4H_9)_3$	69	14.8	80	3	2.2
$(\pi\text{-}C_3H_5)_2Ni$	$P(cyclo\text{-}C_6H_{11})_3$	52	19	5	70	6

^{a} From refs. 4 and 89.

The aryl-substituted bisalkyne, 1,2-bis[1-oxophenylpropargyl]benzene (**9**), in which the alkyne groups are able to approach each other closely, reacts in the presence of nickel tetracarbonyl to give the cyclooctatetraene derivative **10** (133). The role of the nickel is unknown, although the formation of a

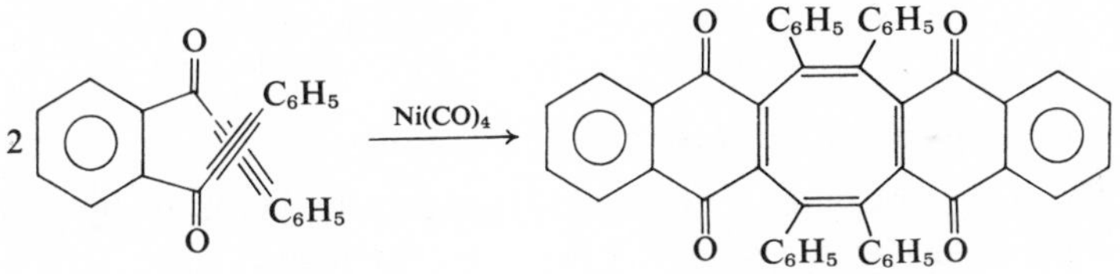

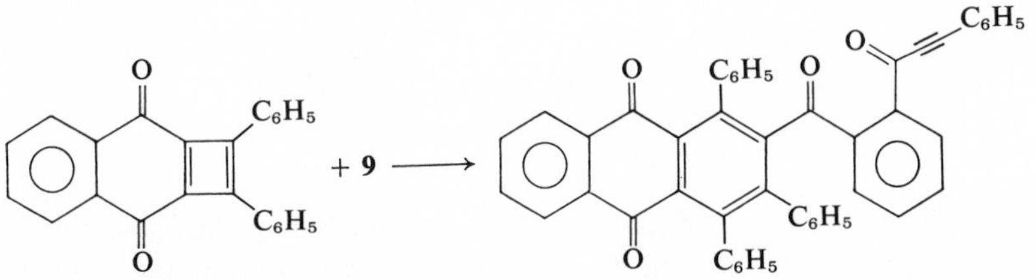

number of intermediates has been observed spectroscopically; interestingly, no carbon monoxide is evolved during the reaction. The postulation of an initial intramolecular dimerization to give a cyclobutadiene derivative that then dimerizes is supported by the isolation of traces of **11**, which is the expected product of a Diels-Alder reaction with unreacted alkyne. The cyclobutadiene intermediate can also be trapped by carrying out the catalytic reaction in the presence of diphenylacetylene or phenylbenzoylacetylene.

A rather unusual reaction has been observed to occur on refluxing nickel acetylacetonate and cyclooctyne in THF. In addition to the cyclic trimer, the benzene derivatives **12** and **13** are formed as a result of reaction between the alkyne and the acetylacetone ligand (51).

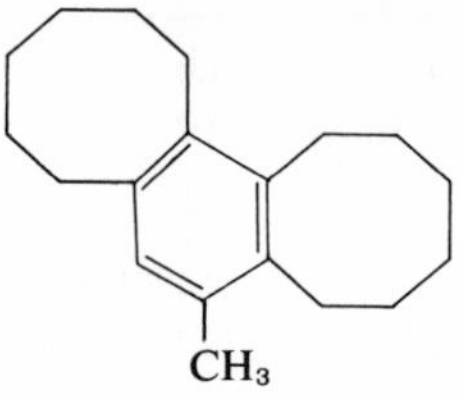

$$CH_3$$

12 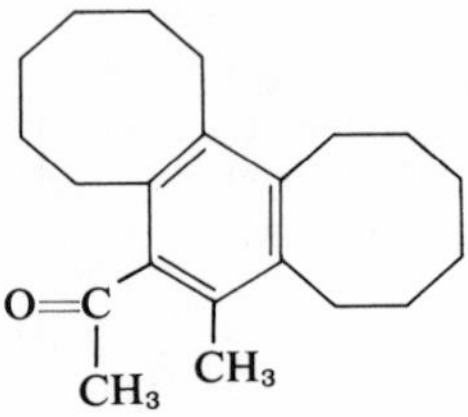 **13**

It is convenient to mention here that benzonitrile is cyclotrimerized in the presence of Raney nickel to triphenyltriazine (23, 49).

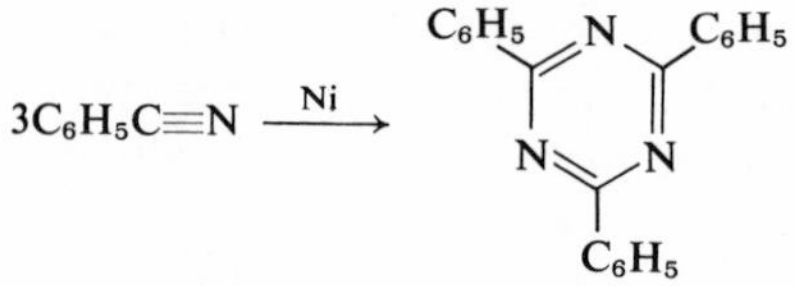

$$3C_6H_5C\equiv N \xrightarrow{\ Ni\ }$$

C. The Co-oligomerization of Alkynes (Table II-4)

Mono- or disubstituted COT derivatives may be prepared by co-oligomerizing acetylene with a mono- or disubstituted alkyne.

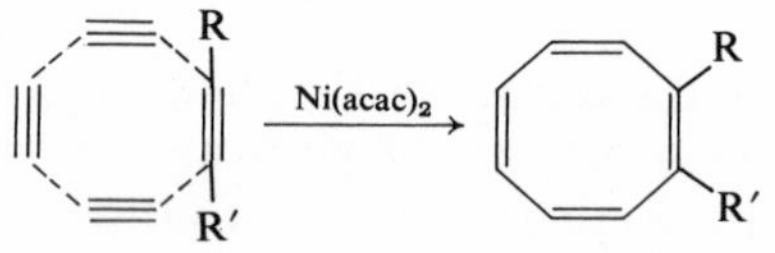

$$\xrightarrow{\ Ni(acac)_2\ }$$

$R = H$, $R' = CH_3$, $HOCH_2$, C_6H_5, etc.
$R = R' = CH_3$, C_6H_5

Similar reactions carried out with a phosphine-modified catalyst produce mono- or disubstituted benzene derivatives, e.g., *o*-divinylbenzene (86, P47).

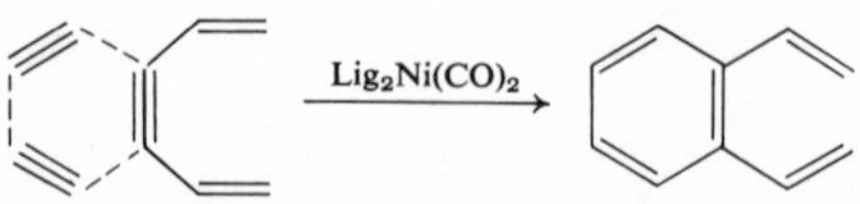

$$\xrightarrow{\ Lig_2Ni(CO)_2\ }$$

A variation of this type of reaction is the dimerization of the α,ω-dialkyne **14** to 1,3-bis(5-indanyl)propane (13, 87) and the reactions of the dialkyne **9** described in the previous section (133).

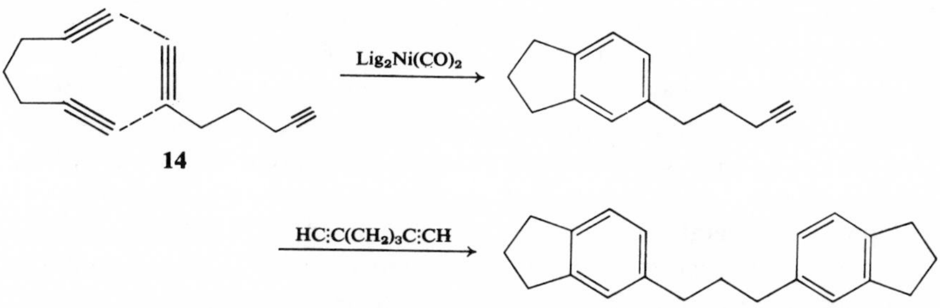

Polycyclotrimerization of diethynylbenzene with phenylacetylene is catalyzed by $Ni(acac)_2$ and $NiCl_2-NaBH_4$ in the presence of triphenylphosphine. The resulting polymer has a molecular weight of 1000 in the first case and 1600 in the second (161).

D. Mechanistic Considerations

Although COT was discovered over 30 years ago, the mechanism of its formation is unknown. Various proposals have been made but concrete evidence is lacking.

The active species is probably the same in all cases even though a variety of different compounds can be used as the catalyst, e.g., $Ni(CN)_2$ or $(COD)_2Ni$; $(R_3P)_2Ni(CO)_2$ or $Ni(acac)_2-PR_3$. Whether this species is formed by reduction of a nickel($2+$) system or by oxidation of a nickel(0) system is an open question. Instinctively one also assumes that the linear oligomers and the cyclic oligomers are produced by different mechanisms.

A mechanism for the formation of the linear oligomers involving insertion into a nickel-hydride or nickel-alkynyl bond is attractive and explains the absence of linear products in reactions involving disubstituted alkynes (33). A possible reaction sequence leading to the dimer is shown below.

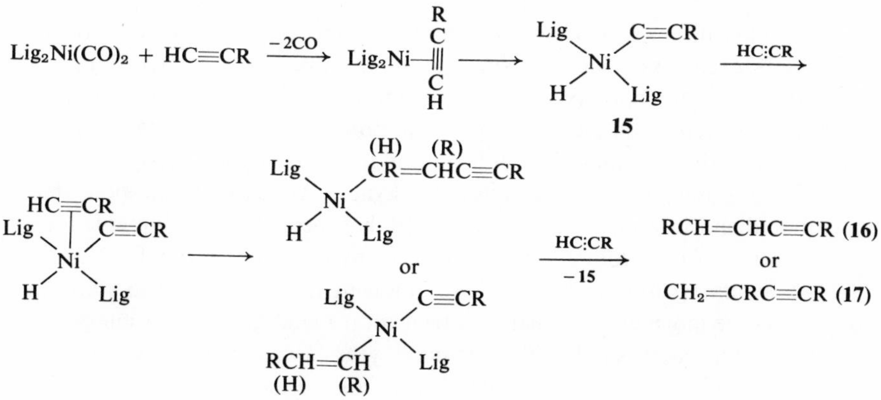

The linear dimers produced in reactions involving 1-butyne, 1-pentyne, or 1-heptyne have been shown to be a mixture of **16** and **17**, whereas only **16** [R = $HOC(CH_3)_2$] has been isolated from the reaction involving HOC-$(CH_3)_2C\!:\!CH$ (see footnote on p. 97). Assuming that a steric effect is important, then in this last case insertion probably occurs into the nickel-hydride bond of **15**.

Support for several of the intermediates postulated above has been obtained by studying the reactions of alkynes with zerovalent nickel complexes. The bistriphenylphosphine adduct corresponding to **15** has been isolated from the reaction of tetrakistriphenylphosphine nickel with phenylacetylene, whereas the coupling of two alkyl fragments has been observed on treating $[(C_6H_5)_3P]_2Ni(tert\text{-}C_4H_9)_2$ with the same alkyne (90, 91, 111, 157). A similar

$$[(C_6H_5)_3P]_4Ni \ + \ HC\!\equiv\!CC_6H_5$$

$$\searrow\ -2P(C_6H_5)_3$$

$$[(C_6H_5)_3P]_2NiH(C\!\equiv\!CC_6H_5)$$

$$HC\!\equiv\!CC_6H_5 \qquad \nearrow\ -2[C_4H_9]$$
$$|$$
$$[(C_6H_5)_3P]_2Ni(tert\text{-}C_4H_9)_2$$

reaction has been reported to occur between various bipyridyl–nickel-dialkyl complexes and a variety of olefins (see Volume I, p. 206). The intermediacy of nickel-hydride species in the reactions involving sodium borohydride is supported by the isolation of such compounds (e.g., **18**) from the reaction with bisphosphine nickel dihalides (107) (see Volume I, p. 141).

$$[(cyclo\text{-}C_6H_{11})_3P]_2NiCl_2 \ \xrightarrow[-\,NaCl/BH_3]{NaBH_4} \ \begin{array}{c} (cyclo\text{-}C_6H_{11})_3P \diagdown \quad \diagup H \\ Ni \\ Cl \diagup \quad \diagdown P(cyclo\text{-}C_6H_{11})_3 \end{array}$$

$$\textbf{18}$$

A detailed discussion of the mechanism of the cyclo-oligomerization of alkynes is even less justified than that of the linear oligomerization. An attempt to explain both processes with one mechanism (33; see also 157) with a final cyclization step, in which a hydrogen atom is transferred from the nickel to the terminal alkynyl group, is ruled out by the ready cyclotrimerization observed for disubstituted alkynes. An early suggestion that cyclobutadiene intermediates are involved has been finally eliminated by investigation of the products from reactions involving $CH_3C\!:\!CCD_3$ (73, 75, 92, 93). Symmetry arguments, moreover, indicate that the concerted addition of two alkyne molecules bonded to a transition metal atom is an energetically unfavored process (94, 139). For many years a mechanism involving an octahedral nickel intermediate in which four alkyne molecules are π bonded to the nickel atom has been discussed. This has the merit of giving a simple explanation for the formation of benzene when one site on the nickel atom is blocked by a donor ligand and for the absence of reaction when two sites are blocked (42, 43, 95, 100, 101, 110). However, this mechanism is no longer universally accepted, partly because of the recognition that a great many, if not all, of the nickel-catalyzed olefin transformations occur by stepwise mechanisms.

A possible mechanism for the formation of a benzene derivative involving a nickelacyclopentadiene intermediate is shown below (33, 100, 106).

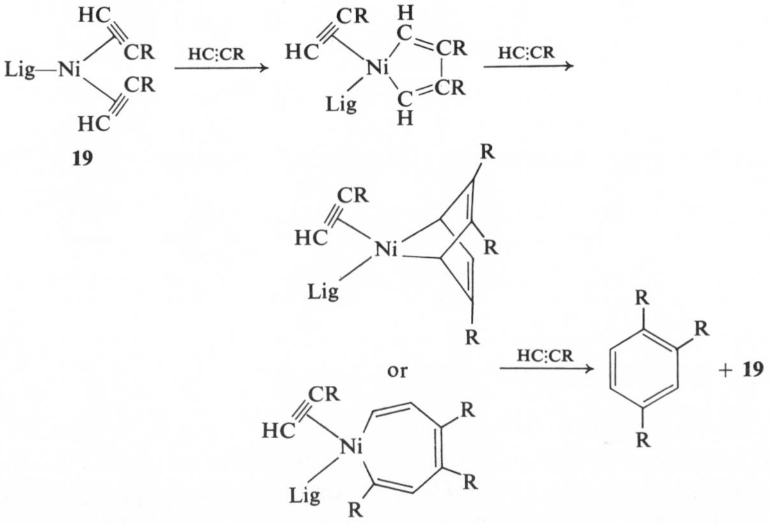

It may be expected that the position of the substituents in the product can be influenced by steric effects originating from both the ligand and from the substituent on the alkyne. In the absence of the ligand, it is possible that three alkyne molecules form a nickelacycloheptatriene intermediate, which reacts with a fourth alkyne molecule to give the cyclooctatetraene derivative. [This suggestion is reminiscent of one made by Reppe himself more than 25 years ago (37).] An alternative mechanism involving a bisnickelacyclopentadiene intermediate, which then couples, is less likely because this does not account for the observed preference for the formation of 1,2,4,6- and 1,2,4,7-tetrasubstituted COT derivatives.

Precedence for two of the proposed intermediates is to be found in the bis(alkyne)nickel-ligand complexes, which, moreover, react with further alkyne to give cyclic trimers (96), and in the nickelacycloheptatriene derivative $[(CH_3O)_3P]_2NiC_6(CF_3)_6$, the structure of which has been confirmed by x-ray crystallography (99, 155). As yet no nickelacyclopentadiene system has been reported, although palladium, platinum, and iridium complexes of this type have been isolated and shown to react with additional alkyne (97, 98).

III. The Co-oligomerization of Alkynes with Olefins (Table II-5)

The same catalysts that trimerize and tetramerize alkynes are able to co-oligomerize alkynes with olefins.

Cyclohexadiene derivatives (e.g., **20** and **21**) have been isolated from reactions involving N-substituted maleimide (77) or butadiene (83–85). The

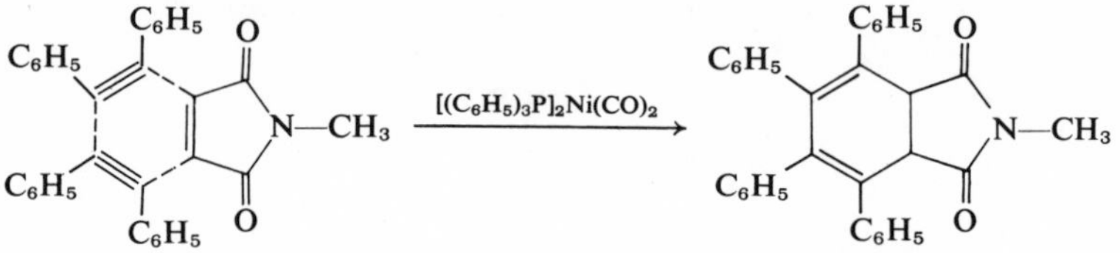

5-vinylcyclohexadiene derivatives (e.g., **21**) are not normally isolated because they undergo a facile internal Diels-Alder reaction to give a tricyclo[$2.2.2.0^{2,6}$]-7-octene derivative (e.g., **22**).* The isolation of 1,2,3,4-tetraphenyl-1,3-cyclohexadiene from the reaction of diphenylacetylene with $NiBr_2$–C_2H_5MgBr may also be the result of a co-oligomerization reaction between two tolan molecules and ethylene (formed by elimination from an intermediate nickel-ethyl species) (46, P32). The mechanism for these reactions is perhaps analogous to that shown on page 103 and involves an intermediate nickelacyclopentadiene species.

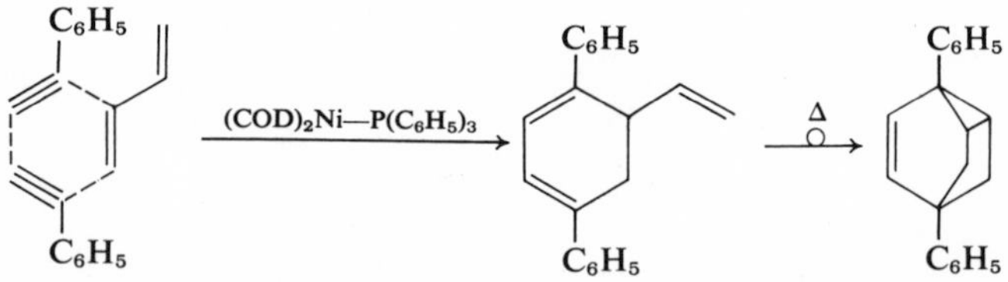

Reppe was the first to observe that acetylene could be co-oligomerized with acrylic acid ester using a triphenylphosphine nickel carbonyl catalyst (38). A reinvestigation of this system showed that the initial product of the reaction with methylacrylate was methyl-2,4,6-heptatrienoate, which at 125° underwent a ring closure reaction to give the product probably isolated by Reppe

---

* The reaction of alkynes with butadiene also produces ten-membered rings (butadiene:alkyne = 2:1) and 12-membered rings (butadiene:alkyne = 2:2); these products are likely to be formed through a bis(π-allyl)C_8–nickel intermediate and discussion is therefore reserved for the chapter devoted to the cyclo-oligomerization of 1,3-dienes (Chapter III).

$$2HC{\equiv}CH + CH_2{=}CHCO_2CH_3 \xrightarrow{[(C_6H_5)_3P]_2Ni(CO)_2}$$

$$CH_2{=}CHCH{=}CHCH{=}CHCO_2CH_3 \xrightarrow{\Delta}$$

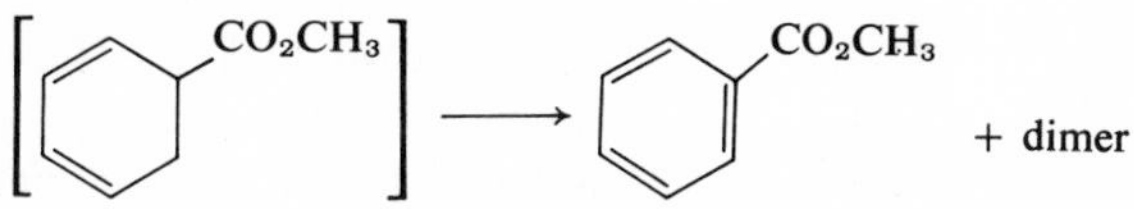

+ dimer

(76, P45, P52; see also 140, P53). A similar reaction occurs with acrylonitrile to give 2,4,6-heptatrienenitrile (23).

No convincing mechanism for these reactions has been proposed. The rather obvious proposal that the acrylonitrile molecule is complexed to the nickel during the reaction is supported by the result of a stoichiometric reaction between bis(acrylonitrile)nickel and acetylene, which produces 23

$$(CH_2{=}CHCN)_2Ni + 2HC{\equiv}CH \xrightarrow{-[NiCH_2{:}CHCN]} CH_2{=}CHCH{=}CHCH{=}CHCN$$
$$\textbf{23}$$

as well as benzene and COT (42, 44). Reactions using labeled acrylonitrile make it improbable that a substituted cyclobutadiene derivative is involved (73, 75). An observation that may have mechanistic implications is that the reaction between the disubstituted alkyne diphenylacetylene and bis(acrylonitrile)nickel [or bis(acrolein)nickel] does not produce a linear product. Instead, 1,2,4,6-tetraphenylbenzonitrile (or 1,2,4,6-tetraphenylbenzaldehyde) is formed (42).

A nickel-catalyzed Diels-Alder reaction has been reported to occur on reacting 1,3,5-hexatriene with various alkynes in the presence of a tricyclohexylphosphine-modified nickel catalyst (156). The product, a vinylcyclohexadiene derivative, is readily aromatized.

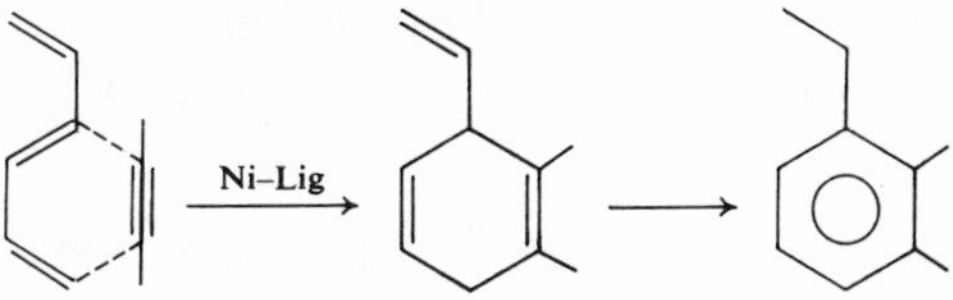

IV. The Hydrosilylation of Alkynes

The hydrosilylation of alkynes has received only slight attention. The reaction of dichloromethylsilane with diphenylacetylene is catalyzed by 1,1'-bis(dimethylphosphino)ferrocene–NiCl$_2$ and preferential cis addition is observed (147). A mechanism related to that proposed for the hydrosilylation of olefins (see p. 70) seems probable and is supported by the observed

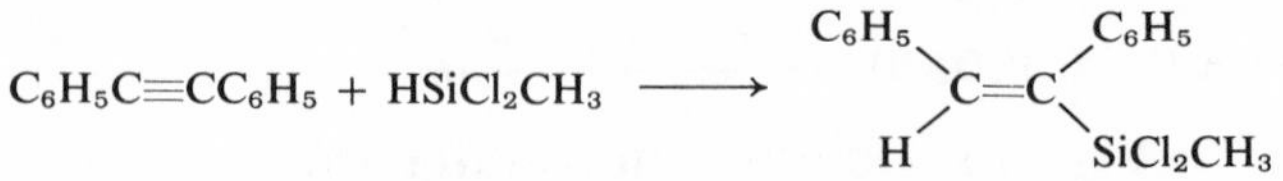

$$C_6H_5C{\equiv}CC_6H_5 + HSiCl_2CH_3 \longrightarrow \underset{H}{\overset{C_6H_5}{\diagdown}}C{=}C\underset{SiCl_2CH_3}{\overset{C_6H_5}{\diagup}}$$

transfer of a trichlorosilyl group from nickel to the alkyne during the reaction of $bipyNi(SiCl_3)_2$ with diphenylacetylene (150).

The reaction of 1-pentyne with triethyl- or triethoxysilane catalyzed by $Ni(acac)_2–Al(C_2H_5)_3$ takes a different course: an isomeric mixture of 1,3-dienes is obtained (e.g., **24**), suggesting that insertion of an alkyne molecule into the intermediate nickel–vinyl system has occurred (148, 170). Related to

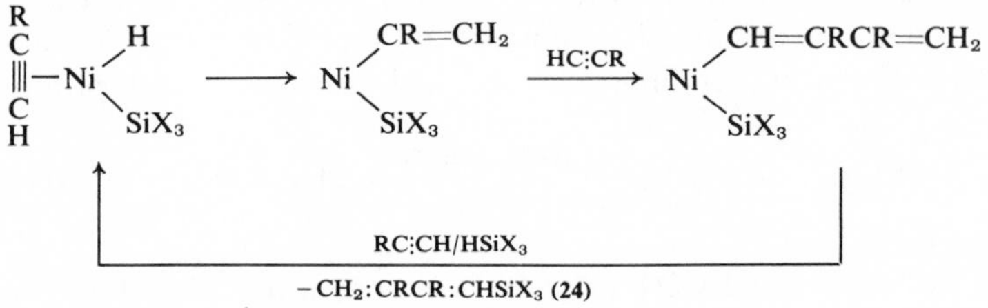

$$\underset{H}{\overset{R}{\underset{C}{\overset{C}{\underset{\|\||}{C}}}}}{-}Ni\underset{SiX_3}{\overset{H}{\diagup}} \longrightarrow Ni\underset{SiX_3}{\overset{CR=CH_2}{\diagup}} \xrightarrow{HC:CR} Ni\underset{SiX_3}{\overset{CH=CRCR=CH_2}{\diagup}}$$

$$RC:CH/HSiX_3$$
$$-CH_2:CRCR:CHSiX_3 \text{ (24)}$$

this reaction is that of tetramethyldisilane with alkynes, which is catalyzed by $[(C_6H_5)_3P]_2NiCl_2$ and in which 1-silacyclopentadiene derivatives are formed (149).

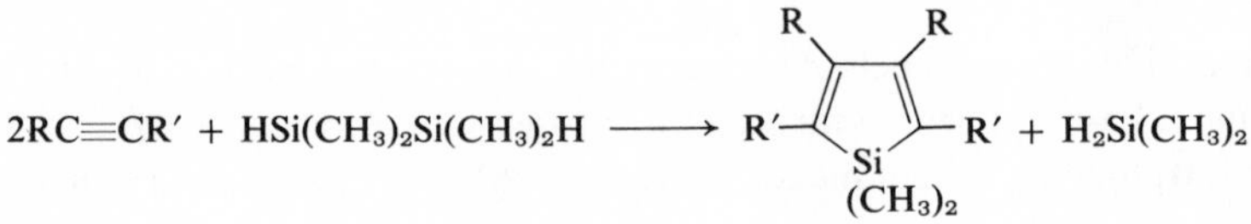

$$2RC{\equiv}CR' + HSi(CH_3)_2Si(CH_3)_2H \longrightarrow R'\overset{R\quad R}{\diagup\diagdown}R' + H_2Si(CH_3)_2$$

A reaction analogous to that of hydrosilylation is the addition of the P–H bond to alkynes, which is catalyzed by $[(C_6H_5)_3P]_2Ni(CO)_2$ (P54).

$$(cyclo\text{-}C_6H_{11})_2PH + C_3H_7C{\equiv}CC_3H_7 \xrightarrow{\text{[Ni]}} (cyclo\text{-}C_6H_{11})_2PC(C_3H_7){=}C(C_3H_7)H$$

A formal relationship exists between these reactions and that of alkyl- and arylmagnesium halides with alkynes catalyzed by $[(C_6H_5)_3P]_2NiCl_2$: cis addition of the Grignard reagent to the alkyne molecule occurs (151, 152).

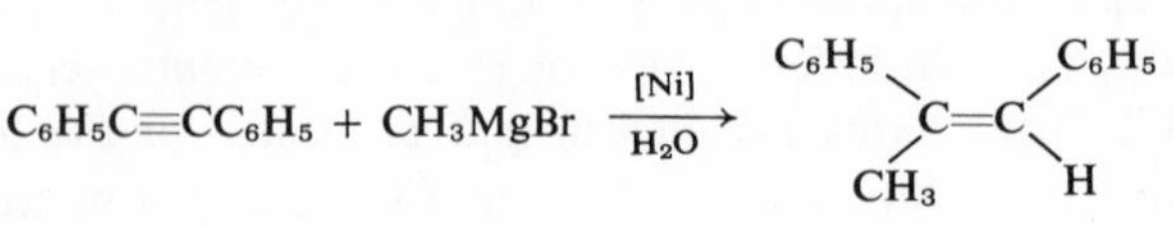

$$C_6H_5C{\equiv}CC_6H_5 + CH_3MgBr \xrightarrow[H_2O]{\text{[Ni]}} \underset{CH_3}{\overset{C_6H_5}{\diagdown}}C{=}C\underset{H}{\overset{C_6H_5}{\diagup}}$$

TABLE II-3

THE OLIGOMERIZATION OF ALKYNES

Alkyne	Catalyst	Dimer	Trimer (benzene deriv.)	Tetramer (COT deriv.)	Higher oligomers and polymer	Ref.
HC⫶CH	Ni(CN)$_2$	—	C$_6$H$_6$	C$_8$H$_8$	Polymer	21, 37, 39, 48, P12, P14, P19, P23, P26, P37
	Ni$_3$[Cr(CN)$_6$]$_2$	—	—	—	—	P13
	Ni(CN)$_2$–P(C$_6$H$_5$)$_3$	—	C$_6$H$_6$	C$_8$H$_8$	—	43
	Ni(CN)$_2$–Ni(acac)$_2$	—	—	—	—	P24
	NiX$_2$ (X = Cl, Br)	—	—	C$_8$H$_8$	Higher oligomer	4, 145, 146, P19
	NiX$_2$–NaBH$_4$	—	—	C$_8$H$_8$	Polymer	P1, P4, P5
	NiX$_2$–Al(*iso*-C$_4$H$_9$)$_3$	—	—	C$_8$H$_8$ (trace)	Polymer	7, 66, 112, 141
	NiX$_2$–e$^-$ (X = Br)	—	—	—	Polymer	P30, P31
	(R$_3$P)$_2$NiX$_2$	—	—	C$_8$H$_8$	Polymer	62, P49
	(R$_3$P)$_2$NiX$_2$–NaBH$_4$	—	—	C$_8$H$_8$	Polymer	17, 19, 28, 29, P2
	Ni(OH)$_2$	—	—	—	Higher oligomer, polymer	P19
	Ni(formate)$_2$	—	—	—	Higher oligomer, polymer	P19
	Ni(acetate)$_2$	CH$_2$⫶CHC⫶CH	C$_6$H$_6$	C$_8$H$_8$	Polymer	5
	Ni(acetate)$_2$–Al(*iso*-C$_4$H$_9$)$_3$	—	—	—	Polymer	7
	Ni(NCS)$_2$	—	—	C$_8$H$_8$	Higher oligomer, polymer	P19
	Ni(acac)$_2$	—	—	C$_8$H$_8$		14–16, 21, 22, 48, 59, 60, 63, P16, P18, P23
	Ni(acac)$_2$–PR$_3$	CH$_2$⫶CHC⫶CH	C$_6$H$_6$	—	Polymer	5, 43
	Ni(acac)$_2$–C$_2$H$_5$Al(OC$_2$H$_5$)H	—	—	C$_8$H$_8$	—	43, P17
	Ni(acac)$_2$–LiAlH$_4$	—	—	C$_8$H$_8$	Polymer	P21
	Ni(acac)$_2$–Al(C$_2$H$_5$)$_3$	—	C$_6$H$_6$	—	Polymer	45, 157
	Ni(acac)$_2$–Al(C$_2$H$_5$)$_3$–2PR$_3$	—	C$_6$H$_6$	—	Polymer	45
	Ni(acac)$_2$–Ni(ethyl-acetoacetate)$_2$	—	—	C$_8$H$_8$	—	P15, P24
	Ni(ethylacetoacetate)$_2$	—	C$_6$H$_6$	C$_8$H$_8$	Polymer	20, 21, P15, P23, P33, P41
	Ni(ethylacetoacetate)$_2$–Lig	CH$_2$⫶CH⫶CH	—	—	Polymer (trace)	5, P40
	Ni(ethylacetoacetate)$_2$–Al(*iso*-C$_4$H$_9$)$_3$	—	C$_6$H$_6$	C$_8$H$_8$	Polymer	52
	Ni(R-acetoacetate)$_2$, R = alkyl, aryl	—	—	C$_8$H$_8$	—	7
	Ni(salicaldehyde)$_2$	—	—	C$_8$H$_8$	—	P39
	Ni(*N*-alkylsalicylaldimine)$_2$	CH$_2$⫶CHC⫶CH	—	C$_8$H$_8$	—	21
	BipyNiR$_2$,	CH$_2$⫶CHC⫶CH	—	—	—	5
	(π-allyl)$_2$Ni–Lig	—	C$_6$H$_6$	—	Polymer	21, 43
	(π-allyl)NiX)$_2$–Lig	—	C$_6$H$_6$	—	Polymer	54, 67; see also 159
	(CH$_2$⫶CHCN)$_2$Ni	—	C$_6$H$_6$, C$_6$H$_5$CH⫶CH$_2$	C$_8$H$_8$	Polymer	8, 9
					—	8, 71, 72
						42, 44

(continued)

TABLE II-3 *(continued)*

Alkyne	Catalyst	Dimer	Trimer (benzene deriv.)	Tetramer (COT deriv.)	Higher oligomers and polymer	Ref.
	$(CH_2:CHCN)_2Ni[P(C_6H_5)_3]_2$	—	$C_6H_6, C_6H_5CH:CH_2$	—	Higher oligomer	42
	$(COD)NiI$	—		—	Polymer	55
	Cl_3P_4Ni	—		—	Oil	P10; see also 27
	$[(C_6H_5)_3P]_2Ni(CO)_2$	—	$C_6H_6, C_6H_5CH:CH_2$	—	—	35, 38, 53, 61, 73, P20 P28, P29; see also 157
$CH_3C:CH$	$(\pi\text{-}C_5H_5Ni)_2HC:CH$	Linear dimer		—	Polymer	P55
	$[(C_6H_5)_3P]_2Ni(CO)_2$		1,2,4 isomer, linear trimer	—		31
	$[(C_6H_5)_3P]_2Ni$	—	1,2,4 isomer 1,3,5 isomer 1,2,3 isomer	tetramer (traces)	—	57
	$(\pi\text{-}C_3H_5NiBr)_2$	—		—	Polymer	P27
	$[(C_6H_5)_3P]_2NiBr_2$	—	1,2,4 isomer, 1,3,5 isomer	—	Polymer	70
	$[(C_6H_5)_3P]_2NiCl_2\text{-}NaBH_4$	—		—		17, P2
$HOCH_2C:CH$	$Ni(acac)_2\text{-}Al(iso\text{-}C_4H_9)_3$	—		—		57
	$Ni(CO)_4$	—	Cyclic trimer	Cyclic tetramer	Polymer	11
	$[(C_6H_5)_3P]_2Ni(CO)_2$	—	1,2,4 Isomer, 1,3,5 isomer	—	—	10, 31, 38, P20, P48
	$[(RO)_3P]_2Ni$	—	1,2,4 Isomer, 1,3,5 isomer	—	—	40, P22
	$(\pi\text{-}C_3H_5)_2Ni\text{-}PR_3$	—	Cyclic trimer	—	Polymer	4
	$NiX_2\text{-}NaBH_4$	—		—	Polymer	P4
	$(R_3P)_2NiCl_2\text{-}NaBH_4$	—	1,2,4 Isomer, 1,3,5 isomer	—		29, P2
	$[(C_6H_5)_3P]_2NiBr_2$	—	1,2,4 Isomer, 1,3,5 isomer	—	Polymer	62, P49
$C_2H_5OCH_2C:CH$	$[(RO)_3P]_2Ni$	—	1,2,4 Isomer, 1,3,5 isomer	—	—	40, P22
$C_6H_5OCH_2C:CH$	$[(C_6H_5)_3P]_2Ni(CO)_2$	—	1,2,4 Isomer, 1,3,5 isomer	—	—	31
$CH_3COCH_2C:CH$	$NiBr_2\text{-}NaBH_4$	—		—	Polymer	P4
$ClCH_2C:CH$	$NiCl_2\text{-}NaBH_4$	—		—	Polymer	P4
$R_2NCH_2C:CH$	$(R_3P)_2NiCl_2\text{-}NaBH_4$	—	1,3,5 Isomer	—	Polymer	P2
	$[(C_6H_5)_3P]_2Ni(CO)_2$	—	1,3,5 Isomer	—		31, P3, P35
$(C_2H_5)_2NCH_2C:CH$	$[(RO)_3P]_2Ni$	—		—		40, P22
	$NiX_2\text{-}NaBH_4$	—		—	Polymer	29, P2
$C_2H_5C:CH$	$(R_3P)_2NiCl_2\text{-}NaBH_4$	Linear dimer	1,2,4 Isomer, 1,3,5 isomer, linear trimer	Linear tetramer	Polymer Linear pentamer, linear heptamer	P4
	$[(C_6H_5)_3P]_2Ni(CO)_2$					31
	$(R_3P)_2NiCl_2\text{-}NaBH_4$	—	Linear trimer	Linear tetramer	Linear pentamer	29

Alkyne	Catalyst	Dimer	Trimer	Tetramer	Higher	References
$HOC_2H_4C{:}CH$	$[(C_6H_5)_3P]_2Ni(CO)_2$	Linear dimer	Cyclic trimer	—	—	31
$HOC(CH_3)HC{:}CH$	$[(C_6H_5)_3P]_2Ni(CO)_2$	Linear dimer	1,2,4 Isomer, 1,3,5 isomer	—	—	31
	$[(RO)_3P]_2Ni$	—	1,3,5 isomer	—	—	40, P22
$(C_2H_5)_2NC(CH_3)HC{:}CH$ $CH_3COC{:}CH$ $C_3H_7C{:}CH$	$[(RO)_3P]_2Ni$	—	1,3,5 Isomer	—	—	31
	$[(C_6H_5)_3P]_2Ni(CO)_2$	Linear dimer	1,2,4 Isomer, 1,3,5 isomer	—	—	31
	$(R_3P)_2Ni(CO)_2$	Linear dimer	1,2,4 Isomer, 1,3,5 isomer, linear trimer	Linear tetramer	Linear pentamer, linear hexamer	31, P3, P7, P11
	$NiCl_2\text{-}NaBH_4$	Linear dimer	Linear trimer	Linear tetramer	Linear pentamer	29
$HOC(CH_3)_2C{:}CH$	$Ni(CO)_4$	—	1,2,4 Isomer, 1,3,5 isomer	1,2,4,6 Isomer, 1,2,4,7 isomer, 1,3,5,7 isomer	—	11
	$(RO)_3PNi(CO)_3$	—	1,3,5 Isomer	—	—	P34
	$[(C_6H_5)_3P]_2Ni(CO)_2$	—	1,2,4 Isomer, 1,3,5 isomer, linear trimer	—	Polymer	10, 31, P36
	$[(RO)_3P]_2Ni$	—	1,3,5 Isomer	—	—	40, P22
	$(COD)_2Ni$	—	1,2,4 Isomer, 1,3,5 isomer	1,2,4,6 Isomer, 1,2,4,7 isomer	—	11
	$(\pi\text{-}C_3H_5)_2Ni$	Linear dimer	1,2,4 Isomer, 1,3,5 isomer	1,2,4,6 Isomer, 1,3,5,7 isomer	Polymer	4, 9
	$(\pi\text{-}C_3H_5)_2Ni\text{-}P(C_6H_5)_3$	Linear dimer	1,2,4 Isomer, 1,2,4 Isomer, 1,3,5 isomer	1,2,4,6 Isomer	Polymer	4, 9
	$NiX_2\text{-}NaBH_4$	—	1,3,5 Isomer, linear trimer	1,2,4,6 Isomer, 1,2,4,7 isomer, 1,3,5,7 isomer	—	10, 11, P4
	$NiX_2\text{-}NaBH_4\text{-}P(C_6H_5)_3$	—	1,3,5 Isomer, linear trimer, 1,2,4 Isomer, 1,3,5 isomer	1,2,4,6 Isomer, 1,2,4,7 isomer, 1,3,5,7 isomer	—	10, 11
	$Ni(acac)_2$	—	—	1,2,4,6 Isomer, 1,2,4,7 isomer, 1,3,5,7 isomer	—	11, 89
	$Ni(N\text{-}phenylsalicyl\text{-}aldimine)_2$	Linear dimer	1,3,5,7 Isomer	—	—	89, 164, 167
	$Ni(acac)_2\text{-}P(C_6H_5)_3$	Linear dimer	1,2,4 Isomer, 1,3,5 isomer	—	Polymer, cryst. polymer	89
	$Ni(N\text{-}phenylsalicyl\text{-}aldimine)_2\text{-}P(C_6H_5)_3$	Linear dimer	1,3,5 Isomer	—	—	10, 89, 164, see also 168
	$Ni(CN)_2$	—	1,2,4 Isomer, 1,3,5 isomer	1,2,4,6 Isomer, 1,2,4,7 isomer, 1,3,5,7 isomer	—	11

(continued)

TABLE II-3 *(continued)*

Alkyne	Catalyst	Dimer	Trimer (benzene deriv.)	Tetramer (COT deriv.)	Higher oligomers and polymer	Ref.
$(CH_3)_2CHCHOHC:CH$	$[(RO)_3P]_2Ni$	—	1,2,4 Isomer, 1,3,5 isomer	—	—	P22
$C_2H_5C(CH_3)OHC:CH$	$[(C_4H_9)_3P]_2NiBr_2$	—	1,3,5-Isomer	—	—	168
$C_3H_7CHOHC:CH$	$[(C_4H_9)_3P]_2NiBr_2$	—	1,3,5-Isomer	—	—	168
$(CH_3)_3CC(CH_3)OHC:CH$	$Ni(ethylacetoacetate)$-py.	Linear dimer	—	—	—	167
$C_4H_9C:CH$	$[(C_6H_5)_3P]_2Ni(CO)_2$	Linear dimer	1,2,4 Isomer, 1,3,5 isomer, linear trimer	—	—	31, P3, P11
	$(\pi\text{-}CH_3CHCHCH_2NiCl_2)_2$	—	1,2,4 Isomer, 1,3,5 isomer	—	—	36
	$(\pi\text{-}CH_3CHCHCH_2NiI)_2$	Linear dimer	1,2,4 Isomer,	—	—	36
	$(\pi\text{-}C_5H_5)_2Ni\cdot2AlBr_3$	Linear dimer	1,3,5 isomer	—	—	36
	$NiCl_2\text{-}NaBH_4$	—	Cyclic trimer,	—	Polymer	58
	$[(C_6H_5)_3P]_2NiBr_2$	—	linear trimer	—	—	62
$C_5H_{11}C:CH$	$(R_3P)_2Ni(CO)_2$	Linear dimer	Cyclic trimer, linear trimer	—	—	31, 32, P7, P9, P11
$cyclo\text{-}C_6H_{11}C:CH$	$(R_3P)_2NiX_2\text{-}NaBH_4$	Linear dimer	Trimer	Tetramer	Polymer	29, 65, P2
$cyclo\text{-}C_6H_{10}OHC:CH$	$[(C_6H_5)_3P]_2Ni(CO)_2$	Linear dimer	—	—	—	31, P3
	$[(C_6H_5)_3P]_2Ni(CO)_2$	Linear dimer	—	—	—	31, P3, see also 167
	$[(C_4H_9)_3P]_2NiBr_2$	—	1,3,5 Isomer	—	—	168
$C_6H_5CH(OH)C:CH$	$[(C_4H_9)_3P]_2NiBr_2$	—	1,3,5 Isomer	—	—	168
	$[(C_6H_5)_3P]_2Ni(CO)_2$	—	1,2,4 Isomer	—	—	31
$CH_2:CHC:CH$	$(\pi\text{-}C_3H_5)_2Ni$	Linear dimer	Cyclic trimer	—	Polymer	9, 26, 163
	$(\pi\text{-}CH_3CHCHCH_2NiX)_2$	—	—	—	Polymer	26
	$NiNO_3\text{-}NaBH_4$	—	—	—	Polymer	P4
$CH_3OCH:CHC:CH$	$[(C_6H_5)_3P]_2Ni(CO)_2$	Linear dimer	1,3,5 Isomer, linear trimer	—	—	31
$CH_2:C(CH_3)C:CH$	$[(C_6H_5)_3P]_2Ni(CO)_2$	—	1,2,4 Isomer, 1,3,5 isomer	—	—	31, 69, P36
	$[(RO)_3P]_2Ni$	—	1,2,4 Isomer,	—	—	40, P22
	$(R_3P)_2NiX_2\text{-}NaBH_4$	—	1,2,4 Isomer, 1,3,5 isomer	—	—	17
	$(\pi\text{-}C_3H_5)_2Ni$	—	1,2,4 Isomer, 1,3,5 isomer	—	Polymer	163
	$Ni(acac)_2\text{-}Al(C_2H_5)_3\text{-}P(C_6H_5)_3$	—	—	—	Polymer	69
$CH_2:CHCH_2C:CH$	$[(iso\text{-}C_3H_7O)_3P]_4Ni$	—	1,2,4 Isomer, 1,3,5 isomer	—	—	P38
$C_6H_5C:CH$	$(R_3P)_2Ni(CO)_2$	—	Cyclic trimer	—	—	31, 33, 34, 41, 104
		—	1,3,5 isomer, linear trimer	—	—	
	$(C_6H_5)_2PC_2H_4P(C_6H_5)_2\text{-}Ni(CO)_2$	—	Linear trimer (?)	—	—	P46
	$[(C_6H_5)_2P]_2\text{-}o\text{-}C_6H_4Ni(CO)_2$	—	1,3,5 Isomer	—	—	P46
	$[(RO)_3P]_2Ni$	—	1,2,4 Isomer	—	—	56, P22
	$(COD)_2Ni\text{-}P(C_6H_5)_3\text{-}C_4H_6$	—	1,2,4 Isomer	—	—	84, see also 166
	$[(C_6H_5)_3P]_2Ni(CH_2:CH_2)$	—	1,2,4 Isomer, linear trimer	—	—	103

Substrate	Catalyst	Linear product	Cyclic / isomer product	Polymer	Reference
	(π-C₃H₅)₂Ni	—	1,2,4 Isomer, 1,3,5 isomer, linear trimer	—	9, 36
	(π-allylNiX)₂	—	1,2,4 Isomer, 1,3,5 isomer	Polymer	8, 36
	(π-allylNiX)₂–PR₃	—	Trimer	Polymer	8
	(π-C₅H₅)₂Ni: 2AlBr₃	—	1,2,4 Isomer, 1,3,5 isomer	Polymer	36
	NiX₂–NaBH₄	Linear dimer	—	Polymer	P4
	(R₃P)₂NiX₂–NaBH₄	—	1,2,4 Isomer, 1,3,5 isomer, linear trimer	—	17, 28, 29, P2, P11
	[(C₆H₅)₃P]₂Ni(C:CC₆H₅)₂	—	1,2,4 Isomer, 1,3,5 isomer	Polymer	12
	(R₃P)₂NiBr₂	—	1,2,4 Isomer, 1,3,5 isomer	Polymer	12, 62
p-NO₂C₆H₄C:CH	Ni(acac)₂–Al(iso-C₄H₉)₃	Linear dimer	Cyclic trimer	Polymer	57
CH₃O₂CC:CH	Ni(AlCl₄)₂·2C₆H₆	—	—	Polymer	169
	[(C₆H₅)₃P]₂Ni(CO)₂	—	1,2,4 Isomer	—	31
	[(C₆H₅)₃P]₂Ni(CO)₂	—	1,2,4 Isomer	—	87
	(R₃P)₄Ni	—	1,2,4 Isomer, 1,3,5 isomer	—	31
	[(RO)₃P]₂Ni	—	1,2,4 Isomer, 1,3,5 isomer; 1,2,4,6 Isomer	—	27, P10
	[(RO)₃P]₂Ni	—	1,2,4 Isomer, 1,3,5 isomer	—	40, P22
C₂H₅O₂CC:CH	[(C₆H₅)₃P]₂Ni(CO)₂	—	1,2,4 Isomer, 1,3,5 isomer	—	31, 33, 129, P9
	(R₃P)₄Ni	—	1,2,4 Isomer, 1,3,5 isomer; 1,2,4,6 Isomer, 1,3,5,7 isomer	—	27, P10
	[(RO)₃P]₂Ni	—	1,2,4 Isomer, 1,3,5 isomer	—	40
CH₂:CHCH₂OCOC:CH	(R₃P)₂NiX₂–NaBH₄	—	1,2,4 Isomer, 1,3,5 isomer	Polymer	65, P2, P11
C₆H₅OC:CH	Ni(CO)₄	—	C₆H₆	Polymer	87
NCC:CH	[(C₆H₅)₃P]₂Ni(CO)₂	—	1,2,4 Isomer	—	31
(π-C₅H₄C:CH)₂Fe	(π-C₄H₇)₂Ni	—	1,2,4 Isomer	Polymer	162
CH₃C:CCH₃	(CH₂:CHCN)₂Ni	—	C₆R₆	Polymer	142
	(COD)₂Ni–PR₅–C₄H₆	—	C₆R₆	—	42, 92
	(COD)NiX	—	C₆R₆	—	3, 84
	(Mesityl)₂Ni	—	C₆R₆	—	55
	NiBr₂–e⁻	—	C₆R₆	Polymer	47, P32
HOCH₂C:CCH₂OH	Ni(acac)₂–Al(iso-C₄H₉)₃	—	—	Polymer	P30
	[(C₆H₅)₃PNi(CO)₃	—	—	Polymer	57
CH₃OCH₂C:CCH₂OCH₃	[(RO)₃P]₂Ni	—	—	—	31, P25
CH₃C:CC₄H₉	(π-CH₃CHCHCH₂NiX)₂	—	—	—	40
	(π-C₅H₅)₂Ni: 2AlBr₃	—	—	—	40, P22
CH₃C:CCO₂CH₃	(COD)₂Ni–Pt(C₆H₅)₃–C₄H₆	—	1,2,4-(CH₃)₃, 3,5,6-(CO₂CH₃)₃C₆	—	36
		—	Trimer	—	36
		—		—	84
C₂H₅C:CC₂H₅	NiCl₂–NaBH₄	Dimer	Tetramer	—	29
C₃H₇C:CC₃H₇	(π-CH₃CHCHCH₂NiCl)₂	—	C₆H₆	—	36

(continued)

TABLE II-3 (*continued*)

Alkyne	Catalyst	Dimer	Trimer (benzene deriv.)	Tetramer (COT deriv.)	Higher oligomers and polymer	Ref.
cyclo-Octyne	(CH₂:CHCN)₂Ni	—	C₆R₆	—	—	50
	Ni(acac)₂	—	C₆R₆	—	—	51
	NiX₂	—	C₆R₆	Tetramer	—	50, 51
CF₃C:CCF₃	[(C₆H₅)₂CH₃P]₄Ni	—	—	—	Polymer	2
	[CH₂:CHCN)₂Ni	—	C₆R₆	—	Polymer	1
	(R₃P)₂NiCl₂—NaBH₄	—	—	—	Polymer	1
C₆H₅C:CC₆H₅	Ni(CO)₄	—	C₆H₆	—	—	25, 68; see also 160
	(CH₂:CHCN)₂Ni	—	C₆R₆	—	—	42
	(Mesityl)₂Ni	—	C₆R₆	—	—	46, 47, P32
	[(C₆H₅)₃P]₂Ni(CH₂:CH₂)	—	C₆R₆	—	—	18
C₆H₅C:CCO₂CH₃	(COD)₂Ni–P(C₆H₅)₃–C₄H₆	—	1,2,4-(C₆H₅)₃–3,5,6-(CO₂CH₃)₃C₆	—	—	84
CH₃CO₂C:CCO₂CH₃	K₄Ni₂(CN)₆	—	C₆R₆	—	—	30
C₂H₅CO₂C:CCO₂C₂H₅`	Ni(CO)₄	Cyclic dimer	C₆R₆	—	—	102
HC:CCO₂CH₂C:CH	Ni(CO)₄	—	C₆R₆	—	Polymer	32
HC:C(CH₂)ₙCO₂CH₂C:CH (n = 2–5)	Ni(CO)₄	—	C₆R₆	—	Polymer	87
	[(C₆H₅)₃P]₂Ni(CO)₂	—	Cyclic trimer (n = 3)	—	Polymer	13, 87, P6, P8
HC:C-C:CH	(R₃P)₂NiCl₂—NaBH₄	—	1,2,4 Isomer	—	Polymer	29, P2
CH₃C:C-C:CCH₃	[(C₆H₅)₃P]₂PNi(CO)₃	—	1,2,4 Isomer	—	Polymer	24
HOCH₂C:C-C:CC₆H₅	[(C₆H₅)₃P]₂Ni(CO)₂	—	1,2,4 Isomer	—	—	6
CH₃OCH₂C:C-C:CC₆H₅	C₆H₅)₃PNi(CO)₃	—	1,2,4 Isomer	—	—	24
C₆H₅C:C-C:CC₆H₅	[(C₆H₅)₃P]₂Ni(CO)₂	—	1,2,4 Isomer	—	—	24
[2,2-bis(phenylethynyl)-1,3-indandione structure, with C₆H₅—C≡C groups]	Ni(CO)₄	—	Cyclic trimer (11)	Cyclic dimer (10)	—	133
[2,2-bis(phenylethynyl)benz[f]indene-1,3-dione structure, with C₆H₅—C≡C groups]	Ni(CO)₄	—	Cyclic trimer	Cyclic dimer	—	133
C₆H₅C≡N	Raney nickel	—	1,3,5-Cyclic trimer	—	—	23, 49

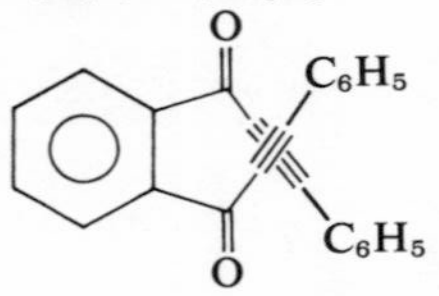

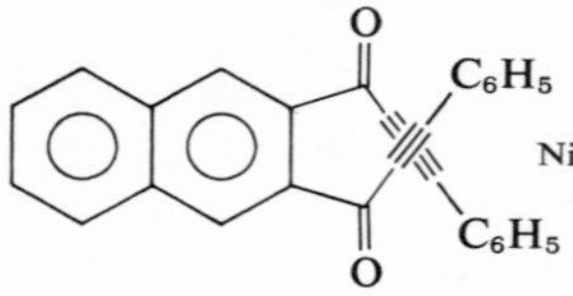

TABLE II-4

THE CO-OLIGOMERIZATION OF ALKYNES

Alkyne		Catalyst	Trimer (benzene deriv.)	Tetramer (COT deriv.)	Oligomer	Ref.
$HC{\vdots}CH$	$CH_3C{\vdots}CH$	$NiCl_2$–$NaBH_4$	—	—	Polymer	P5
		$[(C_6H_5)_3P]_2NiCl_2$–$NaBH_4$	$CH_3C_6H_5$ *m*-$(CH_3)_2C_6H_4$	—	Polymer	17, P2
		$Ni(acac)_2$	—	CH_3–C_8H_7	—	78, 79, P43
$HC{\vdots}CH$	$HOCH_2C{\vdots}CH$	$Ni(acac)_2$	—	$HOCH_2$–C_8H_7	—	81
$HC{\vdots}CH$	$C_6H_5OCH_2C{\vdots}CH$	$NiBr_2$–$NaBH_4$	—	—	Polymer	P4
$HC{\vdots}CH$	$(C_2H_5)_2NCH_2C{\vdots}CH$	$NiCl_2$–$NaBH_4$	—	—	Polymer	P4
$HC{\vdots}CH$	$C_2H_5C{\vdots}CH$	$NiCl_2$–$NaBH_4$	—	—	Polymer	P5
$HC{\vdots}CH$	$HOC_2H_4C{\vdots}CH$	$Ni(acac)_2$	—	HOC_2H_4–C_8H_7	—	81
$HC{\vdots}CH$	$(CH_3)_2NC_2H_4C{\vdots}CH$	$Ni(acac)_2$	—	$(CH_3)_2NC_2H_4$–C_8H_7	—	81
$HC{\vdots}CH$	$CH_3COC{\vdots}CH$	$Ni(acac)_2$	—	CH_3CO–C_8H_7	—	82
$HC{\vdots}CH$	$CH_3C(OH)HC{\vdots}CH$	$Ni(acac)_2$	—	$CH_3CH(OH)$–C_8H_7	—	82
$HC{\vdots}CH$	$C_3H_7C{\vdots}CH$	$Ni(acac)_2$	—	C_3H_7–C_8H_7	—	P43
$HC{\vdots}CH$	$HO(CH_2)_3C{\vdots}CH$	$Ni(acac)_2$	—	$HO(CH_2)_3$–C_8H_7	—	82
$HC{\vdots}CH$	$CN(CH_2)_3C{\vdots}CH$	$Ni(acac)_2$	—	$CN(CH_2)_3$–C_8H_7	—	82
$HC{\vdots}CH$	$HO(CH_3)_2CC{\vdots}CH$	$Ni(NO_3)_2$–$NaBH_4$	—	—	Polymer	P4
		$Ni(acac)_2$	—	$HO(CH_3)_2C$–C_8H_7	—	82
$HC{\vdots}CH$	$C_4H_9C{\vdots}CH$	$Ni(acac)_2$	—	C_4H_9–C_8H_7	—	79, P43
$HC{\vdots}CH$	$CH_2{:}CHC{\vdots}CH$	$[(C_6H_5)_3P]_2Ni(CO)_2$	C_6H_6, styrene, divinylbenzene	—	—	53, P42
		Ni(ethylacetoacetate)$_2$	—	$CH_2{:}CH$–C_8H_7	—	21, 22
$HC{\vdots}CH$	$C_6H_5C{\vdots}CH$	$Ni(acac)_2$	—	C_6H_5–C_8H_7	—	78, 79, P43
$HC{\vdots}CH$	$CH_3C{\vdots}CCH_3$	$[(C_6H_5)_3P]_2Ni(CO)_2$	C_6H_6, *o*-$(CH_3)_2C_6H_4$, 1,2,3,4-$(CH_3)_4C_6H_2$, styrene	—	—	31, 73, 75
		$Ni(acac)_2$	—	1,2-$(CH_3)_2$–C_8H_6	—	78, 79, P43
$HC{\vdots}CH$	$CH_2{:}CHC{\vdots}CCH{:}CH_2$	$(C_6H_5)_3PNi(CO)_3$	*o*-$(CH_2{:}CH)_2C_6H_4$, styrene	—	—	86, P47
$HC{\vdots}CH$	$C_6H_5C{\vdots}CC_6H_5$	$Ni(acac)_2$	*o*-$(C_6H_5)_2C_6H_4$	1,2-$(C_6H_5)_2$–C_8H_6	—	80
$C_3H_7C{\vdots}CH$	$C_5H_{11}C{\vdots}CH$	$(R_3P)_2NiX_2$–$NaBH_4$	Trimers	Tetramers	—	29, 31
$C_3H_7C{\vdots}CH$	$HC{\vdots}C(CH_2)_nC{\vdots}CH$	$[(C_6H_5)_3P]_2Ni(CO)_2$	—	—	Polymer	13, P6, P8
$C_3H_7C{\vdots}CH$	$C_6H_5C{\vdots}CH$	$[(C_6H_5)_3P]_2Ni(CO)_2$	—	—	—	31
$C_5H_{11}C{\vdots}CH$	$HC{\vdots}C(CH_2)_nC{\vdots}CH$	$[(NCC_2H_4)_3P]_2Ni(CO)_2$	—	—	Polymer	13, P6, P8
$CH_3CO_2C{\vdots}CH$	$C_2H_5CO_2C{\vdots}CH$	$(Cl_3P)_4Ni$	—	$(CH_3CO_2)_3C_2H_5CO_2$–C_8H_4	—	27, P10

TABLE II-5
THE CO-OLIGOMERIZATION OF ALKYNES WITH OLEFINS

Alkyne	Olefin	Catalyst	Product	Ref.
HC:CH	$CH_2:CHCN$	$[(C_6H_5)_3P]_2Ni(CO)_2$	$CH_2:CHCH:CHCH:CHCN$	38, 73, 75, 76, P20, P44
HC:CH	$CH_2:CHCO_2CH_3$	$(CH_2:CHCN)_2NiP(C_6H_5)_3$ $[(C_6H_5)_3P]_2Ni(CO)_2$	$CH:CHCH:CHCH:CHCN$ $CH_2:CHCH:CHCH:CHCO_2CH_3$	42, 44 76, P44
HC:CH	$CH_2:CHCO_2C_2H_5$	$\{[(C_6H_5)_3P]_2Cu\}[Ni(CN)_4]$	$CH_2:CHCH:CHCH:CHCO_2C_2H_5$	P45, P52, P53
HC:CH	$CH_2:CHCH:CH_2{}^a$	$Ni(acac)_2-Al(C_2H_5)_3-PR_3$	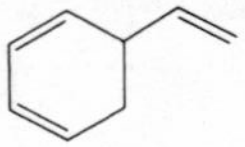	88
HC:CC_6H_5	$CH_2:CHCH:CH_2{}^a$	$(COD)_2Ni-P(C_6H_5)_3$	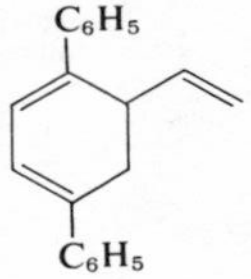	3, 83, 84
$CH_3C:CCH_3$	$CH_2:CHCH:CH_2{}^a$	$(COD)_2Ni$	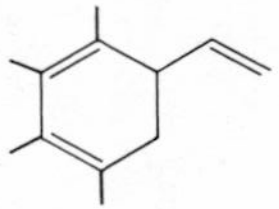	3, 83, 84
$CH_3OCH_2C:CCH_2OCH_3$	$CH_2:CHCH:CH_2{}^a$	$(COD)_2Ni-P(cyclo\text{-}C_6H_{11})_3$	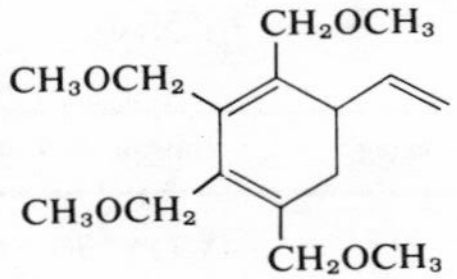	83, 84
$CH_3C:CCO_2CH_3$	$CH_2:CHCH:CH_2{}^a$	$(COD)_2Ni-P(C_6H_5)_3$	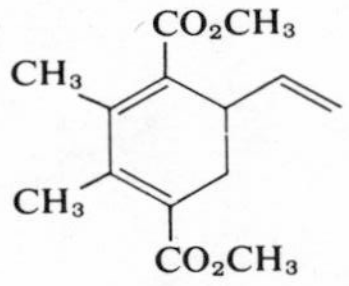 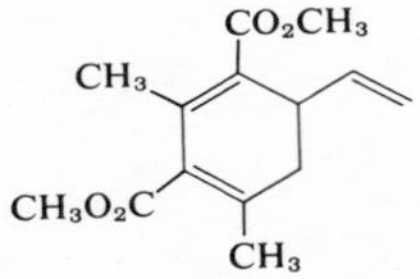	83, 84

114

$C_4H_9C:CCO_2C_2H_5$	$CH_2:CHCH:CH_2{}^a$	$(COD)_2Ni-P(C_6H_5)_3$	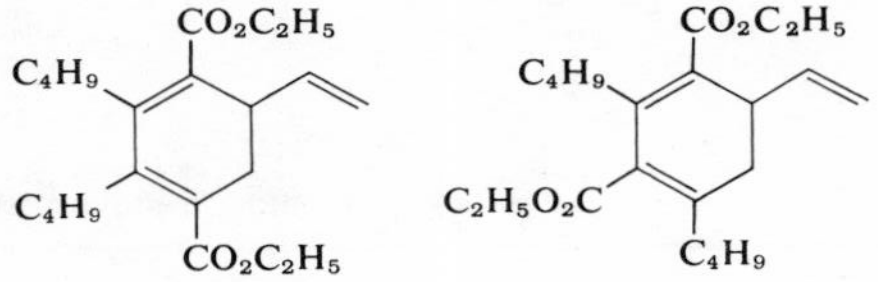 83, 84
$C_6H_5C:CCO_2CH_3$	$CH_2:CHCH:CH_2{}^a$	$(COD)_2Ni-P(C_6H_5)_3$	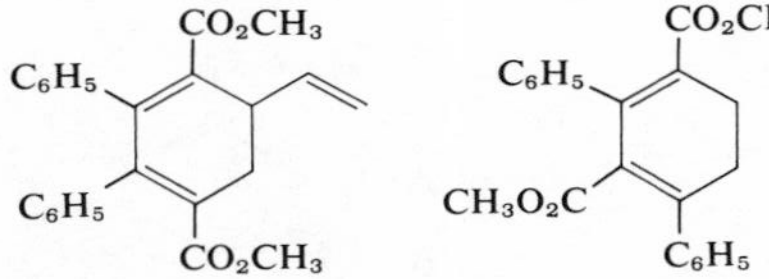 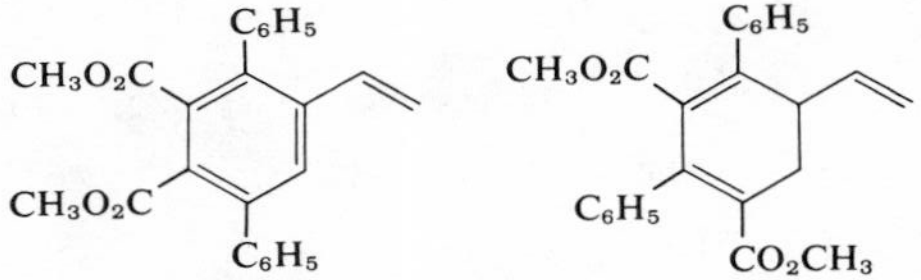 83, 84
$CH_3C:CCH_3$		$(COD)_2Ni-P(cyclo\text{-}C_6H_{11})_3$	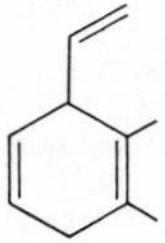 156
$C_2H_5C:CC_2H_5$		$(COD)_2Ni-P(cyclo\text{-}C_6H_{11})_3$	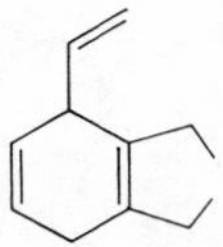 156
$CH_3C:CC_2H_5$		$(COD)_2Ni-P(cyclo\text{-}C_6H_{11})_3$	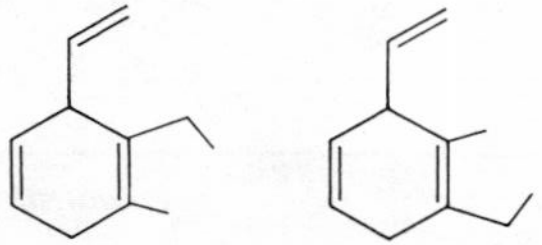 156

(continued)

115

TABLE II-5 (*continued*)

Alkyne	Olefin	Catalyst	Product	Ref.
$CH_3C{:}CCH_3$		$(COD)_2Ni{-}P(cyclo{-}C_6H_{11})_3$	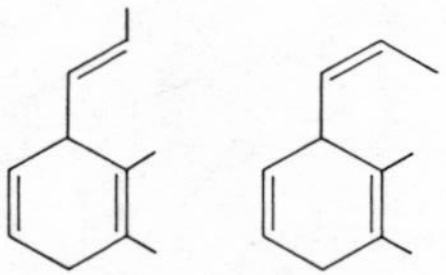	156
$C_4H_9C{:}CH$	N-Methylmaleimide	$[(C_6H_5)_3P]_2Ni(CO)_2$	Diels-Alder adduct to 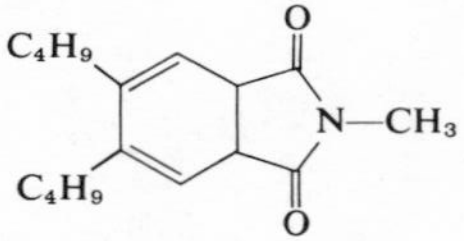	77
$C_6H_5C{:}CC_6H_5$	N-Methylmaleimide	$[(C_6H_5)_3P]_2Ni(CO)_2$	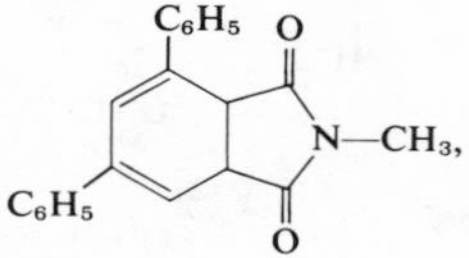 $N{-}CH_3$, Diels-Alder adduct	77
$CH_3C{:}CC{:}CCH_3$	N-Phenylmaleimide	$[(C_6H_5)_3P]_2Ni(CO)_2$	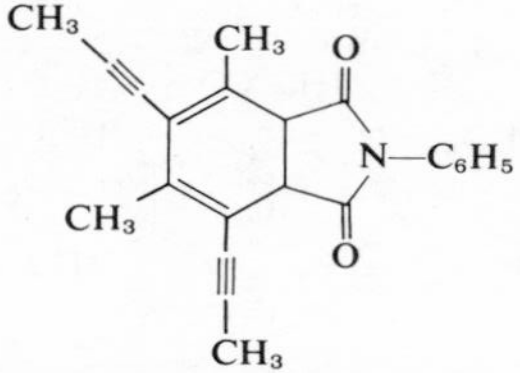 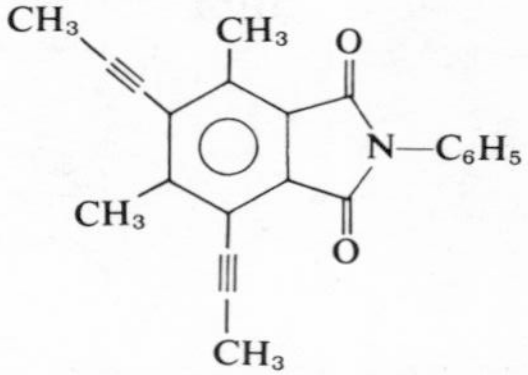	77

^a The reaction between butadiene and alkynes to give 10- and 12-membered rings is discussed in Chapter III.

116

V. The Oligomerization of 1,2-Dienes

A. The Oligomerization of Allene ($CH_2:C:CH_2$) (*Table II-6*)

Allene is readily cyclo-oligomerized by a variety of nickel catalysts to compounds containing *exo*-methylene groups (e.g., **25–31**). Polymeric material is also formed.

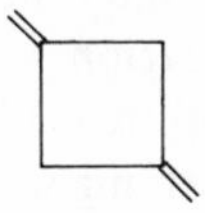

| **25** | 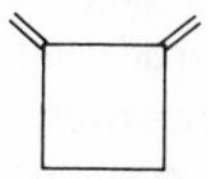   **26** | 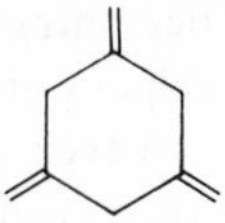   **27** | 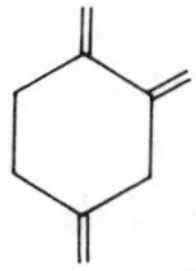   **28** | 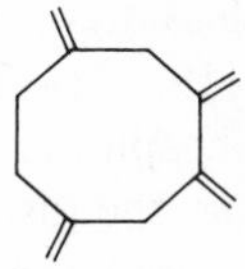   **29** |

| **30** | 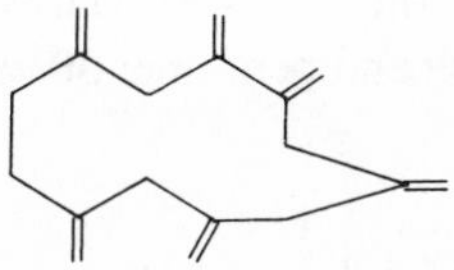   **31** |

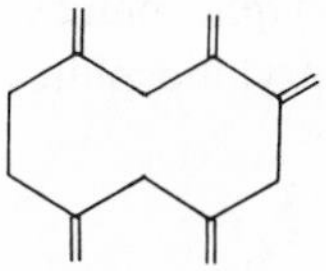

The reported composition of the resulting mixtures is somewhat misleading because those researchers interested in polymer formation have tended to neglect the oligomers and vice versa. The dimers **25** and **26** have only been isolated from reactions conducted in the vapor phase (114) and are also produced in the thermal vapor phase oligomerization of allene, although in neither the same proportions nor under the same conditions (131). The trimer 1,2,4-trimethylenecyclohexane (**28**) is frequently the main product of the catalytic reaction, whereas the symmetrical isomer **27** is observed only in traces. The tetramer **29**, originally thought to be the 1,3,5,7-tetramethylene compound, has been reformulated as the 1,2,4,7 isomer with the help of its ^{13}C NMR spectrum (115, 120). The pentamer **30** is produced in high yield if allene is cyclo-oligomerized in the presence of bis(cyclooctadiene)nickel with the rigorous exclusion of air; under other conditions mainly polymer formation has been reported (121, 126). The hexamer **31** has only been identified among the products formed using bistriphenylphosphine nickel as catalyst (115).

The cyclo-oligomerization reaction is generally accompanied by formation of a polymer and this is in many cases the main product. The polymer chain is composed of 1,2-allene units that may be arranged in a –1,2,1,2– or in a –1,2,2,1– sequence (**32, 33**) or in a random combination of the two. A highly regular 1,2,1,2-polyallene (benzene soluble, m.p. 113–122°) has been obtained using $Ni(acac)_2$–$Al(iso$-$C_4H_9)_3$ as the catalyst (128) [the use of aluminum triethyl in place of aluminum tri*iso*butyl is reported to give an insoluble gel

$$-CH_2-\underset{\underset{1}{C}}{\overset{\overset{CH_2}{\|}}{\underset{2}{C}}}-CH_2-\underset{2}{\overset{\overset{CH_2}{\|}}{C}}- \qquad\qquad -CH_2-\underset{2}{\overset{\overset{H_2C}{\|}}{\underset{2}{C}}}-\underset{2}{\overset{\overset{CH_2}{\|}}{C}}-CH_2$$

$$\underset{\mathbf{32}}{1\quad 2\quad 1\quad 2} \qquad\qquad \underset{\mathbf{33}}{1\quad 2\quad 2\quad 1}$$

(121)], whereas a somewhat less regular but still crystalline polymer (benzene soluble, m.p. 60–80°) is produced in the presence of π-allylnickel derivatives; this latter probably contains a small percentage of –1,2,2,1– ordered molecules (119, 121–123, see also 158). The waxlike polymer produced by bistriphenyl-phosphine nickel is assumed to have an even more disordered structure (115), and the insolubility of the polymer produced using (COD)NiI is thought to indicate that cross-linking has here occurred (125).

The formation of polymer is apparently suppressed by the presence of donor ligands in the catalyst: trialkyl- or triarylphosphines favor formation of the tetramer **29** and pentamer **30** whereas triarylphosphites favor formation of the trimer **28**.

B. *The Oligomerization of Substituted 1,2-Dienes*

Relatively little attention has been given to the oligomerization of substituted allenes using nickel catalysts. 1,2-Butadiene reacts in the presence of $(COD)_2Ni$, $(\pi\text{-}C_3H_5NiX)_2$, or $\pi\text{-}C_3H_5NiCl[P(C_6H_5)_3]$ to give an irregular polymer that probably consists of a random arrangement of –1,2– and –2,3– units (119, 125, P51). 1-Methoxyallene (as a mixture with butadiene) is converted completely to higher oligomers on interaction with a $Ni\text{-}P(OC_6H_4\text{-}o\text{-}C_6H_5)_3$ catalyst (134). The reaction of 1,1′-dimethylallene is more encouraging and from reactions with a catalyst containing $P(OC_6H_4\text{-}o\text{-}C_6H_5)_3$ or $P(C_6H_5)_3$ a dimer **34** and a tetramer probably having structure **35** have been isolated (116, 134).

$$\mathbf{34} \qquad\qquad \mathbf{35}$$

1,1′-Dimethylallene is polymerized by $(COD)_2Ni$, $(\pi\text{-}C_3H_5NiBr)_2$, and (COD)NiX (119, 125, P51). The product (benzene soluble, m.p. 175°) is thought to have the regular –1,2,1,2– structure **36**, and this arrangement is also believed to be present in the polymer formed by treating optically active 1,3-dimethylallene with $(\pi\text{-}C_3H_5NiI)_2$ (137). The product in this last example is optically active and is therefore believed to contain stereoregular sequences having structure **37**. Catalytic oligomerization of 1,2-cyclononadiene to a

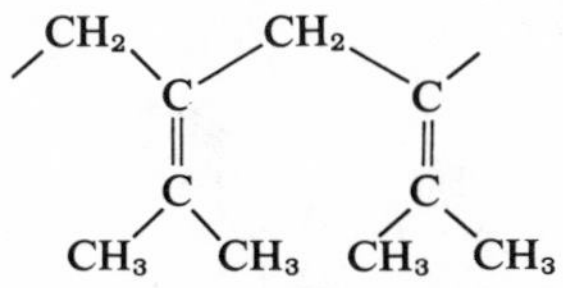

36 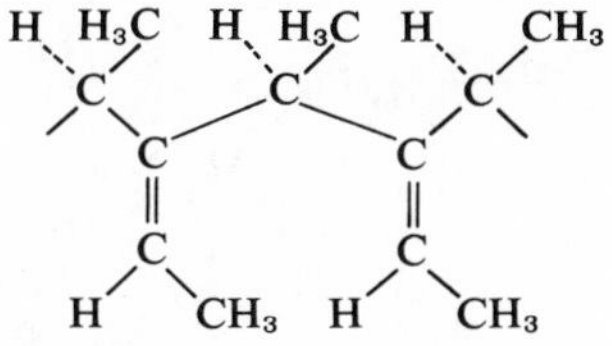   37

mixture containing trimers, tetramers, pentamers, and hexamers has been observed on heating the monomer with $[(C_6H_5\text{-}o\text{-}C_6H_4O)_3P]_2Ni$ (116).

1,3-Diphenylallene is dimerized in the presence of π-allylnickel iodide and triethylphosphite to give *anti*-1,2-bisphenylmethylene-3,4-diphenylcyclo-butane (**38**) (143).

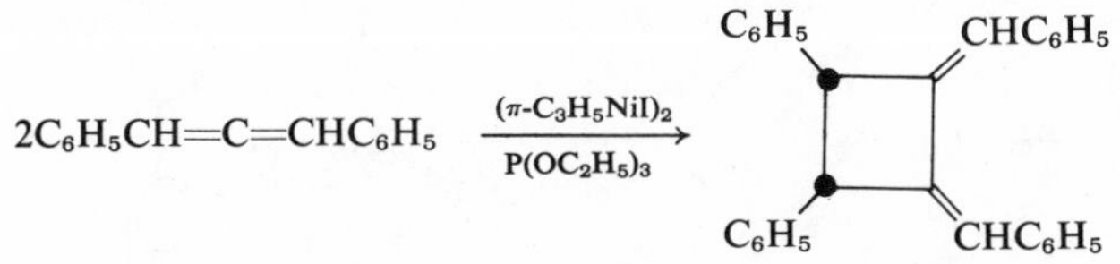

$$2C_6H_5CH{=}C{=}CHC_6H_5 \xrightarrow[P(OC_2H_5)_3]{(\pi\text{-}C_3H_5NiI)_2}$$

38

C. Mechanistic Considerations

The isolation of various intermediates allows a plausible mechanism for the cyclo-oligomerization of allene to be postulated. Initial coordination of two allene molecules to the nickel may be supposed to lead to three isomeric nickelacyclopentane derivatives (**41–43**) that may react with a third allene molecule, either by ring closure to give the cyclic dimers **25** and **26**, or by insertion to give the precursor (**44** and **45**) of the cyclic trimers (for simplicity this is shown only for **41**). Stepwise insertion of further allene molecules generates the other products.

A mono–allene complex corresponding to **39** [Lig = $P(C_6H_5)_3$, $n = 2$] has been isolated from the reaction between allene and bistriphenylphosphine nickel (115; see also 144), whereas the bisallene adduct **40** has precedence in the bis(ethylene)nickel phosphine complexes (see Volume I, p. 263). Although no evidence exists for the nickelacyclopentane intermediates, both ligand-free and ligand-stabilized complexes corresponding to the trimeric and tetrameric precursors **44** and **46** have been isolated, and an x-ray structural determination has been published for **44** [Lig = $P(cyclo\text{-}C_6H_{11})_3$] (116, 117, 120, 132, 144). The catalytic reaction can be simulated by treating **44** and **46** with carbon monoxide or carbon disulfide, whereby coupling of the π-allyl groups is accompanied by elimination of **28** and **29**, respectively. An NMR study of the reaction of **46** [Lig = $P(C_6H_5)_3$] with allene shows that the formation of **29** is accompanied by reformation of the trimeric precursor **44** (120, 144). The absence of the tetramer **47** in the product mixture indicates

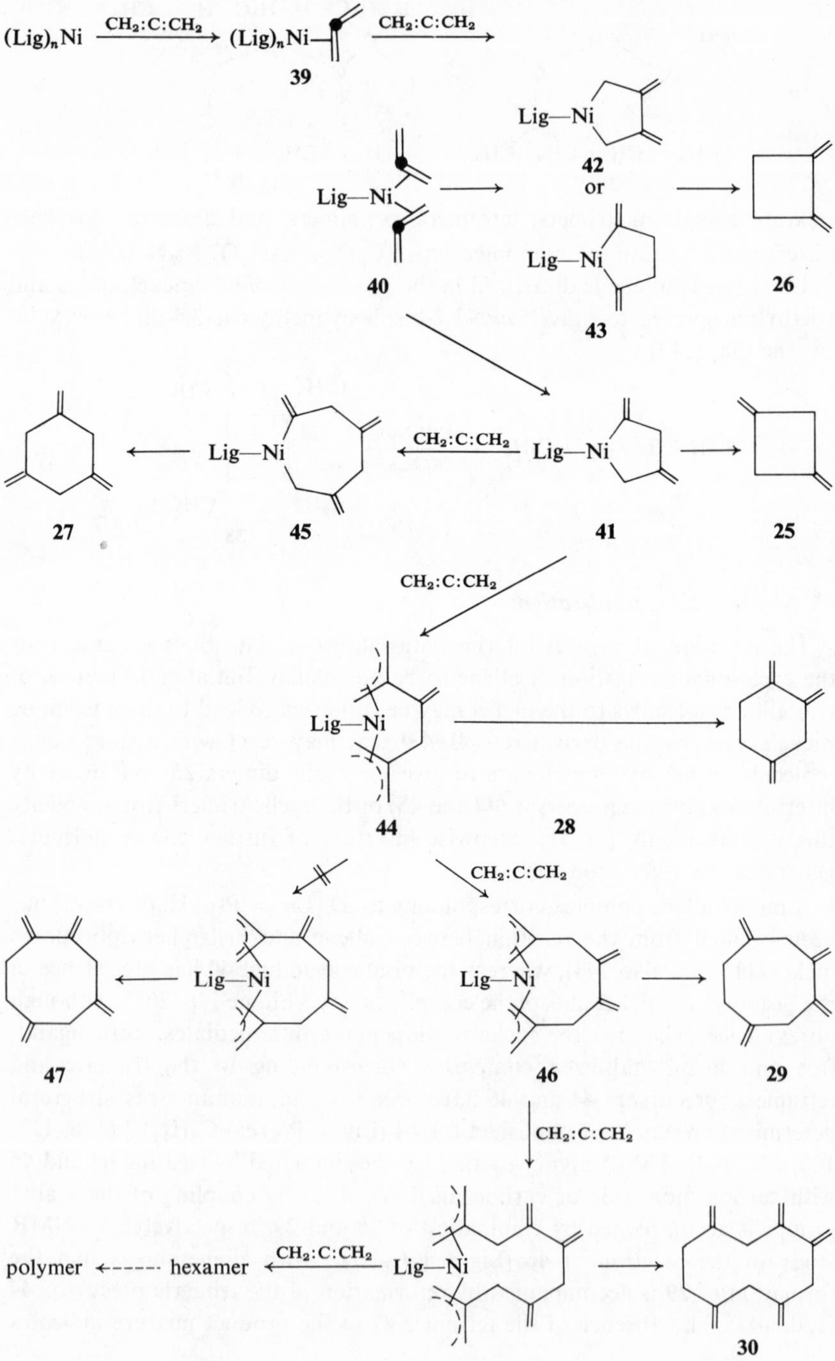
(Lig)$_n$Ni
CH$_2$:C:CH$_2$
(Lig)$_n$Ni
CH$_2$:C:CH$_2$
39
Lig—Ni
42
or
Lig—Ni
43
Lig—Ni
40
26
Lig—Ni
45
CH$_2$:C:CH$_2$
Lig—Ni
41
27
25
CH$_2$:C:CH$_2$
Lig—Ni
44
28
CH$_2$:C:CH$_2$
Lig—Ni
47
Lig—Ni
46
29
CH$_2$:C:CH$_2$
polymer
hexamer
CH$_2$:C:CH$_2$
Lig—Ni
30

that the reaction of **44** with further allene occurs exclusively by insertion into that π-allyl group adjacent to the *exo*-methylene group, which for the tricyclohexylphosphine adduct has been shown to be the less distorted of the two (132).

VI. The Co-oligomerization of 1,2-Dienes with Alkynes and Olefins
(Table II-7)

The same catalysts that are active in the cyclo-oligomerization of acetylene are able to co-oligomerize alkynes with allene to *exo*-methylenecyclohexene and -cyclooctene derivatives. Nickel acetylacetonate or nickel cyanide

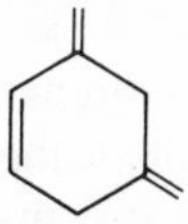

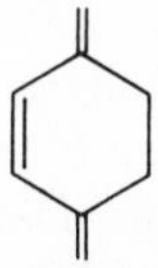

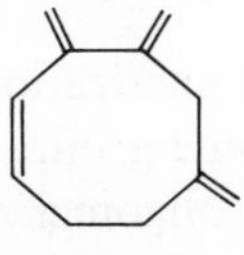

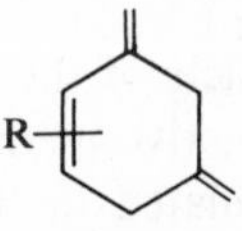

$R = CH_3, C_6H_5, C_6H_5CH:CH_2$

48   **49**   **50**   **51**

converts a mixture of acetylene and allene into the cotrimer **48** and the tetramer **50**. The formation of tetramer is not observed when a phosphine-modified catalyst is used; instead, a mixture of **48** and **49** is isolated. Substituted alkynes react similarly to give a mixture of 1- and 2-substituted 3,5-dimethylenecyclohexene derivatives (**51**) (74, P56).

The reaction between allene and ethylene is also catalyzed by phosphine–nickel systems and the linear co-oligomer 3,5-dimethylene-1-hexene (**52**) results. Methylacrylate reacts similarly to give **53** as well as two isomeric forms of the 2:2 co-oligomer **54** (58).

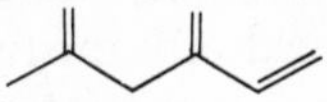

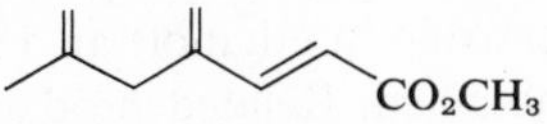

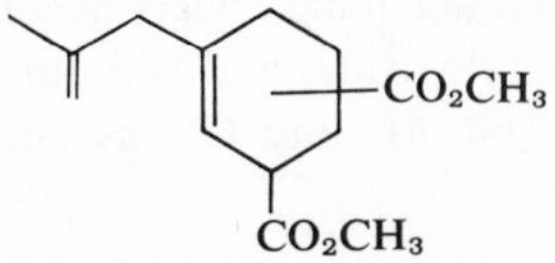

52   **53**   **54**

A cotrimer (**55**) has been isolated from the reaction of methoxyallene with butadiene in the presence of a zerovalent nickel catalyst modified by tri-(2-biphenylyl)phosphite* (134).

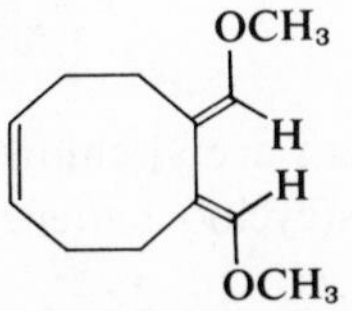

55

* The reaction of allene with two or more molecules of butadiene is discussed in Chapter III.

A variety of olefins (e.g., methylacrylate and butadiene) react with allene in the presence of $[(C_6H_5)_3P]_2Ni(CO)_2$ to give octahydronaphthalene derivatives (e.g., **56**); these, however, are believed to be formed by a Diels-Alder reaction between the catalytically formed 1,2,4-trimethylenecyclohexane and the dienophile (113, P57).

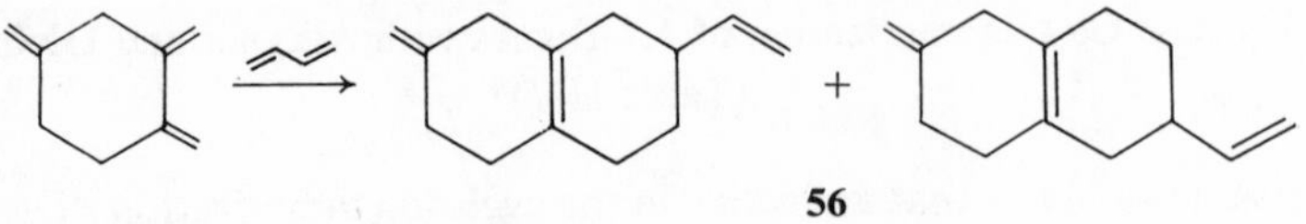

56

VII. The Telomerization of Allene

The structure of the products of the telomerization of allene with amines and compounds containing active methylene groups suggests that the same intermediates as those postulated in the oligomerization of allene are involved here also (153, 154). For example, the reaction of butylamine with allene in the presence of a catalyst prepared from nickel bromide and phenyl-diiso-propoxyphosphine produces a mixture of alkylated amines in which **57**, **58**, and **59** predominate. Active methylene compounds are less reactive.

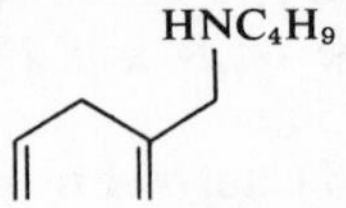

57 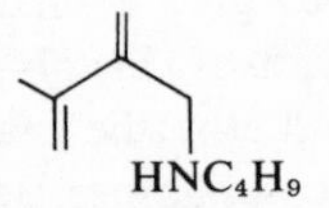   **58** 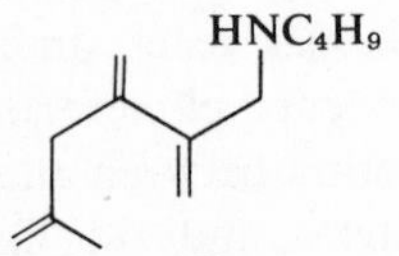   **59**

From the reaction with benzyl methyl ketone (using a catalyst prepared by treating nickel acetylacetonate with phenydiiso-propoxyphosphine, sodium boryhydride, and sodium phenoxide in ethanol) an 18% yield of a mixture of **60**, **61**, and **62** has been obtained. Related products have been isolated

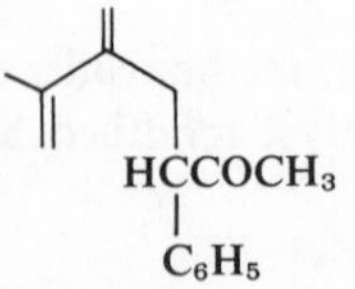

60 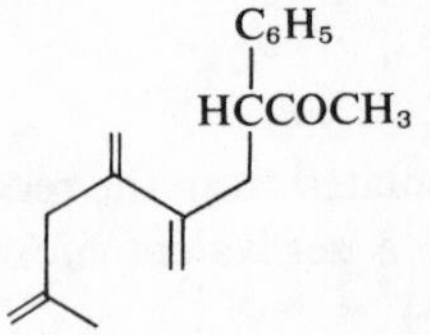   **61** 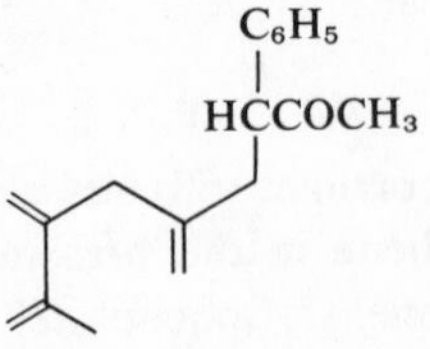   **62**

from the reaction of allene with acetyl chloride (e.g., **63**) and acetaldehyde (e.g., **64**) in the presence of bis(cyclooctadiene)nickel (154).

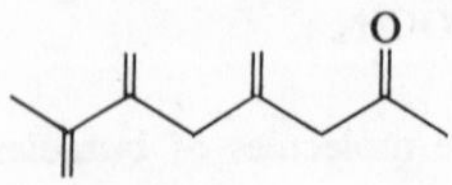

63 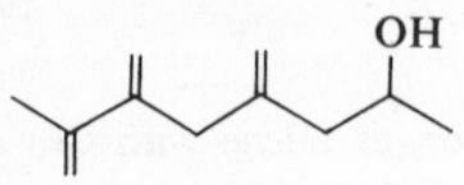   **64**

TABLE II-6

THE CYCLO-OLIGOMERIZATION OF ALLENE

Catalyst	Dimer	Trimer	Tetramer	Pentamer	Hexamer	Polymer	Ref.
$[(C_6H_5O)_3P]_{4-n}Ni(CO)_n$	—	27, 28	29	—	Higher oligomers	Higher oligomers	113, P50
$[(C_6H_5)_3P]_2Ni(CO)_2$	25, 26	28	—	—	—	—	114
	—	27, 28	29	—	—	—	113, P50
$[(C_6H_5)_2PC_6H_4P(C_6H_5)_2Ni(CO)_2]_n$	25, 26	28	—	30	—	—	114
	—	—	29	—	Higher oligomers	Higher oligomers	114
$[(C_6H_5)_3P]_2Ni$	—	27, 28	29	30	31	Wax	115
$[(C_6H_5-o-C_6H_4O)_3P]_2Ni$	—	28	—	—	—	Solid	116
$(Cl_3P)_4Ni$	—	27, 28	29	—	—	—	27, P10
$[(C_6H_5)_3P]_4Ni$	—	27, 28	29	30	—	—	117, 118, 120, 144
$[(C_6H_5O)_3P]_4Ni$	—	27, 28	29	30	—	Cryst.	120, 144
$(tert-C_4H_9NC)_4Ni$	—	—	—	30	—	Cryst.	117–119, P51
$(COD)_2Ni$	—	—	—	30	—	Cryst.	119, 121, 126, 130, P51
$(COD)NiX$	—	—	—	—	—	Amorph.	125
$(CH_2\!:\!CHCN)_2Ni$	—	—	—	—	—	Cryst.	119, 121, P51
$(\pi-C_3H_5)_2Ni$	—	—	—	—	—	Cryst.	119, 121, 158, P51
$(R_3P)_2NiBr_2$	—	—	—	—	—	Gel	119, 121, 127, 136
$Lig_2NiX_2-NaBH_4$	—	27 (?)	39	—	—	—	135
$[(C_6H_5)_3P]_2Ni-C_6H_5OH$	—	—	—	—	—	Cryst.	115
NiX_2-AlR_3	—	—	—	—	—	Gel	57, 119, 121, 124, 128, 138
$(Cl_3P)_4Ni-AlBr_3-LiC_4H_9$	—	—	—	—	—	Gel	138

TABLE II-7

The Co-oligomerization of Allene with Alkynes and Olefins[a]

Allene	Alkyne	Catalyst	Cotrimer	Cotetramer	Misc.	Ref.
$CH_2:C:CH_2$	$HC:CH$	$Ni(acac)_2$	48	50	—	74, P56
$CH_2:C:CH_2$	$CH_3C:CH$	$[(C_6H_5O)_3P]_2Ni(CO)_2$	48, 49	—	—	74, P56
$CH_2:C:CH_2$		$Ni(acac)_2$	51	—	—	74, P56
			$(R = CH_3)$			
$CH_2:C:CH_2$	$C_6H_5C:CH$	$Ni(acac)_2$	51	—	—	74, P56
			$(R = C_6H_5)$			
$CH_2:C:CH_2$	$CH_2:CHC:CH$	$Ni(acac)_2$	51	—	—	74, P56
			$(R = CH:CH_2)$			
$CH_2:C:CH_2$	$CH_2:CH_2$	$[(C_6H_5)_3P]_2Ni$	52	—	Allene	58
					oligomers	
$CH_2:C:CH_2$	$CH_2:CHCO_2CH_3$	$[(C_6H_5)_3P]_2Ni$	53	—	54	58
$CH_2:C:CHOCH_3$	$CH_2:CHCH:CH_2$	$Ni(acac)_2-$ $P(OC_6H_4-o-C_6H_5)_3-$ $(C_2H_5)_2AlOC_2H_5$	55	—	—	134

[a] The reaction of allene with butadiene is discussed in Chapter III.

References

1. J. L. Boston, D. W. A. Sharp, and G. Wilkinson, *J. Chem. Soc., London* p. 3488 (1962).
2. J. Browning, C. S. Cundy, M. Green, and F. G. A. Stone, *J. Chem. Soc., A* p. 448 (1971).
3. W. Brenner, P. Heimbach, K.-J. Ploner, and P. Thömel, *Angew. Chem.* **81**, 744 (1969).
4. G. A. Chukhadzhyan, E. L. Sarkisyan, L. M. Davtyan, and I. M. Rostomyan, *J. Org. Chem. USSR* **8**, 481 (1972).
5. G. A. Chukhadzhyan, Z. I. Abramyan, and M. A. Manukyan, *J. Org. Chem. USSR* **7**, 916 (1971).
6. A. J. Chalk and R. A. Jerussi, *Tetrahedron Lett.* p. 61 (1972).
7. G. A. Chukhadzhyan, Z. I. Abramyan, and V. G. Grigoryan, *Arm. Khim. Zh.* **23**, 608 (1970); *Chem. Abstr.* **74**, 76730 (1971).
8. G. A. Chukhadzhyan, Z. I. Abramyan, and G. A. Gevorkyan, *Vysokomol. Soedin., Ser. A* **12**, 2462 (1970); *Chem. Abstr.* **74**, 54260 (1971).
9. G. A. Chukhadzhyan, Z. I. Abramyan, E. L. Markosyan, L. M. Davotyan, and G. A. Gevorkyan, *Arm. Khim. Zh.* **22**, 1039 (1969); *Chem. Abstr.* **72**, 79206 (1970).
10. P. Chini, A. Santambrogio, and N. Palladino, *J. Chem. Soc., C* p. 830 (1967).
11. P. Chini, N. Palladino, and A. Santambrogio, *J. Chem. Soc., C* p. 836 (1967).
12. D. Cordischi, A. Furlani, P. Bicev, M. V. Russo, and P. Carusi, *Gazz. Chim. Ital.* **101**, 526 (1971).
13. E. C. Colthup and L. S. Meriwether, *J. Org. Chem.* **26**, 5169 (1961).
14. A. C. Cope and L. L. Estes, *J. Amer. Chem. Soc.* **72**, 1128 (1950).
15. A. C. Cope and S. W. Fenton, *J. Amer. Chem. Soc.* **73**, 1195 (1951).
16. L. E. Craig and C. E. Larrabee, *J. Amer. Chem. Soc.* **73**, 1191 (1951).
17. A. F. Donda and G. Moretti, *J. Org. Chem.* **31**, 985 (1966).
18. E. O. Greaves, C. J. L. Lock, and P. M. Maitlis, *Can. J. Chem.* **46**, 3879 (1968).
19. M. L. H. Green, M. Nehmé, and G. Wilkinson, *Chem. Ind. (London)* p. 1136 (1960).
20. N. Hagihara, *J. Chem. Soc. Jap.* **73**, 237 (1952); *Chem. Abstr.* **47**, 9954 (1953).
21. N. Hagihara, *J. Chem. Soc. Jap.* **73**, 323 (1952); *Chem. Abstr.* **47**, 10490 (1953).
22. N. Hagihara, *J. Chem. Soc. Jap.* **73**, 373 (1952); *Chem. Abstr.* **47**, 10491 (1953).
23. W. Z. Heldt, *J, Organometal. Chem.* **6**, 292 (1966).
24. W. Hübel and R. Merenyi, *Chem. Ber.* **96**, 930 (1963).
25. W. Hübel and C. Hoogzand, *Chem. Ber.* **93**, 103 (1960).
26. V. A. Kormer, L. A. Churlyaeva, and T. L. Yufa, *Vysokomol. Soedin., Ser B* **12**, 483 (1970); *Chem. Abstr.* **73**, 88225 (1970).
27. J. R. Leto and M. F. Leto, *J. Amer. Chem. Soc.* **83**, 2944 (1961).
28. L. B. Luttinger, *Chem. Ind. (London)* p. 1135 (1960).
29. L. B. Luttinger, *J. Org. Chem.* **27**, 1591 (1962).
30. J. P. Martella and W. C. Kaska, *Tetrahedron Lett.* p. 4889 (1968).
31. L. S. Meriwether, E. C. Colthup, G. W. Kennerly, and R. N. Reusch, *J. Org. Chem.* **26**, 5155 (1961).
32. E. R. H. Jones, T. Y. Shen, and M. C. Whiting, *J. Chem. Soc., London* p. 48 (1951).
33. L. S. Meriwether, M. F. Leto, E. C. Colthup, and G. W. Kennerly, *J. Org. Chem.* **27**, 3930 (1962).
34. C. G. Overberger and J. M. Whelan, *J. Org. Chem.* **24**, 1155 (1959).
35. L. Pichat and C. Baret, *Tetrahedron* **1**, 269 (1957).
36. V. O. Reikhsfel'd, B. I. Lein, and K. L. Makovetskii, *Proc. Acad. Sci. USSR* **190**, 31 (1970).

37. W. Reppe, O. Schlichting, K. Klager, and T. Toepel, *Justus Liebigs Ann. Chem.* **560**, 1 (1948).
38. W. Reppe and W. J. Schweckendiek, *Justus Liebigs Ann. Chem.* **560**, 104 (1948).
39. W. Reppe, O. Schlichting, and H. Meister, *Justus Liebigs Ann. Chem.* **560**, 93 (1948).
40. W. Reppe, N. von Kutepow, and A. Magin, *Angew. Chem.* **81**, 717 (1969).
41. J. D. Rose and F. S. Statham, *J. Chem. Soc.*, *London* p. 69 (1950).
42. G. N. Schrauzer, *Chem. Ber.* **94**, 1403 (1961).
43. G. N. Schrauzer and S. Eichler, *Chem. Ber.* **95**, 550 (1962).
44. G. N. Schrauzer, *J. Amer. Chem. Soc.* **81**, 5310 (1959).
45. F. K. Shmidt, V. G. Lipovich, and I. V. Kalechits, *Kinet. Catal.* **11**, 206 (1970).
46. M. Tsutsui and H. Zeiss, *J. Amer. Chem. Soc.* **81**, 6090 (1969).
47. M. Tsutsui and H. Zeiss, *J. Amer. Chem. Soc.* **82**, 6255 (1960).
48. S. N. Uskahov and O. F. Solomon, *Bull. Acad. Sci. USSR* p. 593 (1954).
49. G. Oehme and H. Pracejus, *Z. Chem.* **9**, 140 (1969).
50. G. Wittig and S. Fischer, *Chem. Ber.* **105**, 3544 (1972).
51. G. Wittig and P. Fritze, *Justus Liebigs Ann. Chem.* **712**, 79 (1968).
52. H. Yamazaki and N. Hagiwara, *Kogyo Kagaku Zasshi* **61**, 21 (1958); *Chem. Abstr.* **53**, 18885 (1959).
53. K. Yamamoto and M. Oku, *Bull. Chem. Soc. Jap.* **27**, 382 (1954).
54. A. Yamamoto, K. Marifuji, S. Ikeda, T. Saito, Y. Uchida, and A. Misono, *J. Amer. Chem. Soc.* **87**, 4652 (1965).
55. L. Porri, M. C. Gallazzi, and G. Vitulli, *J. Polym. Sci.*, *Part B* **5**, 629 (1967).
56. M. Englert, Ph.D. Thesis, University of Bochum, 1971.
57. J. E. van den Enk and H. J. van der Ploeg, *J. Polym. Sic.*, *Part A-1* **9**, 2403 (1971).
58. R. J. de Pasquale, *J. Organometal. Chem.* **32**, 381 (1971).
59. V. D. Azatyan, *Dokl. Akad. Nauk SSSR* **98**, 403 (1958).
60. V. D. Azatyan and R. S. Gyuli-Kevkhyan, *Izv. Akad. Nauk Arm. SSR, Khim. Nauki* **14**, 451 (1961); *Chem. Abstr.* **58**, 3327 (1963).
61. H. Baddenhausen, H. Götte, and L. Wiesner, *Angew. Chem.* **71**, 444 (1959).
62. W. E. Daniels, *J. Org. Chem.* **29**, 2936 (1964).
63. G. Schröder, "Cyclooctatetraen," p. 70. Verlag Chemie, Weinheim, 1965.
64. R. J. Fredericks, D. G. Lynch, and W. E. Daniels, *J. Polym. Sci.*, *Part B* **2**, 803 (1964).
65. L. B. Luttinger and E. C. Colthup, *J. Org. Chem.* **27**, 3752 (1962).
66. G. A. Chukhadzhyan and Z. I. Abramyan, *Arm. Khim. Zh.* **14**, 329 (1971); *Chem, Abstr.* **75**, 77372 (1971).
67. G. A. Chukhadzhyan, Z. I. Abramyan, and G. A. Gevorkyan, *Arm. Khim. Zh.* **23**. 89 (1970); *Chem. Abstr.* **73**, 133277 (1970).
68. C. W. Bird and J. Hudec, *Chem. Ind. (London)* p. 570 (1959).
69. P. Chini, G. de Venuto, T. Salvatori, and M. de Maldé, *Chim. Ind. (Milan)* **46**, 1049 (1964).
70. T. Okamoto, K. Takagi, Y. Sakakibara, and S. Kunichika, *Bull. Inst. Chem. Res., Kyoto Univ.* **48**, 96 (1970).
71. F. Guerrieri and G. P. Chiusoli, *Chem. Commun.* p. 781 (1967).
72. M. Dubini, F. Montino, and G. P. Chiusoli, *Chim. Ind. (Milan)* **47**, 839 (1965).
73. J. C. Sauer and T. L. Cairns, *J. Amer. Chem. Soc.* **79**, 2659 (1957).
74. R. E. Benson and R. V. Lindsey, *J. Amer. Chem. Soc.* **81**, 4250 (1959).
75. T. I. Bieber, *Chem. Ind. (London)* p. 1126 (1957).
76. T. L. Cairns, V. A. Engelhardt, H. L. Jackson, G. H. Kalb, and J. C. Sauer, *J. Amer. Chem. Soc.* **74**, 5636 (1952).

77. A. J. Chalk, *J. Amer. Chem. Soc.* **94**, 5928 (1972).
78. A. C. Cope and H. C. Campbell, *J. Amer. Chem. Soc.* **73**, 3536 (1951).
79. A. C. Cope and H. C. Campbell, *J. Amer. Chem. Soc.* **74**, 179 (1952).
80. A. C. Cope and D. S. Smith, *J. Amer. Chem. Soc.* **74**, 5136 (1952).
81. A. C. Cope and D. F. Rugen, *J. Amer. Chem. Soc.* **75**, 3215 (1952).
82. A. C. Cope and R. M. Pike, *J. Amer. Chem. Soc.* **75**, 3220 (1953).
83. P. Heimbach, K.-J. Ploner, and F. Thömel, *Angew. Chem.* **83**, 285 (1971).
84. F. Thömel, Ph.D. Thesis, University of Bochum, 1970.
85. K.-J. Ploner, Ph.D. Thesis, University of Bochum, 1969.
86. F. W. Hoover, O. W. Webster, and C. T. Handy, *J. Org. Chem.* **26**, 2234 (1961).
87. S. Isaoka, K. Kogami, and J. Kumanotani, *Makromol. Chem.* **135**, 1 (1970).
88. D. R. Fahey, *J. Org. Chem.* **37**, 4471 (1972).
89. G. A. Chukhadzhyan, E. L. Sarkisyan, and T. S. Elbakyan, *J. Org. Chem. USSR* **8**, 1133 (1972).
90. G. A. Chukhadzhyan and Z. K. Evoyan, *Arm. Khim. Zh.* **24**, 530 (1971); *Chem. Abstr.* **75**, 118393 (1971).
91. G. A. Chukhadzhyan, Z. K. Evoyan, and G. A. Kevorkian, *Abstr., Int. Conf. Organometal. Chem., 5th, 1971*, Vol. 2, p. 187 (1971).
92. G. M. Whitesides and W. J. Ehmann, *J. Amer. Chem. Soc.* **91**, 3800 (1969).
93. H. C. Longuett-Higgins and L. E. Orgel, *J. Chem. Soc., London* p. 1969 (1956).
94. F. D. Mango and J. H. Schachtschneider, *J. Amer. Chem. Soc.* **91**, 1030 (1969).
95. G. N. Schrauzer, P. Glockner, and S. Eichler, *Angew. Chem.* **76**, 28 (1964).
96. B. Büssemeier, Ph.D. Thesis, University of Bochum, 1973.
97. J. P. Collman, J. W. Kang, W. F. Little, and M. F. Sullivan, *Inorg. Chem.* **7**, 1298 (1968).
98. K. Moseley and P. M. Maitlis, *Chem. Commun.* p. 1604 (1971).
99. J. Browning, M. Green, B. R. Penfold, J. L. Spencer, and F. G. A. Stone, *Chem. Commun.* p. 31 (1973).
100. E. A. Reinsch, *Theor. Chim. Acta* **11**, 296 (1968).
101. E. A. Reinsch, *Theor. Chim. Acta* **17**, 309 (1970).
102. R. M. MacDonald and P. M. Maitlis, unpublished results, quoted in Maitlis (106).
103. G. Herrman, Ph.D. Thesis, Technische Hochschule Aachen, 1963.
104. G. Booth and J. M. Rowe, *Chem. Ind. (London)* p. 661 (1960).
105. L. S. Meriwether, E. C. Colthup, and G. W. Kennerly, *J. Org. Chem.* **26**, 5163 (1961).
106. P. M. Maitlis, *Pure Appl. Chem.* **30**, 427 (1972).
107. M. L. H. Green, T. Saito, and P. J. Tanfield, *J. Chem. Soc., A* p. 152 (1971).
108. R. Willstätter and E. Waser, *Chem. Ber.* **44** 3423 (1911).
109. A. C. Cope and C. G. Overberger, *J. Amer. Chem. Soc.* **70**, 1433 (1948).
110. F. D. Mango and J. H. Schachtschneider, *J. Amer. Chem. Soc.* **89**, 2484 (1967).
111. G. A. Chukhadzhyan, Z. I. Abramyan, V. G. Grigoryan, and D. V. Avetisyan, *Arm. Khim. Zh.* **23**, 860 (1970); *Chem. Abstr.* **74**, 54206 (1971).
112. G. A. Chukhadzhyan and Z. I. Abramyan, *Gomogen Katal.* p. 217 (1970); *Chem. Abstr.* **77**, 5857 (1972).
113. R. E. Benson and R. V. Lindsey, *J. Amer. Chem. Soc.* **81**, 4247 (1959).
114. F. W. Hoover and R. V. Lindsey, *J. Org. Chem.* **34**, 3051 (1969).
115. R. J. de Pasquale, *J. Organometal. Chem.* **32**, 381 (1971).
116. M. Englert, P. W. Jolly, and G. Wilke, *Angew. Chem.* **83**, 84 (1971); **84**, 120 (1972); M. Englert, Ph.D. Thesis, University of Bochum, 1971.
117. S. Otsuka, A. Nakamura, S. Ueda, and K. Tani, *Chem. Commun.* p. 863 (1971),

118. S. Otsuka, A. Nakamura, S. Ueda, and H. Minamida, *Kogyo Kagaku Zasshi* **72**, 1809 (1969).

119. S. Otsuka, K. Mori, T. Suminoe, and F. Imaizumi, *Eur. Polym. J.* **3**, 73 (1967).

120. S. Otsuka, A. Nakamura, T. Yamagata, and K. Tani, *J. Amer. Chem. Soc.* **94**. 1037 (1972).

121. S. Otsuka, K. Mori, and F. Imaizumi, *J. Amer. Chem. Soc.* **87**, 3017 (1965).

122. H. Tadokoro, Y. Takahashi, O. Otsuka, K. Mori, and F. Imaizumi, *J. Polym. Part B* **3**, 697 (1965).

123. H. Tadokoro, M. Kobayashi, K. Mori, Y. Takahashi, and S. Taniyama, *J. Polym. Sci., Part C* **22**, 1031 (1967).

124. W. P. Baker, *J. Polym. Sci., Part A* **1**, 655 (1963).

125. L. Porri, M. C. Gallazzi, and G. Vitulli, *J. Polym. Sci., Part B* **5**, 629 (1967).

126. S. Otsuka, A. Nakamura, K. Tani, and S. Ueda, *Tetrahedron Lett.* p. 297 (1969).

127. S. Kunichika, Y. Sakakibara, and T. Okamoto, *Kogyo Kagaku Zasshi* **72**, 1814 (1969).

128. J. E. van den Enk and H. J. van der Ploeg, *J. Polym. Sci., Part A-1* **9**, 2395 (1971).

129. G. O. Evans, C. U. Pittman, R. McMillan, R. T. Beach, and R. Jones, *J. Organometal. Chem.* **67**, 295 (1974).

130. R. Baker, B. N. Blackett, and R. C. Cookson, *Chem. Commun.* p. 802 (1972).

131. For references to the thermal oligomerization of allene, see B. Weinstein and A. H. Fenselau, *J. Chem. Soc., C* p. 368 (1967); *J. Org. Chem.* **32**, 2278 and 2988 (1971).

132. B. L. Barnett, C. Krüger, and Y.-H. Tsay, *Angew. Chem.* **84**, 121 (1972).

133. F. Wagner and H. Meier, *Tetrahedron* **30**, 773 (1974).

134. H. Selbeck, Ph.D. Thesis, University of Bochum, 1972.

135. L. B. Luttinger, *J. Org. Chem.* **27**, 1591 (1962).

136. T. Okamoto, K. Takagi, Y. Sakakibara, and S. Kunichika, *Bull. Inst. Chem. Res., Kyoto Univ.* **48**, 96 (1970).

137. L. Porri, R. Rossi, and G. Ingrosso, *Tetrahedron Lett.* p. 1083 (1971).

138. C. Dixon, E. W. Duck, and D. K. Jenkins, *Organometal. Chem. Syn.* **1**, 77 (1970).

139. F. D. Mango, *Chem. Technol.* p. 758 (1971).

140. M. Foa and L. Cassar, *Gazz. Chim. Ital.* **102**, 85 (1972).

141. Z. I. Abramyan and G. A. Chukhadzhyan, *Vysokomol. Soedin., Ser. B.* **14**, 489 (1972); *Chem. Abstr.* **77**, 140609 (1972).

142. C. Simionescu, T. Lixandru, I. Negulescu, I. Mazilu, and L. Tataru, *Makromol. Chem.* **163**, 59 (1973).

143. G. Ingrosso, M. Iqbal, R. Rossi, and L. Porri, *Chim. Ind. (Milan)* **55**, 540 (1973).

144. S. Otsuka, K. Tani, and T. Yamagata, *J. Chem. Soc., Dalton Trans.* p. 2491 (1973).

145. G. A. Chukhadzhyan, V. G. Tovmasyan, and Z. I. Abramyan, *Vysokomol. Soedin., Ser. B* **14**, 795 (1972); *Chem. Abstr.* **78**, 124970 (1973).

146. G. A. Chukhadzhyan, Z. I. Abramyan, and M. A. Manukyan, *Khim. Atsetilina, Tr. Vses. Konf., 3rd, 1968* p. 255 (1972); *Chem. Abstr.* **79**, 53869 (1973).

147. Y. Kiso, M. Kumada, K. Tamao, and M. Umeno, *J. Organometal. Chem.* **50**, 297 (1973).

148. M. F. Lappert and S. Takahashi, *Chem. Commun.* p. 1272 (1972).

149. H. Okinoshima, K. Yamamoto, and M. Kumada, *J. Amer. Chem. Soc.* **94**, 9263 (1972).

150. Y. Kiso, K. Tamao, and M. Kumada, *Chem. Commun.* p. 1208 (1972).

151. J. G. Duboudin and B. Jousseaume, *J. Organometal. Chem.* **44**, C1 (1972).

152. J. G. Duboudin and B. Jousseaume, *C.R. Acad. Sci., Ser. C* **276**, 1421 (1973).

153. R. Baker and A. H. Cook, *Chem. Commun.* p. 472 (1973).

154. R. Baker, *Chem. Rev.* **73**, 487 (1973) (see p. 529).
155. J. Browning, M. Green, J. L. Spencer, and F. G. A. Stone, *J. Chem. Soc., Dalton Trans.* p. 97 (1974).
156. K. H. Scholz, Ph.D. Thesis, University of Bochum, 1974.
157. G. A. Chukhadzhyan, Z. I. Abramyan, and E. A. Gevorkyan, *Zh. Obsch. Khim.* **43**, 2012 (1973); *Chem. Abstr.* **80**, 36401 (1974).
158. G. N. Bondarenko, E. M. Kharkova, E. A. Mushina, M. P. Teterina, and B. A. Krentsel, *Bull. Acad. Sci. USSR* p. 1184 (1973).
159. M. L. H. Green and M. J. Smith, *J. Chem. Soc., Dalton Trans.* p. 639 (1971).
160. H. W. Sternberg, R. Markby, and I. Wender, *J. Amer. Chem. Soc.* **82**, 3638 (1960).
161. L. G. Grigor'eva, V. A. Sergeev, V. K. Shitikov, and V. V. Korshak, *Bull. Acad. Sci. USSR* p. 1446 (1973).
162. G. A. Chukhadzhyan, A. E. Kalaidzhyan, and E. M. Nazaryan, *Vysokomol. Soedin., Ser. B* **15**, 715 (1973); *Chem. Abstr.* **80**, 37495 (1974).
163. G. A. Chukhadzhyan, E. L. Sarkisyan, and I. M. Rostomyan, *Vysokomol. Soedin., Ser. B* **15**, 203 (1973); *Chem. Abstr.* **79**, 66875 (1973).
164. G. A. Chukhadzhyan, E. L. Sarkisyan, and T. S. Elbakyan, *Zh. Obshch. Khim.* **43**, 2302 (1973); *Chem. Abstr.* **80**, 70486 (1974).
165. W. Reppe, "Chemie und Technik der Acetylen-Druckreaktionen," Verlag Chemie, Weinheim, 1952.
166. K. G. Allum, R. D. Hancock, S. McKenzie, and R. C. Pitkethly, *Catal. Proc. Int. Congr. 5th* **1**, 477 (1972).
167. L. A. Akopyan, S. G. Grigoryan, G. A. Chukhadzhyan, and S. G. Matsoyan, *J. Org. Chem. USSR* **9**, 2020 (1973).
168. P. Bicev, A. Furlani, and G. Sartori, *Gazz. Chim. Ital.* **103**, 849 (1973).
169. G. A. Chukhadzhyan, G. A. Gevorkyan, and Z. I. Abramyan, *Arm. Khim. Zh.* **27**, 355 (1974).
170. M. F. Lappert, T. A. Nile, and S. Takahashi, *J. Organometal. Chem.* **72**, 425 (1974).

Patents

P1. American Cyanamid Co., Brit. Patent 897,099 (1962); *Chem. Abstr.* **57**, 11387 (1963).
P2. L. B. Luttinger (American Cyanamid Co.), U.S. Patent 3,131,155 (1964); Conc. Brit. Patent 889,730 (1962); *Chem. Abstr.* **58**, 5513 (1963).
P3. L. S. Meriwether (American Cyanamid Co.), U.S. Patent 2,961,330 (1960); *Chem. Abstr.* **55**, 7907 (1961).
P4. G. W. Kennerly, D. S. Hoffenberg, and J. S. Noland (American Cyanamid Co.), U.S. Patent 3,092,613 (1963); *Chem. Abstr.* **59**, 5280 (1963).
P5. L. B. Luttinger (American Cyanamid Co.), U.S. Patent 3,174,956 (1965); *Chem. Abstr.* **62**, 11934 (1965).
P6. L. S. Meriwether and E. C. Colthup (American Cyanamid Co.), U.S. Patent 2,998,463 (1961); *Chem. Abstr.* **56**, 595 (1962).
P7. L. S. Meriwether and M. L. Fiene (American Cyanamid Co.), U.S. Patent 3,051,694 (1962); *Chem. Abstr.* **58**, 4423 (1963).
P8. L. S. Meriwether and E. C. Colthup (American Cyanamid Co.), U.S. Patent 3,066,119 (1962); *Chem. Abstr.* **59**, 11685 (1963).
P9. L. S. Meriwether (American Cyanamid Co.), U.S. Patent 3,117,952 (1964); *Chem. Abstr.* **61**, 13444 (1964).
P10. J. R. Leto and M. L. Fiene (American Cyanamid Co.), U.S. Patent 3,076,016 (1963); *Chem. Abstr.* **59**, 6276 (1963).

P11. K. G. Allum and R. D. Hancock (British Petroleum Co. Ltd.), U.S. Patent 3,658,884 (1972); Conc. Ger. Offen. 2,022,488 (1971); *Chem. Abstr.* **74**, 126337 (1971).

P12. T. Fujisaki and T. Ogawara (Board Industrial Technol.), Jap. Patent 5077('56) (1956); *Chem. Abstr.* **52**, 11886 (1958).

P13. T. Fujisaki and T. Ogawara (Bureau of Industrial Technics), Jap. Patent 4329('52) (1952); *Chem. Abstr.* **48**, 5214 (1954).

P14. D. V. N. Hardy (British Oxygen Co. Ltd.), Brit. Patent 706,629 (1954); *Chem. Abstr.* **49**, 9032 (1955).

P15. N. von Kutepow, C. Berding, and W. Pfab (B.A.S.F.), Ger. Patent 1,102,141 (1961); *Chem. Abstr.* **56**, 10000 (1962).

P16. H. Pirzer, R. Stadler, and F. Becke (B.A.S.F.), Ger. Patent 1,138,763 (1962); *Chem. Abstr.* **58**, 10104 (1963).

P17. N. von Kutepow and H. Reis (B.A.S.F.), Ger. Patent 1,138,762 (1962); *Chem. Abstr.* **58**, 10104 (1963).

P18. W. Reppe, W. Pfab, N. von Kutepow, and W. Büche (B.A.S.F.), Ger. Patent 1,029,369 (1958); *Chem. Abstr.* **54**, 22538 (1960).

P19. W. Reppe and T. Toepel (B.A.S.F.), Ger. Patent 859,464 (1952); *Chem. Abstr.* **50**, 7852 (1956).

P20. W. Reppe, W. Schweckendiek, A. Magin, and K. Klager (B.A.S.F.), Ger. Patent 805,692 (1951); *Chem. Abstr.* **47**, 602 (1953).

P21. H. Pirzer and F. Beck (B.A.S.F.), Ger. Patent 1,092,908 (1960); *Chem. Abstr.* **56**, 3375 (1962).

P22. N. von Kutepow and F. Meier (B.A.S.F.), Ger. Patent 1,159,951 (1963); *Chem. Abstr.* **60**, 9198 (1964).

P23. W. Reppe, F. Reicheneder, K. Dury, and H. Suter (B.A.S.F.), Ger. Patent 1,019,297 (1957); *Chem. Abstr.* **54**, 1365 (1960).

P24. W. Reppe, O. Schlichting, and H. Schweter (B.A.S.F.), Ger. Patent 1,039,059 (1958); *Chem. Abstr.* **54**, 22538 (1960).

P25. R. F. Kleinschmidt (General Anilin and Film Corp.), U.S. Patent 2,542,417 (1951); *Chem. Abstr.* **45**, 7594 (1951).

P26. C. E. Barnes (General Anilin and Film Corp.), U.S. Patent 2,579,106 (1951); *Chem. Abstr.* **46**, 6671 (1952).

P27. H. Mori, H. Ikeda, I. Nagaoka, and S. Hirayanagi (Japan Syn. Rubber Co. Ltd.), Jap. Patent 70 09,925 (1970); *Chem. Abstr.* **73**, 56605 (1970).

P28. Mitsui Chem. Ind. Co., Brit. Patent 730,038 (1955); *Chem. Abstr.* **50**, 7854 (1956).

P29. M. Tanaka and K. Yamamoto (Mitsui Chem. Ind. Co), Jap. Patent 3819('53) (1953); *Chem. Abstr.* **48**, 8821 (1954).

P30. W. A. Kornicker (Monsanto Co.), U.S. Patent 3,578,626 (1971); *Chem. Abstr.* **75**, 49875 (1971).

P31. W. A. Kornicker (Monsanto Co.), U.S. Patent 3,474,012 (1969); *Chem. Abstr.* **72**, 22145 (1970).

P32. H. Zeiss and M. Tsutsui (Monsanto Chem. Co), U.S. Patent 2,980,741 (1961); *Chem. Abstr.* **55**, 17583 (1961).

P33. F. Nagasawa, K. Matsusawa, G. Hashizume, and K. Yoshida (Mitsubishi Chem. Ind. Co.), Jap. Patent 3931('55) (1955); *Chem. Abstr.* **51**, 15561 (1957).

P34. R. A. Bafford (Pennwalt Corp.), Ger. Offen. 2,046,200 (1971); *Chem. Abstr.* **74**, 125135 (1971).

P35. J. O. van Hook and W. J. Croxall (Rohm and Haas Co.), U.S. Patent 2,613,208 (1952); *Chem. Abstr.* **48**, 2772 (1954).

P36. C. H. McKeever and J. O. van Hook (Rohm and Haas Co.), U.S. Patent 2,542,551 (1951); *Chem. Abstr.* **45**, 7591 (1951).

P37. A. J. Canale and J. F. Kincaid (Rohm and Haas Co.), U.S. Patent 2,613,231 (1952); *Chem. Abstr.* **47**, 7547 (1953).

P38. R. T. Holm, L. G. Cannell, and W. de Acetis (Shell Oil Co.), U.S. Patent 3,277,198 (1966); *Chem. Abstr.* **65**, 20004 (1966).

P39. S. Murakashi, N. Hagihara, and H. Yamazaki (Univ. of Osaka), Jap. Patent 4814('59) (1959); *Chem. Abstr.* **54**, 14155 (1960).

P40. E. C. Herrick and J. C. Sauer (E.I. du Pont), U.S. Patent 2,667,520 (1954); *Chem. Abstr.* **48**, 6162 (1954).

P41. K. Wintersberger and G. Zirger (B.A.S.F.), Ger. Patent 1,025,870 (1958); *Chem. Abstr.* **54**, 18393 (1960).

P42. F. Nagasawa and K. Yoshida (Mitsubishi Chem. Ind. Co.), Jap. Patent 2428('54) (1954); *Chem. Abstr.* **49**, 14803 (1955).

P43. H. C. Campbell and A. C. Cope (Research Corp.), U.S. Patent 2,772,314 (1956); *Chem. Abstr.* **51**, 7409 (1957).

P44. G. H. Kalb and J. C. Sauer (E.I. du Pont), U.S. Patent 2,540,736 (1951); *Chem. Abstr.* **45**, 5712 (1951).

P45. H. Friederich, W. Schweckendiek, and K. Sepp (B.A.S.F.), Ger. Patent 1,005,954 (1957).

P46. J. Chatt, F. A. Hart, and G. A. Rowe (I.C.I. Ltd), Brit Patent 882,400 (1961); *Chem. Abstr.* **57**, 12540 (1962).

P47. A. C. Cope and C. T. Handy (E.I. du Pont), U.S. Patent 2,950,334 (1960); *Chem. Abstr.* **55**, 1527 (1961).

P48. O. F. Hecht and A. P. Castaldi (General Aniline and Film Corp.), U.S. Patent 3,405,110 (1968); *Chem. Abstr.* **70**, 4806 (1969).

P49. General Aniline and Film Corp., Brit. Patent 1,021,948 (1966); *Chem. Abstr.* **64**, 17738 (1966).

P50. E. I. du Pont, Brit. Patent 812,902 (1959); *Chem. Abstr.* **54**, 7591 (1960).

P51. S. Otsuka, H. Hiroshi, T. Suminoe, F. Imaizumi, and T. Taketome (Japan Syn. Rubber Co. Ltd.), U.S. Patent 3,405,112 (1968); *Chem. Abstr.* **69**, 107279 (1968).

P52. H. Friedrich and H. Hoffmann (B.A.S.F.), Ger. Patent 1,042,572 (1959); *Chem. Abstr.* **55**, 2486 (1961).

P53. W. Reppe and A. Magin (B.A.S.F.), U.S. Patent 3,396,191 (1968); *Chem. Abstr.* **69**, 76662 (1968).

P54. K. C. Lin (Hooker Chem, Corp.), U.S. Patent 3,673,285 (1972); *Chem. Abstr.* **77**, 101890 (1972).

P55. M. Dubeck and A. H. Filbey (Ethyl Corp.), U.S. Patent 3,256,260 (1966); *Chem. Abstr.* **65**, 7307 (1966).

P56. R. E. Benson (E.I. du Pont), U.S. Patent 2,943,116 (1960); *Chem. Abstr.* **55**, 2522 (1961).

P57. R. E. Benson (E.I. du Pont), U.S. Patent 2,894,936 (1959); *Chem. Abstr.* **54**, 16011 (1960).

Reviews

A selection of general review articles relevant to the material discussed in this chapter is listed below.

R. Baker, π-Allylmetal derivatives in organic synthesis. *Chem. Rev.* **73**, 487 (1973).

C. W. Bird, "Transition Metal Intermediates in Organic Synthesis," Academic Press, New York, 1967.

M. G. Chauser, V. D. Ermakova, and M. I. Cherkashin, Structural course of the polymerization of diynes. *Russ. Chem. Rev.* **41**, 687 (1972).

C. Hoogzand and W. Hübel, Cyclic polymerization of acetylenes by metal carbonyl compounds. *Org. Syn. Metal Carbonyls* **1**, 343 (1968).

P. M. Maitlis, The oligomerization of acetylenes induced by metals of the nickel triad. *Pure Appl. Chem.* **30**, 427 (1972).

P. M. Maitlis, The palladium catalyzed cyclotrimerization of acetylenes. *Pure Appl. Chem.* **33**, 489 (1972).

V. O. Reikhsfeld and K. L. Makovetskii, The cyclization of alkynes and their derivatives, *Russ. Chem. Rev.* **35**, 510 (1966).

W. Reppe, N. von Kutepow, and A. Magin, Cyclisierung acetylenischer Verbindungen. *Angew. Chem.* **81**, 717 (1969).

G. N. Schrauzer, P. Glockner, and S. Eichler, Koordinations-Chemie und Katalyse. *Angew. Chem.* **76**, 28 (1964).

B. L. Shaw and A. J. Stringer, Transition metal-allene complexes. *Inorg. Chim. Acta Rev.* **7**, 1 (1973).

L. P. Yur'eva, Oligomerization of acetylenes on transition-metal compounds. *Usp. Khim.* **43**, 95 (1974).

CHAPTER III

The Oligomerization and Co-oligomerization of Butadiene and Substituted 1,3-Dienes

I. Introduction

The resurgence of interest in the organic chemistry of nickel that has occurred during the last 15 years is directly related to the discovery that butadiene can be cyclotrimerized to 1,5,9-cyclododecatriene by zerovalent nickel catalysts. The isolation of the intermediate in this reaction, a π-allyl-nickel complex, has given a new impulse to research both into the use of nickel in organic syntheses and into the chemistry of organonickel complexes. The scope of the oligomerization reaction has been greatly increased by allowing the butadiene to co-oligomerize with a second monomer and it is possible to prepare cyclic or linear dimers, large ring systems, or telomeric compounds. The addition of a second organometallic component to the catalyst produces systems capable of converting butadiene into macro-cyclic polyenes or polymers. The extent of the literature concerned with the nickel-catalyzed polymerization of butadiene has necessitated a separate treatment and this is to be found in Chapter IV.

It is doubtful whether a mechanistically better understood family of reactions exists in the whole of transition metal catalysis. This insight has been obtained by combining product analysis with the isolation and investi-gation of model compounds. A mass of evidence has been generated, indi-cating that the oligomerization, in particular cyclodimerization, is the result of multistep reactions and not of a one-step concerted process, even though the latter has been shown to be theoretically possible (28–30, 33, 108–114, 166). The intellectual development of these multistep mechanisms can be followed from a series of reviews (8, 15, 54, 97, 102–105, 115, 150) and the mechanisms discussed in this chapter are based on these articles.

As yet no oligomerization or co-oligomerization products of butadiene have been commercially produced by processes using nickel-containing catalysts: 1,5,9-cyclododecatriene is synthesized using titanium-based catalysts and 1,5-cyclooctadiene is not manufactured on a large scale. The situation in the latter case, however, may change with the development of new syntheses of polymers based on the metathesis of cycloolefins.

II. The Cyclotrimerization of 1,3-Dienes

A. The Cyclotrimerization of Butadiene (Table III-1)

Ligand-free* zerovalent nickel catalysts cyclotrimerize butadiene to 1,5,9-cyclododecatriene (CDT). The catalyst is conveniently prepared by reducing a nickel salt in the presence of butadiene; a standard combination is nickel acetylacetonate and an aluminum-alkyl (Table III-1, row 1). Other reducing agents that have been used include metal dispersions (P19, P23), sodium hydride (P124), lithium-alkyls (P75), Grignard reagents (48, P93), lithium aluminum-hydride (27, 69), and electric current (26, 70).

Essentially identical results are obtained by using preformed "ligand-free" zerovalent nickel complexes, e.g., $(COD)_2Ni$, $(CH_2:CHCN)_2Ni$, and $(\pi\text{-}C_3H_5)_2Ni$ (Table III-1, row 2). An active catalyst can even be generated from nickel tetracarbonyl by treating it at elevated temperature with butadiene or cyclooctadiene and venting the carbon monoxide evolved (95, P12). The activity for cyclotrimerization shown by $(\pi\text{-indenyl})_2Ni$ (P92), in contrast to nickelocene, is probably associated with its π-allyl-type structure (see Volume I, p. 432, and ref. 161).

Cyclotrimerization is also the main reaction observed in the presence of zerovalent nickel catalysts combined with certain donor ligands, e.g., arsines (3, 6, P111), stibines (P39, P68, P111), tertiary amines (P38, P43, P45, P61, P103, P111), bipyridyl (41, 55, 57, 59, P111), pyridine (3, 95), Schiff bases (55, P76, P115), carbon monoxide (P38, P45), isonitriles (P60), and certain phosphites (9, 75, P67, P68) (Table III-1, row 3). The addition of dialkyl-cyanamide or various S and N heterocyclic compounds, e.g., 1,4-thioxan, is claimed to suppress the formation of polymer (165). This class of catalyst is normally associated with the cyclodimerization of butadiene and it is an open question whether the ligand remains attached to the metal during the reaction. In some cases the activity for cyclotrimerization is lost on addition of a second equivalent of ligand. The cyclo-oligomerization of butadiene in

---

* A "ligand-free" zerovalent nickel catalyst is defined as a system in which the only ligands associated with the metal are organic groups readily displaced by the reactant.

the presence of olefins or alkynes can also, under certain conditions, produce the cyclic trimer as the main product (5, 13, 73, P111).

Other systems that convert butadiene into CDT include binary catalysts derived from π-cyclopentadienylnickel complexes, e.g., $(\pi\text{-}C_5H_5)_2\text{Ni--}$ $\text{Al}(C_2H_5)_3$; nickel tetracarbonyl, e.g., $\text{Ni}(CO)_4\text{--Al}(C_2H_5)_3$; substituted carbonyl complexes, e.g., $\text{Lig}_2\text{Ni}(CO)_2\text{--}C_4H_9\text{Li}$, and π-allyl complexes, e.g., $(\pi\text{-}C_3H_5)_2\text{Ni--}(C_2H_5)_3\text{Tl}$ or $(iso\text{-}C_4H_9)_2\text{AlH}$ (Table III-1, row 4).

In addition, CDT is observed as a byproduct in most of the oligomerization and co-oligomerization reactions involving butadiene which are discussed in the sections that follow.

The principal product of the cyclotrimerization reaction is *trans,trans,trans*-CDT (**1**). In addition, smaller amounts of the trans,trans,cis and trans,cis,cis isomers are produced (**2** and **3**) as well as cyclic dimers and higher boiling oligomers: none of the fourth CDT isomer, *cis,cis,cis*-CDT, has ever been detected.

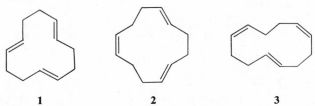

| 1 | 2 | 3 |

A careful study has been made of the effect of the reaction conditions on the product distribution using a catalyst derived from nickel acetylacetonate and diethylaluminum ethoxide. The following conclusions may be drawn (3, 97).

a. The rate of ring formation decreases with increasing conversion.

b. The proportion of the trans,trans,cis isomer depends only on the temperature of the reaction and increases with increasing temperature at the expense of the sum of the trans,trans,trans and trans,cis,cis isomers.

c. The ratio of trans,trans,trans isomer to trans,cis,cis isomer mainly depends on the butadiene concentration. With increasing temperature the proportion of trans,cis,cis isomer increases at the expense of the trans,trans, trans isomer.

d. The proportion of C_8 hydrocarbons produced increases on raising the temperature of the reaction.

The decrease in the rate with increasing conversion indicates that the reaction products, probably the side product cyclooctadiene in particular, compete successfully with butadiene in complexing to the nickel.

The isomeric distribution can also be influenced, within a limited range, by various additives (Table III-2).

A useful review of the chemistry of CDT is to be found in reference 106.

TABLE III-1

THE CYCLOTRIMERIZATION OF BUTADIENE

Catalyst	Product^a				Ref.^{b}
	CDT (%)	COD (%)	VCH (%)	Misc. (%)	
1. Ni(acac)$_2$–(C$_2$H$_5$)$_2$AlOC$_2$H$_5$	93.5	1.5	1.5	3.5	3, 4, 10, 26, 27, 48, 51, 54, 55, 69, 70, 165, P13, P19, P42, P43, P47–P49, P75, P77, P93, P102, P109, P111, P112, P118
2a. (COD)$_2$Ni	93.7	4.6	—	1.7	3, 54, 75, 77, 95, P59, P109
2b. (CH$_2$:CHCN)$_2$Ni	80.4	7.1	6.0	6.5	9, 35, 36, P68, P98
2c. (π-C$_3$H$_5$)$_2$Ni	86.8	13.1	Trace	—	38, 52, 53, 58, 98, 99
2d. Ni(CO)$_4$	69	23	?	?	95, P12
2e. Ni(atomic)	45	55	—	—	174
3a. Ni(acac)$_2$–(C$_2$H$_5$)$_2$AlOC$_2$H$_5$–As(C$_6$H$_5$)$_3$	80.5	10.6	5.3	3.6	3, 6, P111
3b. Ni(CO)$_4$–N(C$_2$H$_5$)$_3$	55	20	?	?	P61
3c. (C$_2$H$_5$NC)$_2$Ni	69.5	16	14.5	—	P60
3d. Ni(acac)$_2$–(C$_2$H$_5$)$_2$AlOC$_2$H$_5$–py	81.6	14.2	—	1,1	3
3e. [Ni pyridine complex CH:NCH$_3$]–Al(C$_2$H$_5$)$_3$	87.5	6.7	4.9	1.3	55, P76, P115
4a. (π-C$_5$H$_5$)Ni$^+$B(C$_6$H$_5$)$_4$$^-$–Al(C$_2H_5$)$_3$	48.8	29.8	12.6	13.8	45, 46
4b. Ni(CO)$_4$–Al(C$_2$H$_5$)$_3$	54.5	28.8	13.8	—	38, 87, P28
4c. [(C$_6$H$_5$)$_3$P]$_2$Ni(CO)$_2$–C$_4$H$_9$Li	54	37	7	—	P26
4d. (π-C$_3$H$_5$)$_2$Ni–(C$_2$H$_5$)$_3$Tl	92	Trace	—	8	101, P88

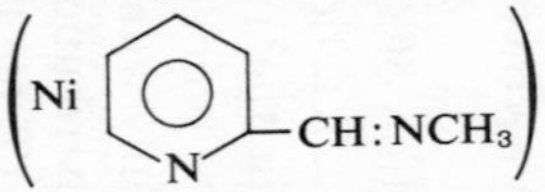

^{a} CDT = mixture of *trans,trans,trans*-, *trans,trans,cis*-, and *trans,cis,cis*-1,5,9-cyclododecatriene; COD = 1,5-cyclooctadiene; VCH = 4-vinylcyclohexene.

^{b} The references refer not only to the catalyst quoted in the table but also to related systems.

TABLE III-2

THE EFFECT OF ADDITIVES ON THE ISOMER DISTRIBUTION IN THE
CYCLOTRIMERIZATION OF BUTADIENE[a]

Catalyst	*trans,trans, trans*-CDT (%)	*trans,trans, cis*-CDT (%)	*trans,cis, cis*-CDT (%)
$(COD)_2Ni$	84.7	4.9	10.4
$(COD)_2Ni$–py	94.7	4.9	0.2
$Ni(acac)_2$–$(C_2H_5)_2AlOC_2H_5$–$As(C_6H_5)_3$	88.4	11.6	0

[a] From refs. 3, 6, P111.

B. The Cyclotrimerization of Substituted 1,3-Dienes*

The cyclotrimerization of 1,3-dienes other than butadiene has received relatively little attention. Initial results, which indicated that perdeutero-butadiene cyclotrimerizes much less readily than butadiene, have been disproved by competition experiments using a mixture of d_6-butadiene and d_0-butadiene: both dienes react at practically the same rate (100; see also 38).

Alkyl-substituted dienes only reluctantly take part in cyclo-oligomerization reactions—this is probably the result of the increased stability of the inter-mediate alkyl substituted π-allyl systems—and in the case of isoprene the rate of reaction has been shown to be three times slower than that for buta-diene. Both cyclo-co-oligomerization with butadiene and cyclotrimerization of the substituted 1,3-diene alone have been investigated.

The principal cotrimers formed in the reaction of isoprene with butadiene are 1-methyl-CDT and, to a lesser extent, dimethyl-CDT. The 1-methyl-CDT is formed as a mixture of isomers the distribution of which parallels that observed in CDT itself; 1-methyl-*trans,trans,trans*-CDT is the main com-ponent and the various 1-methyl-*trans,trans,cis* and 1-methyl-*trans,cis,cis* isomers are formed in smaller amounts. 1-Methyl-CDT has also been isolated from the reaction of *cis*-1,2-divinylcyclobutane with isoprene. 2,3-Dimethyl-butadiene is less reactive than isoprene: the main co-oligomers formed with butadiene are 1,2-dimethyl-*trans,trans,trans*-CDT and 1,2-dimethyl-*trans, trans,cis*-CDT. Only traces of the co-oligomer 3-methyl-CDT have been isolated from the reaction involving piperylene (8, 11, 15, 81, 96, 102, P104).

* Both methylene insertion and gas chromatographic techniques have been used to identify the various methyl-substituted dimers and trimers, and the reader is referred to reference 102 for a detailed account.

1,2-Dimethylenecyclohexane co-oligomerizes readily with butadiene giving a mixture of the two bicyclic trienes 4 and 5 (81). For completeness it should

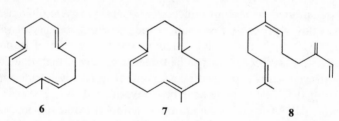

be mentioned that a patent describes the preparation of 3-(1-methyl-2-propenyl)-cyclododecatriene by co-oligomerization of butadiene and 3-methyl-1,4,6-heptatriene using a catalyst consisting of a mixture of salts of cobalt and nickel or iron and nickel combined with a trialkylaluminum (P63).

A recent significant development is the co-oligomerization of butadiene with dialkylidenehydrazine derivatives ($R_2C:NN:CR_2$) which leads to 1,2-diaza-1,5,9-cyclododecatriene derivatives (P135).

The study of the cyclotrimerization of substituted 1,3-dienes alone has apparently been limited to isoprene and piperylene. Isoprene is not cyclotrimerized by the "ligand-free" zerovalent nickel catalysts (55); it is claimed in patents that cyclodimerization occurs (P48, P96). However, use of the "nickel–ligand" catalyst results in both dimerization and trimerization (65, 78, P111). The trimeric product has been shown to consist mainly of 1,5,9-trimethyl-CDT (6), 1,5,10-trimethyl-CDT (7), and the linear trimer farnessene

(8). The distribution of the products is controlled by the ligand attached to the nickel; particularly effective for producing the CDT derivatives is a catalyst containing trishydroxymethylpropane phosphite, and the highest yield of linear trimer is obtained in the presence of triphenylphosphine (the formation of the dimers is discussed in Section III,B). The reaction of piperylene with a nickel–ligand system produces almost exclusively dimers (78; see also 169).

C. Mechanistic Considerations

Insight into the mechanism of the cyclotrimerization of butadiene was obtained with the isolation of the key intermediate, an α,ω-bis(π-allyl)-C_{12}-nickel complex (9) formed by the reaction of three butadiene molecules at a nickel atom. Further reaction of this intermediate with butadiene leads to ring closure to give CDT and is accompanied by the regeneration of **9** probably via an intermediate π-allyl-C_8-nickel system (**10**).

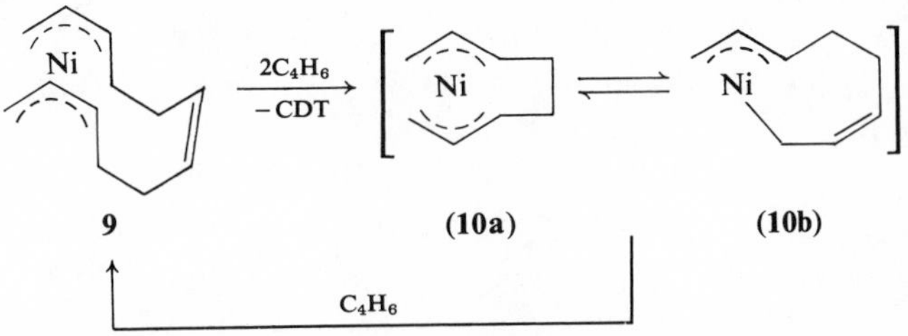

The catalytic reaction may be simulated in a stoichiometric manner by treating **9** with a ligand such as $P(C_2H_5)_3$; ring closure first occurs, to give a (CDT)Ni–Lig complex (**11**) and this is followed by the displacement of the CDT with the formation of a tetrakisligand nickel system (3, 52, 99).

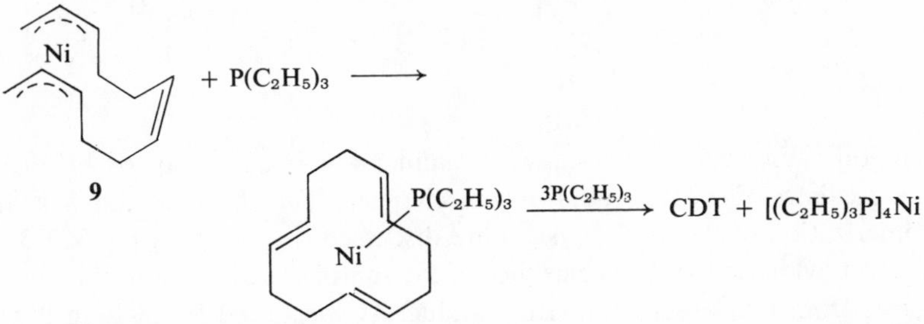

The bis(π-allyl)-C_{12}-Ni complex **9** is stable at room temperature (in the absence of butadiene) and the configuration of the C_{12} chain has been investigated by NMR techniques (3, 97). Although 12 isomers are possible, **12–17** (formal rotation of one π-allyl group relative to the other gives a further six— this is illustrated for **18**, which is derived from **12**), the NMR spectrum indicates that only two are present in solution, both of which contain an uncomplexed trans double bond. One of these two isomers has been assigned with some certainty the cis,anti,anti structure **12** and it is assumed that the second is the trans form **13**. It is debatable whether these two isomers represent the final step in the catalytic reaction because they seem predestined to produce *trans,cis,cis*-CDT, which normally accounts for less than 10% of the product. It is conceivable that under the influence of further butadiene isomerization of the chain occurs, and that *trans,trans,trans*-CDT is formed by coupling of the isomers having a syn,syn arrangement of the π-allyl groups (i.e., **16** and **17**) and *trans,trans,cis*-CDT from the isomers having a syn,anti configuration (i.e., **14** and **15**). It is of course not improbable that the coupling

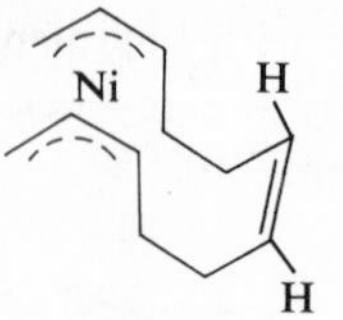

12(=9) 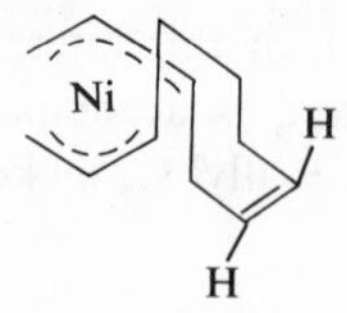   13 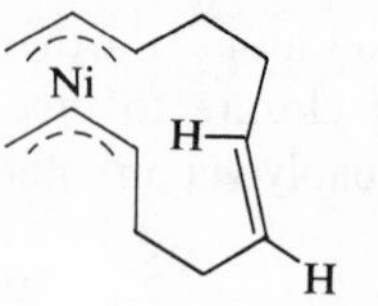   14

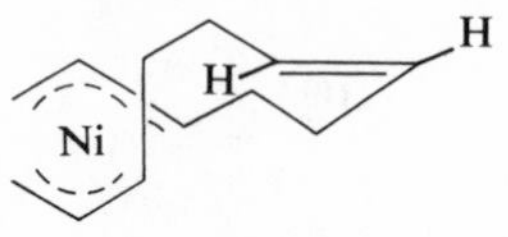

15 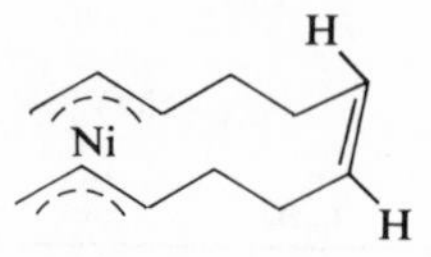   16 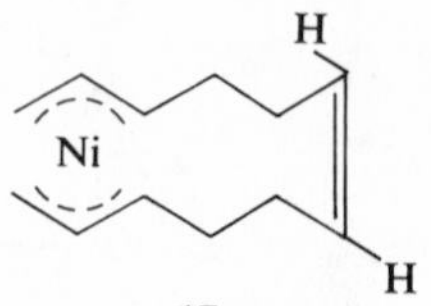   17

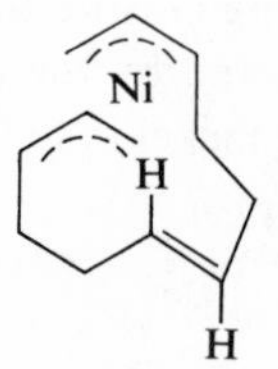

18 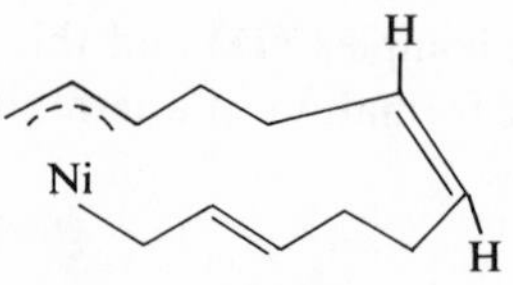   19

reaction itself occurs in a stepwise manner involving mono- or bis-σ-allyl intermediates (e.g., **19**). (Various mechanisms for the σ–π and syn–anti isomerization of the π-allyl system are discussed in Volume I, pp. 360–362.) Indirect evidence for the occurrence of syn–anti isomerization in this system comes from two observations: the adduct **11**, produced by treating **9** with triethylphosphine, contains *trans,trans,trans*- and not *trans,cis,cis*-CDT, while the reaction of **9** with excess triphenylphosphine gives, besides $[(C_6H_5)_3P]_4Ni$, a mixture of *trans,trans,trans*-, *trans,trans,cis*-, and *trans,cis, cis*- CDT in the ratio 94.4:5.0:0.6 (similar results are obtained using COD as ligand.) Syn–anti isomerization also accounts for the increase in the proportion of *trans,trans,trans*-CDT on addition of pyridine to the reaction mixture (Table II-2), pyridine being known to facilitate the σ–π interconversion in π-allyl complexes (Volume I, p. 342).

The reaction sequence leading to the generation of the C_{12} chain has been less well elucidated. However, it is probable that this reaction also occurs in a stepwise manner, the initial formation of nickel-olefin complexes (e.g., **20–23**) being followed by the coupling of two molecules of butadiene to give the π-allyl-C_8-nickel species (**10**). The trans configuration of the uncomplexed double bond in the bis(π-allyl)-C_{12} intermediate **9**, as well as the complete absence of *cis,cis,cis*-CDT among the cyclotrimerization products (even though this isomer forms the most stable complex with nickel—see Volume I, p. 252), suggests that the π-allyl-C_8-nickel intermediate from which **9** is formed has a structure in which at least one of the π-allyl groups adopts a

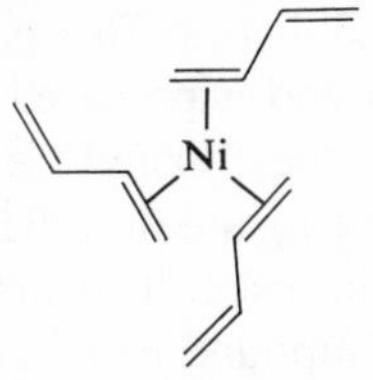

$$\underset{\mathbf{20}}{} \qquad \underset{\mathbf{21}}{} \qquad \underset{\mathbf{22}}{} \qquad \underset{\mathbf{23}}{}$$

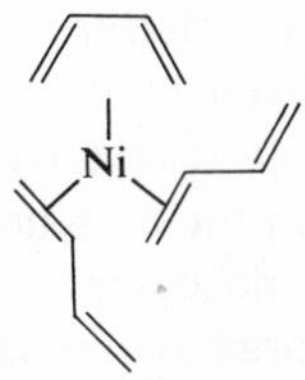

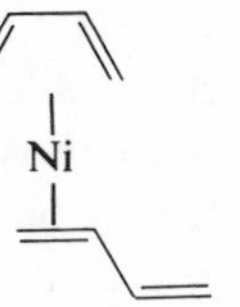

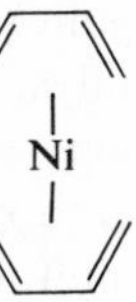

syn configuration, i.e., **10a** or **10b**. It may be surmised that further reaction occurs through the π-allyl,σ-allyl form whereby the reacting butadiene molecule adopts a single-cis configuration and inserts exclusively into the π-allylnickel bond (102). Phosphine stabilized analogs to **10a** and **10b** have been characterized and are discussed in Section III,C. [The volatile product formed in traces during the reaction of butadiene with atomic nickel at $-78°$ and originally identified as $(C_4H_6)_2Ni$ has recently been shown to be bis(π-crotyl)nickel (87).]

One piece of evidence conflicting with the above mechanism is the observation that protonolysis of **9** with 1 equivalent of HCl produces the syn-substituted π-allylnickel chloride species **24** (3): however, it is conceivable that the kinetically preferred anti isomer is formed initially and isomerizes to the thermodynamically preferred product [such behavior has been well documented for a series of π-crotylnickel complexes (143)].

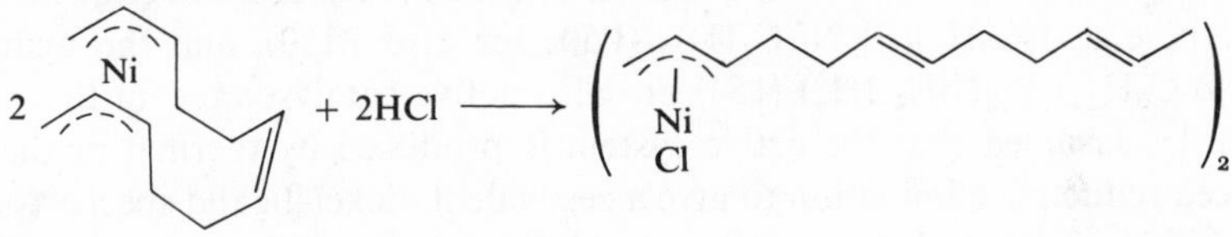

$$2 \quad \underset{\mathbf{9}}{} \quad + \text{ 2HCl} \longrightarrow \left(\underset{\mathbf{24}}{} \right)_2$$

Although the general outline of the mechanism formulated above is probably acceptable, alternative proposals can be (and have been) made for the detailed course of reaction. It has been suggested that the ratio of single-cis butadiene:single-trans butadiene in the free diolefin may have a direct influence on the isomeric distribution of the products. For instance, it is possible that the bis(π-allyl)-C_{12} isomers **12–17** are produced directly by insertion into **10a** and **10b** of single-trans or single-cis butadiene molecules. However, this concept is not easy to verify because not only can the butadiene isomerize on complexation but also the intermediates produced will differ in their reactivity (97).

III. The Cyclodimerization of 1,3-Dienes

The catalysts described in Section II as being active in the cyclotrimerization of 1,3-dienes can be modified by adding a ligand (normally a phosphine or phosphite) to give a system capable of cyclodimerizing 1,3-dienes. This type of catalyst is frequently called a "nickel–ligand" catalyst and is prepared in reactions analogous to those shown in Table III-1 for the "ligand-free" system: a nickel salt may be reduced in the presence of the ligand and diene or a preformed zerovalent nickel–ligand complex may be used. The most commonly used reducing agent is an organoaluminum compound or related organometallic compounds (3, 6, 45, 56, 72, 167, P2, P11, P13, P14, P20, P24, P48–P51, P54, P73, P83–P86, P102, P105, P107, P111). Other reducing agents that have been used include metals, e.g., sodium or aluminum powder (P15, P23, P74); metal hydrides (27, P1, P11, P72, P100, P111); and electric current (17, 26, 70, P70). The most frequently used preformed nickel–ligand complex has been a substituted nickel-carbonyl derivative (2, 40, 49, 95, P16, P26, P27, P31–P37, P39, P41, P44, P46, P57, P58, P66–P69). Other systems are based on (acrylonitrile)nickel–ligand complexes and related systems (9, 35, 36, 61, 95, P8, P17, P65, P67, P76, P89, P99), (olefin)nickel–ligand complexes (6, 12, P105, P107, P108, P110), and tetrakisligand nickel complexes and related systems (25, 36, 92, 95, P4, P17, P29, P30, P53, P79, P91, P92, P94, P110) as well as atomic nickel (174). The activity of the substituted nickel-carbonyl complexes is increased by adding certain activators, e.g., diphenylmethane (P69); alkynes (40, P57, P58); cyclic dienes, e.g., COD, cyclopentadiene (95, P34), or isobutylene (P36): these probably assist in generating free coordination sites at the nickel. A number of nickel-dialkyls, e.g., $[(C_6H_5)_3P]_2Ni(C_4H_9)_2$ (P50; see also P130), and the hydride $[(cyclo\text{-}C_6H_{11})_3P]_2NiH(BH_4)$ (154) are also active catalysts and in this case it can be assumed that the active system is produced by thermal or diene-induced reductive elimination to give a zerovalent nickel–ligand species (both possibilities are well documented and are discussed in Volume I, p. 206). In addition, it is reported in the patent literature that the cyclodimerization is catalyzed by nickelocene (P95), a nickel formate–triphenylphosphine mixture (P13) and nickel cyanide (P52). The catalytic activity of the nickel(1+) complex $[(C_6H_5)_3P]_3NiCl$ in the presence of benzaldehyde or ethylene oxide is probably the result of the formation of a zerovalent nickel species by disproportionation (117). Cyclic dimers are also produced in varying amounts during the cyclotrimerization and co-oligomerization reactions and in certain cases these can be the main products (see Section II).

A. The Cyclodimerization of Butadiene (*Table III-3*)

The catalysts described above cyclodimerize butadiene to *cis*-1,2-divinylcyclobutane (DVCB, **25**), 1,5-cyclooctadiene (COD, **26**) and 4-vinylcyclohexene (VCH, **27**) (the formation of a fourth dimer, 1-methylene-2-vinylcyclopentane, is discussed in Section VI).

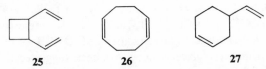

25	**26**	**27**

The distribution of the products can be influenced by varying the reaction conditions and the ligand. A particularly detailed study is reported in reference 6 from which Table III-3 has been extracted. Maximum rates of conversion are obtained with a nickel:ligand ratio of 1:1. The yield of COD and its ratio to VCH is ligand dependent and, as can be seen from Table III-3, both electronic and steric effects are involved (compare in the table examples 1, 11, 12, and 13 and 7, 8, and 9). Other compounds, in addition to phosphines and phosphites, that increase the proportion of COD in the product include pyrrole and related compounds (P66) or aromatic hydrocarbons, e.g., diphenylmethane or fluorene (P69).

Cyclooctadiene is obtained in highest yields from the system involving $P(OC_6H_4\text{-}o\text{-}C_6H_5)_3$ with a reaction temperature of 80°. If the reaction is carried out at a lower temperature and terminated before complete conversion of the butadiene has occurred, then DVCB may be isolated in up to

TABLE III-3
THE CYCLODIMERIZATION OF BUTADIENE^{a,b}

		Product distribution			
Ligand	Rate (gm BD/gm Ni/hr)	COD (%)	VCH (%)	CDT (%)	Misc. (%)
1. $P(OC_6H_4\text{-}o\text{-}C_6H_5)_3$	780	96	3.1	0.2	0.2
2. $P(OC_6H_4\text{-}2\text{-}tert\text{-}C_4H_9,4\text{-}CH_3)_3$	600	96	2.6	1.0	0.2
3. $P(OC_6H_4\text{-}2\text{-}CH_3,4\text{-}tert\text{-}C_4H_9)_3$	315	91	6.8	1.5	—
4. $P(OC_6H_4\text{-}o\text{-}CH_3)_3$	400	92	5.7	1.4	0.6
5. $P(OC_6H_4\text{-}m\text{-}CH_3)_3$	110	74	8.1	10.8	7.1
6. $P(OC_6H_4\text{-}p\text{-}CH_3)_3$	70	73	8.8	9.7	8.2
7. $P(OC_6H_4\text{-}o\text{-}OCH_3)_3$	325	90	5.8	3.7	0.3
8. $P(OC_6H_4\text{-}m\text{-}OCH_3)_3$	140	83	7.5	9.3	—
9. $P(OC_6H_4\text{-}p\text{-}OCH_3)_3$	90	78	10.1	9.6	1.8
10. $P(OC_6H_5)_2OC_6H_4\text{-}o\text{-}C_6H_5$	190	88	5.8	4.4	2.4
11. $P(OC_6H_5)_3$	100	81	7.4	9.2	2.3
12. $P(C_6H_5)_3$	180	64	27	6.0	2.8
13. $P(cyclo\text{-}C_6H_{11})_3$	35	41	40	14	4.8

^{a} Taken from ref. 6.
^{b} Ni:Lig = 1:1; Temp. = 80°.

40% yield. The yield of DVCB is limited by the facile catalytic Cope rearrangement to COD and VCH (6, 11, 12, 71, 72). An interesting observation is that this nickel-catalyzed rearrangement of DVCB is retarded by the presence of bistrimethylsilylacetylene; it has been suggested that this may be the result of the formation of a π complex the strength of which is intermediate between that of the nickel complexes formed by DVCB and butadiene (82).

Vinylcyclohexene is the main product of the thermal Diels-Alder dimerization of butadiene (116) and therefore any "catalyzed" reaction carried out at excessively high temperature and producing mainly VCH is suspect. It is perhaps this thermal reaction which has been observed in the experiments conducted in the presence of $[(C_6H_5)_3P]_2Ni(C_6H_4CH_3)_2$ at 180° (P50), $[(C_6H_5O)_3P]_4Ni-CO_2$ at 120° (25), $Ni(CO)_4-Al(C_2H_5)_3$ at 165° (P28), nickelocene at 130°, or $\pi-C_5H_5NiNO$ at 170° (P95). In the last two cases this suggestion is supported by the isolation of DVCB (presumably the trans isomer) as a minor product—*trans*-DVCB is formed in ca. 10% yield in the thermal dimerization.

B. The Cyclodimerization of Substituted 1,3-Dienes (Tables III-6 and III-7)

Much of our understanding of the mechanism of the oligomerization of 1,3-dienes at a nickel atom has been derived from a systematic study of the products formed during the oligomerization and co-oligomerization of substituted 1,3-dienes. A detailed discussion of the mechanistic implications of this work is to be found in references 102 and 150.

The cyclodimerization reaction has been extended to include *cis*- and *trans*-piperylene, isoprene, 2,3-dimethylbutadiene, 1,3-hexadiene, 1,3,5-hexatriene, 2,4,6-octatriene, and (in patents) chloroprene. The cyclo-codimerization of butadiene with a substituted 1,3-diene or of two differently substituted 1,3-dienes has also been studied; the undesired cyclodimerization of butadiene

TABLE III-4
THE CYCLODIMERIZATION OF METHYL-SUBSTITUTED 1,3-DIENES[a]

1,3-Diene	Rate (gm dim/gm Ni/hr)	C_6 Ring (%)	C_8 Ring (%)
Butadiene	220	2.3	97.2
Piperylene	31	5.3	90.9
Isoprene	14	34.8	55.1
2,3-Dimethylbutadiene	0.6	86.3	6.1
2,4-Hexadiene	0	—	—

[a] From refs. 74 and 97.

TABLE III-5

THE CODIMERIZATION OF METHYL-SUBSTITUTED 1,3-DIENES
WITH BUTADIENE[a]

1,3-Diene	C_6 Ring (%)	C_8 Ring (%)	Dimer of subst. diene
Piperylene	1.6	86.4	11.5
Isoprene	5.5	84.0	9.8
2,3-Dimethylbutadiene	7.6	92.3	Trace

[a] From ref. 97.

with itself, which is the main reaction in the first case, can be suppressed by keeping the butadiene concentration to a minimum.

Most of the serious investigations have been carried out with a $Ni-P(OC_6H_4-o-C_6H_5)_3$ catalyst that is conveniently prepared by adding 1 equivalent of the phosphite to bis(cyclooctadiene)nickel. In general the substituted 1,3-dienes react far less readily than butadiene itself; this is illustrated in Table III-4 for the methyl-substituted 1,3-dienes.

The product from the cyclodimerization reaction usually consists of a mixture of substituted DVCB, COD, or VCH derivatives, e.g. **28–32**,

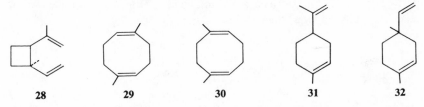

28   **29**   **30**   **31**   **32**

derived from isoprene. The distribution of the products, as well as their structure, depends on both the reaction conditions and the nature of the substituent. The effect of methyl substituents on the cyclodimerization and cyclo-codimerization of methyl-substituted 1,3-dienes is shown in Tables III-4 and III-5: mainly COD derivatives are obtained from the cyclo-codimerization with butadiene, whereas the product from the cyclodimerization depends on the position and number of substituents. The data shown in Tables III-4 and III-5 have been obtained by allowing the diene to react completely. As a result, the intermediate DVCB derivatives formed rearrange to the corresponding substituted COD compounds. Isolation of the DVCB derivatives is possible if the reaction is terminated after a fraction of the diene has been dimerized. The mechanistic implications of the configuration of the substituents in the four-membered ring are discussed in the following section.

The reactions involving *cis*- and *trans*-piperylene are complicated by the rapid nickel-catalyzed isomerization that occurs, and products derived from both isomers are invariably isolated. A similar effect, although less pronounced, is observed in the reactions involving the trienes.

The cyclodimerization of conjugated trienes produces principally six-membered ring systems, e.g., **33** and **34**, derived from 1,3,5-hexatriene.

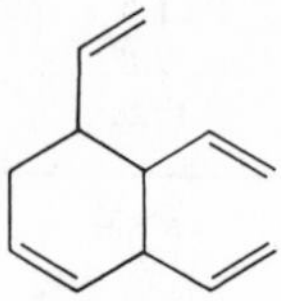

33 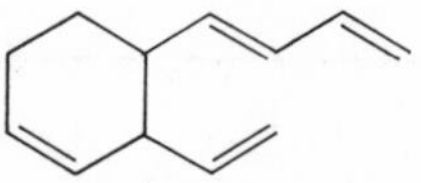   **34** 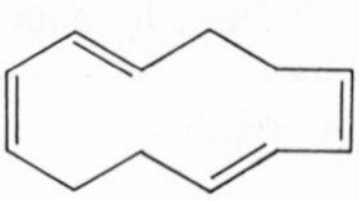   **35**

An interesting byproduct from the cyclodimerization of 1,3-*trans*-5-hexatriene is the cyclododecatetraene **35** in which all three double bonds of the triene have been incorporated into the ring. The product of the co-oligomerization of a triene with a 1,3-diene depends on the nature of the catalyst used: co-oligomerization of one molecule of triene with one of diene to give a divinyl-cyclohexene derivative, e.g., **36**, is the principal reaction in the presence of a "nickel–ligand" catalyst, whereas co-oligomerization of one molecule of triene with two molecules of diene to give a 14-membered ring, e.g., **37**, derived from 1,3,5-hexatriene and butadiene, is the main product of the reaction catalyzed by a "ligand-free" catalyst (150, 151).

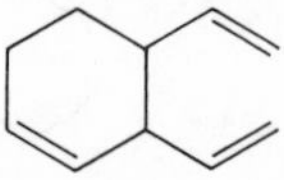

36 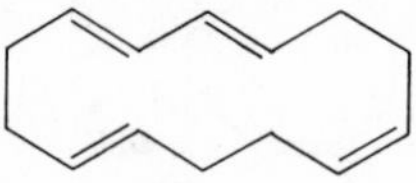   **37**

C. Mechanistic Considerations

Both product analysis and the study of model compounds have been used to investigate the mechanism of the cyclodimerization of 1,3-dienes and for no other system has the effect of varying the ligand and the 1,3-diene been studied in such detail. We shall initially develop the mechanism using butadiene as the example and then discuss the implications of the results using methyl-substituted 1,3-dienes.

The cyclodimerization of butadiene using a "nickel–ligand" catalyst is believed to occur in a stepwise manner: two butadiene molecules react at a nickel atom to form a π-allyl-C_8-nickel–ligand intermediate (e.g., **38–40**), which reacts further through the corresponding bis-σ-allyl form (e.g., **41–43**) to give the cyclic product. It is suggested that the course of the reaction is controlled by the configuration of the π-allyl-C_8 chain, VCH originating

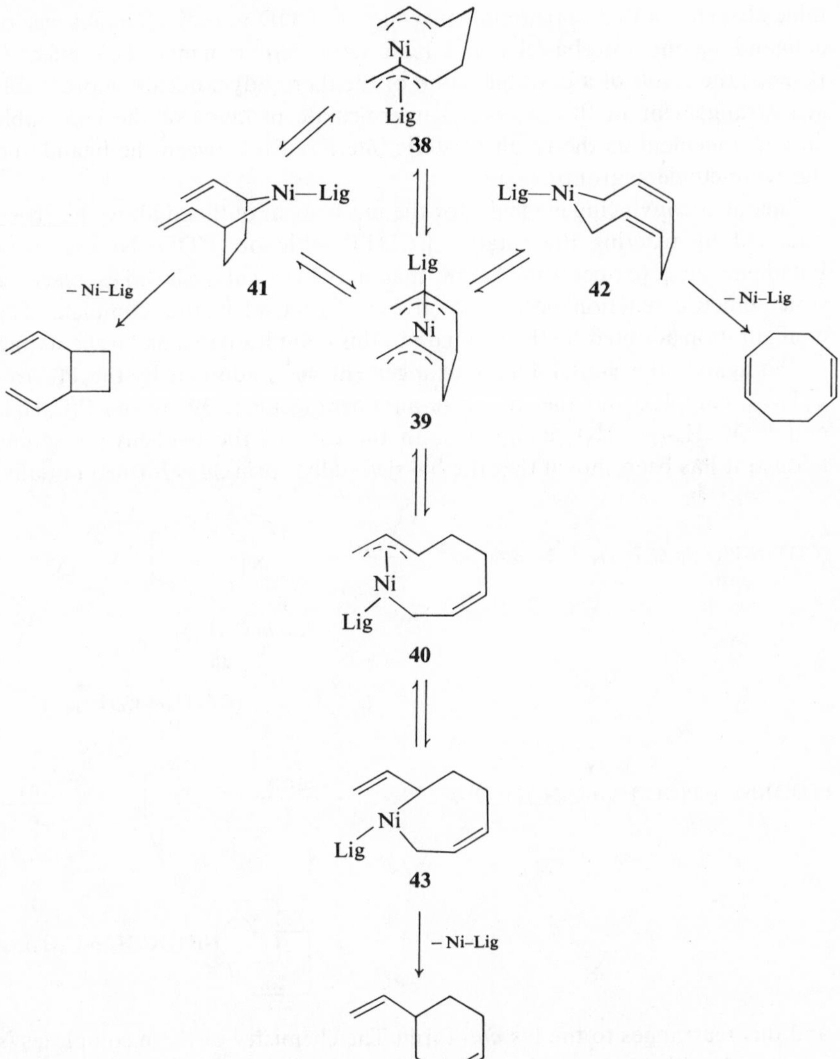

from the syn-π-allyl, σ-allyl form **40**, COD from the bis-$anti$-π-allyl form **39**, and DVCB from either **39** or the bis-syn-π-allyl form **38** (6, 97). The distribution of the products is determined by the concentration and reactivity of these intermediates which, in turn, is influenced by the reaction conditions and the steric and electronic properties of the ligand. Basic ligands, e.g., P($cyclo$-C_6H_{11})$_3$, stabilize the π-allyl,σ-allyl form **40**, and less basic ligands stabilize the bis-π-allyl forms. As can be seen by inspection of Table III-3, the experimental results are in agreement with this hypothesis. The same table also shows that maximum conversion to COD requires the presence of a ligand having low basicity and high steric requirements. This effect is perhaps the result of a destabilization of the thermodynamically more stable syn arrangement in the bis-π-allyl intermediate in favor of the less stable anti arrangement as the result of steric interference between the ligand and the *syn*-methylene groups.

The most convincing evidence for the mechanism outlined above has been obtained by reacting the catalyst [(CDT)Ni–Lig or (COD)$_2$Ni–Lig] with butadiene at a temperature below that at which catalysis takes place: a stoichiometric reaction occurs to give a C_8-nickel–ligand complex. The configuration adopted by the C_8 chain in this complex depends on the nature of the ligand; the π-allyl-σ-allyl arrangement **40** is adopted by the P(*cyclo*-C_6H_{11})$_3$ complex and the bis-*anti*-π-allyl arrangement **39** by the P(C_6H_5)$_3$ and P(OC_6H_4-*o*-C_6H_5)$_3$ complexes. In the case of the triphenylphosphine adduct, it has been shown that the bis-*syn*-π-allyl form **38** is formed initially,

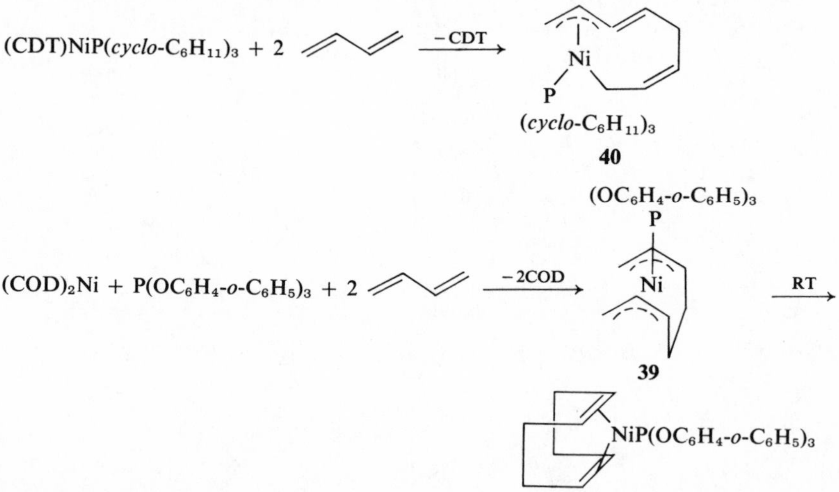

$$(CDT)NiP(cyclo\text{-}C_6H_{11})_3 + 2 \; \diagup\diagdown\diagup \; \xrightarrow{-CDT}$$

40

$$(COD)_2Ni + P(OC_6H_4\text{-}o\text{-}C_6H_5)_3 + 2 \; \diagup\diagdown\diagup \; \xrightarrow{-2COD} \qquad \xrightarrow{RT}$$

39

NiP(OC_6H_4-*o*-C_6H_5)$_3$

and this rearranges to the bis-*anti* form. The chemistry of these complexes is in accord with the postulated mechanism for the cyclodimerization: **40** [Lig = P(*cyclo*-C_6H_{11})$_3$] reacts with CO at $-78°$ to give VCH and **39** [Lig = P(OC_6H_4-*o*-C_6H_5)$_3$] rearranges spontaneously at room temperature to give the corresponding COD complex. The formation of considerable quantities of butadiene (in addition to the three cyclic dimers) on treating **39** or **40** with additional ligands, e.g., P(C_6H_5)$_3$ or COD, at room temperature or above has also been observed during the catalytic conversion of DVCB to COD and VCH (7, 20, 21, 54, 107).

The mechanism outlined above for the formation of COD is not the only possibility and it has also been shown that the "nickel–ligand" catalyst is capable of rearranging DVCB to COD (6, 11, 71, 72, 97). The intermediacy of a π-allyl-C_8-nickel species in this reaction is suggested by the conversion of (DVCB)NiP($cyclo$-C_6H_{11})$_3$ into the π-allyl complex **40** in solution at room

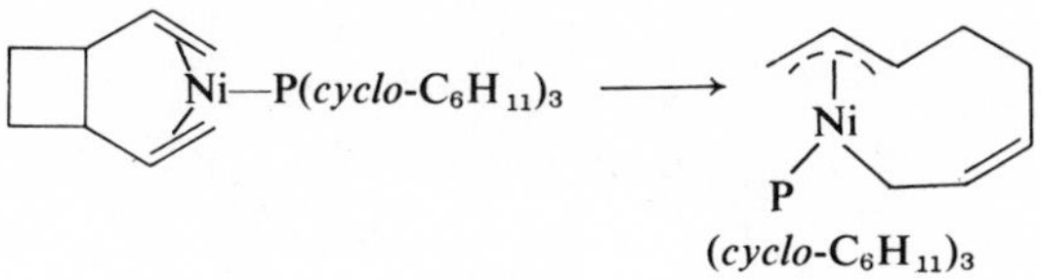

$(cyclo$-$C_6H_{11})_3$

40

temperature (20). The observed facility with which **39** [Lig = P(OC$_6$H$_4$-o-C$_6$H$_5$)$_3$] undergoes ring closure to give the COD complex (107), suggests, however, that COD is not formed exclusively from DVCB. The induction period of up to 1h that is observed in the formation of dimethylcyclooctadiene but not in the formation of the various dimethyl-DVCB derivatives on dimerizing piperylene, however, suggests strongly that in this case the four-membered ring is the primary product (71, 97).

In the case of the cyclodimerization of butadiene, the use of product analysis or the study of model compounds cannot give detailed information on either the mechanism of the formation of the C_8 chain [in spite of an early claim (20), a bis(butadiene)nickel–ligand system has still to be isolated] or the mechanism of the ring closure step. The situation, however, is different for reactions involving methyl-substituted 1,3-dienes, and here valuable information on the detailed course of the reaction has been obtained by determining the stereochemistry of the products. The relevance of the mechanism described above for butadiene to the cyclodimerization of methyl-substituted 1,3-dienes has been demonstrated with the isolation of intermediates analogous to **40**. The structures of two of these complexes—**44** and **45**, derived from isoprene and *cis*-piperylene—have been confirmed by x-ray studies (68, 84, 107). Both of these species react with CO at low temperature to give mainly cyclic products.

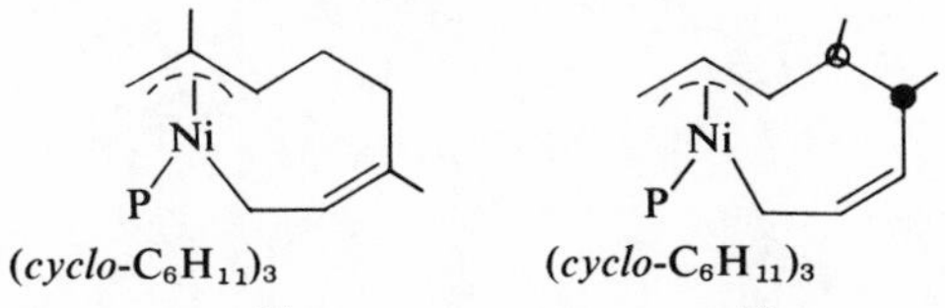

$(cyclo$-$C_6H_{11})_3$ $(cyclo$-$C_6H_{11})_3$

44   **45**

A careful study has been made of the stereochemistry of the DVCB derivatives formed in the reactions of *cis*- and *trans*-piperylene with the Ni–P(OC$_6$H$_4$-o-C$_6$H$_5$)$_3$ catalyst (16, 71, 97, 102). The product of the reaction

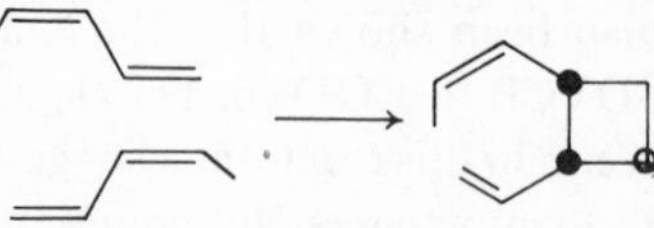

46

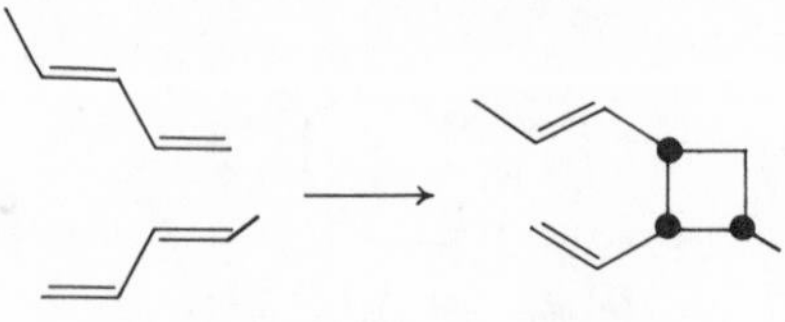

47 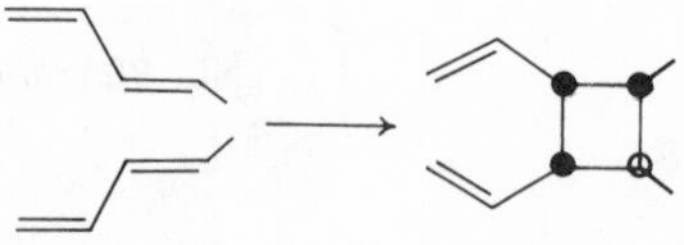   **48**

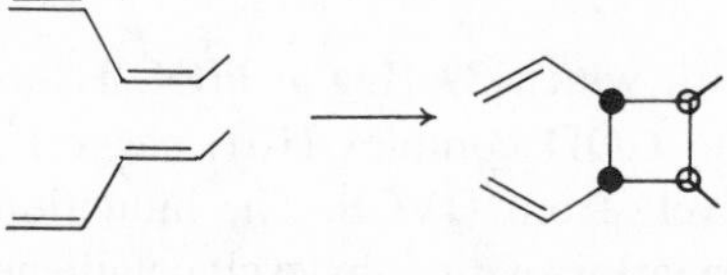

49
                                                                              50 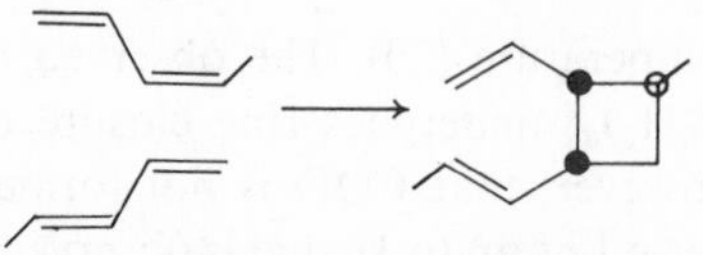

involving *cis*-piperylene is **46**, whereas **47** and **48** have been isolated from the dimerization of *trans*-piperylene. The codimerization of *cis*-piperylene with *trans*-piperylene produces two new isomers, **49** and **50**. The five other possible isomers (**51–55**) are not formed. The bis(allyl)-C_8-nickel intermediates from

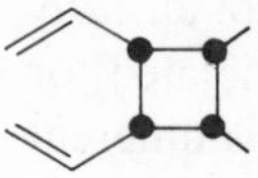

51 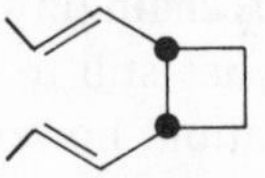   **52** 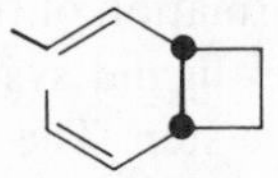   **53** 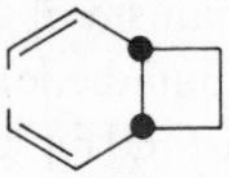   **54** 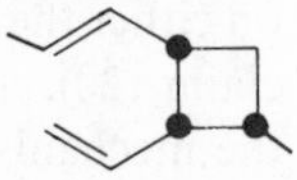   **55**

which it may be assumed that **46–50** are formed are generated by the head-to-tail or tail-to-tail coupling of two piperylene molecules (e.g., **56** and **57**). This, combined with the absence of products formed by head-to-head coupling (e.g., **58**), suggests that before coupling the two diene molecules are complexed through the least substituted double bond; this is in accord with the known stability of the nickel-olefin complexes (see Volume I, p. 268).

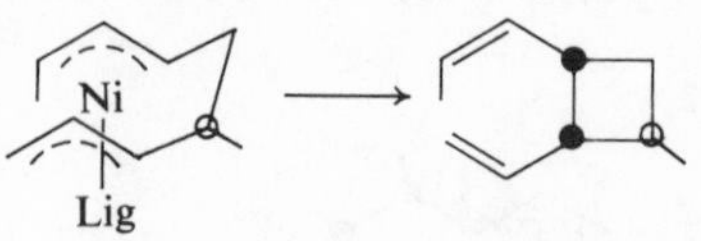

56   **46** 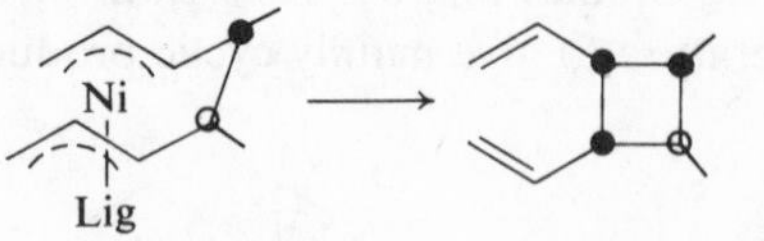   **57**   **48**

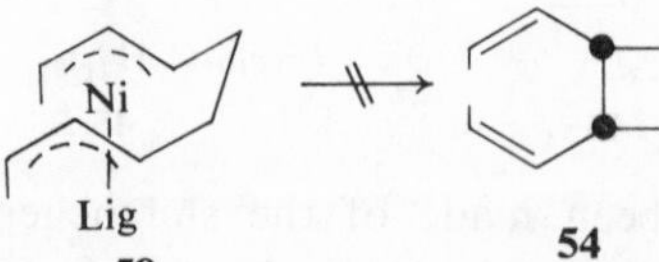

58   **54**

Similar arguments can be applied to the reactions of other substituted 1,3-dienes. The stereochemistry of the products of the reaction involving isoprene or 1,3-hexadiene also arise by exclusive head-to-tail or tail-to-tail coupling (in the case of isoprene additional support is found in the x-ray structural study of the intermediate **44**), whereas the substituted DVCB derivatives that have been isolated from the cyclo-codimerization of butadiene with piperylene are the result of tail-to-tail coupling (71, 74).

The preferred formation of six-membered ring compounds from the reactions involving hexatrienes suggests that here also a π-allyl,σ-allyl intermediate is involved (e.g., **59**) and that interaction of the vinyl group with the metal atom instigates ring closure.

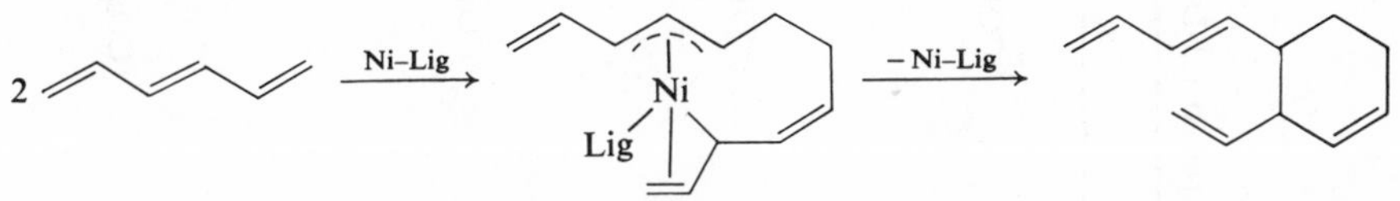

59

The nature and distribution of the products obtained by the cyclo-codimerization of a substituted 1,3-diene with butadiene or a second substituted diene may be controlled by the relative stability of the intermediate π-allylnickel complexes. A clear example of this is found in the structure of the cyclohexene derivatives **61** and **63** obtained by codimerizing butadiene with sorbic acid ester or with 2,3-dimethylbutadiene: in both cases the probable intermediates, **60** and **62**, involve the most stable π-allyl systems (80, 97).

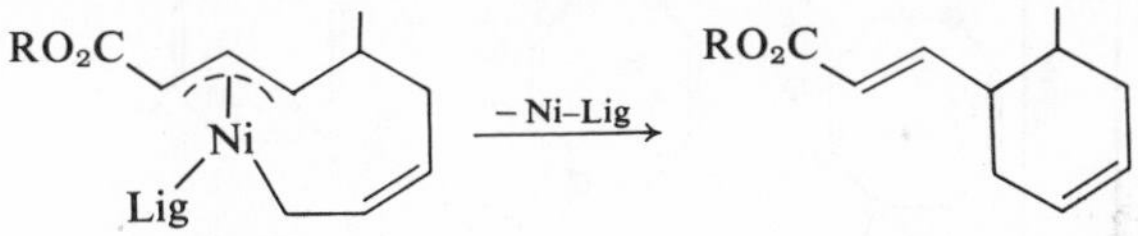

60 **61**

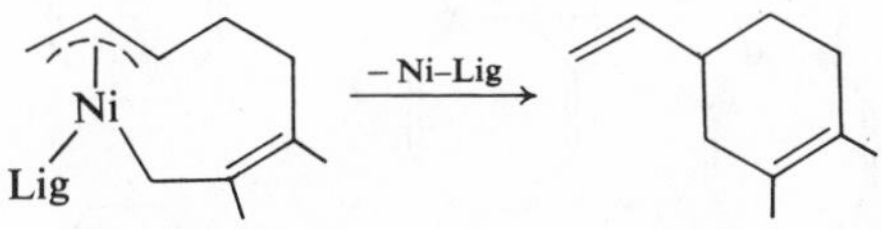

62 **63**

The discussion given above has been essentially limited to the further reaction of the substituted bis(allyl)-C_8 chain. The mechanism of its formation remains unknown. One possibility is that a bis(diene)nickel–ligand complex (analogous to the bisolefin complexes) is first formed, in which the diene molecules adopt a single-trans arrangement. Coupling of the two diene molecules to form the C_8 chain is then the result of interaction of one of the free double bonds in this complex with the nickel atom; the stereochemistry of the resultant C_8 chain may therefore have an electronic or steric origin. An attempt has been made to apply the Woodward-Hoffmann rules to these steps, and the reader is referred to references 8, 102, and 150 for a detailed account.

TABLE III-6

The Cyclodimerization and Cyclo-oligomerization of Substituted 1,3-Dienes

Product[a]	Diene	Catalyst	Ref.
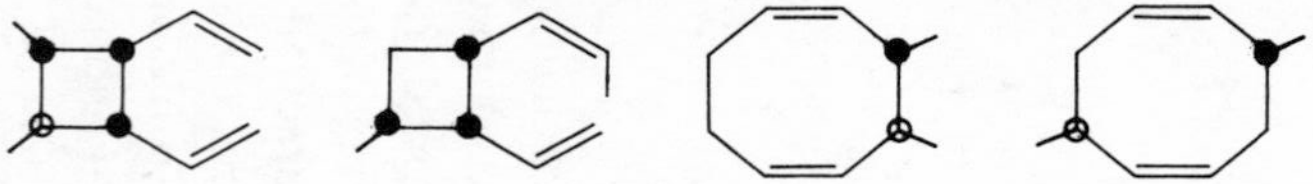	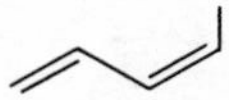	$(COD)_2Ni-Lig$	8, 16, 71, 78, 83, 96, P54, P103, P105, P111
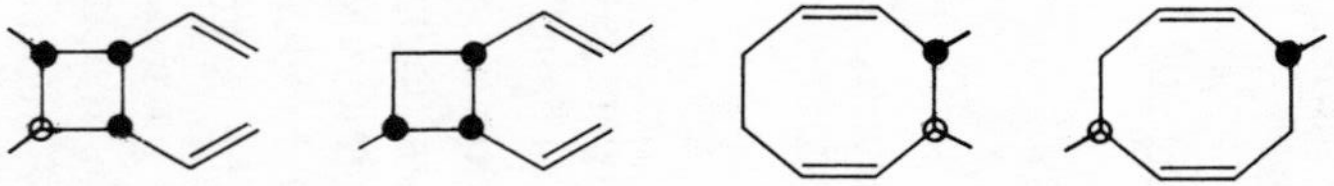	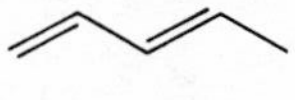	$(COD)_2Ni-Lig$	8, 16, 71, 78, 83, 96, 169, P54, P103, P105, P111
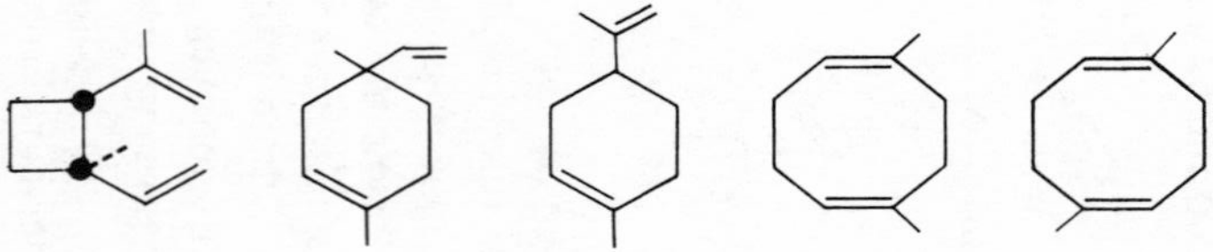	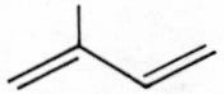	$(COD)_2Ni-Lig$	50, 55, 62, 65, 74, 78, 91, 158, P23, P34, P35, P48, P96, P103, P111
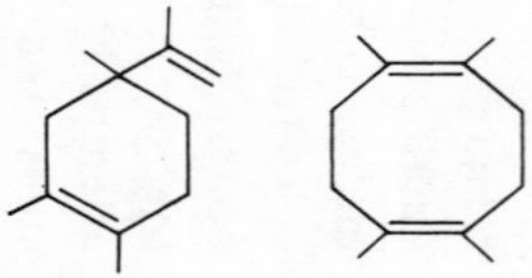	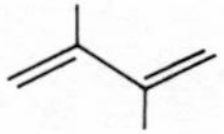	$(COD)_2Ni-Lig$	15, 118, P23

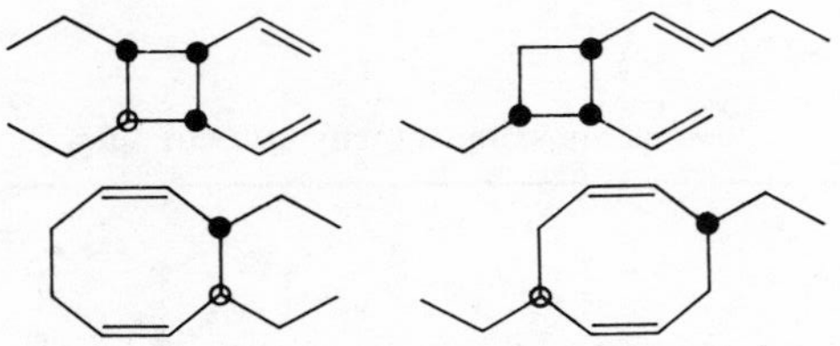

8, 74, 83

$(COD)_2Ni–Lig$

P34–P36

$Lig_2Ni(CO)_2$

8, 74, 150, 151

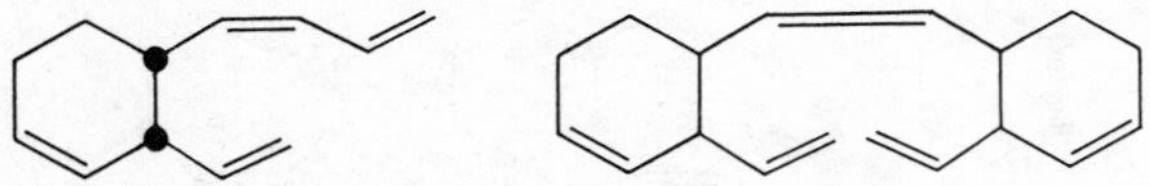

$(COD)_2Ni \pm Lig$

8, 74

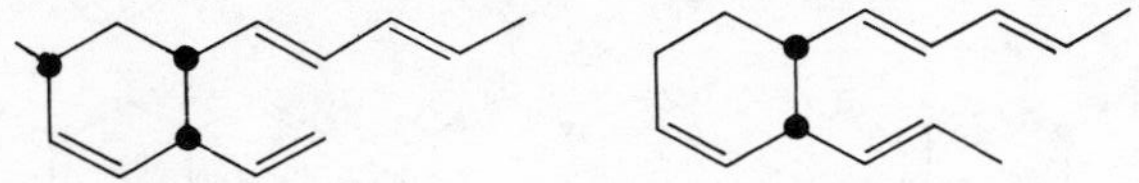

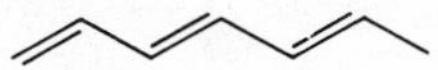

$(COD)_2Ni–Lig$

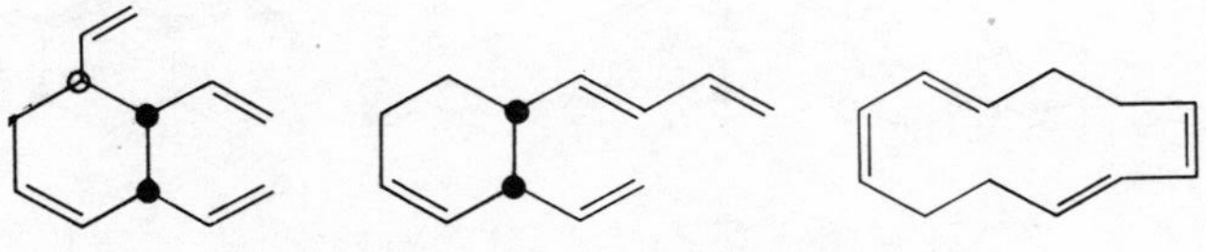

151

$(COD)_2Ni–Lig$

(continued)

Dichlorocyclooctadiene

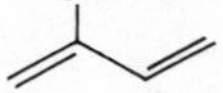

TABLE III-6 (continued)

Product[a]			Diene	Catalyst	Ref.
				$(COD)_2Ni–Lig$	151
				$(COD)_2Ni–Lig$	151
				$(COD)_2Ni–Lig$	151
				$(COD)_2Ni–Lig$	8,74

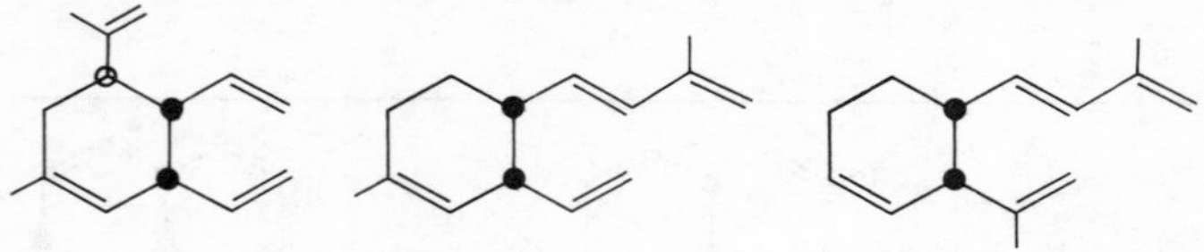

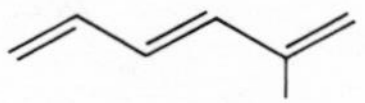

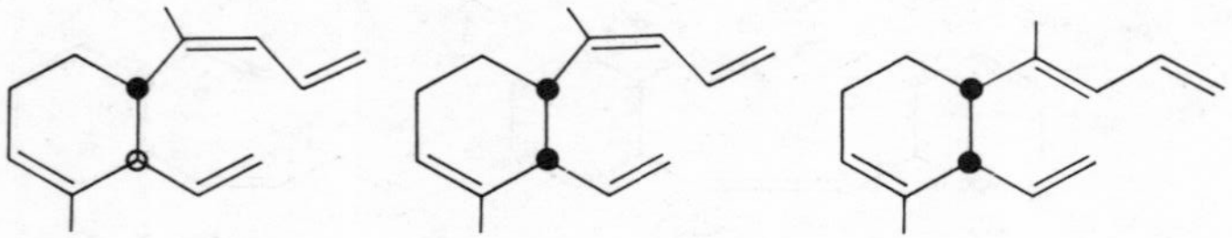

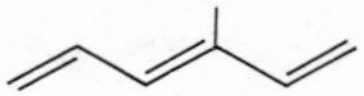

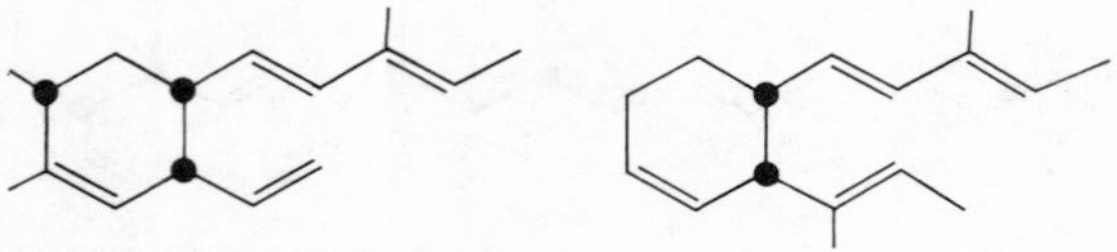

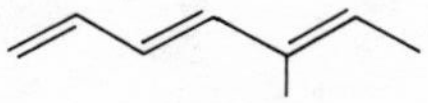

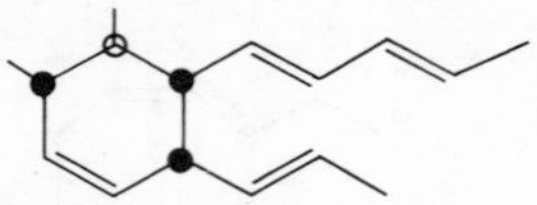

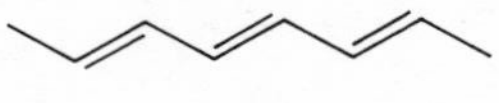

[a] Only the principal products are shown.

TABLE III-7

THE CYCLO-CODIMERIZATION AND CO-OLIGOMERIZATION OF SUBSTITUTED 1,3-DIENES

Product^a		Diene		Catalyst	Ref.
				$(COD)_2Ni-Lig$	71, 97, P104
				$(COD)_2Ni-Lig$	71, 97, P104
				$(COD)_2Ni-Lig$	97
				$Ni(acac)_2-Lig-$ $(C_2H_5)_2AlOC_2H_5$	83
				$(COD)_2Ni-Lig$	14, 97, 117, P103, P104
				$(COD)_2Ni-Lig$	97, P63, P104
				$(COD)_2Ni-Lig$	97, P104

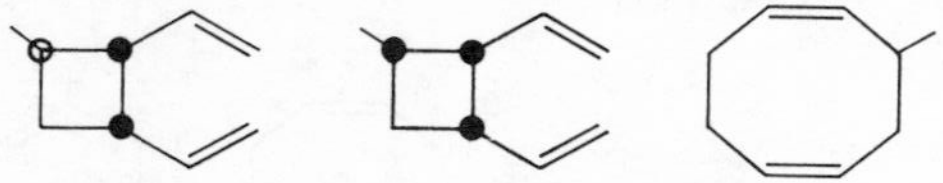

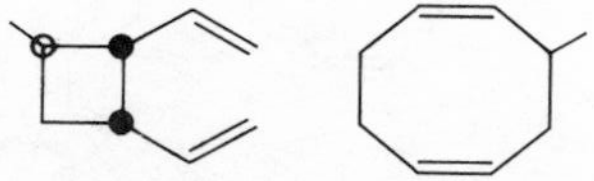

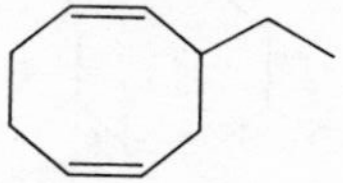

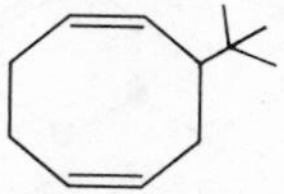

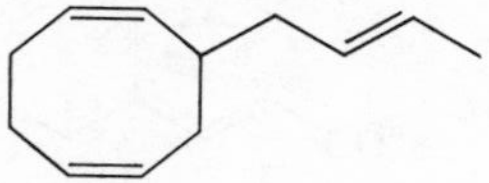

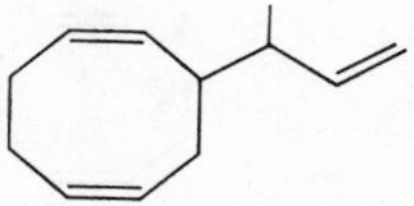

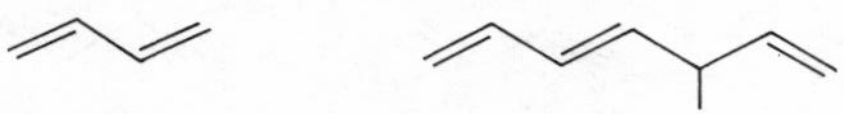

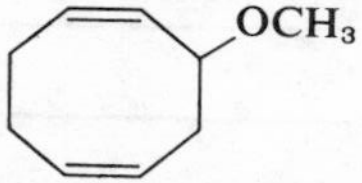

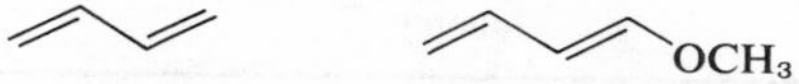

(*continued*)

155

TABLE III-7 (*continued*)

Product[a]	Diene		Catalyst	Ref.
		C_6H_5	$Ni(acac)_2-Lig-(C_2H_5)_2AlOC_2H_5$	P104
		CO_2R	$Ni(acac)_2-Lig-(C_2H_5)_2AlOC_2H_5$	80
		CO_2R	$Ni(acac)_2-Lig-(C_2H_5)_2AlOC_2H_5$	80, 83, 97, 159, P104
			$Ni(acac)_2-(C_2H_5)_2AlOC_2H_5$	81, 97, P62, P104
			$Ni(acac)_2-Lig-(C_2H_5)_2AlOC_2H_5$	97, P104

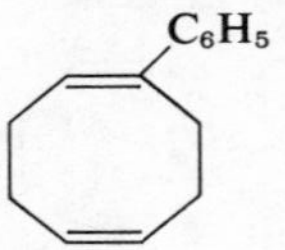

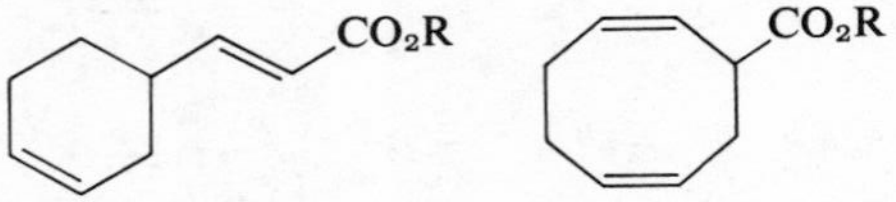

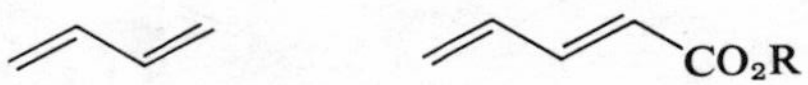

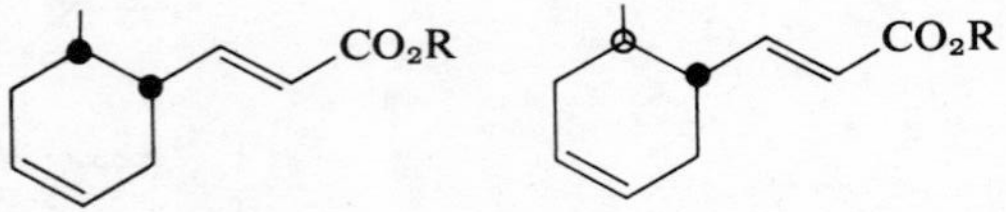

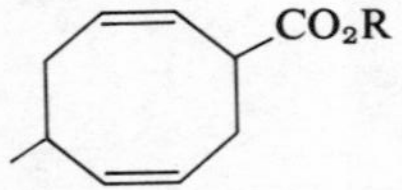

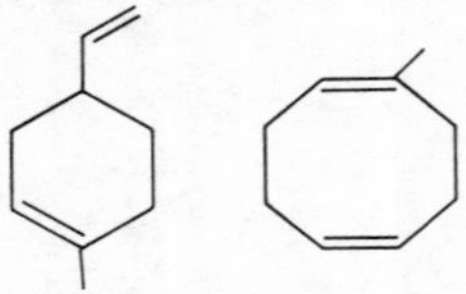

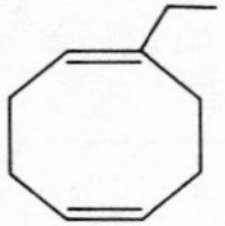

156

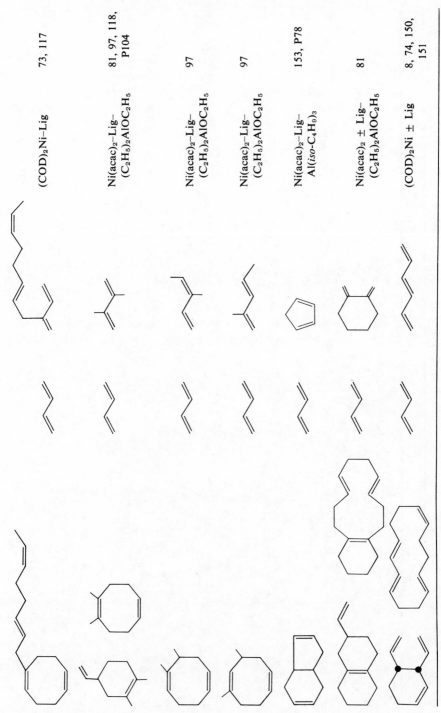

$(COD)_2Ni–Lig$	73, 117
$Ni(acac)_2–Lig–$ $(C_2H_5)_2AlOC_2H_5$	81, 97, 118, P104
$Ni(acac)_2–Lig–$ $(C_2H_5)_2AlOC_2H_5$	97
$Ni(acac)_2–Lig–$ $(C_2H_5)_2AlOC_2H_5$	97
$Ni(acac)_2–Lig–$ $Al(iso\text{-}C_4H_9)_3$	153, P78
$Ni(acac)_2 \pm Lig–$ $(C_2H_5)_2AlOC_2H_5$	81
$(COD)_2Ni \pm Lig$	8, 74, 150, 151

(continued)

157

TABLE III-7 (*continued*)

Product[a]	Diene		Catalyst	Ref.
			$(COD)_2Ni \pm Lig$	8, 74, 151
			$(COD)_2Ni \pm Lig$	151
			$(COD)_2Ni–Lig$	151
			$(COD)_2Ni–Lig$	151
			$(COD)_2Ni–Lig$	151
			$(COD)_2Ni \pm Lig$	74, 151

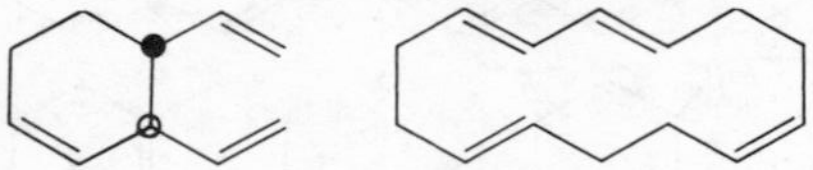

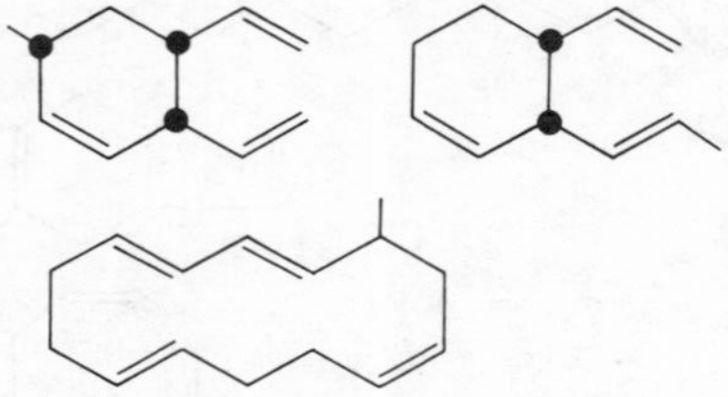

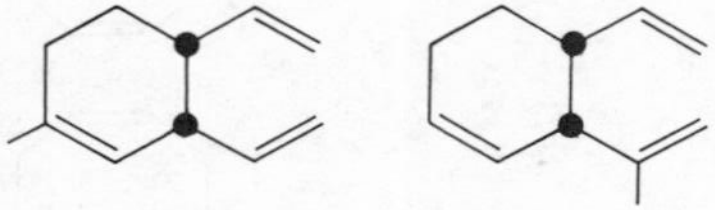

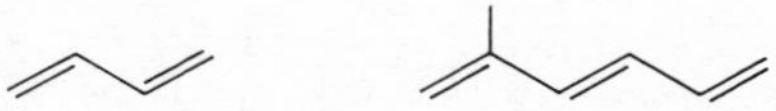

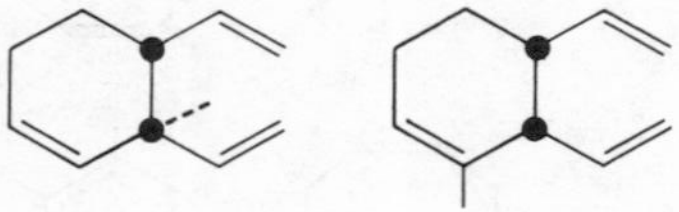

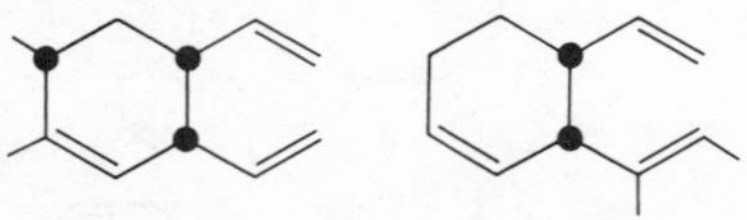

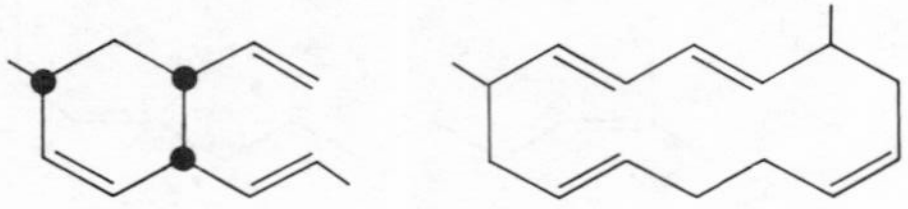

158

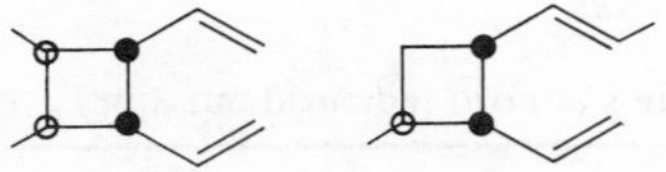

8, 16, 71, 83

(COD)$_2$Ni-Lig

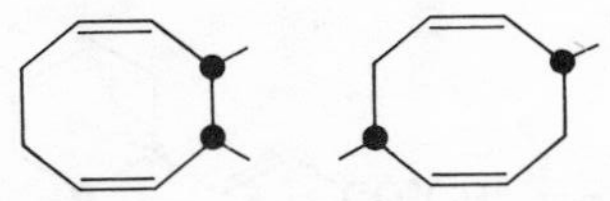

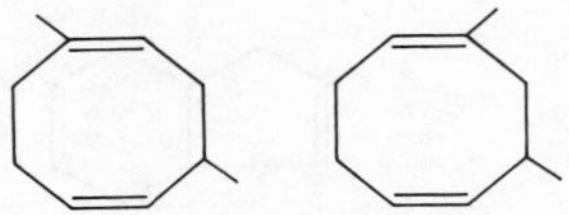

Ni(acac)$_2$-Lig-
(C$_2$H$_5$)$_2$AlOC$_2$H$_5$

P104

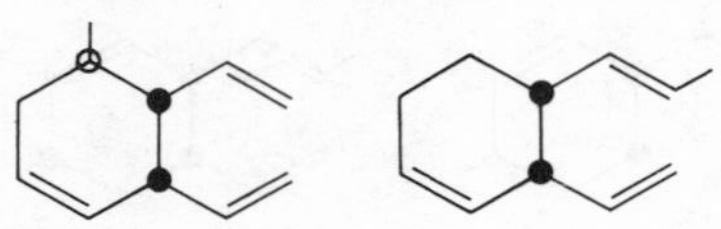

(COD)$_2$Ni-Lig

151

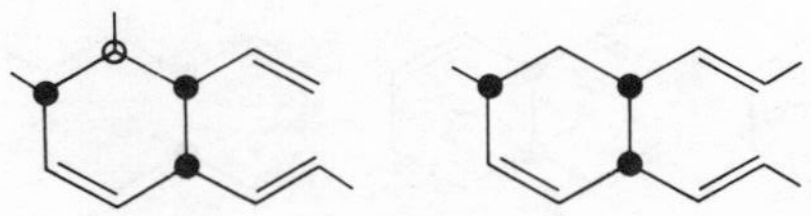

(COD)$_2$Ni-Lig

8, 74

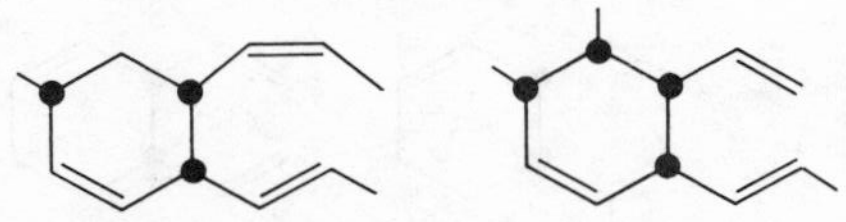

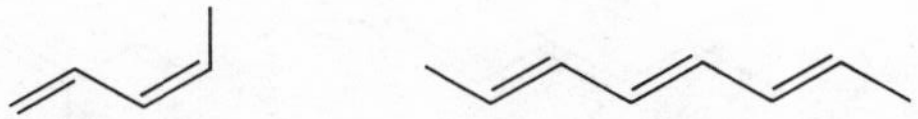

(COD)$_2$Ni-Lig

8, 74

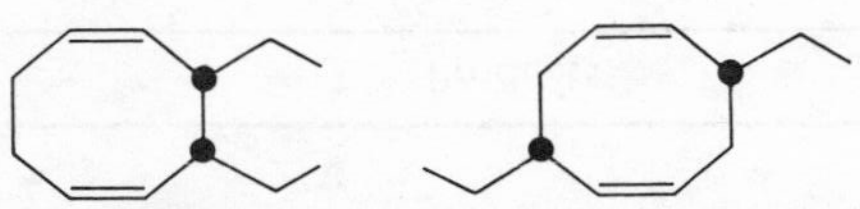

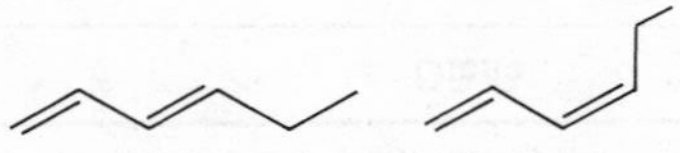

(COD)$_2$Ni-Lig

83

(continued)

159

TABLE III-7 (*continued*)

Product[a]		Diene		Catalyst	Ref.
				$(COD)_2Ni-Lig$	151
				$(COD)_2Ni-Lig$	151
				$(COD)_2Ni-Lig$	8, 74
				$Ni(acac)_2-Lig-$ $Al(iso-C_4H_9)_3$	153
			CO_2CH_3	$Ni(acac)_2-P(C_6H_5)_3-$ $Al(C_2H_5)_3$	159

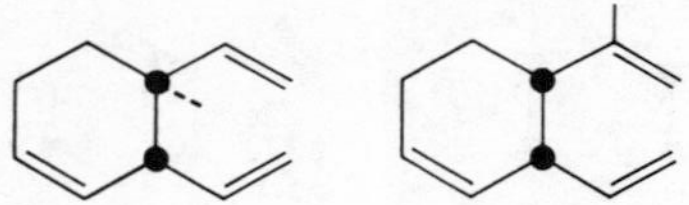

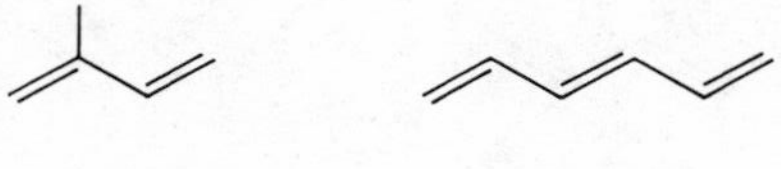

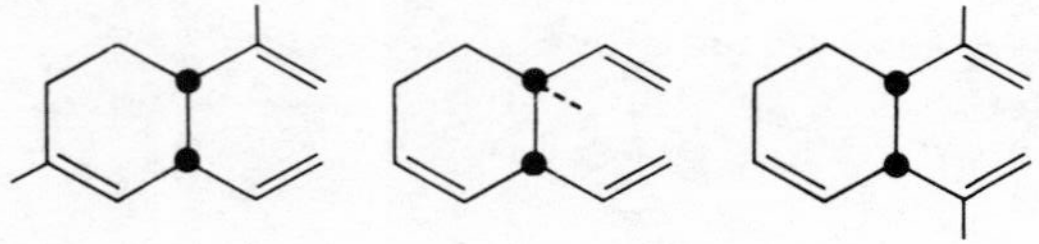

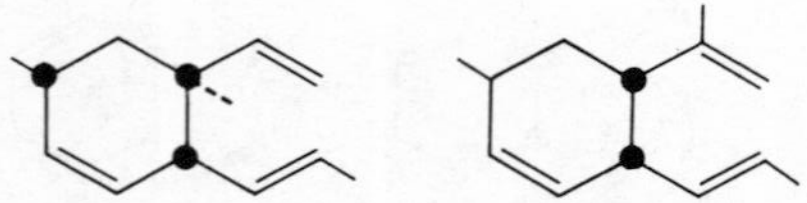

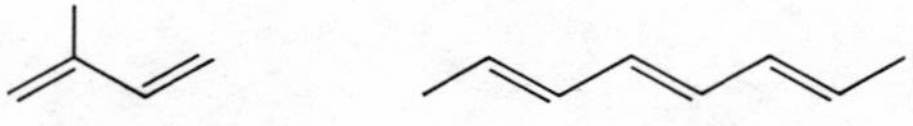

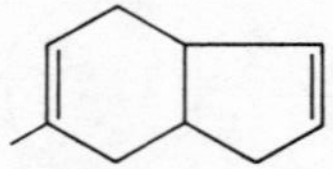

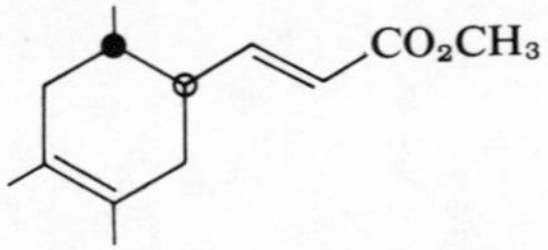

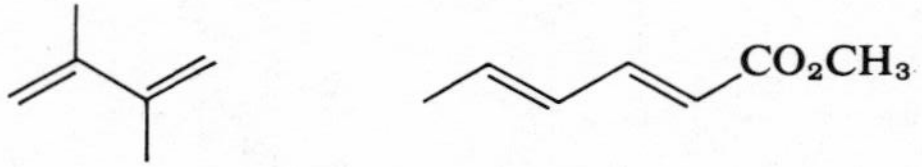

[a] Only the principal products are shown.

160

IV. The Preparation of Macrocyclic Polyenes

An important extension of the work described in the preceding sections is the conversion of butadiene into a mixture of macrocyclic polyenes by a two-component catalyst system, e.g., $(\pi\text{-}C_3H_5)_2Ni\text{-}(\pi\text{-}C_3H_5NiCl)_2$ (34, 101, P82, P87, P88, P119–P123, P136, P137). The product has a wide molecular weight distribution, with $C_8\text{-}C_{32}$ polyenes predominating. Two basic structures are adopted (**64** and to a lesser extent **65**) and these contain exclusively

64   **65**

trans double bonds. The second component can be varied widely and active catalysts include bis(π-allyl)nickel in conjunction with organic halides, e.g., $HC\vdots CCH_2Cl$, $(C_6H_5)_3CCl$, and CH_3COCl; organic acids, e.g., CF_3CO_2H; organometallic compounds of the Group I–III elements, e.g., $(C_2H_5)_2AlCl$, $Al(C_2H_5)_3$, C_2H_5ZnCl, and C_2H_5Li; and organotransition metal compounds, e.g., $\pi\text{-}C_3H_5PdCl$, $(\pi\text{-}C_5H_5)_2TiBr$; and lithium bromide. Yet another active combination is $(\pi\text{-}C_3H_5NiBr)_2\text{-}(\pi\text{-}C_5H_5)_2M$ (M = Ti, V). Typical product distributions are shown in Fig. III-1.

Because $(\pi\text{-}C_3H_5)_2Ni$ alone cyclotrimerizes butadiene, whereas $(\pi\text{-}C_3H_5NiCl)_2$ is an active polymerization catalyst, the formation of macrocyclic compounds on combining the two suggests that both components act

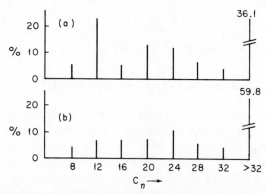

Fig. III-1. The distribution of the monocyclic polyene in the cyclo-oligomerization of butadiene catalyzed by (a) $(\pi\text{-}C_3H_5)_2Ni\text{-}(\pi\text{-}C_3H_5NiCl)_2$ and (b) $(\pi\text{-}C_3H_5)_2Ni\text{-}(C_2H_5)_2Zn$ (101).

together. A possible mechanism is that shown below, in which growth of the oligomer chain occurs on a binuclear species and ring closure after dissociation into a mononuclear complex.

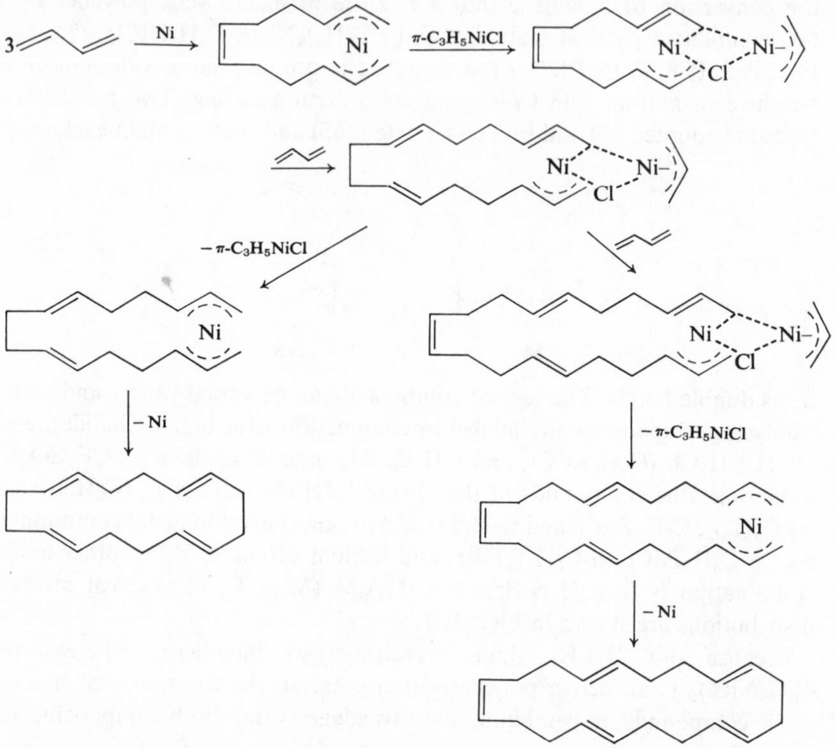

Some of the binary catalysts mentioned above are very similar to systems described in Chapter IV that are used to polymerize butadiene and one suspects that in some cases the formation of *trans*-1,4-polybutadiene is accompanied by the generation of macrocyclic polyenes.

V. The Co-oligomerization of 1,3-Dienes with Olefins and Alkynes

In the presence of olefins or alkynes the cyclodimerization or cyclotrimerization of butadiene is suppressed and instead a co-oligomerization reaction occurs to give cyclic or linear products, depending on the conditions. The reactions are catalyzed by both the "ligand-free" zerovalent nickel catalyst and the ligand-modified catalyst and, as discussed in the previous sections, the catalyst may be formed *in situ* either by reducing a nickel salt alone or in the presence of a donor ligand, or alternatively, by using a preformed nickel complex that may either be "ligand-free", e.g., $(COD)_2Ni$, or contain donor ligands, e.g., $Lig_2Ni(CO)_2$. The most detailed studies have made use of the catalysts $Ni(acac)_2-(C_2H_5)_2AlOC_2H_5$ and $Ni(acac)_2-(C_2H_5)_2-AlOC_2H_5-Lig$. The discussion in this section is limited to the formation of co-oligomers containing two or more 1,3-diene molecules. The formation of co-oligomers (e.g., hexadiene or vinylcyclohexadiene) involving only one molecule of 1,3-diene is discussed in Chapter I (Section III,B) and Chapter II (Section III)*.

A. The Co-oligomerization of 1,3-Dienes with Olefins (*Table III-8*)

The co-oligomerization of a large variety of olefins has been studied. Most attention has been given to the co-oligomerization of butadiene with ethylene and this will be discussed in some detail (13, 97). The product of the reaction, a mixture of *cis,trans*-1,5-cyclodecadiene (**66**) and 1-*trans*-4,9-decatriene (**67**), is formed by the reaction of two molecules of butadiene with one of ethylene.

Selective formation of cyclodecadiene can be achieved using the "ligand-free" catalyst at low temperature (ca. 80% yield at 20°) but the reaction is rather slow (ca. 2.2 gm product/gm Ni/hr). The rate can be increased by raising the temperature, but this increase is accompanied by a loss in selectivity and at 80° decatriene is the main product. The temperature at which the catalytic reaction can be carried out is limited to 80–100° because above this the thermal Cope rearrangement to *cis*-1,2-divinylcyclohexane (**68**) becomes significant (120). The rate of reaction using the "nickel–ligand" catalyst is higher

but has the disadvantage that COD and VCH are also produced. The proportion of cyclic to linear co-oligomerization product is ligand dependent, and cyclodecadiene essentially free of decatriene can be obtained using a catalyst involving triphenyl phosphite. Similar results are obtained using DVCB in place of butadiene (11, 15). The addition of aryl halides or sulfones is reported to lead to preferential formation of the decatriene (146, P5–P7).

Alkyl-substituted straight-chain olefins are less reactive than ethylene and

* The elucidation of the structure of the cyclodecadiene and cyclodecatriene derivatives described in this section has been realized using a combination of thermal Cope rearrangement and methylene insertion reactions and is described in detail in references 97, 102, and 147.

only in the case of propylene has a cyclo-oligomerization reaction been observed. The propylene molecule can be incorporated in two different orientations and a mixture of the two isomeric methylcyclodecatrienes **69** and **71** and the corresponding linear trienes (**70** and **72**) is obtained. The third possible linear triene **73** is not observed; migration of a hydrogen atom from the methyl group to give 1,5-*trans*-10-undecatriene (**74**) occurs instead.

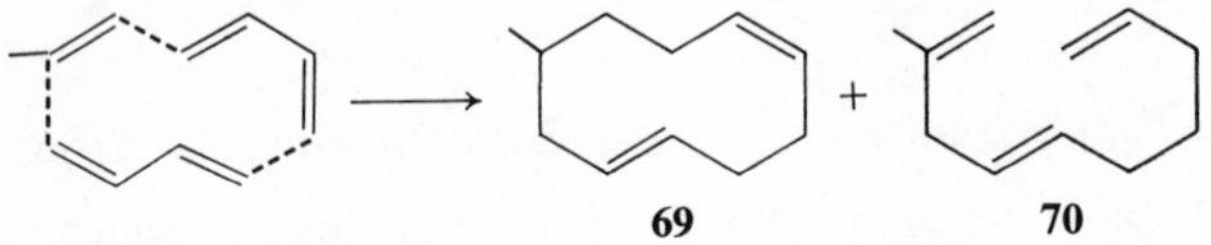

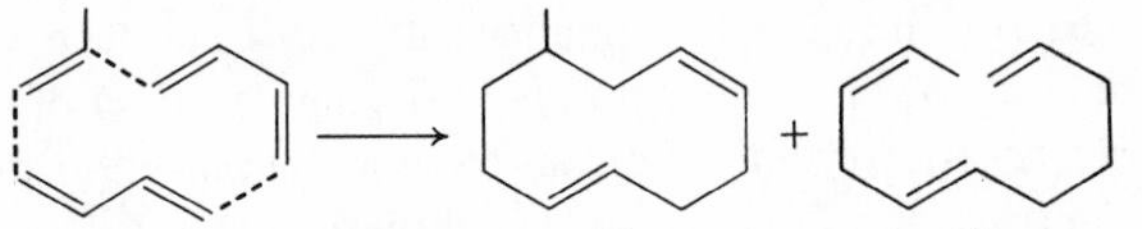

Although simple dialkyl-substituted olefins, e.g., 2-butene or isobutene, do not co-oligomerize with butadiene (73), it has been reported that both cyclodecadiene and *trans,trans,trans*-CDT can be co-oligomerized with butadiene if these compounds are used as the solvent for the reaction: the product is a C_8 alkylated ring (13, 97).

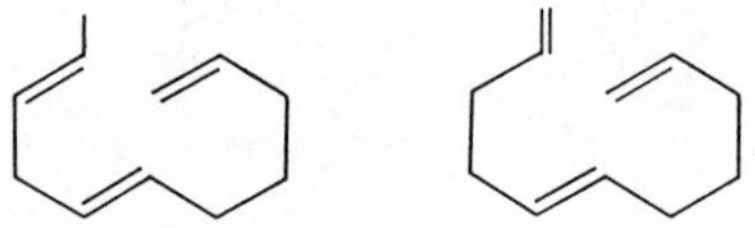

The reactivity of the olefinic double bond can be considerably increased by incorporation into a strained ring and the co-oligomerization of butadiene has been studied with cyclic olefins, e.g., 3,3-dimethylcyclobutene, bicycloheptene, and methylenecyclopropane. The reactions involving bicycloheptene and related compounds give exclusively cyclic products, e.g., **75** and **76**, derived from bicycloheptadiene and dicyclopentadiene, respectively. Fusion of the ten-membered ring occurs exclusively to the exo side of the bicycloheptane skeleton.

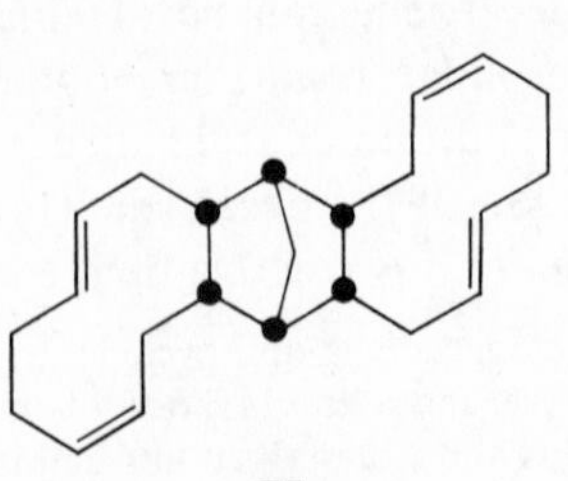

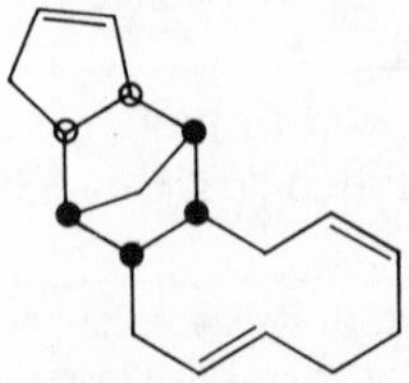

The reactions involving methylenecyclopropane and related compounds are characterized by the facile opening of the strained ring. As a result a number of interesting byproducts are obtained. Methylenecyclopropane itself co-oligomerizes with butadiene to give a mixture of the expected spirodecadienes (**77** and **78**) and the linear oligomer **79**. The product distribution

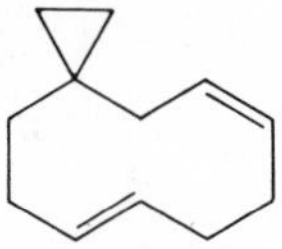

77 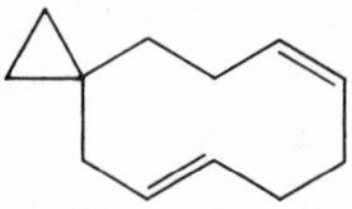 **78** 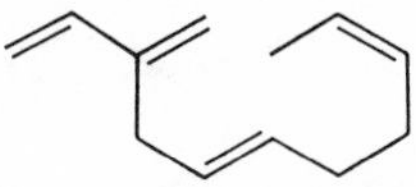 **79**

using the "nickel–ligand" catalyst system can be influenced by varying the ligand: $P(OC_6H_4\text{-}o\text{-}C_6H_5)_3$ favors the formation of the cyclic oligomers and $P(cyclo\text{-}C_6H_{11})_3$ favors the linear oligomer. Byproducts of the reactions involving 2,2'-dimethylmethylenecyclopropane have been identified as the 11-membered ring systems **80** and **81**, whereas the pyrrole derivative **83** is formed in modest yield in the reaction between butadiene and N-isopropyl-2-methyleneaziridine (**82**) in addition to the expected co-oligomers.

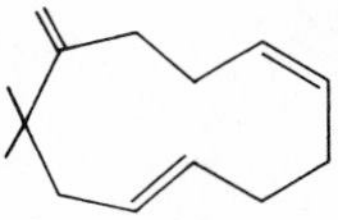

80 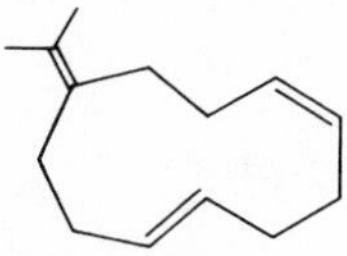 **81**

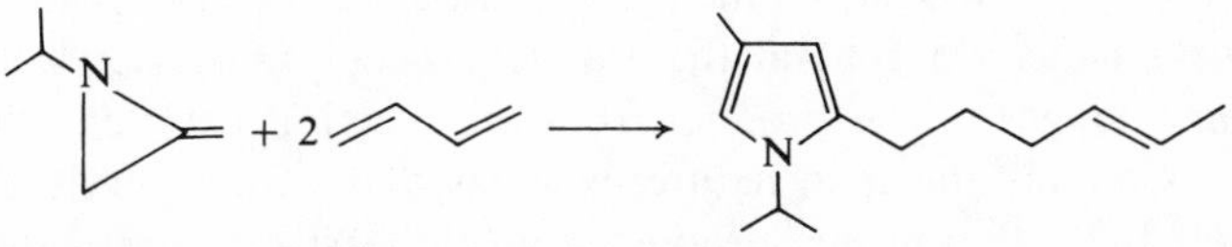

82 **83**

The reactivity of the olefin can also be increased by the introduction of electron-withdrawing substituents. Investigation of the co-oligomerization of such olefins with butadiene is apparently limited to styrene and related compounds, itaconic acid, and the acrylic acid esters. The course of the reaction with styrene is very temperature dependent; linear products are formed almost exclusively at higher temperature, whereas the proportion of the cyclic co-oligomer increases on lowering the temperature. Interestingly, of the two possible isomers of phenylcyclodecadiene, only that isomer having the phenyl group in the 9 position has been isolated. The reactions involving the acrylic acid esters are characterized by the ease with which incorporation of additional reactant molecules occurs: in addition to the expected cyclic and linear co-oligomers, methyl acrylate yields compounds formed by the incorporation of a second olefin molecule (e.g., **84**) or further butadiene molecules (e.g., **85**).

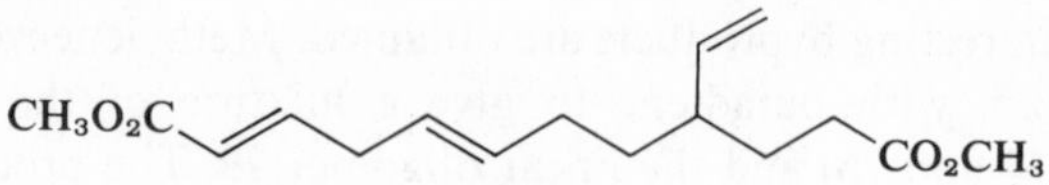

84

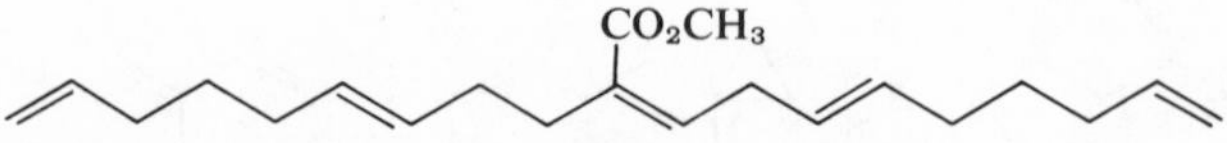

85

The main product from the reaction of methacrylate esters with butadiene is the 1,5,10-undecatriene derivative **86**, which is formed as the result of a hydrogen abstraction from the methyl group. The alternative, hydrogen abstraction from the methylene group, which would lead to a product (**87**) containing a tertiary olefin, is not observed, and it has been suggested that this is evidence that complexation of the reacting olefin to the nickel plays a significant role in the hydrogen transfer process (80, 97). Similar behavior has been observed in the reaction between itaconic acid and butadiene (90, P117).

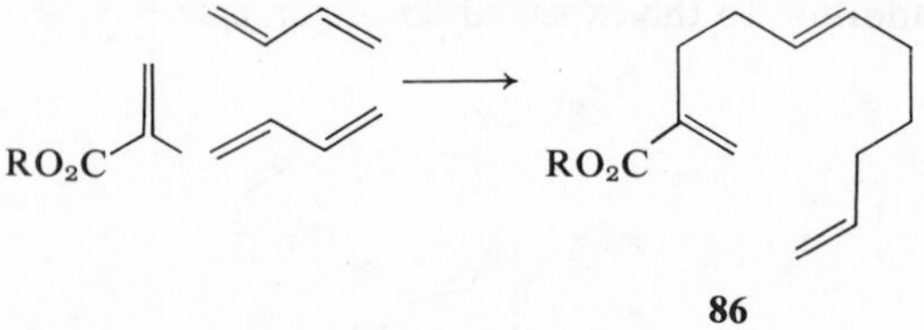

86 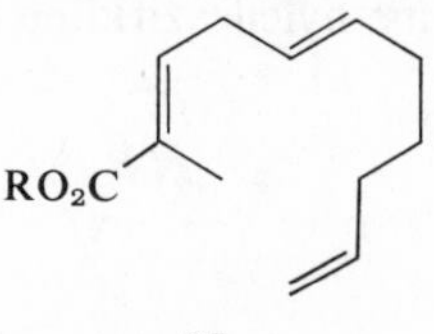   **87**

The co-oligomerization of methyl-substituted 1,3-dienes with olefins has also been studied. In general the reactions occur more slowly than those involving butadiene. The product from the reaction between ethylene and isoprene is mainly linear in nature, whereas that with piperylene is mainly cyclic (73, 102). An unusual product from this last reaction is the 12-membered cyclic diene (**88**), which is formed in the presence of a "ligand-free" catalyst as a result of the co-oligomerization of two molecules of ethylene with two of *trans*-piperylene.

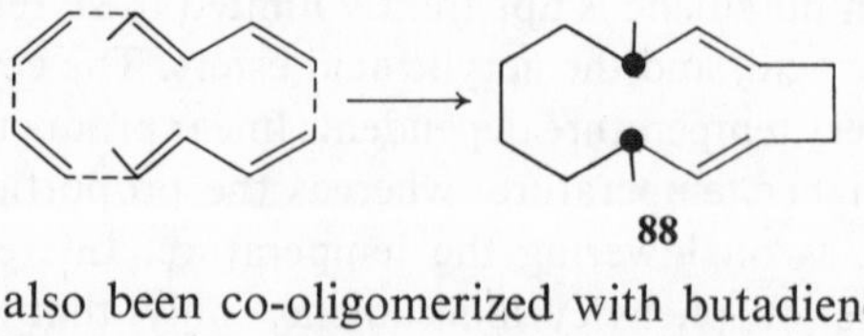

88

1,2-Dienes have also been co-oligomerized with butadiene (81, 124). The main co-oligomer from the reaction with allene is a mixture of the two expected methylenecyclodecadiene compounds. Minor components have been identified as a 12-membered cyclic tetraene (probably **89**–allene:butadiene = 2:2) and the methyltetradecapentaene **90**(allene:butadiene = 1:3).

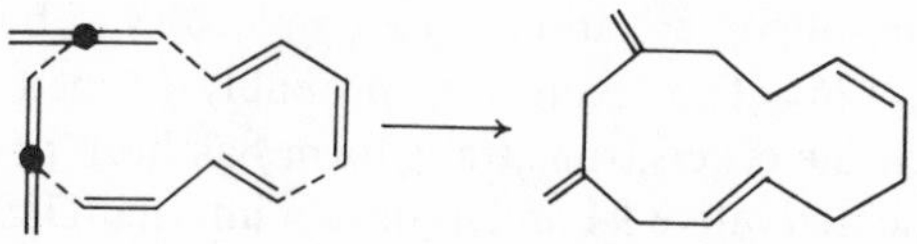

89

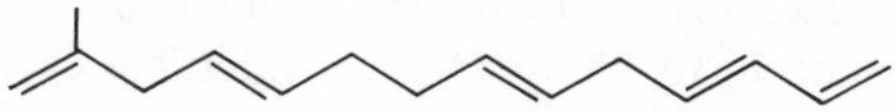

90

B. The Co-oligomerization of 1,3-Dienes with Alkynes (Table III-9)

The co-oligomerization of butadiene with alkynes differs from the reaction involving olefins in the preferential formation of ring products and in the marked tendency to incorporate more than one alkyne molecule. The reaction involving 2-butyne has received the most attention. Both the "nickel–ligand" and the "ligand-free" catalysts are active but whereas in the first case the main product from the reaction of 2 moles of butadiene with 1 mole of butyne is 2,5-dimethyl-*cis,cis,trans*-1,4,7-cyclodecatriene (**91**), in the second case considerable quantities of higher oligomers are formed. The preferred ligands are triphenylphosphine and tris-*o*-biphenylyl phosphite; the highest yield of cyclodecatriene (ca. 90%) is obtained with the first, whereas the relatively small amount of higher boiling material obtained with the second facilitates the workup. The temperature at which the reaction is carried out must be kept as low as possible (< 40°) in order to suppress the thermal Cope rearrangement, which leads to 1,2-dimethyl-*cis*-4,5-divinyl-cyclohexene (**92**) (5, 72, 97, 120, 123, 147). An excess of butadiene has no

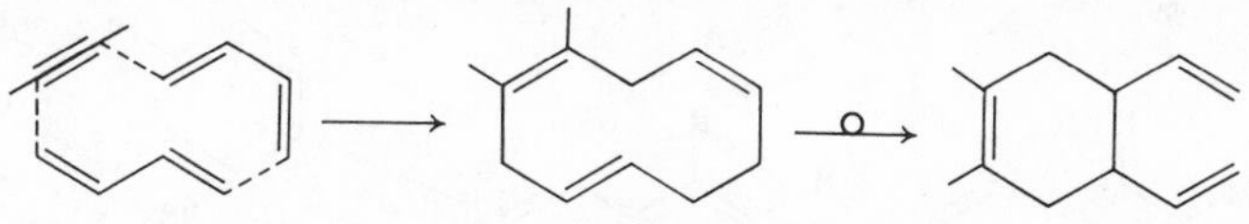

91   **92**

significant effect on the co-oligomerization but a deficiency encourages the formation of products containing more than one alkyne molecule. The isolation of the higher oligomers is hindered by the ease with which thermal rearrangement occurs; identification is limited to that of a 12-membered tetraene (**93**) (76, 96, 149), formed by co-oligomerization of two molecules of butadiene with two of butyne, and that of a vinylcyclohexadiene derivative

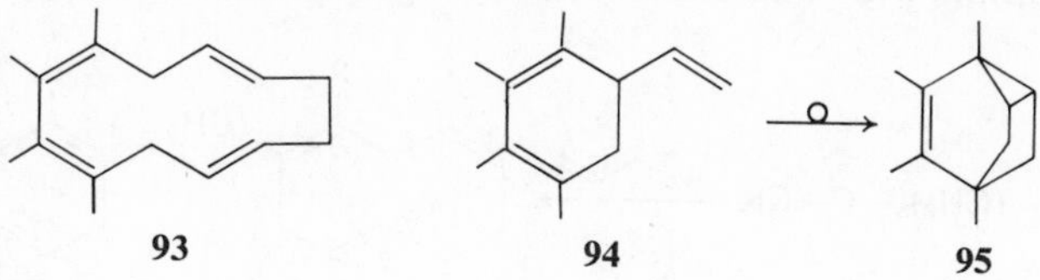

93   **94**   **95**

94, formed by the co-oligomerization of two molecules of butyne with one of butadiene (82, 96, 126). The tetraene is probably formed as a mixture of isomers from which the cis,cis,trans,trans form has been obtained pure. The vinylcyclohexadiene derivative **94** undergoes an internal Diels-Alder reaction under very mild conditions to give the tricyclic compound **95**. Further details of this and related reactions are to be found in Chapter II (Section III). In addition to the formation of **93** and **94**, considerable cyclotrimerization of butyne to hexamethylbenzene occurs.

The co-oligomerization of acetylene with butadiene has not been investigated in detail. In an early publication (40) dealing with the dimerization of butadiene catalyzed by $(R_3P)_2Ni(CO)_2$ and activated by acetylene, it is suggested that the acetylene reacts with butadiene in a side reaction to give divinylcyclohexene [a reinvestigation has confirmed this (168)]. In addition, the product related to **94**, e.g., 5-vinyl-1,3-cyclohexadiene, has been isolated from a reaction catalyzed by the "nickel–ligand" system (125).

Alkynes containing substituents in which functional groups are separated from the triple bond by at least two methylene groups, e.g., $CH_3O(CH_2)_2C\vdots$ $C(CH_2)_2OCH_3$, react to form ten-membered ring systems in high yield. In contrast, considerable amounts of 12-membered ring compounds are produced in the reactions involving 1,4-dimethoxy-2-butyne and 1-methoxy-2-pentyne. Unsymmetrically substituted alkynes generally produce both of the possible isomeric substituted cyclodecatrienes (**96** and **97**). In many cases the isolation and separation of these isomers has been avoided and the product either partially hydrogenated to the substituted cyclodecene (**98**) or rearranged thermally to the divinylcyclohexene derivative (**99**).

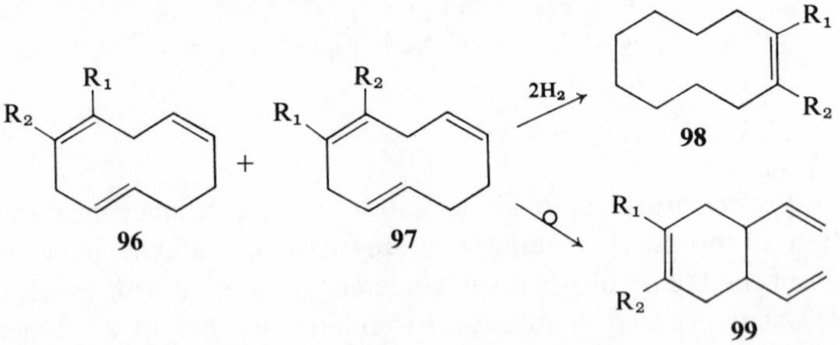

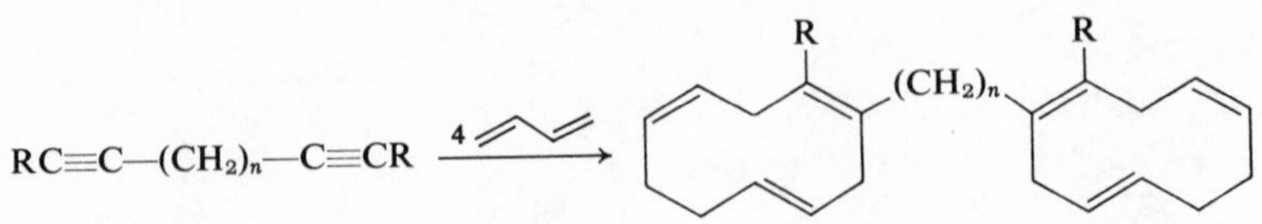

Both of the alkyne groups present in dialkynes are available for co-oligomerization with butadiene. The co-oligomerization of butadiene with cycloalkynes occurs in an analogous manner and forms the basis of an elegant and simple synthesis of large ring compounds. Another synthetically valuable

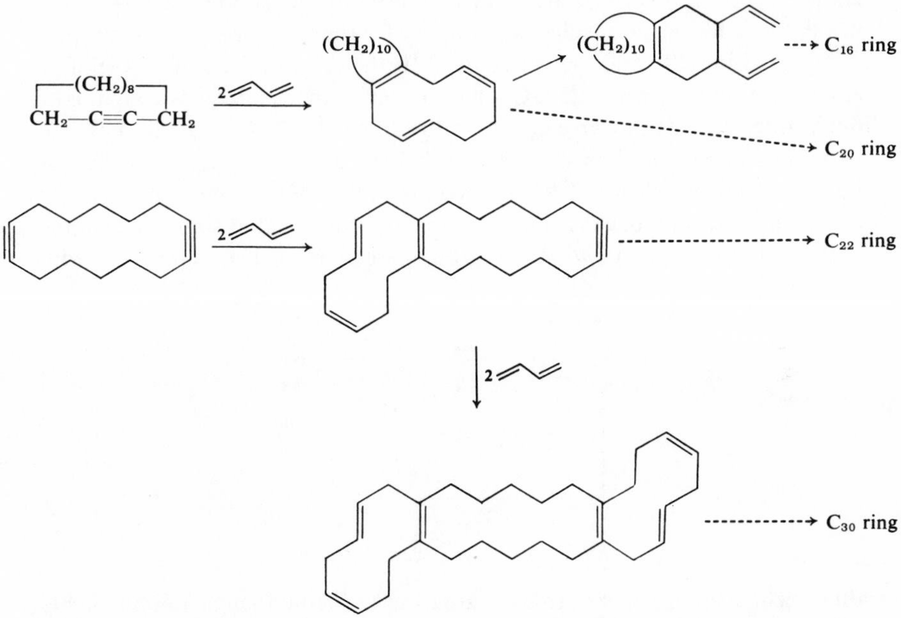

reaction is the essentially quantitative syntheses of substituted benzene derivatives by isomerization of the disubstituted divinylcyclohexene intermediate with potassium *tert*-butoxide (121).

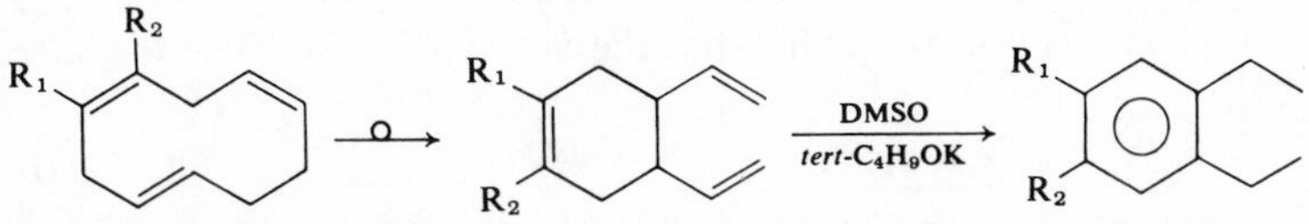

Substituted alkynes in which the substituent contains an alcohol, amine, ketone, or acidic group cannot be co-oligomerized directly: hydrogen transfer reactions leading to open-chain products or destruction of the catalyst occur. The 2-tetrahydropyranyl group has been found to be particularly effective for protecting alcohols (76, 149).

The co-oligomerization of alkynes with 1,3-dienes other than butadiene is limited to a study of the reaction between butyne and isoprene or piperylene (82). Both reactions yield a mixture of ten-membered ring products (which were not isolated but thermally converted into the corresponding divinyl-cyclohexene derivatives) and 12-membered ring compounds.

C. Mechanistic Considerations

The mechanism of the co-oligomerization of butadiene with olefins or alkynes is not completely understood. It is attractive to suppose that the same bis(allyl)-C_8-nickel intermediates postulated for the cyclodimerization reaction are formed initially and that these react further by insertion of an olefin or alkyne molecule to generate a C_{10}-nickel species that undergoes coupling or β-hydrogen transfer to give cyclic or linear products. One of the possible variations of this mechanism is shown below for the co-oligomerization of butadiene with ethylene (13, 97). The configuration of the double bonds in the ten-membered ring (one cis and one trans) indicates that at some stage in the reaction both *syn*- and *anti*-π-allyl groups are generated. In agreement with

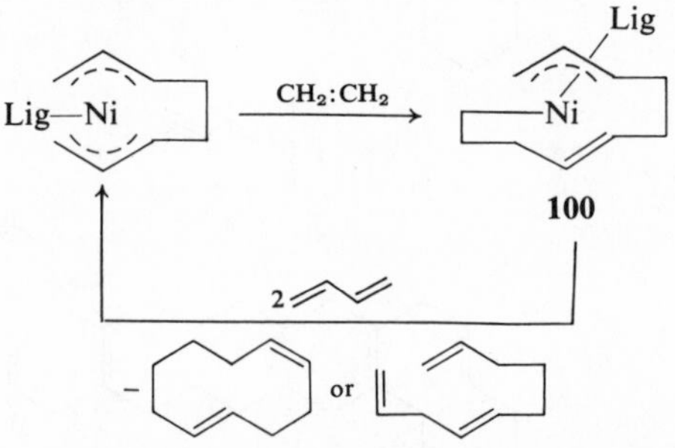

this mechanism, no linear products are observed from the reaction involving alkynes or olefins in which β-hydrogen atoms are absent [β elimination from the intermediate involving a strained olefin (e.g., bicycloheptene) would either involve trans elimination or generate a bridgehead olefin].

The plausibility of this basic concept is underlined by the isolation of **101** (a ligand-free analog to **100**) from the reaction of **39** [Lig $=$ P(C_6H_5)$_3$] with the diethyl ester of acetylenedicarboxylic acid (107, 171). The displacement

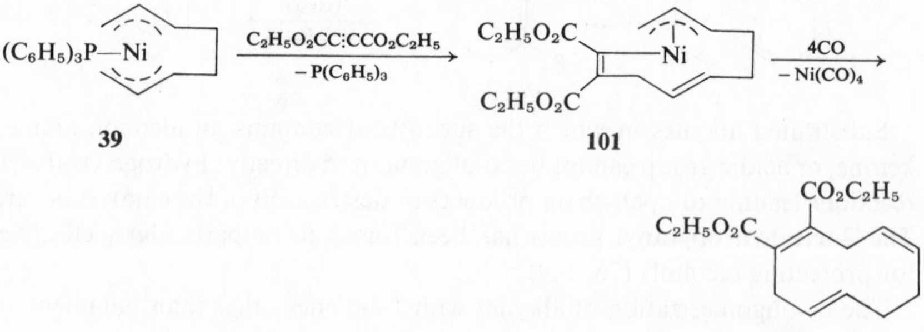

of the ligand during this reaction is not regarded as typical but is a property of this particular alkyne, which is known to react readily with triphenylphosphine (85). Treatment of **101** with CO causes ring closure and formation of the diethyl ester of 4,5-dicarboxylic acid-*cis,cis,trans*-1,4,7-cyclodecatriene (**102**). The syn–anti isomerization of the π-allyl group that necessarily accompanies this reaction has precedence in the isomerization of the π-pentenyl system **103**, which has been observed in solution (141).

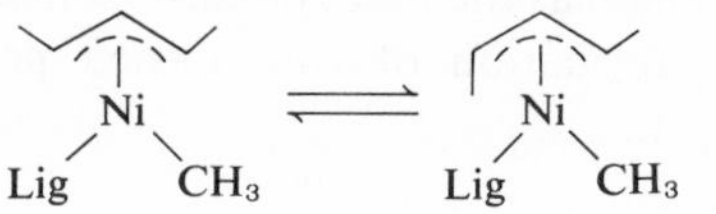

$$\underset{\text{Lig}\qquad \text{CH}_3}{\overset{\text{Ni}}{}} \rightleftharpoons \underset{\text{Lig}\qquad \text{CH}_3}{\overset{\text{Ni}}{}}$$

103

An interesting example of the role played by the ligand is to be found by contrasting the reaction involving ethylene with that involving butyne. Optimum yield of ten-membered ring is obtained in the first case in the presence of triphenyl phosphite and in the second case in the presence of triphenylphosphine. Assuming that both reactions operate under optimal conditions with similar electronic density at the nickel, it can be seen that apparently the better acceptor ligand (triphenyl phosphite) is paired with the better organic π donor (ethylene) (5, 6, 97).

The mechanism discussed above is supported by an analysis of the stereochemistry of the products of the co-oligomerization of ethylene with piperylene, which indicates that the configuration of the methyl groups in the ten-membered ring is identical to that observed in the substituted DVCB derivative produced by cyclodimerization (see p. 150). This is illustrated below for one of the isomers formed from *trans*-piperylene.

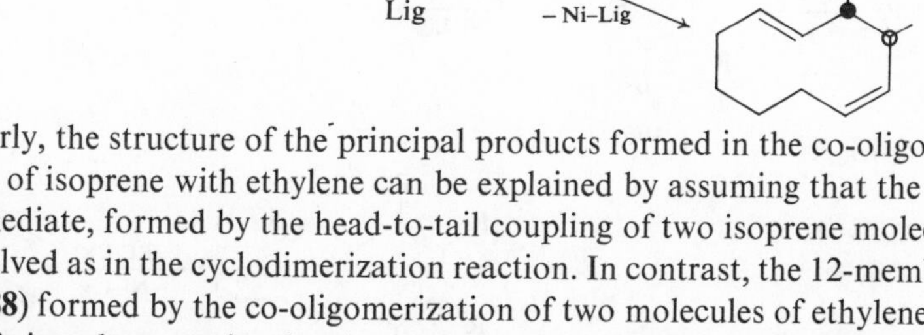

Similarly, the structure of the principal products formed in the co-oligomerization of isoprene with ethylene can be explained by assuming that the same intermediate, formed by the head-to-tail coupling of two isoprene molecules, is involved as in the cyclodimerization reaction. In contrast, the 12-membered ring (**88**) formed by the co-oligomerization of two molecules of ethylene with two of piperylene requires head-to-head coupling if the reaction is to proceed through a C_8 intermediate.

The reactions involving methylenecyclopropane are unusual in that, besides the expected spiro product, opening of the three-membered ring is also observed. Ring opening can occur either by rearrangement of an intermediate spiro complex (e.g., **104**) to give linear products, or by direct insertion of the cyclopropane group into the C_8 chain (e.g., **105**) to give an 11-membered ring. Illustrative reaction mechanisms are shown below. The formulation of the C_8 chain as the π-allyl,σ-allyl isomer is justified here by the observed preferential formation of ring-opened products using the Ni–P($cyclo$-C_6H_{11})$_3$ catalyst.

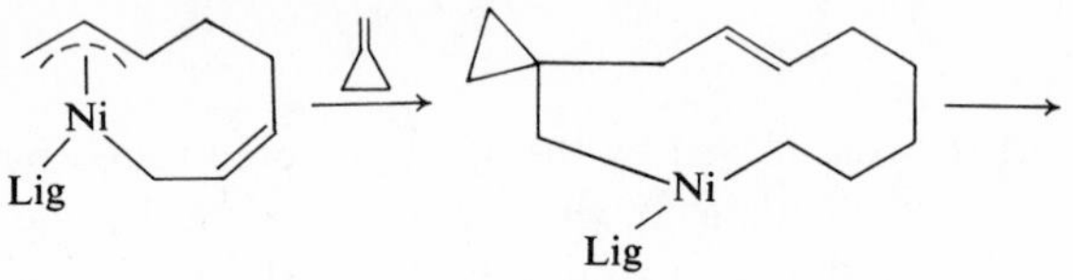

104

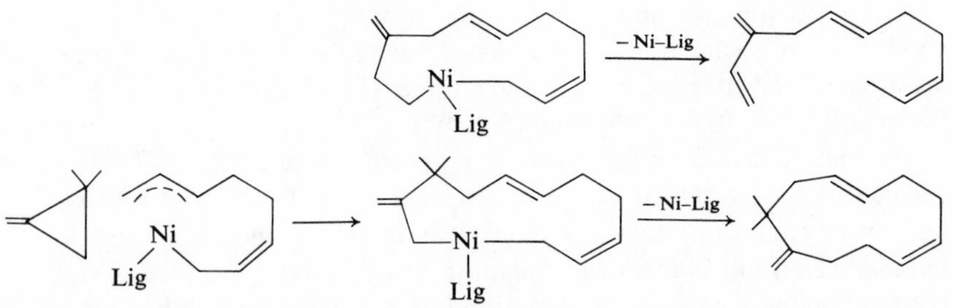

105

The formation of 12-membered rings by the co-oligomerization of butadiene with an excess of alkyne can be equally well explained as either the insertion of butadiene into a metallacyclopentadiene system (**106**) or the insertion of alkyne molecules into the C_8–nickel system (76, 82). A distinction

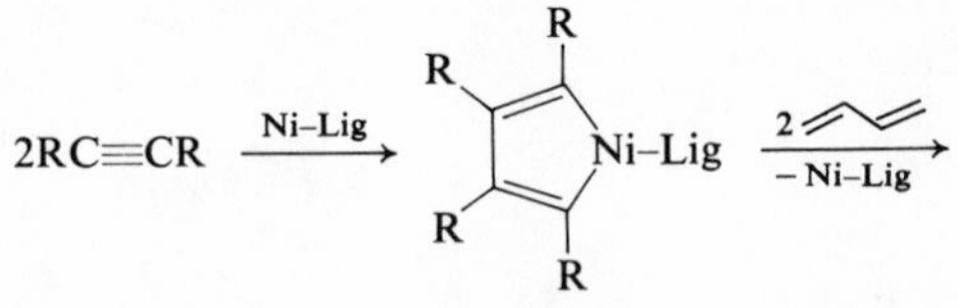

106

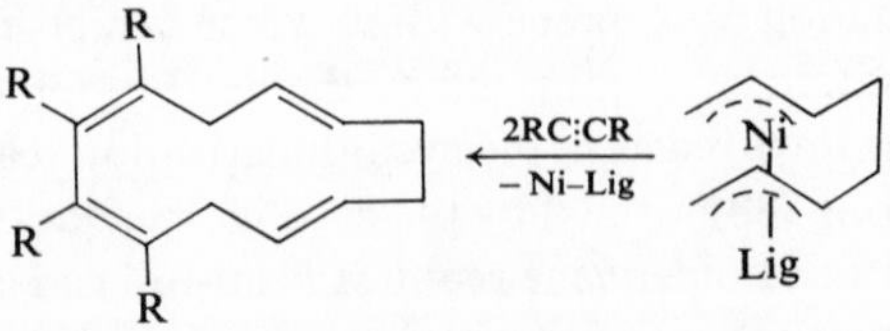

between these two possibilities cannot at present be made and may well depend on the electronic properties of the particular alkyne involved.

TABLE III-8

THE CO-OLIGOMERIZATION OF 1,3-DIENES WITH OLEFINS

Product	Diene	Olefin	Catalyst	Ref.
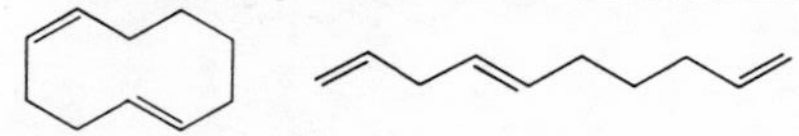		$CH_2:CH_2$	$Ni(acac)_2 \pm Lig-$ $(C_2H_5)_2AlOC_2H_5$	1, 5, 11, 13, 18, 19, 35, 54, 95, 97, 146, 160, P3, P5–P7, P12, P25, P38, P40, P43, P45, P76, P80, P92, P97, P106, P116
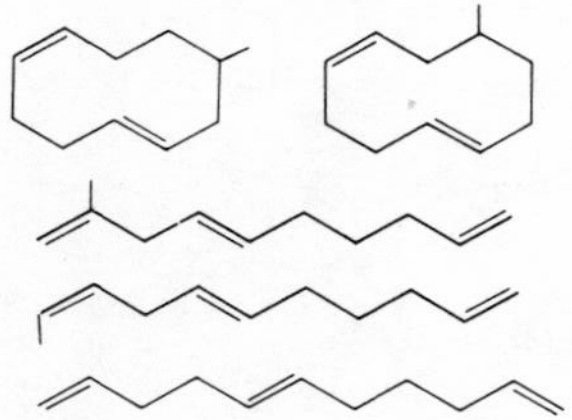		$CH_3CH:CH_2$	$Ni(acac)_2-Lig-$ $(C_2H_5)_2AlOC_2H_5$	8, 73, 102, 146, P18
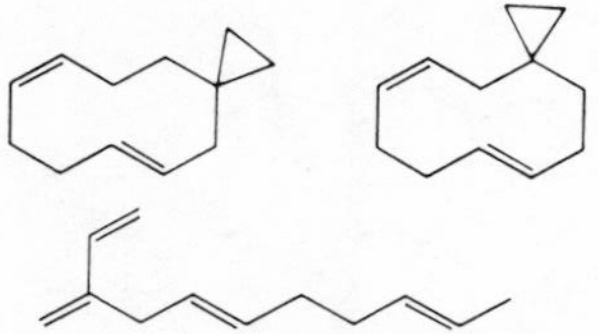		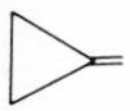	$Ni(acac)_2-Lig-$ $(C_2H_5)_2AlOC_2H_5$	73, 102, 117
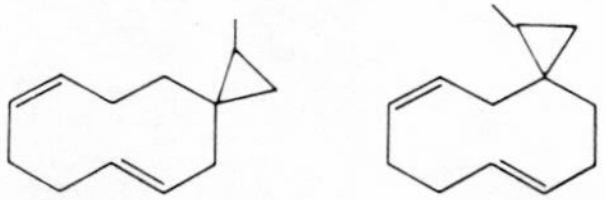	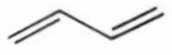	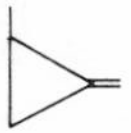	$Ni(acac)_2-Lig-$ $(C_2H_5)_2AlOC_2H_5$	117

(continued)

173

TABLE III-8 (*continued*)

Product	Diene	Olefin	Catalyst	Ref.
			Ni(acac)$_2$–Lig– (C$_2$H$_5$)$_2$AlOC$_2$H$_5$	117
			Ni(acac)$_2$–Lig– (C$_2$H$_5$)$_2$AlOC$_2$H$_5$	117

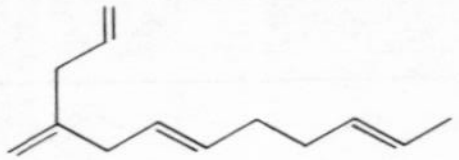

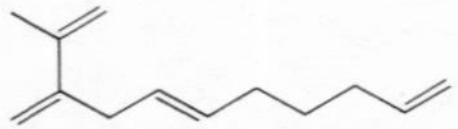

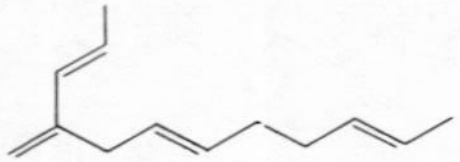

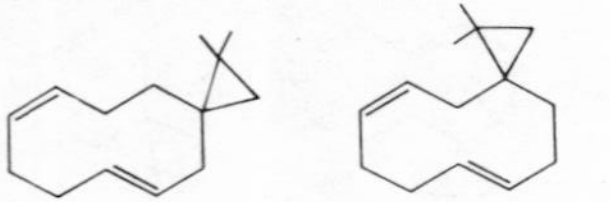

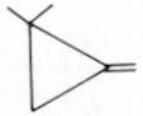

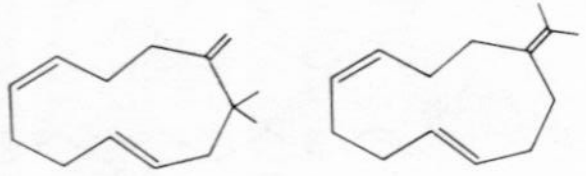

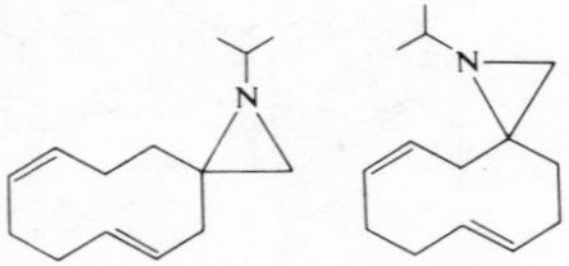

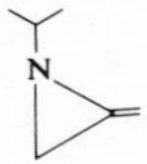

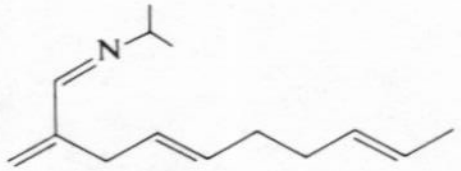

174

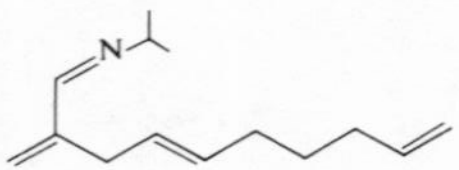

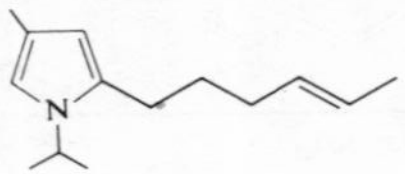

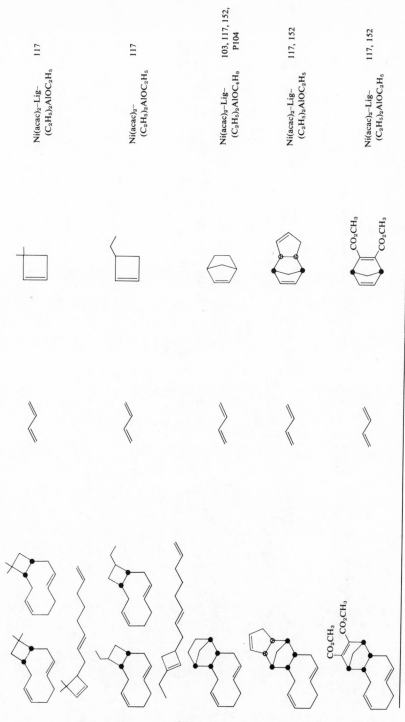

Ni(acac)₂–Lig–(C₂H₅)₂AlOC₂H₅   117

Ni(acac)₂–(C₂H₅)₂AlOC₂H₅   117

Ni(acac)₂–Lig–(C₂H₅)₂AlOC₄H₉   103, 117, 152, P104

Ni(acac)₂–Lig–(C₂H₅)₂AlOC₂H₅   117, 152

Ni(acac)₂–Lig–(C₂H₅)₂AlOC₂H₅   117, 152

(continued)

175

TABLE III-8 (*continued*)

Product	Diene	Olefin	Catalyst	Ref.
		CO_2CH_3 CO_2CH_3	$Ni(acac)_2$–Lig– $(C_2H_5)_2AlOC_2H_5$	117, 152
		CH_2OCOCH_3	$Ni(acac)_2$–Lig– $(C_2H_5)_2AlOC_2H_5$	117, 152
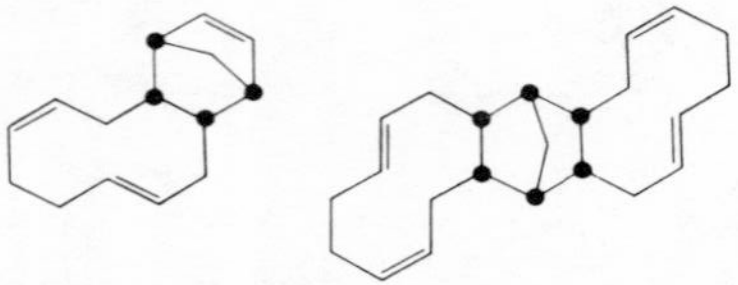			$Ni(acac)_2$–Lig– $(C_2H_5)_2AlOC_2H_5$	117, 152
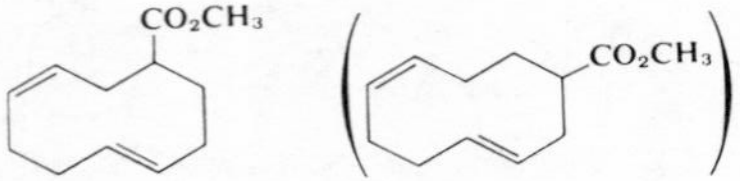 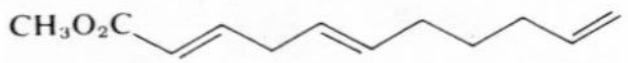 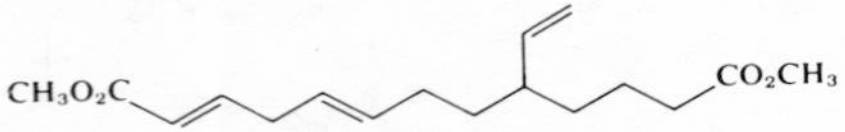 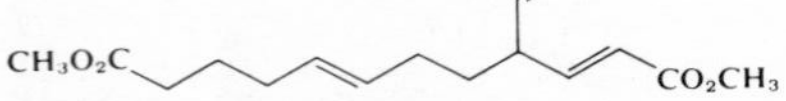		$CH_2\!:\!CHCO_2CH_3$	$Ni(acac)_2$–Lig– $(C_2H_5)_2AlOC_2H_5$	80, 97, P55, P56, P92, P101, P104, P114
		$CH_2\!:\!C(CH_3)CO_2CH_3$	$Ni(acac)_2$–Lig– $(C_2H_5)_2AlOC_2H_5$	44, 80, 97, P101
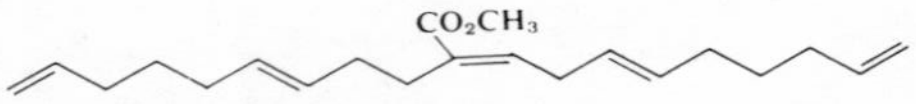		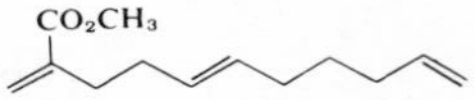 CO_2CH_3	$Ni(acac)_2$–Lig– $(C_2H_5)_2AlOC_2H_5$	44, 80, 97

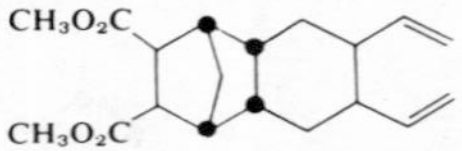

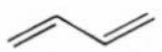

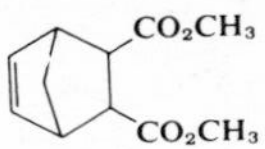

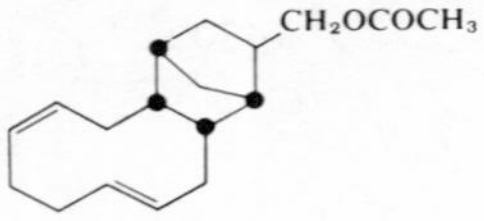

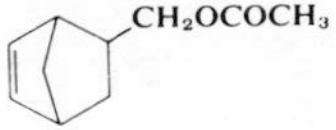

176

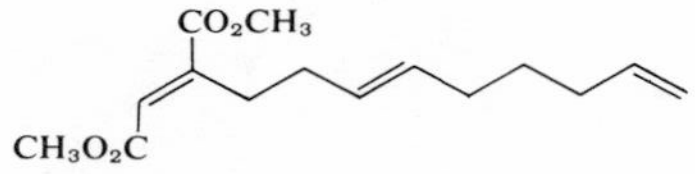  $CH_2:C(CO_2CH_3)$—$CH_2CO_2CH_3$	$Ni(acac)_2$–Lig–$Al(C_2H_5)_3$	90, P117,
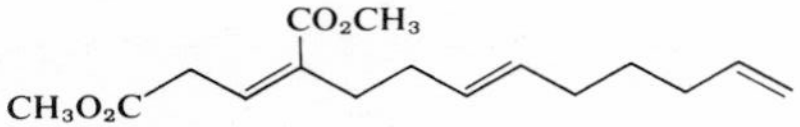 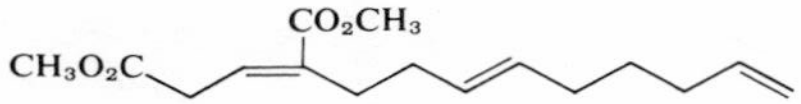  $CH_2:C(CO_2CH_3)$—$(CH_2)_2CO_2CH_3$	$Ni(acac)_2$–Lig–$Al(C_2H_5)_3$	90, P117
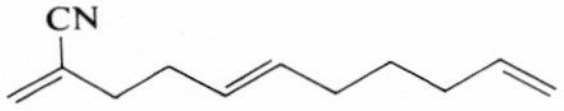  $CH_2:C(CH_3)CN$	$Ni(acac)_2$–Lig–$Al(C_2H_5)_3$	90
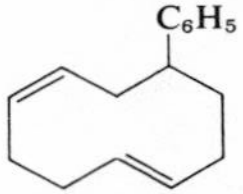 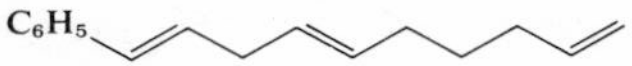 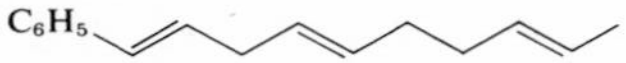 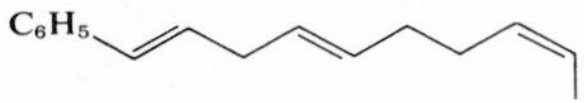  $C_6H_5CH:CH_2$	$Ni(acac)_2$–Lig–$(C_2H_5)_2AlOC_2H_5$	31, 67, 95, 97, P10, P21, P22, P38, P45, P90, P92
 p-$CH_3OC_6H_4CH:CH_2$	$(COD)_2Ni$–$Al(C_2H_5)_3$	P10
 Divinylbenzene	$(COD)_2Ni$–$Al(C_2H_5)_3$	P10
 Vinyl, ethylbenzene	$(COD)_2Ni$–$Al(C_2H_5)_3$	P10
	 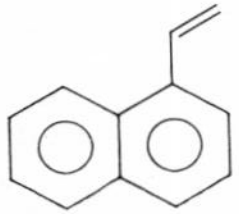 $(COD)_2Ni$–$Al(C_2H_5)_3$	P10

Methoxyphenyldecatriene

Vinyldecatrienylbenzene

Ethyldecatrienylbenzene

Naphthyldecatriene

Nonadecapentaenenitrile

177

(continued)

TABLE III-8 (*continued*)

Product	Diene	Olefin	Catalyst	Ref.
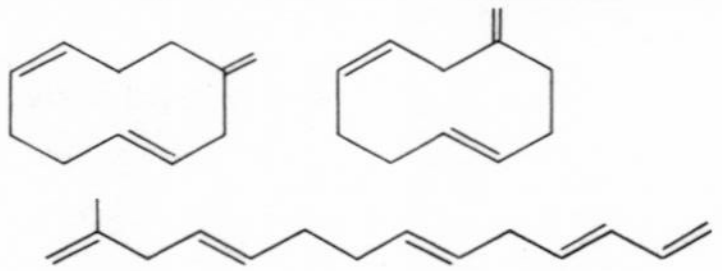		$CH_2:C:CH_2$	Ni(acac)$_2$–Lig–(C$_2$H$_5$)$_2$AlOC$_2$H$_5$	81, 102, 124
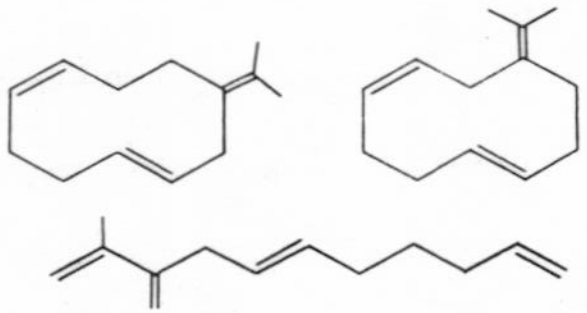		$(CH_3)_2C:C:CH_2$	Ni(acac)$_2$–Lig–(C$_2$H$_5$)$_2$AlOC$_2$H$_5$	81, 124
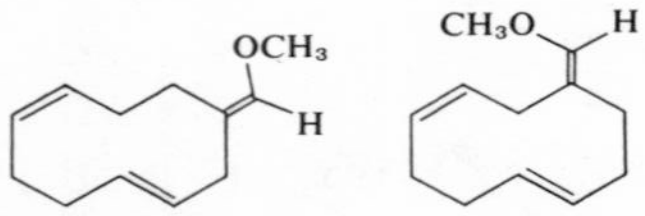		$CH_3OCH:C:CH_2$	Ni(acac)$_2$–Lig–(C$_2$H$_5$)$_2$AlOC$_2$H$_5$	81, 124
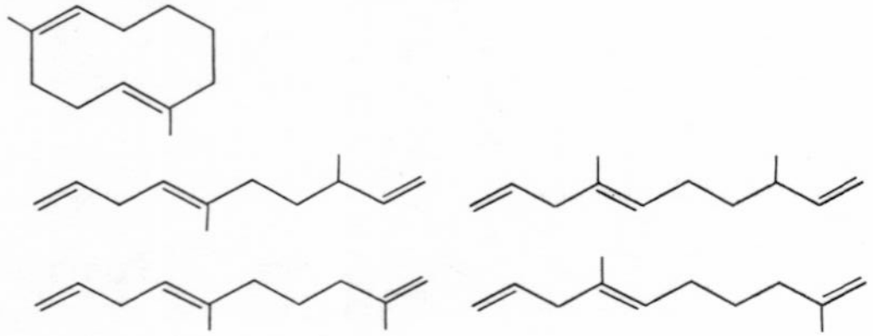	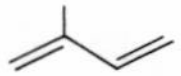	$CH_2:CH_2$	Ni(acac)$_2$–Lig–(C$_2$H$_5$)$_2$AlOC$_2$H$_5$	73, 102, 146, P5
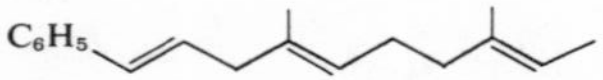	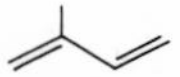	$C_6H_5CH:CH_2$	Ni(acac)$_2$–Lig–Al(C$_2$H$_5$)$_3$	95, P10
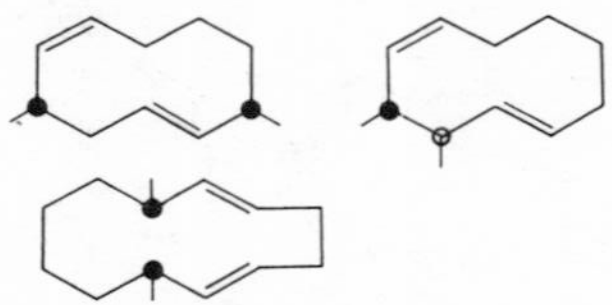		$CH_2:CH_2$	Ni(acac)$_2$–Lig–(C$_2$H$_5$)$_2$AlOC$_2$H$_5$	8, 73, 102

178

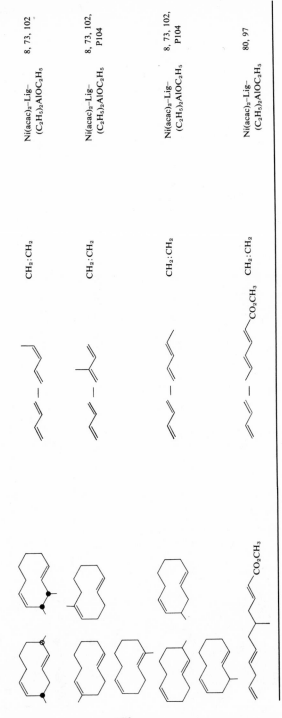

		$CH_2:CH_2$	$Ni(acac)_2-Lig- (C_2H_5)_2AlOC_2H_5$	8, 73, 102
		$CH_2:CH_2$	$Ni(acac)_2-Lig- (C_2H_5)_2AlOC_2H_5$	8, 73, 102, P104
		$CH_2:CH_2$	$Ni(acac)_2-Lig- (C_2H_5)_2AlOC_2H_5$	8, 73, 102, P104
		$CH_2:CH_2$	$Ni(acac)_2-Lig- (C_2H_5)_2AlOC_2H_5$	80, 97

179

TABLE III-9

THE CO-OLIGOMERIZATION OF 1,3-DIENES WITH ALKYNES

Product[a]	Diene	Alkyne	Catalyst	Ref.
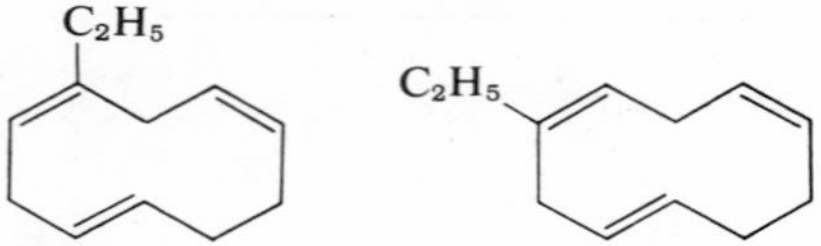	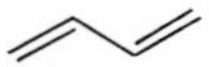	$HC:CCH_2CH_3$	$Ni(acac)_2-Lig-(C_2H_5)_2AlOC_2H_5$	76
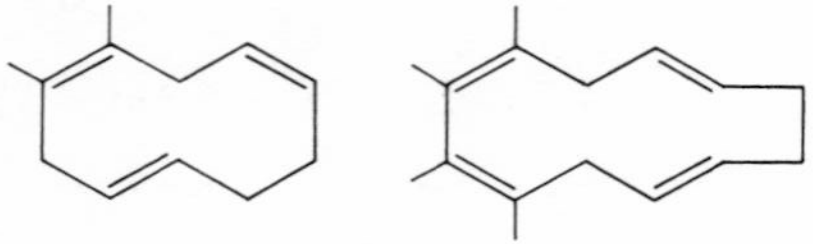	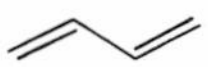	$CH_3C:CCH_3$	$Ni(acac)_2-Lig-(C_2H_5)_2AlOC_2H_5$	5, 54, 72, 76, 82, 96, 97, 123, 147, P104, P106
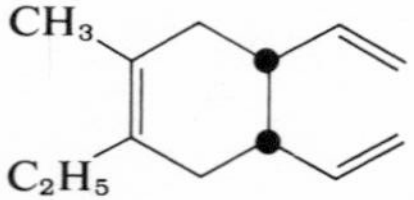		$CH_3C:CCH_2CH_3$	$(COD)_2Ni-Lig$	82, 147
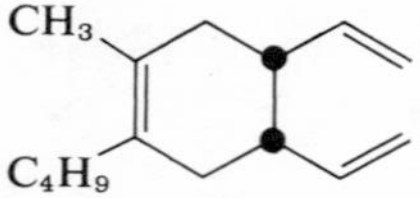		$CH_3C:CC_4H_9$	$(COD)_2Ni-Lig$	82, 147
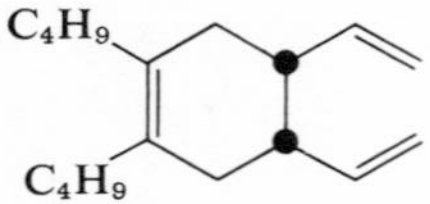	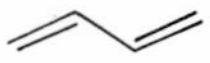	$C_4H_9C:CC_4H_9$	$(COD)_2Ni-Lig$	82, 147
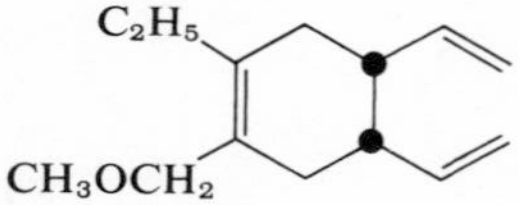		$C_2H_5C:CCH_2OCH_3$	$Ni(acac)_2-Lig-(C_2H_5)_2AlOC_2H_5$	76, 82, 96, 149

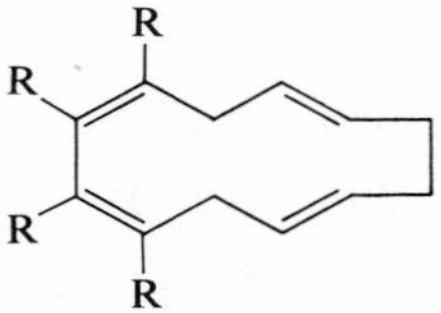

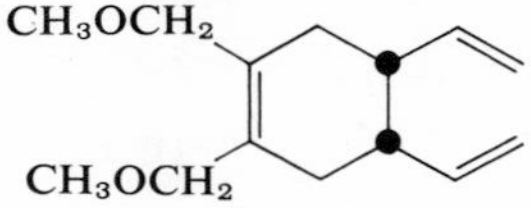

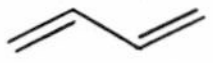

$CH_3OCH_2C:CCH_2OCH_3$ $Ni(acac)_2$–Lig– $(C_2H_5)_2AlOC_2H_5$ 76, 82, 96, 149

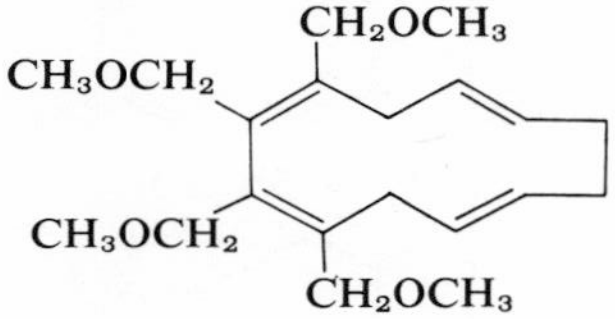

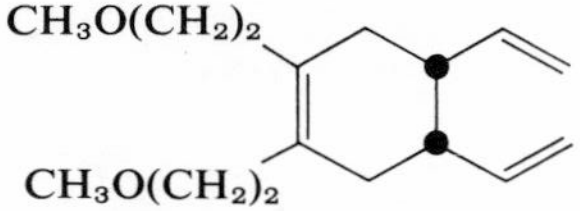

$CH_3O(CH_2)_2C:C-(CH_2)_2OCH_3$ $(COD)_2Ni$–Lig   82

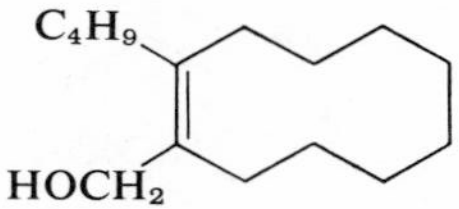

$C_4H_9C:CCH_2Opyr^b$ $Ni(acac)_2$–Lig– $(C_2H_5)_2AlOC_2H_5$ 76, 149

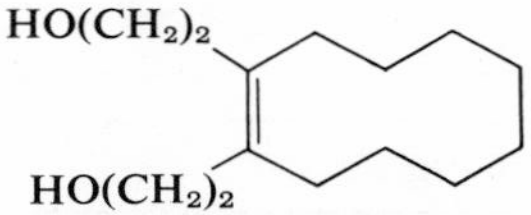

$pyrO(CH_2)_2C:C-(CH_2)_2Opyr^b$ $Ni(acac)_2$–Lig– $(C_2H_5)_2AlOC_2H_5$ 76, 149

$R = C_2H_5$ or CH_2OCH_3

(continued)

TABLE III-9 (continued)

Product[a]	Diene	Alkyne	Catalyst	Ref.
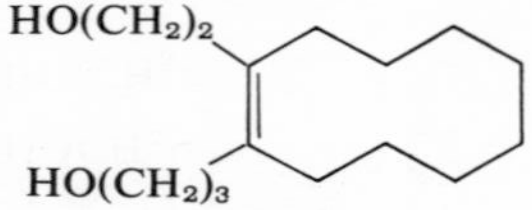		$pyrO(CH_2)_2C:C-(CH_2)_3Opyr^b$	$Ni(acac)_2-Lig-(C_2H_5)_2AlOC_2H_5$	76, 149
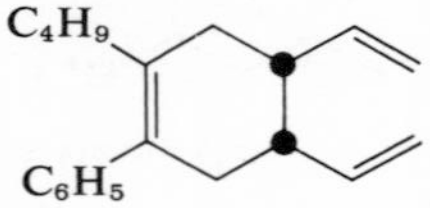	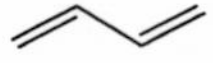	$C_4H_9C:CC_6H_5$	$(COD)_2Ni-Lig$	82
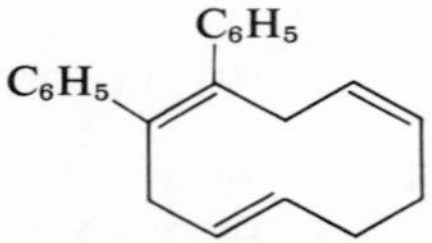	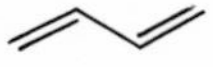	$C_6H_5C:CC_6H_5$	$(COD)_2Ni-Lig$	82
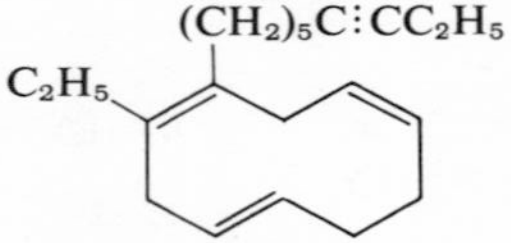 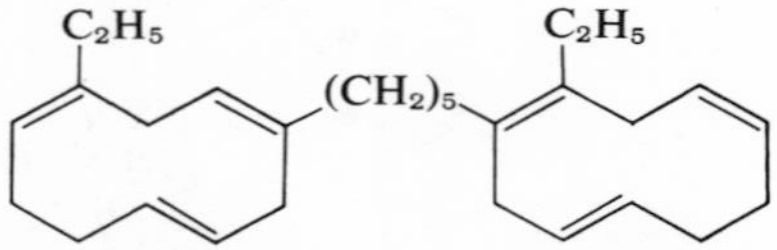		$C_2H_5C:C(CH_2)_5C:C-C_2H_5$	$(COD)_2Ni-Lig$	72, 147
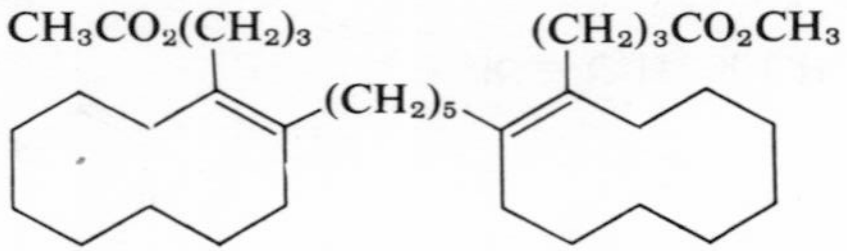		$CH_3CO_2(CH_2)_3C:C-(CH_2)_5C:C(CH_2)_3CO_2CH_3$	$Ni(acac)_2-Lig-(C_2H_5)_2AlOC_2H_5$	76, 149

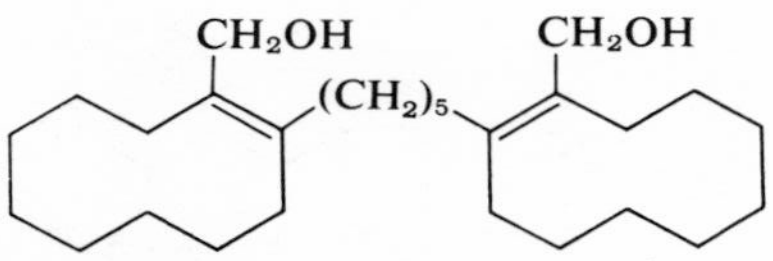

76, 149

Ni(acac)$_2$–Lig–
(C$_2$H$_5$)$_2$AlOC$_2$H$_5$

pyrOCH$_2$C:C(CH$_2$)$_5$–
C:CCH$_2$Opyr

72, 122, 148,
P104

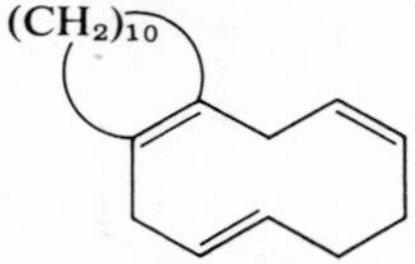

Ni(acac)$_2$–Lig–
(C$_2$H$_5$)$_2$AlOC$_2$H$_5$

$\overset{\text{(CH}_2)_8}{\underset{\text{CH}_2\text{C:CCH}_2}{\rule{0pt}{0pt}}}$

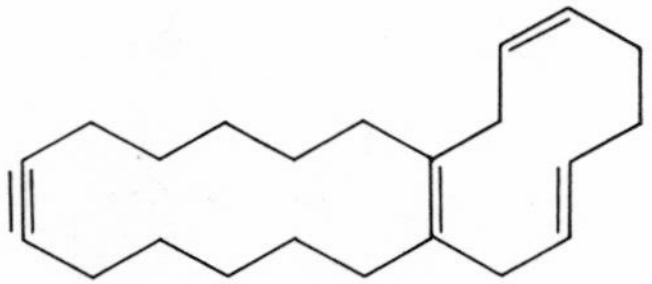

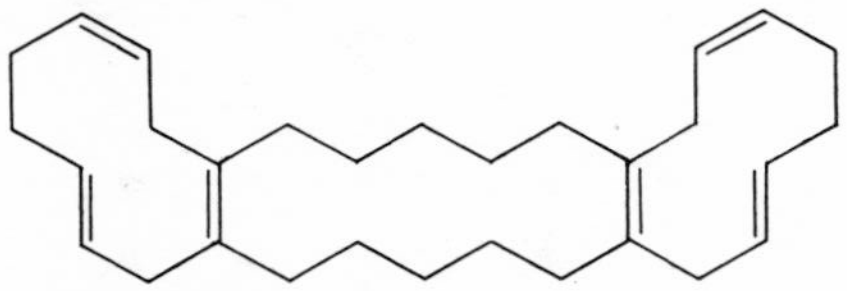

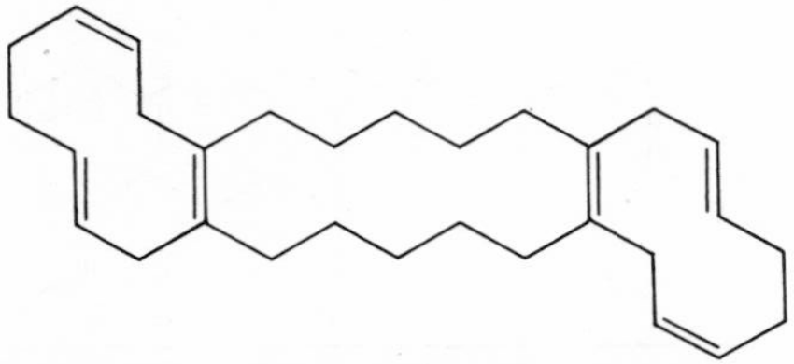

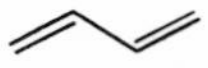

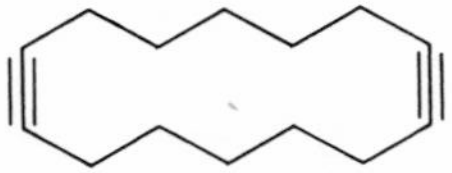

72, 122, 148,
P104

Ni(acac)$_2$–Lig–
(C$_2$H$_5$)$_2$AlOC$_2$H$_5$

(*continued*)

TABLE III-9 (*continued*)

Product[a]	Diene	Alkyne	Catalyst	Ref.
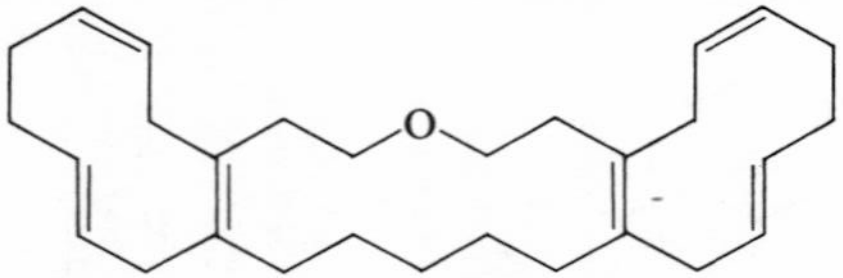	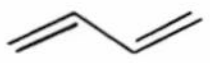	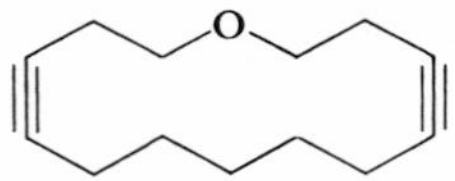	$Ni(acac)_2$–Lig–$(C_2H_5)_2AlOC_2H_5$	72, 148
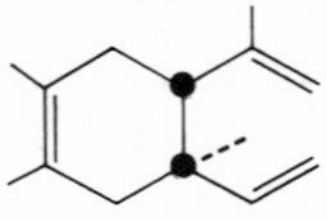 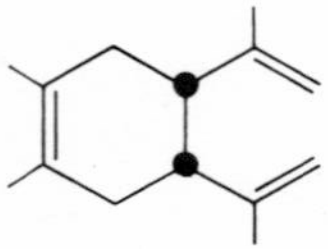 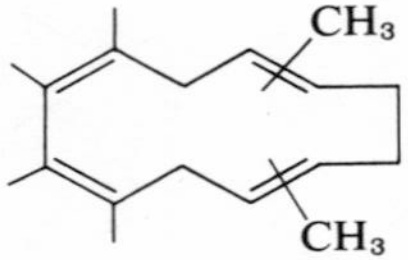	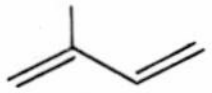	$CH_3C{:}CCH_3$	$(COD)_2Ni$–Lig	82
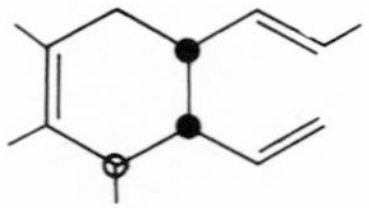 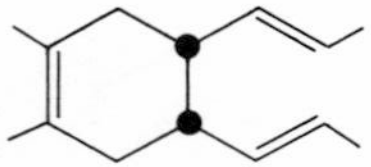 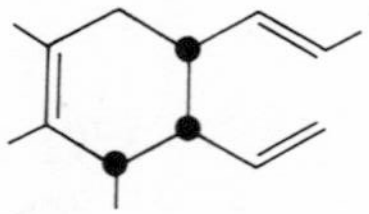 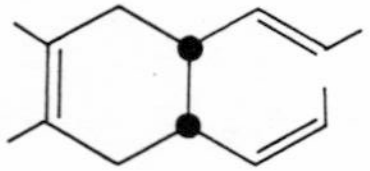 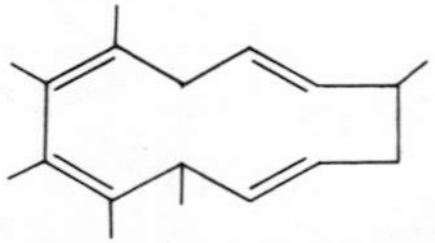	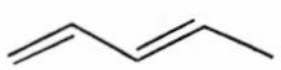	$CH_3C{:}CCH_3$	$(COD)_2Ni$–Lig	82

[a] In some cases the isolated product has been obtained by hydrogenation or thermal rearrangement.   [b] pyr = pyranyl.

VI. The Linear Oligomerization and Telomerization of 1,3-Dienes

The same catalysts that cyclo-oligomerize butadiene can be modified to produce mainly linear oligomers by adding a cocatalyst. This reaction is accompanied by telomerization with the cocatalyst and under certain conditions this can become the principal reaction. Cocatalysts particularly effective in promoting linear oligomerization include alcohols, phenols, and secondary amines, whereas telomerization reactions have been observed, in addition, with compounds containing active methylene groups, trimethylsilane, and hydrogen cyanide.

A. The Formation of Linear Oligomers

In general, the catalyst is based on the "nickel–ligand" system and has been prepared by using either preformed zerovalent nickel–ligand complexes, e.g., $Lig_2Ni(CO)_2$, $(CH_2:CHCN)_2NiLig_2$ (9, 95, 161, P9, P64, P65, P128, P129), and $[(C_6H_5)_3P]_4Ni$ (39, 42), or hydridonickel complexes, e.g., $HNiCl[P(cyclo-C_6H_{11})_3]_2$ (42, 154), or arylnickel complexes, e.g., o-tolylNiNCS-$[P(C_2H_5)_3]_2$ (23, 24). Reduction of a nickel salt in the presence of the ligand has also been used; the reducing agents employed include aluminum-alkyls (14, 79, 97, P103), lithium aluminum-hydride and sodium borohydride (64, 69, 88, P81), lithium-alkyls (24, 130, 163, 164, P71, P132), electric current (39, 127, 128), or sodium methoxide (88, 161). It is important that the reducing agent be present in greater than equimolar concentrations. The activity shown by the nickel($1+$) complex $[(C_6H_5)_3P]_2NiCl$ is probably the result of a disproportionation reaction to give a zerovalent nickel complex (117). Catalysts that may perhaps be regarded as "ligand-free" have been prepared by the electrolysis of nickel chloride in ammonia (39, 127, 128; see also 86).

The product of the reaction consists mainly of linear dimers and trimers (in addition to cyclic dimers) and the composition can be influenced by varying both the cocatalyst and the ligand. The reactions in which a secondary amine (e.g., morpholine or piperidine) is used as cocatalyst produce a mixture of *trans,trans*- and *cis,trans*-1,3,6-octatriene (**107** and **108**) and the three dodecatetraenes **109**, **110**, and **111**. The tetraenes have been shown to be formed by reaction of the octatriene with butadiene (14, 79, 97; see also 164).

107   108

109   110   111

Varying the donor ligand effects the distribution and rate of formation of the product but not its nature. Catalysts involving triethyl phosphite and triphenylphosphine are particularly active. Similar effects have been observed using phenol or a substituted phenol (e.g., m-cresol) as cocatalyst but in this case the principal linear oligomer formed is 1,3,7-octatriene, irrespective of the ligand present (9, P64, P65). In contrast, a remarkable ligand effect has been observed using the system $(CH_2:CHCN)_2Ni$–Lig–C_2H_5OH: with $P(OC_2H_5)_3$ as ligand the product is mainly 1,3,6-octatriene; with $P(C_4H_9)_3$, 1,3,7-octatriene; and with trismorpholinophosphine, 2,4,6-octatriene (95, P9, P129). The 1,3,7 isomer is also the main product from the reactions catalyzed by $[(C_6H_5)_3P]_4Ni$–C_2H_5OH (39) and $[(cyclo$-$C_6H_{11})_3P]_2NiH(BH_4)$ (154).

The catalysts prepared by the electrolytic reduction of nickel chloride in methanol in the presence of pyridine, or by reduction of a $NiCl_2$–ligand system with $LiAlH_4$ or $NaBH_4$, convert butadiene into a mixture of dimers, trimers, tetramers, and telomers; the reaction is accompanied by partial hydrogenation of the product. The dimers have been identified as 1,6-octadiene and 1,7-octadiene; the trimers as 5-vinyl-2,8-decadiene, 3-methyl-1,5,10-undecatriene, and 1,6,10-dodecatriene; and the tetramers as 5-vinyl-1,8,13-tetradecatriene and 1,6,10,15-hexadecatetraene (39, 69, 127, 128; see also 86).

Investigations of the linear oligomerization of 1,3-dienes other than butadiene are confined to isoprene, piperylene, and 1,3,6-octatriene using morpholine as the cocatalyst (79, P103). In all cases dimers are produced. Interestingly, the carbon skeletons of the C_{16} hydrocarbons produced by hydrogenation of the octatriene dimers, i.e., **112** and **113**, are different from those of the

                    112   **113**

tetramers mentioned above. The linear co-oligomerization of butadiene with isoprene, as well as with 1,3,6-octatriene and 1,3,7-octatriene, has also been studied (79, 102, P134).

We should also mention that the presence of a cocatalyst is not always a necessary requirement for the linear oligomerization of isoprene: linear dimers are reported to be the main products formed during the "ligand-free" catalyzed reaction (P48, P96), as well as from the reaction catalyzed by $[(C_6H_5)_3P]_2NiX_2$–C_2H_5MgBr (50), and linear trimers (mainly farnessene, **114**) are obtained from reactions catalyzed by the "nickel–ligand" systems (65).

A curiosity can be conveniently included here. It is claimed in a patent that

114   **115**

the product of the reaction of butadiene with a catalyst formed by reducing a nickel halide with phenylmagnesium bromide is a mixture of 3-methyl-1,5-heptadiene, 1,3,5-cyclooctatriene, and a trimer having the formula $C_{12}H_{16}$ (P93).

An interesting byproduct observed in the reactions catalyzed by $(CH_2:CHCN)_2NiLig_2$ is the five-membered ring compound 1-methylene-2-vinylcyclopentane (**115**) (9, 95, P9, P64, P65). This dimer is the main product from the reaction catalyzed by a variety of ligand-modified nickel complexes in the presence of a cocatalyst and an alcohol, e.g., $[(C_2H_5O)_3P]_4Ni–CF_3CO_2H$ (42), $[(C_4H_9)_3P]_2NiBr_2–NaBH_4$ (24, 88), $NaOCH_3$ (88) or C_4H_9Li (22, 24, 63, 163, P132), o-tolylNiBr$[P(C_2H_5)_3]_2$ (23, 24, 63), and $(\pi\text{-}C_3H_5NiBr)_2–$ $P(C_4H_9)_3$ (22, 24, P138). It is important that the molar concentration of the cocatalyst does not exceed that of the nickel component; otherwise linear dimers predominate. The reaction also proceeds (less efficiently) in the presense of amines (23, 64). In some cases minor changes in the composition of the nickel component, e.g., substitution of a halide by NO_2 or NCS or of $P(C_4H_9)_3$ by $P(C_6H_5)_3$ or $P(cyclo\text{-}C_6H_{11})_3$, result in the suppression of the formation of methylenevinylcyclopentane and linear dimers are formed instead. An optically active form of methylenevinylcyclopentane has been obtained from the reaction of methylenecyclopropane with a bisphosphine nickel dibromide–C_4H_9Li catalyst containing the optically active phosphine $(-)$-methylphenyl-n-propylphosphine. The methylenecyclopropane first undergoes a nickel-catalyzed rearrangement to butadiene (see for example 92), which is then dimerized (60). The optical purity of the product is not known. Conversion of 1,3,7-octatriene into methylenevinylcyclopentane has been carried out with a $[(C_4H_9)_3P]_2NiBr_2$ catalyst in the presence of sodium ethoxide (88). The reaction is accompanied by considerable isomerization to 2,4,6-octatriene. Related to this is the observation that in the presence of the catalyst $[(C_4H_9)_3P]_2Ni(NO_3)_2–KO\text{-}tert\text{-}C_4H_9$, 2,7-octadienylisopropyl ether is converted into the cyclopentane derivative (155).

B. *The Telomerization of 1,3-Dienes* (Table III-10)

The linear dimerization of butadiene described in the preceding section is accompanied by the telomerization of the diene with the cocatalyst and in certain cases this can become the principal reaction. The reported investigations cover the reaction of butadiene with compounds containing active methylene groups, alcohols, phenol, amines, silanes, and hydrogen cyanide.

Although "ligand-free catalysts", e.g., $(COD)_2Ni$ (47), $Ni(CO)_4$ (47, P125), nickelocene (173) and bis(π-allyl)-C_{12}-Ni (14), are active in the telomerization reaction, most attention has been given to the "nickel–ligand" catalyst, which may be prepared *in situ* by reducing a nickel salt in the presence of the ligand or by using preformed zerovalent nickel–ligand complexes. The most frequently used reducing agent has been $NaBH_4$ (64, 66, 129, 131–133, P81, P127); other reagents that are effective include butyl lithium (130, 163), electric current (39), aluminum-alkyls (94, 132, 172), and the ligand itself (66, 129, 131, 135). The preformed catalysts used are of the type Lig_4Ni (39, 42, 43, 134, 175, P113, P126, P133), $Lig_2Ni(CO)_2$ (43, 47), $(CH_2:CHCN)_2$-NiLig (47), $(COD)_2Ni$–Lig (14, 43, 47, 89, 97, 135), and [π-allylNi(Lig)$_2$]$^+$ (175). The hydrosilylation reaction is also catalyzed by Lig_2NiX_2 complexes (93, 119, see also 173). It has recently been shown that the addition of tris-alkoxyaluminum compounds to a Lig_4Ni catalyst facilitates the telomerization of morpholine with butadiene (170).

The product is a mixture of the butene and octadiene derivatives **116–119**.

116   **117**

118   **119**

The straight-chain octadiene derivative formed is most frequently identified as the 2,7 isomer **118** but the 2,6 isomer **120** has been isolated in some cases from the reaction involving morpholine (14, 97), trimethylsilane (47), and ethanol (P81). In two cases trimeric species have also been isolated: 1-morpholino-2,6,10-dodecatriene has been detected in the product of the reaction catalyzed by a "ligand-free" catalyst (14) and methoxy-2,6,11-dodecatriene is formed during the electrolysis of $NiCl_2$–pyridine in the presence of butadiene and methanol (39). Traces of bisbutenyl and bisoctadienyl derivatives (e.g., **121**) are produced in the reaction of butadiene with the primary amine $C_4H_9NH_2$ and with diethylmalonate (66, 129).

120 $C_2H_5O_2C$ $CO_2C_2H_5$
                 121

Mechanistically important are the observations that the ternary system $Ni(acac)_2-P(C_4H_9)_3-C_4H_9Li$ catalyzes the reaction of butadiene with the 1-methoxy-2-butene (116, X = OCH_3) to give 118 (X = OCH_3) and with 3-methoxy-1-butene (117, X = OCH_3) to give 119 (X = OCH_3) (130). Analogous behavior however, has, been ruled out for the corresponding phenoxy compounds (134).

Methyl-substituted 1,3-dienes take part less readily than butadiene in telomerization reactions and give predominantly butene derivatives (129, 135). For example, the reaction of benzylmethyl ketone with isoprene gives a mixture of the butene and octadiene derivatives 122–125, of which 122 and

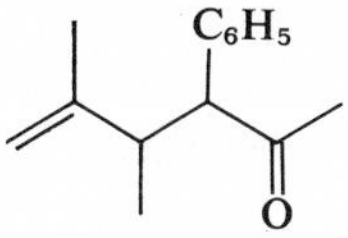

122 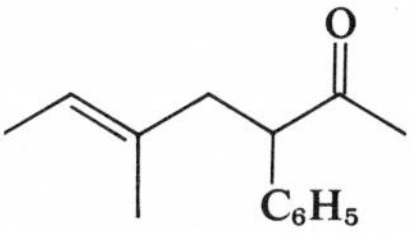 123

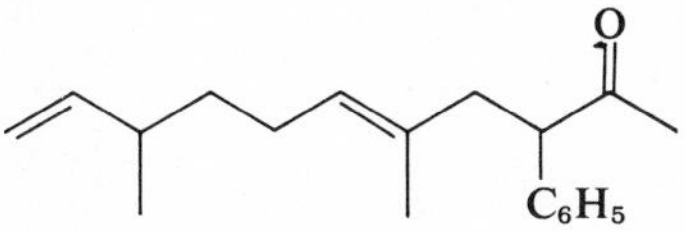

124 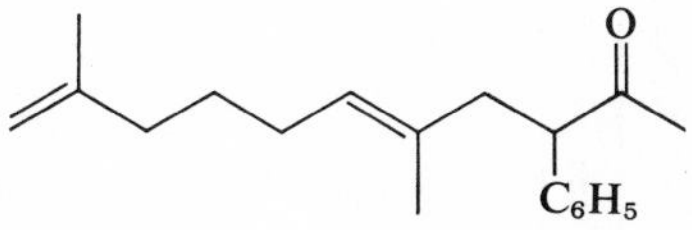 125

123 constitute 77%. The reaction of methoxybutene with isoprene has also been studied and a compound tentatively assigned as 1-methoxymethyl-2,7-octadiene obtained in a small yield (130).

The reaction of isoprene with phenol using the catalyst $(\pi\text{-}C_5H_5)_2Ni-C_6H_5MgBr$, with or without additional ligands, is quite different from the corresponding reaction involving butadiene: instead of the (formal) addition of phenol to the diene, alkylation of the aromatic ring occurs. The main products are 126–128; lesser amounts of dialkylated compounds are also

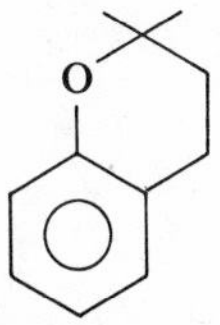

126 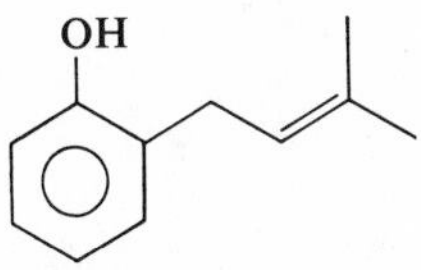 127 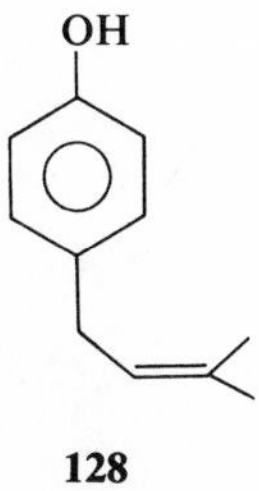 128 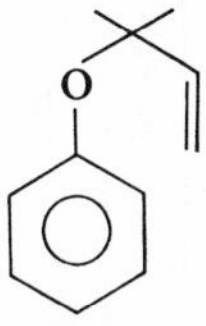 129

formed (139). Although this reaction is highly reminiscent of the Claisen rearrangement, its logical precursor, 129 has not been detected.

TABLE III-10

THE TELOMERIZATION OF 1,3-DIENES

Product	Diene	Telomer	Catalyst	Ref.^{a}
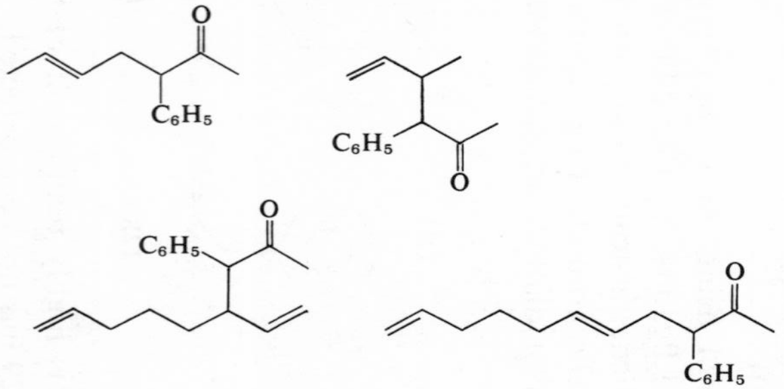		$C_6H_5CH_2COCH_3$	$Ni(acac)_2-C_6H_5P(O\text{-}iso\text{-}C_3H_7)_2-NaBH_4-NaOC_6H_5$	129, 131
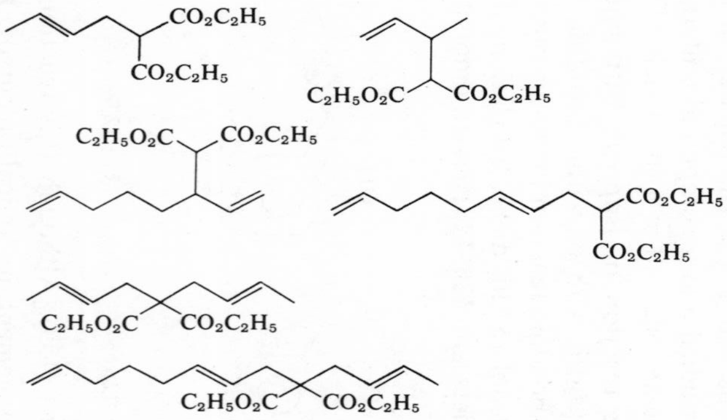	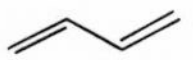	$(C_2H_5OCO)_2CH_2$	$Ni(acac)_2-C_6H_5P(O\text{-}iso\text{-}C_3H_7)_2-NaBH_4-NaOC_6H_5$	129
		$CH_3COCH_2CO_2C_2H_5$	$Ni(acac)_2-C_6H_5P(O\text{-}iso\text{-}C_3H_7)_2-NaBH_4-NaOC_6H_5$	129

190

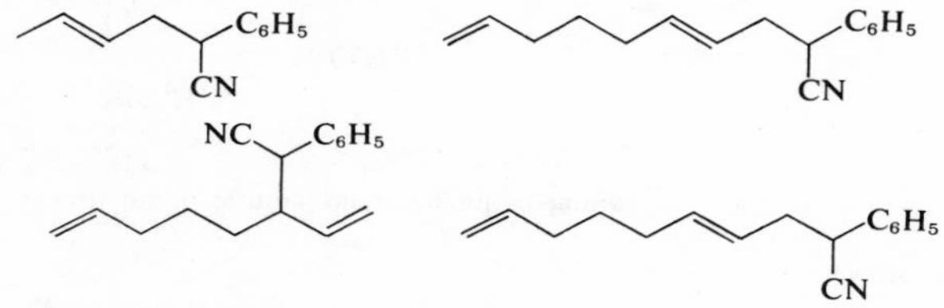	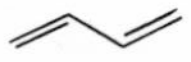	$C_6H_5CH_2CN$	$Ni(acac)_2-C_6H_5P(O\text{-}iso\text{-}C_3H_7)_2-NaBH_4-NaOC_6H_5$	129
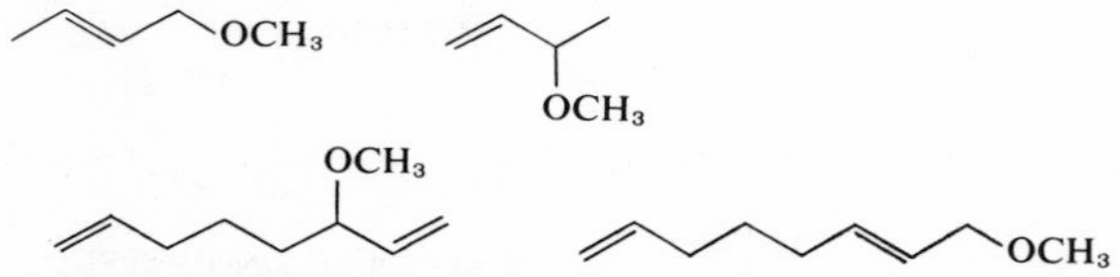	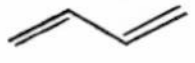	CH_3OH	$Ni(acac)_2-P(C_4H_9)_3-C_4H_9Li$	39, 130, 133, 135, 161, 162, 163
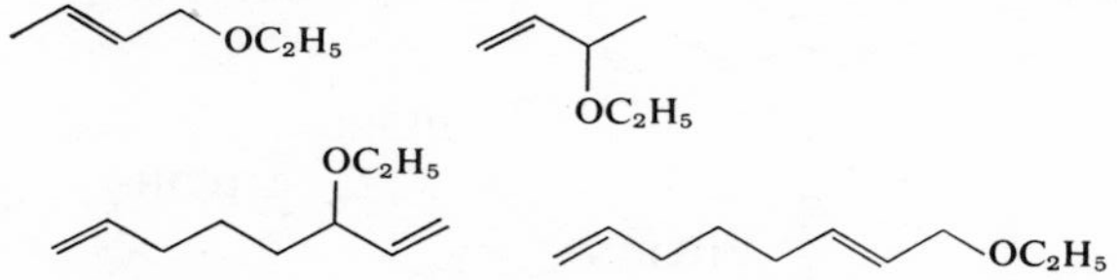		C_2H_5OH	$Ni(acac)_2-C_6H_5P(O\text{-}iso\text{-}C_3H_7)_2-NaBH_4$	39, 133, 163, P81
	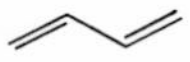	C_3H_7OH	$Ni(acac)_2-P(C_4H_9)_3-C_4H_9Li$	130
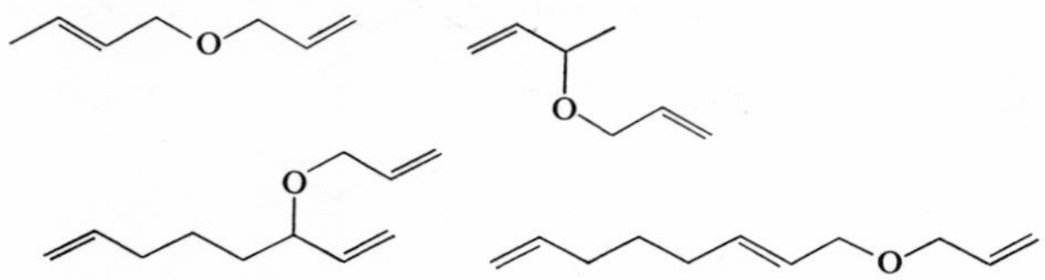		$CH_2{:}CHCH_2OH$	$Ni(acac)_2-P[N(CH_3)_2]_3-C_4H_9Li$	163
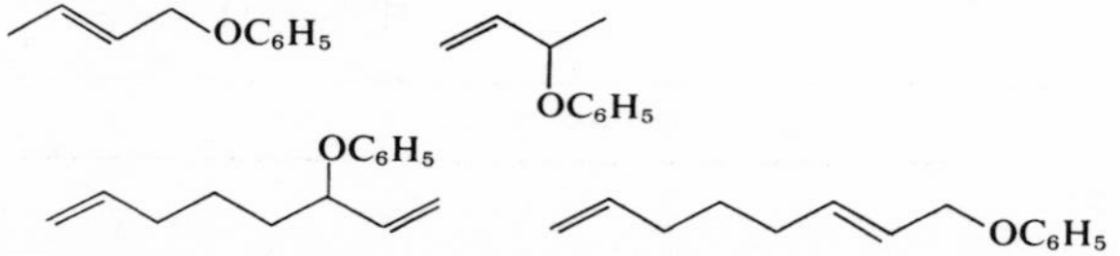	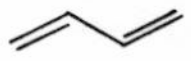	C_6H_5OH	$(R_3P)_4Ni$	43, 133, 134, 163, P113

n-propoxybutene, n-propoxyoctadiene

(continued)

TABLE III-10 (*continued*)

Product	Diene	Telomer	Catalyst	Ref.^{a}
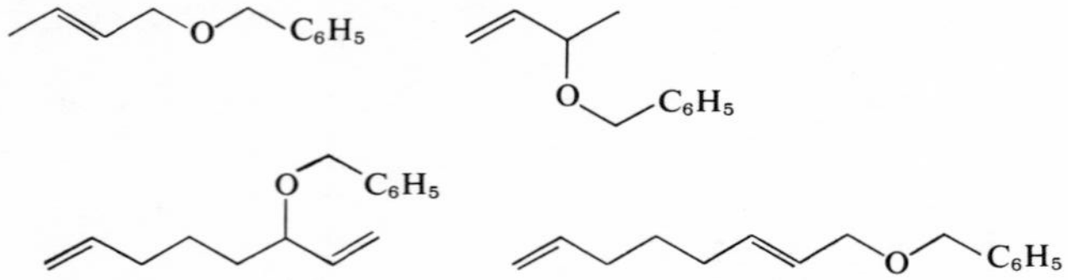		$C_6H_5CH_2OH$	$Ni(acac)_2-P[N(CH_3)_2]_3-$ C_4H_9Li	163
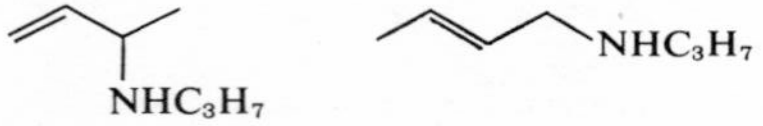		$C_3H_7NH_2$	$[(C_4H_9)_3P]_2NiBr_2-NaBH_4$	64
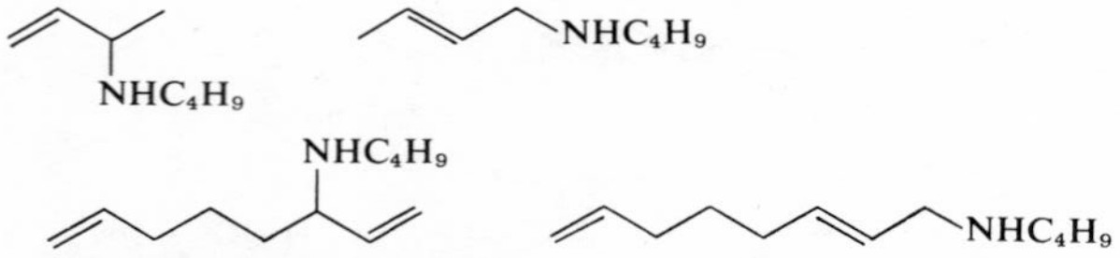		$C_4H_9NH_2$	$NiBr_2-C_6H_5P(O\text{-}iso\text{-}C_3H_7)_2-$ $NaBH_4-NaOC_6H_5$	66, 131
octadienyl-N(C$_4$H$_9$)-butenyl				
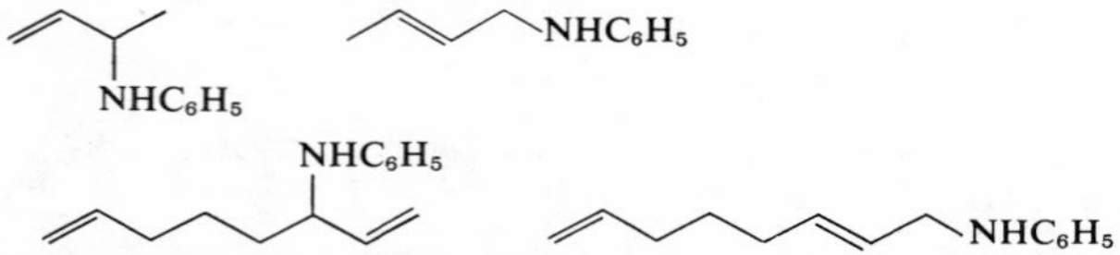		$C_6H_5NH_2$	$Ni(acac)_2-C_6H_5P(O\text{-}iso\text{-}$ $C_3H_7)_2-NaBH_4$	66
Diethylbutenylamine, diethyloctadienylamine		$(C_2H_5)_2NH$	$NiX_2-PR_3-NaBH_4$	64, 132, P127
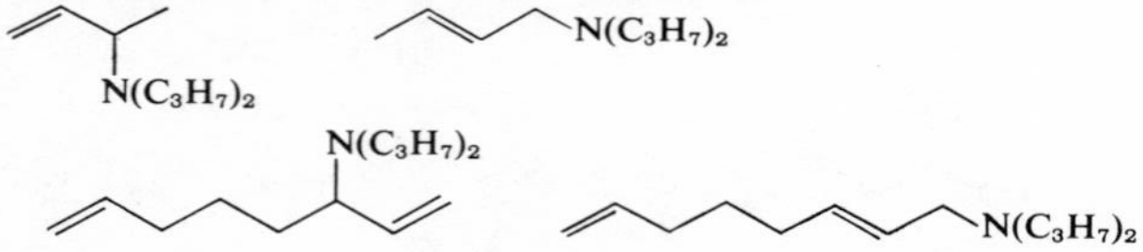		$(C_3H_7)_2NH$	$Ni(acac)_2-C_6H_5P(O\text{-}iso\text{-}$ $C_3H_7)_2-NaBH_4$	66, 131, P127

192

193

TABLE III-10 (*continued*)

Product	Diene	Telomer	Catalyst	Ref.[a]
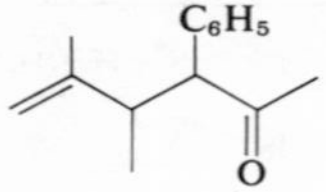 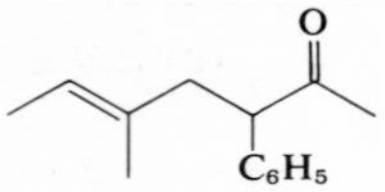	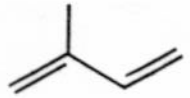	$C_6H_5CH_2COCH_3$	$Ni(acac)_2-C_6H_5P(O\text{-}iso\text{-}C_3H_7)_2-NaBH_4-NaOC_6H_5$	129
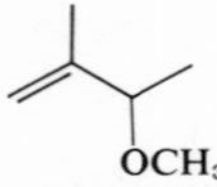	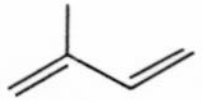	CH_3OH	$(COD)_2Ni-P(C_6H_5)_3$	130, 135
	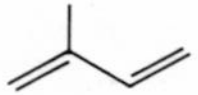	C_6H_5OH	$(\pi\text{-}C_5H_5)_2Ni-C_6H_5MgBr$	139
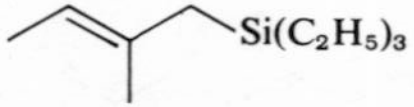 Alkylated phenols	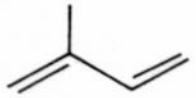	$HSi(C_2H_5)_3$	$Ni(acac)_2-Al(C_2H_5)_3$	94, 172
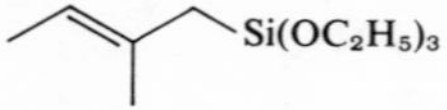	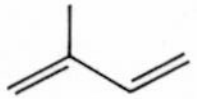	$HSi(OC_2H_5)_3$	$Ni(acac)_2-Al(C_2H_5)_3$	94, 172
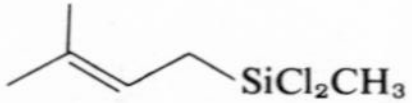	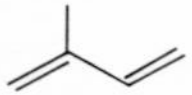	$HSiCl_2CH_3$	Lig_2NiCl_2	93
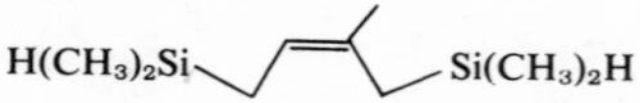	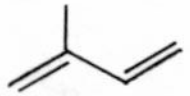	$HSi(CH_3)_2Si(CH_3)_2H$	$[(C_2H_5)_3P]_2NiCl_2$	119

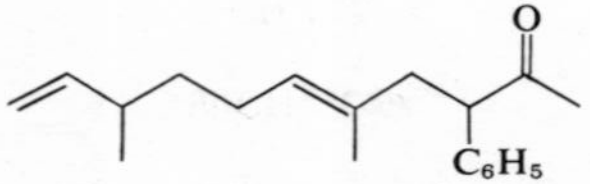

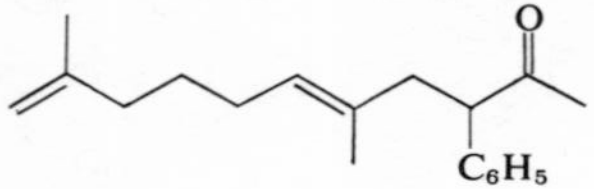

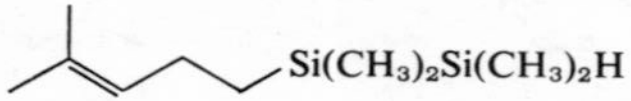

194

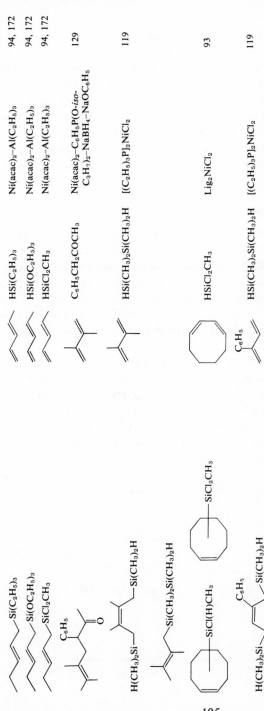

$HSi(C_2H_5)_3$	$Ni(acac)_2-Al(C_2H_5)_3$	94, 172
$HSi(OC_2H_5)_3$	$Ni(acac)_2-Al(C_2H_5)_3$	94, 172
$HSiCl_2CH_3$	$Ni(acac)_2-Al(C_2H_5)_3$	94, 172
$C_6H_5CH_2COCH_3$	$Ni(acac)_2-C_6H_5P(O\text{-}iso\text{-}C_3H_7)_2-NaBH_4-NaOC_6H_5$	129
$HSi(CH_3)_2Si(CH_3)_2H$	$[(C_2H_5)_3P]_2NiCl_2$	119
$HSiCl_2CH_3$	Lig_2NiCl_2	93
$HSi(CH_3)_2Si(CH_3)_2H$	$[(C_2H_5)_3P]_2NiCl_2$	119

[a] The references refer not only to the catalyst shown in the table but also to related systems.

195

A systematic investigation of the effect of varying the ligand on the product distribution has been made for the telomerization of butadiene with $(CH_3)_3SiH$ (47) and C_6H_5OH (134). In the first case, the formation of **116** [X = $Si(CH_3)_3$] is favored by such diverse ligands as $P(C_2H_5)_3$ and $P(OC_6H_5)_3$, whereas $As(C_6H_5)_3$ favors formation of **120** [X = $Si(CH_3)_3$]. In the second case both $P(CH_3)_3$ and $P(C_6H_5)_3$ favor formation of **117** (X = OC_6H_5), whereas $P(OC_6H_5)_3$ favors formation of **118** (X = OC_6H_5).

A reaction that may be related to these is that of chloroprene with benzene. This is catalyzed by the combination diphosNi$(NO_3)_2$–$(C_2H_5)_2AlCl$ and 3-chloro-1-phenyl-*cis*-butene is formed (156).

C. Mechanistic Considerations

The fact that telomers and linear dimers are generally formed in the same reaction suggests that they are mechanistically related. Our knowledge of the mechanism, however, is mainly speculative and one suspects that different mechanisms are operative for different telomers and cocatalysts. The method of forming the catalysts may also be important. Furthermore, some of the catalyst systems are known to be active for the isomerization of olefins (see Chapter I, Section VI), so the identity of the primary product is not always known with certainty.

One of the suggested mechanisms involves the initial formation of the bis-(π-allyl)-C_8-nickel complex and its reaction with the nucleophilic cocatalyst. The formation of a linear dimer is then seen to be the result of protonation of one end of the C_8 chain and β-hydrogen abstraction from the other, whereas the telomer is a result of a coupling reaction. One variation of this mechanism is shown below; others are possible depending on the geometry of the C_8 chain in the intermediate complex (8, 14, 97)

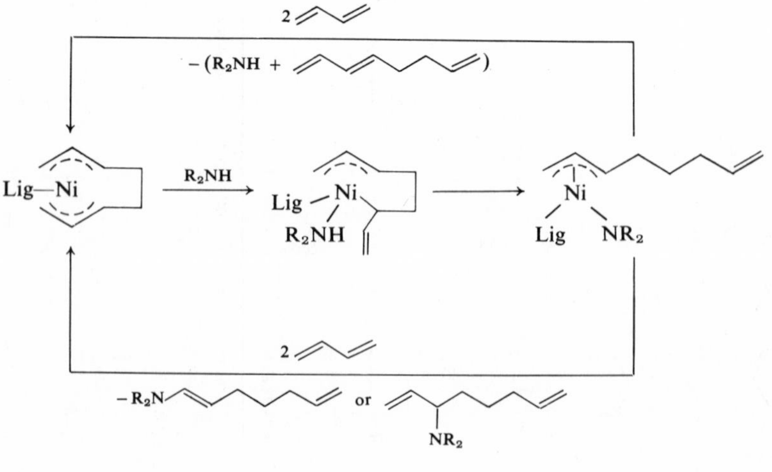

Indirect evidence in support of this mechanism is to be found in the facile π–σ conversion that is observed in the NMR spectra of π-allyl complexes on the addition of amines (79, 140; see Volume I, p. 342), in the ready elimination of methane from the amine complex **130** (141), in the elimination of

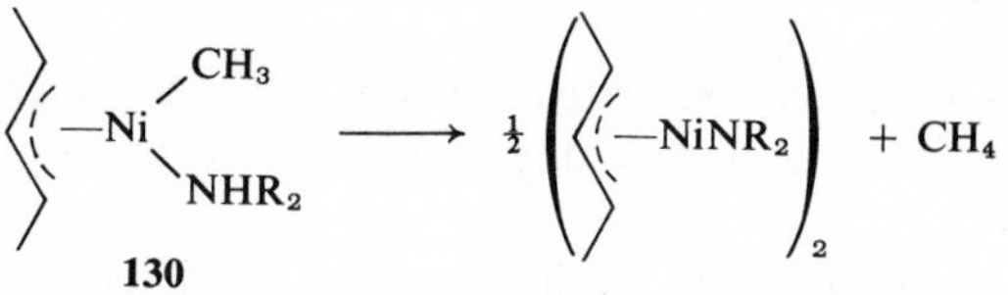

N-allyl-N-methylaniline from the reaction of the amide **131** with excess triethyl phosphite (79), and in the generation of piperylene on treatment of the amide **132** with excess phosphite (79)

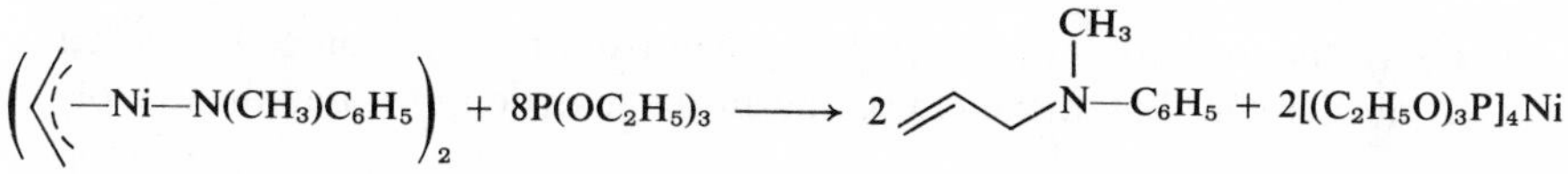

An alternative mechanism involves the addition of a nickel-hydride species to butadiene and provides a route to the butene derivatives (47, 134). This

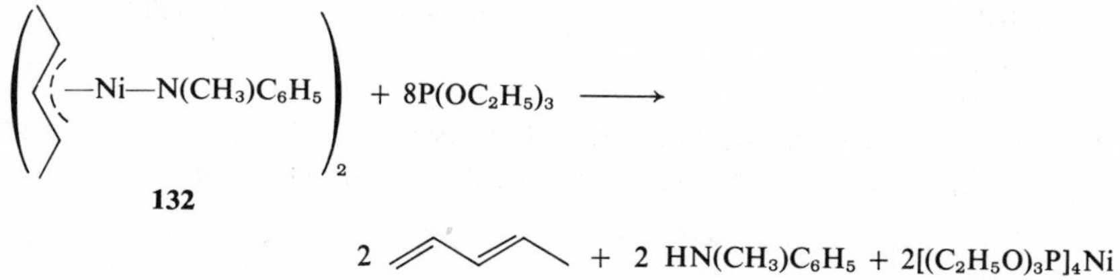

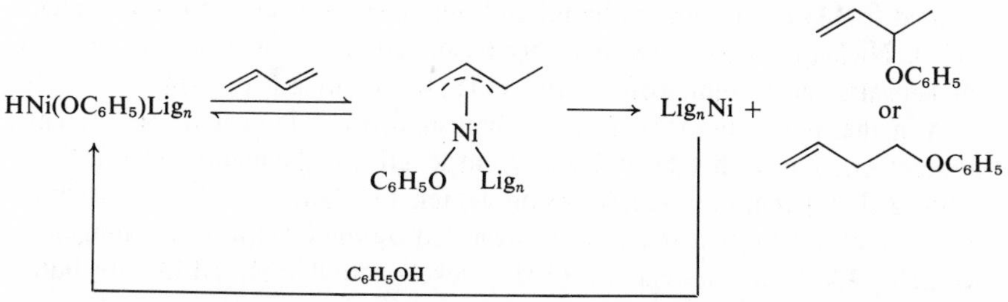

mechanism can be extended to include the formation of octadiene derivatives by supposing that a second molecule of butadiene inserts into the π-crotylnickel bond. Although this step is perhaps irrelevant to the reaction involving phenol (**120**, $X = OC_6H_5$, is not formed) it does, however, provide a simple explanation for the formation of **120** [$X = Si(CH_3)_3$] from the reaction with trimethylsilane (47) (the second possible pathway, i.e., elimination of octatriene, is not observed in this particular example). The suppression of the

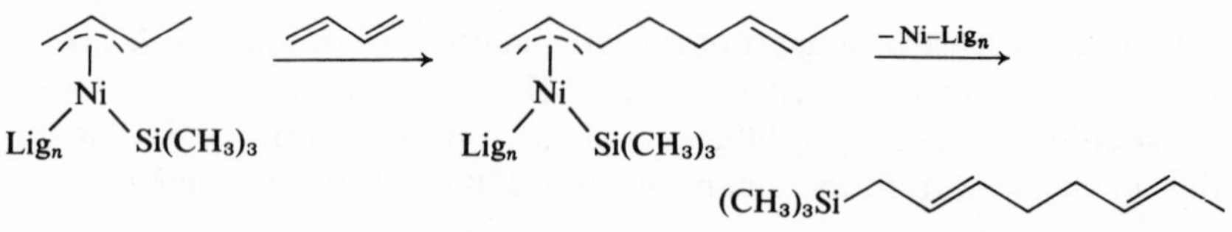

120

formation of octadiene derivatives in favor of the butene derivatives on addition of excess ligand is readily understood in terms of this mechanism (66, 132, 134).

Indirect support for this mechanism is to be found in the reactions of phenol with various ligand–nickel complexes to form hydrides, e.g., **133**

$$[(cyclo\text{-}C_6H_{11})_3P]_2Ni + C_6H_5OH \longrightarrow [(cyclo\text{-}C_6H_{11})_3P]_2Ni(OC_6H_5)H$$

133

(142), in the formation of a π-crotyl system from the reaction of the nickel-hydride **134** with butadiene (143), and in the insertion of a butadiene mole-

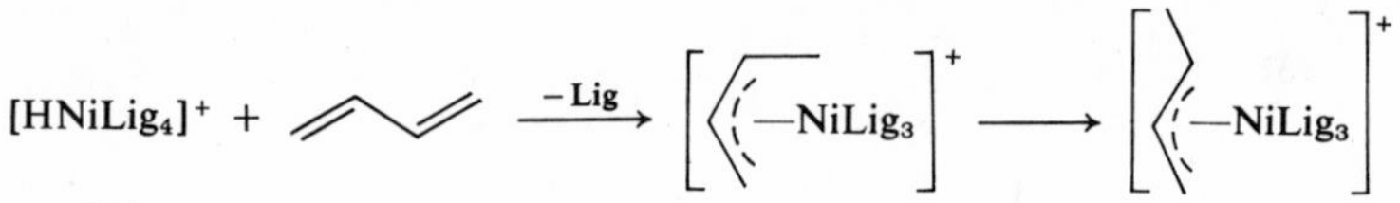

134

cule into the π-crotylnickel bond of $(\pi\text{-}C_4H_7NiI)_2$, which has been followed spectroscopically (144). It has been mentioned briefly that compounds

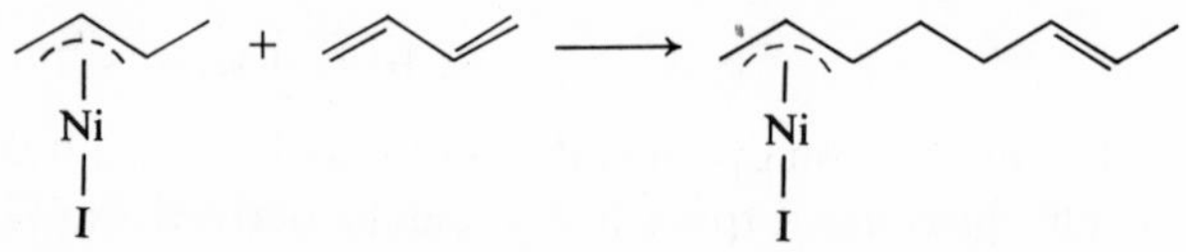

related to **134** react with methanol and butadiene to give methoxyoctadiene (138). Nickel-hydrides have also been isolated from the reaction between nickel salts and sodium borohydride (145; see Volume I, p. 141).

A further possibility is that the mechanism involves binuclear species. This type of mechanism has been discussed at length for the analogous reactions catalyzed by palladium (see, for example, ref. 137), and its possible relevance to the nickel-catalyzed reaction is suggested by the isolation of a binuclear complex **136** from the reaction of the nickel–methyl species **135** with buta-

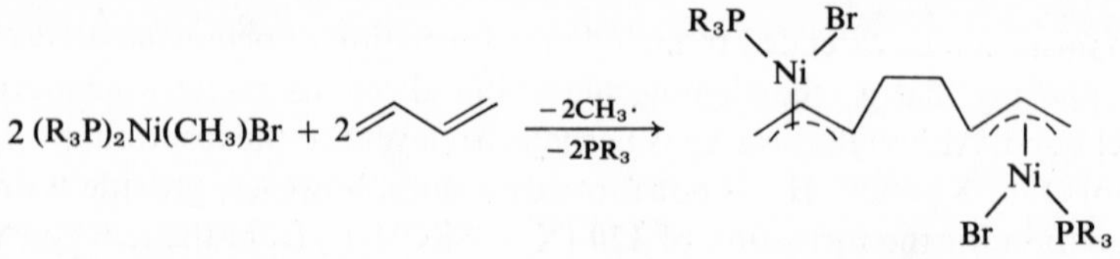

$$2\,(R_3P)_2Ni(CH_3)Br + 2 \xrightarrow[-2PR_3]{-2CH_3\cdot}$$

135　　　　　　　　**136**

$$R = iso\text{-}C_3H_7$$

diene (136). Related binuclear species have been postulated in the reaction of π-allylnickel halides with butadiene (32, 37).

Our present knowledge does not justify an attempt to rationalize the effect of the ligand on the telomerization reactions because it is necessary to take into account the steric and electronic effect of two (or more) ligands, i.e., the donor ligand and the nucleophile.

1-Methylene-2-vinylcyclopentane (115) is presumably the product of an intramolecular ring closure reaction of the type familiar from the chemistry of the analogous main group element alkyls. It can be derived from either of the mechanisms discussed above, e.g.,

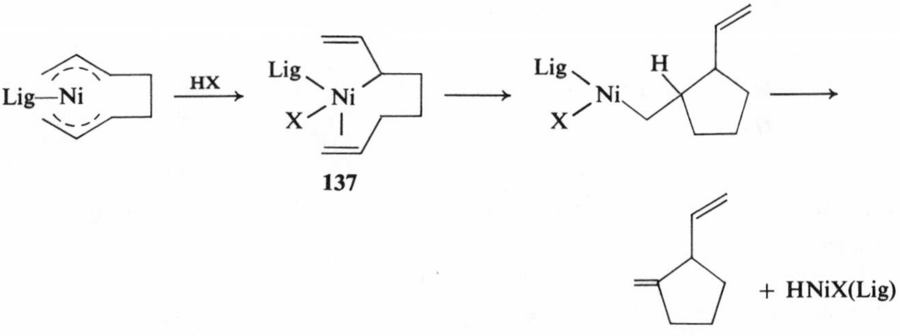

137

115

$+ \text{HNiX(Lig)}$

The formation of a cyclic dimer in preference to a linear dimer is perhaps the result of the electronic and steric environment around the nickel atom in 137, which promotes intramolecular interaction between the nickel atom and the terminal double bond. It may be expected that this interaction is hindered by the presence of the bulky phosphine $P(cyclo\text{-}C_6H_{11})_3$ and consequently octatriene is formed instead. The activity of o-tolylNiBr$[P(C_2H_5)_3]_2$ for the formation of 115 has been shown to be associated with the cleavage of the o-tolyl group, and it is possible that the initial reaction with butadiene is related to the formation of 136 described above. A similar mechanism may be involved in the reaction catalyzed by $(\pi\text{-}C_3H_5\text{NiBr})_2\text{-}P(C_4H_9)_3$, the π-allyl system being converted into a σ-allyl system under the influence of the basic phosphine. 1,3,7-Octatriene may be converted by $[(C_4H_9)_3P]_2\text{NiBr}_2\text{-}$ $NaOC_2H_5$ into methylenevinylcyclopentane (88): experiments carried out with the perdeuterated triene in ethanol lead to a product containing a hydrogen atom in the 4 or the 5 position (in contrast to the result using C_4D_6 and CH_3OH, in which case the hydrogen atom occupies the expected 3 position) and suggests that a different mechanism to that described above must be operative here. Further speculation on the nature of the intermediates involved in these complex reactions is to be found in references 42 and 88.

D. The Reactions of 1,3-Dienes with Aldehydes and Ketones (Table III-11)

A potentially useful reaction for preparing long-chain polyenes has been introduced with the discovery that 1,3-dienes react readily with aldehydes or ketones in the presence of stoichiometric amounts of bis(cyclooctadiene)-nickel to give long-chain diols (97, 105, 157, P131). These can be easily dehydrated to the corresponding polyene. For example, the reaction of acetone with butadiene produces 2,7-dimethyl-*trans*-4-octene-2,7-diol (138) as the result of the coupling of two molecules of acetone with one of butadiene. The water molecules needed to decompose the intermediate nickelate are produced by the formation of mesityl oxide. Aldehydes react similarly, with

$$(COD)_2Ni + 6CH_3COCH_3 + CH_2{=}CHCH{=}CH_2 \longrightarrow$$

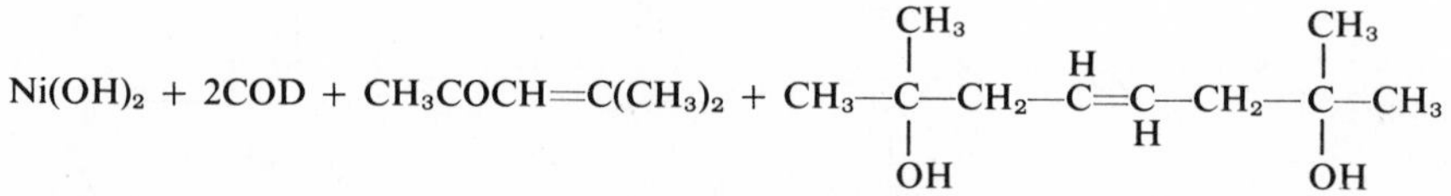

$$Ni(OH)_2 + 2COD + CH_3COCH{=}C(CH_3)_2 + CH_3-\underset{\underset{OH}{|}}{\overset{\overset{CH_3}{|}}{C}}-CH_2-\overset{H}{\underset{H}{C}}{=}C-CH_2-\underset{\underset{OH}{|}}{\overset{\overset{CH_3}{|}}{C}}-CH_3$$

138

the difference that water must be added. Substituted 1,3-dienes have also been reacted.

In certain cases the intermediate complexes involved have been isolated. The reaction of cinnamaldehyde is typical. In benzene a bis(cinnamaldehyde)-nickel complex that is believed to be related to bis(acrolein)nickel may be isolated. This intermediate reacts with butadiene to form a paramagnetic

$$(COD)_2Ni + 2C_6H_5CH{:}CHCHO \longrightarrow (C_6H_5CH{:}CHCHO)_2Ni + 2COD$$

species that, on the strength of spectroscopic evidence, is thought to have structure 139. Compound 139 reacts immediately with water to form 1,10-diphenyl-1,5,9-decatriene-3,8-diol.

$$(C_6H_5CH{=}CHCHO)_2Ni + CH_2{=}CHCH{=}CH_2 \longrightarrow$$

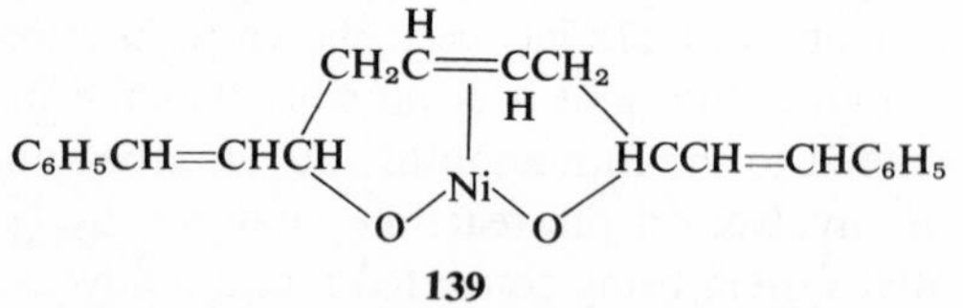

$$C_6H_5CH{=}CHCH \underset{\diagdown O \diagup Ni \diagdown O \diagup}{\overset{CH_2C{=}CCH_2}{\diagup\qquad\diagdown}} HCCH{=}CHC_6H_5$$

139

Another possible intermediate, a π-allylnickelate 140, has been isolated from the reaction of bis(cyclooctadiene)nickel with butadiene in the presence of a carbonyl compound.

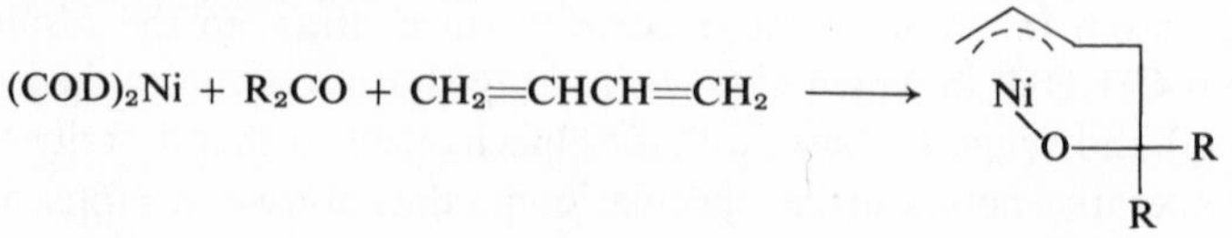

$$(COD)_2Ni + R_2CO + CH_2{=}CHCH{=}CH_2 \longrightarrow \quad Ni$$

140

TABLE III-11

THE REACTION OF 1,3-DIENES WITH KETO COMPOUNDS[a]

Product, RCH:CHR (R =)	Reactant	
	1,3-Diene	Keto compound
$(CH_3)_2COCHCH_2-$	$CH_2:CHCH:CH_2$	CH_3COCH_3
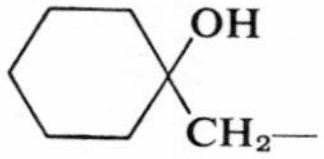	$CH_2:CHCH:CH_2$	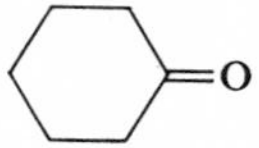 (cyclohexanone)
$C_6H_5-C(CH_3)-CH_2-$ with OH	$CH_2:CHCH:CH_2$	$C_6H_5C(CH_3)O$
$C_6H_5CH(OH)CH_2-$	$CH_2:CHCH:CH_2$	C_6H_5CHO
$C_6H_5CH:CHCH(OH)CH_2-$	$CH_2:CHCH:CH_2$	$C_6H_5CH:CHCHO$
$C_6H_5CH:CHCH:CHCH(OH)CH_2-$	$CH_2:CHCH:CH_2$	$C_6H_5CH:CHCH:CHCHO$
$p-CH_3OC_6H_4CH(OH)CH_2-$	$CH_2:CHCH:CH_2$	$p-CH_3OC_6H_4CHO$
$p(CH_3)_2NC_6H_4CH(OH)CH_2-$	$CH_2:CHCH:CH_2$	$p-(CH_3)_2NC_6H_4CHO$
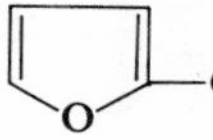 (furyl)—CH(OH)CH_2-	$CH_2:CHCH:CH_2$	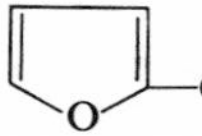 (furyl)—CHO
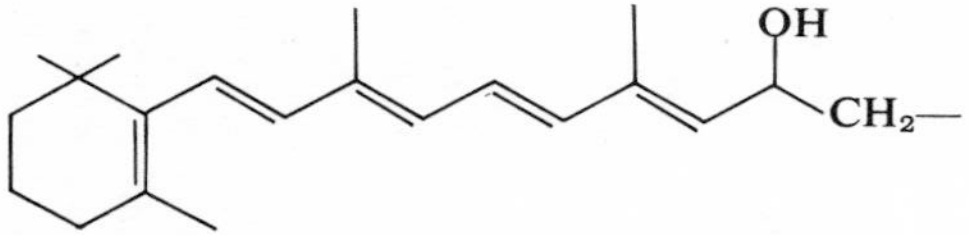 (vitamin A product structure)	$CH_2:CHCH:CH_2$	Vitamin A aldehyde
$R'C(CH_3):CHR'; [R' = C_6H_5CH(OH)CH_2-]$	$CH_2:C(CH_3)CH:CH_2$	C_6H_5CHO
$R'C(CH_3):C(CH_3)R'; [R' = C_6H_5CH(OH)CH_2-]$	$CH_2:C(CH_3)C(CH_3):CH_2$	C_6H_5CHO
$R'C(C_6H_5):C(C_6H_5)R'; [R' = C_6H_5CH(OH)CH_2-]$	$CH_2:C(C_6H_5)C(C_6H_5):CH_2$	C_6H_5CHO

[a] Taken from refs. 57–59 and P8.

References

1. V. M. Akhmedov, M. A. Mardanov, A. V. Khanmetov, and L. I. Zakharkin, *J. Org. Chem. USSR* 7, 2610 (1971).
2. G. Bosmajian, R. E. Burks, C. E. Feazel, and J. Newcombe, *Ind. Eng. Chem., Prod. Res. Develop.* 3, 117 (1964).
3. B. Bogdanović, P. Heimbach, M. Kröner, G. Wilke, E. G. Hoffmann, and J. Brandt, *Justus Liebigs Ann. Chem.* 727, 143 (1969).
4. H. Breil, P. Heimbach, M. Kröner, H. Müller, and G. Wilke, *Makromol. Chem.* 69, 18 (1963).
5. W. Brenner, P. Heimbach, and G. Wilke, *Justus Liebigs Ann. Chem.* 727, 194 (1969).
6. W. Brenner, P. Heimbach, H. Hey, E. W. Müller, and G. Wilke, *Justus Liebigs Ann. Chem.* 727, 161 (1969).
7. J. M. Brown, B. T. Golding, and M. J. Smith, *Chem. Commun.* p. 1240 (1971).
8. H. Buchholz, P. Heimbach, H. J. Hey, H. Selbeck, and W. Wiese, *Coord. Chem. Rev.* 8, 129 (1972).
9. J. Feldman, O. Frampton, B. Saffer, and M. Thomas, *Amer. Chem. Soc., Div. Petrol. Chem., Prepr.* 9, A55 (1964).
10. R. Giezynski, S. Giezynski, and S. Malinowski, *Przem. Chem.* 47, 269 (1968); *Chem. Abstr.* 29, 18644 (1968).
11. P. Heimbach and W. Brenner, *Angew. Chem.* 79, 814 (1967).
12. P. Heimbach and W. Brenner, *Angew. Chem.* 79, 813 (1967).
13. P. Heimbach and G. Wilke, *Justus Liebigs Ann. Chem.* 727, 183 (1969).
14. P. Heimbach, *Angew. Chem.* 80, 967 (1968).
15. P. Heimbach, *Mod. Chem. Ind., Proc. IUPAC Symp., 1968* p. 223 (1968).
16. P. Heimbach and H. Hey, *Angew. Chem.* 82, 550 (1970).
17. W. B. Hughes, *J. Org. Chem.* 36, 4073 (1971).
18. Y. Inoue, T. Kagawa, and H. Hashimoto, *Tetrahedron Lett.* p. 1099 (1970).
19. Y. Inoue, T. Kagawa, Y. Uchida, and H. Hashimoto, *Bull. Chem. Soc. Jap.* 45. 1996 (1972).
20. P. W. Jolly, I. Tkatchenko, and G. Wilke, *Angew. Chem.* 83, 329 (1971).
21. P. W. Jolly, I. Tkatchenko, and G. Wilke, *Angew. Chem.* 83, 328 (1971).
22. J. Kiji, K. Masui, and J. Furukawa, *Tetrahedron Lett.* p. 2561 (1970).
23. J. Kiji, K. Masui, and J. Furukawa, *Chem. Commun.* p. 1310 (1970).
24. J. Kiji, K. Masui, and J. Furukawa, *Bull. Chem. Soc. Jap.* 44, 1956 (1971).
25. J. F. Kohnle, L. H. Slaugh, and K. L. Nakamaye, *J. Amer. Chem. Soc.* 91, 5904 (1969).
26. H. Lehmkuhl and W. Leuchte, *J. Organometal. Chem.* 23, C30 (1970).
27. L. B. Luttinger, *J. Org. Chem.* 27, 1591 (1962).
28. F. D. Mango, *Tetrahedron Lett.* p. 4813 (1969).
29. F. D. Mango and J. H. Schachtschneider, *J. Amer. Chem. Soc.* 93, 1123 (1971).
30. F. D. Mango, *Tetrahedron Lett.* p. 505 (1971).
31. M. A. Mardanov, V. M. Akhmedov, L. I. Zakharkin, and A. A. Khanmetov, *J. Org. Chem. USSR* 6, 1772 (1970).
32. T. Arakawa and K. Saeki, *Kogyo Kagaku Zasshi* 71, 1028 (1968).
33. F. D. Mango, *Chem. Technol.*, p. 758 (1971).
34. A. Miyake, H. Kondo, and M. Nishino, *Angew. Chem.* 83, 851 (1971).
35. S. Miyamoto and T. Arakawa, *Kogyo Kagaku Zasshi* 74, 1394 (1971).
36. I. Ono and K. Kihara, *Hydrocarbon Process. Petrol. Refiner* 46, 147 (1967).
37. V. P. Nechiporenko, A. D. Treboganov, A. I. Kadantseva, A. A. Glazkov, G. I. Myagkova, and N. A. Preobrazhenskii, *J. Org. Chem. USSR* 6, 2633 (1970).

38. S. Otsuka and K. Taketomi, *Eur. Polym. J.* **2**. 289 (1966).
39. T. Ohta, K. Ebina, and N. Yamazaki, *Bull. Chem. Soc. Jap.* **44**, 1321 (1971).
40. H. W. B. Reed, *J. Chem. Soc., London* p. 1931 (1954).
41. T. Saito, Y. Uchida, A. Misono, A. Yamamoto, K. Morifuji, and S. Ikeda, *J. Amer. Chem. Soc.* **88**, 5198 (1966).
42. J. Furukawa, J. Kiji, H. Konishi, K. Yamamoto, S. I. Mitani, and S. Yoshikawa, *Makromol. Chem.* **174**, 65 (1973).
43. E. J. Smutny, H. Chung, K. C. Dewhirst, W. Keim, T. M. Shryne, and H. E. Thyret, *Amer. Chem. Soc., Div. Petrol Chem., Prepr.* **14**, B100 (1969).
44. H. Singer, W. Umbach, and M. Dohr, *Synthesis* p. 42 (1972).
45. Y. Tajima and E. Kunioka, *J. Polym. Sci., Part B* **5**, 221 (1967).
46. Y. Tajima and E. Kunioka, *Bull. Chem. Soc. Jap.* **40**, 697 (1967).
47. S. Takahashi, T. Shibano, H. Kojima, and N. Hagihara, *Organometal. Chem. Syn.* **1**, 193 (1970/1971).
48. M. Tsutsui and J. Ariyoshi, *J. Polym. Sci., Part A* **3**, 1729 (1965).
49. A. D. Treboganov, B. I. Mitsner, E. P. Zinkevich, A. A. Kraevskii, and N. A. Preobrazhenskii, *Zh. Org. Khim.* **1**, 1583 (1965); *Chem. Abstr.* **64**, 611 (1966).
50. S. Watanabe, K. Suga, and H. Kikuchi, *Aust. J. Chem.* **23**, 385 (1970).
51. G. Wilke, *Angew. Chem.* **72**, 581 (1960).
52. G. Wilke, M. Kröner, and B. Bogdanović, *Angew. Chem.* **73**, 755 (1961).
53. G. Wilke and B. Bogdanović, *Angew. Chem.* **73**, 756 (1961).
54. G. Wilke *et al.*, *Angew Chem.* **75**, 10 (1963).
55. C.-Y. Wu and H. E. Swift, *J. Catal.* **24**, 510 (1972).
56. G. Wilke, E. W. Müller, and M. Kröner, *Angew. Chem.* **73**, 33 (1961).
57. A. Yamamoto, K. Morifuji, S. Ikeda, T. Saito, Y. Uchida, and A. Misono, *J. Amer. Chem. Soc.* **87**, 4652 (1965).
58. V. A. Yakovlev, E. I. Tinyakova, and B. A. Dolgoplosk, *Bull. Acad. Sci. USSR* p. 1350 (1968).
59. A. Yamamoto and S. Ikeda, *Bull. Tokyo Inst. Technol.* **97**, 1 (1970).
60. H. Takaya, N. Hayashi, T. Ishigami, and R. Noyori, *Chem. Lett.* p. 813 (1973).
61. L. I. Zakharkin and G. G. Zhigareva, *Izv. Akad. Nauk SSSR, Otd. Khim. Nauk* p. 386 (1963); *Chem. Abstr.* **59**, 13810 (1963).
62. L. I. Zakharkin and G. G. Zhigareva, *Izv. Akad. Nauk SSSR, Ser. Khim.* p. 168 (1964); *Chem. Abstr.* **60**, 11883 (1964).
63. H. Hashimoto, *Senryo To Yakuhin* **17**, 198 (1972).
64. J. Furukawa, J. Kiji, S. Mitani, S. Yoshikawa, K. Yamamoto, and E. Sasakawa, *Chem. Lett.* p. 1211 (1972).
65. K. Suga, S. Watanabe, T. Fujita, and T. Shimada, *J. Appl. Chem. Biotechnol.* **23**, 131 (1973).
66. R. Baker, D. E. Halliday, and T. N. Smith, *Chem. Commun.* p. 1583 (1971).
67. Y. Inoue, T. Yamaguchi, Y. Uchida, and H. Hashimoto, *Yukagaku* **21**, 512 (1972); *Chem. Abstr.* **78**, 4602 (1973).
68. B. Barnett, B. Büssemeier, P. Heimbach, P. W. Jolly, C. Krüger, I. Tkatchenko, and G. Wilke, *Tetrahedron Lett.* p. 1457 (1972).
69. N. Yamazaki and T. Ohta, *J. Macromol. Sci., Chem.* **3**, 1571 (1969).
70. H. W. Leuchte, Ph.D. Thesis, University of Bochum, 1971.
71. H. Hey, Ph.D. Thesis, University of Bochum, 1969.
72. W. Brenner, Ph.D. Thesis, Technische Hochschule Aachen, 1967.
73. H. A. Buchholz, Ph.D. Thesis, University of Bochum, 1971.
74. W. Wiese, Ph.D. Thesis, University of Bochum, 1972.
75. H. F. Zimmermann, Ph.D. Thesis, Technische Hochschule Aachen, 1967.

76. K.-J. Ploner, Ph.D. Thesis, University of Bochum, 1969.
77. M. Kröner, Ph.D. Thesis, Technische Hochschule Aachen, 1961.
78. H.-J. Kaminsky, Ph.D. Thesis, Technische Hochschule Aachen, 1962.
79. W. Fleck, Ph.D. Thesis, University of Bochum, 1971.
80. C. Delliehausen, Ph.D. Thesis, University of Bochum, 1968.
81. H. Selbeck, Ph.D. Thesis, University of Bochum, 1972.
82. F. Thömel, Ph.D. Thesis, University of Bochum, 1970.
83. M. Molin, Ph.D. Thesis, University of Bochum, 1973.
84. C. Krüger and Y. H. Tsay, unpublished work (1973).
85. See, for example, N. E. Waite, J. C. Tebby, R. S. Ward, M. A. Shaw, and D. H. Williams, *J. Chem. Soc.*, C p. 1620 (1971).
86. N. Yamazaki and T. Ohta, *Polym. J.* 4, 616 (1973).
87. P. S. Skell, J. J. Havel, D. L. Williams-Smith, and M. J. McGlinchey, *Chem. Commun.* p. 1098 (1972); P. S. Shell, private communication (1974).
88. J. Kiji, K. Yamamoto, S. Mitani, S. Yoshikawa, and J. Furukawa, *Bull. Chem. Soc. Jap.* 46, 1791 (1973).
89. J. Kiji, K. Yamamoto, E. Sasakawa, and J. Furukawa, *Chem. Commun.* p. 770 (1973).
90. H. Singer, *Synthesis* p. 189 (1974).
91. W. E. Billups, J. H. Cross, and C. V. Smith, *J. Amer. Chem. Soc.* 95, 3438 (1973).
92. M. Englert, P. W. Jolly, and G. Wilke, *Angew. Chem.* 83, 84 (1971).
93. Y. Kiso, M. Kumada, K. Tamao, and H. Umeno, *J. Organometal. Chem.* 50, 297 (1973).
94. M. F. Lappert and S. Takahashi, *Chem. Commun.* p. 1272 (1972).
95. H. Müller, D. Wittenberg, H. Seibt, and E. Scharf, *Angew. Chem.* 77, 318 (1965).
96. W. Brenner, P. Heimbach, K.-J. Ploner, and F. Thömel. *Angew. Chem.* 81, 744 (1969).
97. P. Heimbach, P. W. Jolly, and G. Wilke, *Advan. Organometal. Chem.* 8, 29 (1970).
98. E. I. Tinyakova, A. V. Alferov, T. G. Golenko, B. A. Dolgoplosk, I. A. Oreshkin, O. K. Sharaev, G. N. Chernenko, and V. A. Yakovlec, *J. Polym. Sci., Part C* 16, 2625 (1967).
99. B. Bogdanović, Ph.D. Thesis, Technische Hochschule Aachen, 1962.
100. P. Heimbach, D. Henneberg, E. Janssen, H. Lehmkuhl, G. Schomburg, and G. Wilke, in press (1975).
101. A. Miyake, A. Kondo, M. Nishino, and S. Tokizane, *Spec. Lec., Int. Congr. Pure Appl. Chem., 23rd, 1971* Vol. 6, p. 201 (1971).
102. P. Heimbach, *Aspects Homogen. Catal.* 2 (in press) (1975).
103. B. Bogdanović, P. Heimbach, W. Oberkirch, K. Tanaka, and G. Wilke, *Conf. Chem. Chem. Process. Petrol. Natur. Gas, Plenary Lect., 1965* p. 32 (1968).
104. G. Wilke, B. Bogdanović, P. Heimbach, M. Kröner, and E. W. Müller, *Advan. Chem. Ser.* 34, 137 (1962).
105. G. Wilke, *Pure Appl. Chem.* 17, 179 (1968).
106. P. Rona, *Intra-Sci. Chem. Rep.* 5, 105 (1971).
107. B. Büssemeier, Ph.D. Thesis, University of Bochum, 1973.
108. G. L. Caldow and R. A. MacGregor, *Inorg. Nucl. Chem. Lett.* 6, 645 (1970).
109. G. L. Caldow and R. A. MacGregor, *J. Chem. Soc., A* p. 1654 (1971).
110. F. D. Mango, *Advan. Catal.* 20, 291 (1969).
111. W. T. A. M. van der Lugt, *Tetrahedron Lett.* p. 2281 (1970).
112. F. D. Mango and J. H. Schachtschneider, *J. Amer. Chem. Soc.* 89, 2484 (1967).
113. D. R. Eaton, *J. Amer. Chem. Soc.* 90, 4272 (1968).

114. T. H. Whitesides, *J. Amer. Chem. Soc.* **91**, 2395 (1969).
115. G. Wilke *et al.*, *Angew. Chem.* **78**, 157 (1966).
116. See, for example, W. J. Bailey, *High Polym.* **24**, 757 (1971) (see p. 817).
117. R.-V. Meyer, Ph.D. Thesis, University of Bochum, 1973.
118. G. Wilke and P. Heimbach, unpublished results (1966).
119. H. Okinoshima, K. Yamamoto, and M. Kumada, *J. Amer. Chem. Soc.*, **94**, 9263 (1972).
120. P. Heimbach, *Angew. Chem.* **76**, 859 (1964).
121. P. Heimbach and R. Schimpf, *Angew. Chem.* **80**, 704 (1968).
122. P. Heimbach and W. Brenner, *Angew. Chem.* **78**, 983 (1966).
123. P. Heimbach, *Angew. Chem.* **78**, 983 (1966).
124. P. Heimbach, H. Selbeck, and E. Troxler, *Angew. Chem.* **83**, 731 (1971).
125. D. R. Fahey, *J. Org. Chem.* **37**, 4471 (1972).
126. P. Heimbach, K.-J. Ploner, and F. Thömel, *Angew. Chem.* **83**, 285 (1971).
127. N. Yamazaki and S. Murai, *Chem. Commun.* p. 147 (1968).
128. N. Yamazaki, S. Murai, K. Ebina, T. Ohta, and S. Nakahama, *Kinet. Mech. Polyreact.* **2**, 411 (1969).
129. R. Baker, D. E. Halliday, and T. N. Smith, *J. Organometal. Chem.* **35**, C61 (1972).
130. J. Beger, C. Duschek, and H. Füllbier, *Z. Chem.* **13**, 59 (1973).
131. R. Baker, A. H. Cook, and T. N. Smith, *Tetrahedron Lett.* p. 503 (1973).
132. D. Rose, *Tetrahedron Lett.* p. 4197 (1972).
133. T. C. Shields and W. E. Walker, *Chem. Commun.* p. 193 (1971).
134. F. J. Weigert and W. C. Drinkard, *J. Org. Chem.* **38**, 335 (1973).
135. Y. Inoue, M. Hidai, and Y. Uchida, *Chem. Lett.* p. 1119 (1972).
136. T. S. Cameron and C. K. Prout, *Acta Crystallogr.*, *Sect. B* **28**, 2021 (1972); T. S. Cameron, M. L. H. Green, H. Munakata, C. K. Prout, and M. J. Smith, *J. Coord. Chem.* **2**, 43 (1972).
137. W. Keim, *in* "Transition Metals in Homogeneous Catalysis" (G. N. Schrauzer, ed.), p. 77, Dekker, New York, 1971.
138. M. L. H. Green and H. Munakata, *Chem. Commun.* p. 549 (1971).
139. K. Suga, S. Watanabe, H. Kikuchi, and K. Hijikata, *J. Appl. Chem.* **20**, 175 (1970).
140. U. Birkenstock, Ph.D. Thesis, Technische Hochschule Aachen, 1966.
141. H. Schenkluhn, Ph.D. Thesis, University of Bochum, 1971.
142. K. Jonas and G. Wilke, *Angew. Chem.* **81**, 534 (1969).
143. C. A. Tolman, *J. Amer. Chem. Soc.* **92**, 6777 and 6785 (1970).
144. M. I. Lobach, V. A. Kormer, I. Y. Tsereteli, G. P. Kondratenkov, B. D. Babitskii, and V. I. Klepikova, *J. Polym. Sci.*, *Part B* **9**, 71 (1971); M. I. Lobach, V. A. Kormer, I. Y. Tsereteli, G. P. Kondratenkov, B. D. Babitskii, and V. I. Klepikova, *Proc. Acad. Sci. USSR* **196**, 22 (1971).
145. M. L. H. Green, H. Munakata, and T. Saito, *J. Chem. Soc.*, *A* p. 469 (1971).
146. W. Schneider, *Amer. Chem. Soc.*, *Div. Chem.*, *Prepr.* **17**, B105 (1972).
147. W. Brenner, P. Heimbach, K.-J. Ploner, and F. Thömel, *Justus Liebigs Ann. Chem.*, p. 1882 (1973).
148. W. Brenner and P. Heimbach, *Justus Liebigs Ann. Chem.* (in press) (1975).
149. K.-J. Ploner and P. Heimbach, *Justus Liebigs Ann. Chem.* (in press) (1975).
150. P. Heimbach, *Angew. Chem.* **85**, 1035 (1973).
151. K. H. Scholz, Ph.D. Thesis, University of Bochum, 1974.
152. P. Heimbach, R. V. Meyer, and G. Wilke, *Justus Liebigs Ann. Chem.* (in press) (1975).
153. N. M. Seidov and M. A. Geidarov, *Dokl. Akad. Nauk Azerb. SSR* **28**, 33 (1972); *Chem. Abstr.* **79**, 32623 (1973).

154. M. L. H. Green and H. Munakata, *J. Chem. Soc., Dalton Trans.* p. 269 (1974).
155. J. Furukawa, J. Kiji, K. Yamamoto, and T. Tojo, *Tetrahedron* 29, 3149 (1973).
156. M. Iwamoto, *Bull. Chem. Soc. Jap.* 40, 1721 (1967).
157. C. de Ortueta Spiegelberg, Ph.D. Thesis, Technische Hochschule Aachen, 1965.
158. S. Watanabe, K. Suga, T. Fujita, and N. Takasaka, *Yukagaku* 23, 24 (1974); *Chem. Abstr.* 80, 82192 (1974).
159. P. J. Garratt and M. Wyatt, *Chem. Commun.* p. 251 (1974).
160. V. M. Akhmedov, A. A. Khanmetov, M. A. Mardanov, and L. I. Zakharkin, *J. Org. Chem. USSR* 9, 442 (1973).
161. F. H. Köhler, *Chem. Ber.* 107, 570 (1974).
162. D. Commereuc and Y. Chauvin, *Bull. Soc. Chim. Fr.* p. 652 (1974).
163. J. Beger, C. Duschek, H. Füllbier, and W. Gaube, *J. Prakt. Chem.* 316, 26 (1974).
164. J. Beger, C. Duschek, H. Füllbier, and W. Gaube, *J. Prakt. Chem.* 316, 43 (1974).
165. J. Beger, H. Füllbier, and W. Gaube, *J. Prakt. Chem.* 316, 174 (1974).
166. F. D. Mango, *Intra-Sci. Chem. Rep.* 6, 171 (1972).
167. H. J. Rauh, W. Geyer, H. Schmidt, and G. Geiseler, *Z. Phys. Chem. (Leipzig)* 253, 43 (1973).
168. P. Heimbach, unpublished work (ca. 1962).
169. G. E. Ivanov and V. P. Yur'ev, *Khim. Vysokomol. Soedin. Neftekhim.* p. 46 (1973); *Chem. Abstr.* 80, 132879 (1974).
170. P. Heimbach and W. Scheidt, unpublished results (1973).
171. B. Büssemeier, P. W. Jolly, and G. Wilke, *J. Amer. Chem. Soc.* 96, 4726 (1974).
172. M. F. Lappert, T. A. Nile, and S. Takahashi, *J. Organometal. Chem.* 72, 425 (1974).
173. Y. Kiso, K. Tamao, and M. Kumada, *J. Organometal. Chem.* 76, 95 (1974).
174. V. M. Akhmedov, M. T. Anthony, M. L. H. Green, and D. Young, *Chem. Commun.*, p. 777 (1974).
175. J. Kiji, E. Sasakawa, K. Yamamoto, and J. Furukawa, *J. Organometal. Chem.* 77, 125 (1974).

Patents

P.1 L. B. Luttinger (American Cynamid Co.), U.S. Patent 3,148,224 (1964); *Chem. Abstr.* 61, 13444 (1964).
P2. K. Saotome and O. K. Nakamura (Asahi Chem. Ind. Co. Ltd.), Jap. Patent 72 08,293 (1972); *Chem. Abstr.* 76, 153239 (1972).
P3. J. E. Bozik, C. Y. Wu, and H. E. Swift (Ameripol Inc.), Ger. Offen. 1,965,047 (1970); *Chem. Abstr.* 73, 67042 (1970).
P4. K. G. Allum and I. V. Howell (British Petroleum Co. Ltd.), Brit. Patent 1,259,856 (1972); *Chem. Abstr.* 77, 6947 (1972).
P5. W. Schneider (B.F. Goodrich Co.), U.S. Patent 3,376,359 (1968); see *Chem. Abstr.* 67, 63697 (1967).
P6. W. Schneider (B.F. Goodrich Co.), U.S. Patent 3,392,208 (1968); see *Chem. Abstr.* 68, 2569 (1968).
P7. W. Schneider (B.F. Goodrich Co.), U.S. 3,376,358 (1968); see *Chem. Abstr.* 68, 2569 (1968).
P8. B.A.S.F., Brit. Patent 944,440 (1963); *Chem. Abstr.* 60, 13164 (1964).
P9. H. Seibt and N. von Kutepow (B.A.S.F.), U.S. Patent 3,277,099 (1966); see *Chem. Abstr.* 61, 11891 (1964).

P10. H. Lautenschlager, E. Scharf, D. Wittenberg, and H. Müller (B.A.S.F.), Belg. Patent 622,195 (1963); *Chem. Abstr.* **59**, 9879 (1963).

P11. D. Wittenberg, H. Lautenschlager, N. von Kutepow, F. Meier, and H. Seibt (B.A.S.F.), U.S. Patent 3,219,716 (1965); see *Chem. Abstr.* **59**, 11292 (1963).

P12. H. Müller, E. Scharf, and D. Wittenberg (B.A.S.F.), Ger. Patent 1,194,410 (1966); see *Chem. Abstr.* **62**, 9007 (1965).

P13. D. Wittenberg (B.A.S.F.), Ger. Patent 1,244,770 (1968); see *Chem. Abstr.* **67**, 90459 (1967).

P14. D. Wittenberg (B.A.S.F.), Ger. Patent 1,244,172 (1968); see *Chem. Abstr.* **67**, 90454 (1967).

P15. H. Müller and D. Wittenberg (B.A.S.F.), Ger. Patent 1,251,751 (1968); *Chem. Abstr.* **68**, 39197 (1968).

P16. H. Müller, D. Wittenberg, and E. Scharf (B.A.S.F.), Ger. Patent 1,204,669 (1966); *Chem. Abstr.* **64**, 3380 (1966).

P17. N. von Kutepow, H. Seibt, and F. Meier (B.A.S.F.), Ger. Patent 1,144,268 (1968); see *Chem. Abstr.* **59**, 3790 (1963).

P18. D. Wittenberg (B.A.S.F.), Ger. Offen, 1,443,465 (1968).

P19. H. Müller and D. Wittenberg (B.A.S.F.), Ger. Offen. 1,443,461 (1968).

P20. D. Wittenberg (B.A.S.F.), Ger. Aus. 1,244,178 (1967); *Chem. Abstr.* **67**, 90457 (1967).

P21. H. Müller and E. Scharf (B.A.S.F.), Ger. Patent 1,263,735 (1968); *Chem. Abstr.* **70**, 3502 (1969).

P22. H. Müller and E. Scharf (B.A.S.F.), Ger. Patent 1,201,328 (1965); *Chem. Abstr.* **63**, 18298 (1965).

P23. H. Müller (B.A.S.F.), Ger. Aus. 1,126,864 (1962); *Chem. Abstr.* **57**, 8461 (1962).

P24. U. Hochmuth, N. Wilke, and R. Streck (Chemische Werke Hüls AG), Ger. Offen. 1,804,017 (1970); *Chem. Abstr.* **73**, 3540 (1970).

P25. U. Hochmuth and H. Weber (Chemische Werke Hüls AG), Ger. Aus. 1,226,095 (1966); see *Chem. Abstr.* **65**, 8758 (1966).

P26. C. D. Storrs, R. F. Clark, and H. G. Jackson (Cities Service Research and Develop. Co.), Fr. Patent 1,358,348 (1964); see *Chem. Abstr.* **61**, 593 (1964).

P27. R. Levine (Cities Service Research and Develop. Co.), Fr. Patent 1,321,454 (1963); *Chem. Abstr.* **59**, 11293 (1963).

P28. Cities Service Research and Develop. Co., Brit. Patent 947,656 (1964); *Chem. Abstr.* **60**, 14406 (1964).

P29. Cities Service Research and Develop. Co., Brit. Patent 1,000,477 (1965); see *Chem. Abstr.* **59**, 11342 (1963).

P30. Cities Service Research and Develop. Co., Brit. Patent 971,755 (1964); see *Chem. Abstr.* **57**, 16662 (1962).

P31. Cities Service Research and Develop. Co., Brit. Patent 944,574 (1963); *Chem. Abstr.* **60**, 10569 (1964).

P32. Cities Service Research and Develop. Co., Brit. Patent 979,553 (1965); *Chem. Abstr.* **63**, 632 (1964).

P33. G. Bosmajian (Cities Service Research and Develop. Co.), U.S. Patent 3,004,081 (1961); *Chem. Abstr.* **56**, 4640 (1962).

P34. A. A. Sekul and H. G. Sellers (Cities Service Research and Develop. Co.), U.S. Patent 2,964,575 (1960); *Chem. Abstr.* **55**, 14333 (1961).

P35. H. G. Sellers and A. A. Sekul (Cities Service Research and Develop. Co.), U.S. Patent 2,991,317 (1961); *Chem. Abstr.* **56**, 3374 (1962).

P36. R. E. Burks and A. A. Sekul (Cities Service Research and Develop. Co.), U.S. Patent 2,972,640 (1961); *Chem. Abstr.* **56**, 4640 (1962).

P37. H. Shechter (Columbian Carbon Co.), Ger. Patent 1,298,526 (1970); see *Chem. Abstr.* **71**, 80808 (1969).

P38. S. F. Chappell, J. R. Olechowski, and A. A. Arseneaux (Columbian Carbon Co.), U.S. Patent 3,490,745 (1970); *Chem. Abstr.* **72**, 115269 (1970).

P39. C. D. Storrs and R. F. Clark (Columbian Carbon Co.), U.S. Patent 3,247,269 (1966); *Chem. Abstr.* **65**, 5382 (1966).

P40. L. R. Hellwig (Columbian Carbon Co.), U.S. Patent 3,420,904 (1969); *Chem. Abstr.* **70**, 57091 (1969).

P41. R. F. Clark, C. D. Storrs, and G. B. Barnes (Columbian Carbon Co.), U.S. Patent 3,364,273 (1968); *Chem. Abstr.* **68**, 95391 (1968).

P42. R. F. Clark (Columbian Carbon Co.), Ger. Offen. 1,443,571 (1968); see *Chem. Abstr.* **67**, 73243 (1967).

P43. J. R. Olechowski and A. A. Arseneaux (Columbian Carbon Co.), U.S. Patent 3,392,203 (1968); *Chem. Abstr.* **69**, 43391 (1968).

P44. R. Levine (Columbian Carbon Co.), U.S. Patent 3,272,876 (1966); *Chem. Abstr.* **65**, 16881 (1966).

P45. S. F. Chappell, J. R. Olechowski, and A. A. Arseneaux (Columbian Carbon Co.), U.S. Patent 3,390,195 (1968); *Chem. Abstr.* **69**, 43524 (1968).

P46. H. Schechter (Columbian Carbon Co.), U.S. Patent 3,187,062 (1965); *Chem. Abstr.* **63**, 6886 (1965).

P47. R. Tarao and H. Higashi (Chisso Corp.), Jap. Patent 71 16,974 (1971); *Chem. Abstr.* **75**, 35254 (1971).

P48. S. J. Lapporte (Chevron Research Co.), U.S. Patent 3,235,613 (1966); *Chem. Abstr.* **64**, 14110 (1966).

P49. S. J. Lapporte (Chevron Research Co.), U.S. Patent 3,250,817 (1966); *Chem. Abstr.* **65**, 3767 (1966).

P50. M. A. McCall and H. W. Coover (Eastman Kodak Co.), U.S. Patent 3,290,401 (1966); *Chem. Abstr.* **66**, 37498 (1967).

P51. M. A. McCall and H. W. Coover (Eastman Kodak Co.), U.S. Patent 3,414,629 (1968); see *Chem. Abstr.* **68**, 12557 (1968).

P52. R. E. Foster (E.I. du Pont de Nemours and Co.), U.S. Patent 2,504,016 (1950); *Chem. Abstr.* **44**, 7873 (1950).

P53. H. Schott (Farbwerke Hoechst AG), U.S. Patent 3,535,397 (1970); see *Chem. Abstr.* **73**, 14310 (1970).

P54. Farbwerke Hoechst AG, Fr. Demande 2,000,865 (1969); *Chem. Abstr.* **72**, 78537 (1970).

P55. U. Birkenstock and K. Wedemeyer (Farbenfabriken Bayer AG), Ger. Offen. 1,942,016 (1971); *Chem. Abstr.* **74**, 126392 (1971).

P56. M. Dohr, H. Singer, and W. Umbach (Henkel and Cie. GmbH), Ger. Offen. 2,025,830 (1971); *Chem. Abstr.* **76**, 45753 (1972).

P57. H. W. B. Reed (Imperial Chemical Industries Ltd.), U.S. Patent 2,686,209 (1954); *Chem. Abstr.* **49**, 15957 (1955).

P58. H. W. B. Reed (Imperial Chemical Industries Ltd.), Brit. Patent 701,106 (1953); *Chem. Abstr.* **49**, 2495 (1955).

P59. C. Lassau, R. Stern, and L. Sajus (Institut Français du Pétrole), Fr. Patent 2,045,021 (1971); *Chem. Abstr.* **76**, 46686 (1972).

P60. H. Mori, K. Ikeda, I. Nagaoka, S. Hirayanagi, and A. Kihinoki (Japan Synthetic Rubber Co. Ltd), Jap. Patent 70 28,573 (1970); *Chem. Abstr.* **74**, 63984 (1971).

P61. S. Otsuka and S. Muranishi (Japan Synthetic Rubber Co. Ltd.), Japan Patent 7332('66) (1966); *Chem. Abstr.* **65**, 8758 (1966).

P62. G. Sartori and V. Turba (Montecatini), Ital. Patent 736,944 (1967); *Chem. Abstr.* **71**, 80804 (1969).

P63. H. Morikawa (Mitsubishi Petrochem. Co. Ltd.), U.S. Patent 3,658,926 (1972); *Chem. Abstr.* **77**, 36001 (1972).

P64. J. Feldman, B. A. Saffer, and O. D. Frampton (National Distillers and Chem. Corp.), U.S. Patent 3,480,685 (1969); *Chem. Abstr.* **72**, 66350 (1970).

P65. J. Feldman, B. A. Saffer, and O. D. Frampton (National Distillers and Chem. Corp), U.S. Patent 3,284,529 (1966); *Chem. Abstr.* **66**, 28373 (1973).

P66. J. Feldman and B. A. Saffer (National Distillers and Chem. Corp.), U.S. Patent 3,227,767 (1966); *Chem. Abstr.* **64**, 9614 (1966).

P67. J. Feldman, B. A. Saffer, and M. Thomas (National Distillers and Chem. Corp.), U.S. Patent 3,194,848 (1965); *Chem. Abstr.* **63**, 9989 (1965).

P68. J. Feldman, B. A. Saffer, and M. Thomas (National Distillers and Chem. Corp.), U.S. Patent 3,251,893 (1966); *Chem. Abstr.* **65**, 2147 (1966).

P69. J. Feldman and B. A. Saffer (National Distillers and Chem. Corp.), U.S. Patent 3,223,741 (1965); *Chem. Abstr.* **64**, 4969 (1966).

P70. W. B. Hughes (Phillips Petroleum Co.), U.S. Patent 3,668,086 (1972); *Chem. Abstr.* **77**, 69482 (1972).

P71. E. A. Zuech (Phillips Petroleum Co.), U.S. Patent 3,435,088 (1969); *Chem. Abstr.* **70**, 105922 (1969).

P72. E. A. Zuech and R. A. Gray (Phillips Petroleum Co.), U.S. Patent 3,310,591 (1967); *Chem. Abstr.* **66**, 115360 (1967).

P73. E. A. Zuech (Phillips Petroleum Co.), U.S. Patent 3,284,520 (1966); *Chem. Abstr.* **66**, 10654 (1967).

P74. E. A. Zuech (Phillips Petroleum Co.), U.S. Patent 3,243,467 (1966); *Chem. Abstr.* **64**, 17419 (1966).

P75. E. A. Zuech (Phillips Petroleum Co.), U.S. Patent 3,393,245 (1968); *Chem. Abstr.* **69**, 51519 (1968).

P76. Y. Yagi, A. Kobayashi, and I. Hirata (Sumitomo Chem. Co. Ltd.), Jap. Patent 71 06,776 (1971); *Chem. Abstr.* **75**, 63249 (1971).

P77. G. G. Garifzyanov, R. B. Valitov, and I. K. Bikbulativ (Sterlitamak Chemical Plant), U.S.S.R. Patent 263,588 (1970); *Chem. Abstr.* **73**, 14307 (1970).

P78. N. I. Seidov and M. A. Geidarov, U.S.S.R. Patent 282,648 (1970); *Chem. Abstr.* **75**, 50231 (1971).

P79. R. D. Mullineaux (Shell Oil Co.), U.S. Patent 3,290,348 (1966); *Chem. Abstr.* **66**, 65080 (1967).

P80. T. Yoshida and S. Yuguchi (Toyo Rayon Co. Ltd.), Jap. Patent 70 07,522 (1970); *Chem. Abstr.* **73**, 24887 (1970).

P81. H. Yoshida and S. Yuguchi (Toyo Rayon Co. Ltd.), Jap. Patent 70 09,729 (1970); *Chem. Abstr.* **73**, 55612 (1970).

P82. H. Kondo, M. Nishino, and A. Miyake (Toyo Rayon Co. Ltd.), Ger. Offen. 1,906,361 (1969); *Chem. Abstr.* **71**, 123719 (1969).

P83. K. Tani and S. Yuguchi (Toyo Rayon Co. Ltd.), Jap. Patent 21,619('65) (1965); *Chem. Abstr.* **64**, 1984 (1966).

P84. T. Yoshida and S. Yuguchi (Toyo Rayon Co. Ltd), Jap. Patent 7071('66) (1966); *Chem. Abstr.* **65**, 8757 (1966).

P85. K. Tani and S. Yuguchi (Toyo Rayon Co. Ltd.), Jap. Patent 21,618('65) (1965); *Chem. Abstr.* **64**, 1984 (1966).

P86. I. Ono, H. Sasaki, and T. Tanaka (Toyo Soda Manufg. Co. Ltd.), Jap. Patent 72 07,056 (1972); *Chem. Abstr.* **76**, 153241 (1972).

P87. H. Kondo, M. Nishino, and A. Miyake (Toray Industries Inc.), Jap. Patent 71 31,849 (1971); *Chem. Abstr.* **76**, 24784 (1972).

P88. S. Tokizane, A. Miyake, and K. Kamakura (Toray Industries Inc.), Ger. Offen. 2,103,369 (1971); *Chem. Abstr.* **75**, 129410 (1971).

P89. M. Itamura, H. Tanaka, and H. Ito (Toa Gosei Chem. Ind. Co.), Jap. Patent 70 32,256 (1970); *Chem. Abstr.* **74**, 53170 (1971).

P90. S. Izawa, I. Ono, and K. Kihara (Toyo Soda Manufg. Co. Ltd.), Jap. Patent 68 07,934 (1968); *Chem. Abstr.* **70**, 11325 (1969).

P91. M. Izawa, I. Ono, and K. Kihara (Toyo Soda Manufg. Co. Ltd.), Jap. Patent 4332('67) (1967); *Chem. Abstr.* **67**, 21506 (1967).

P92. R. L. Pruett and W. R. Myers (Union Carbide Corp.), U.S. Patent 3,417,130 (1968); *Chem. Abstr.* **70**, 46957 (1969).

P93. E. A. Rick (Union Carbide Corp.), U.S. Patent 3,359,337 (1967); *Chem. Abstr.* **68**, 39072 (1968).

P94. R. L. Pruett (Union Carbide Corp.), U.S. 3,261,875 (1966); *Chem. Abstr.* **65**, 13575 (1966).

P95. W. R. Myers and R. L. Pruett (Union Carbide Corp.), U.S. Patent 3,201,484 (1965); *Chem. Abstr.* **64**, 1979 (1966).

P96. E. L. De Young (Universal Oil Products Co.), U.S. Patent 3,522,321 (1970); *Chem. Abstr.* **73**, 120139 (1970).

P97. J. R. Larson and R. P. Zimmermann (Universal Oil Products Co.), U.S. Patent 3,349,138 (1967); *Chem. Abstr.* **67**, 116644 (1967).

P98. H. Baltz, W. Pritzkow, and H. Schmidt (VEB Leuna Werke), U.S. Patent 3,529,028 (1970); *Chem. Abstr.* **74**, 120194 (1970).

P99. L. I. Zakharkin and G. G. Zhigareva, U.S.S.R. Patent 186,445 (1966); *Chem. Abstr.* **66**, 85536 (1967).

P100. L. I. Zakharkin and G. G. Zhigareva, U.S.S.R. Patent 187,773 (1966); *Chem. Abstr.* **67**, 11274 (1967).

P101. G. Wilke, P. Heimbach, and C. Delliehausen (Studiengesellschaft Kohle mbH), Can. Patent 855,236 (1970); see *Chem. Abstr.* **75**, 5284 (1971).

P102. G. Wilke (Studiengesellschaft Kohle mbH), Swiss Patent 498,061 (1970).

P103. G. Wilke and P. Heimbach (Studiengesellschaft Kohle mbH), Ger. Offen. 1,493,220 (1969); see *Chem. Abstr.* **67**, 73121 (1967).

P104. G. Wilke and P. Heimbach (Studiengesellschaft Kohle mbH), Ger. Offen. 1,493,221 (1969); see *Chem. Abstr.* **67**, 73242 (1957).

P105. G. Wilke, P. Heimbach, and W. Brenner (Studiengesellschaft Kohle mbH), Ger. Aus. 1,643,063 (1972); *Chem. Abstr.* **77**, 20339 (1972).

P106. G. Wilke and P. Heimbach (Studiengesellschaft Kohle mbH), Fr. Patent 1,351,938 (1963); see *Chem. Abstr.* **60**, 15752 (1964).

P107. G. Wilke and E. W. Müller (Studiengesellschaft Kohle mbH), Swiss Patent 459,185 (1968); *Chem. Abstr.* **59**, 11293 (1963).

P108. G. Wilke and E. Müller (Studiengesellschaft Kohle mbH), Ger. Patent 1,297,600 (1970); *Chem. Abstr.* **71**, 80810 (1969).

P109. G. Wilke and E. W. Müller (Studiengesellschaft Kohle mbH), Ger. Patent 1,283,836 (1968); *Chem. Abstr.* **70**, 67757 (1969).

P110. G. Wilke and E. W. Müller (Studiengesellschaft Kohle mbH), Ger. Patent 1,283,838 (1969); *Chem. Abstr.* **70**, 57289 (1969).

P111. G. Wilke and E. W. Müller (Studiengesellschaft Kohle mbH), Ger. Patent 1,140,569 (1962); *Chem. Abstr.* **68**, 11214 (1963).

P112. G. Wilke and E. W. Müller (Studiengesellschaft Kohle mbH), Ger. Aus. 1,468,708 (1972); *Chem. Abstr.* **77**, 164142 (1972).

P113. H. Chung and W. Keim (Shell Oil Co.), U.S. Patent 3,636,162 (1972); *Chem. Abstr.* **76**, 85545 (1972).

P114. P. Chabardes (Soc. Usines Chimiques Rhône-Poulenc), Fr. Patent 1,433,409 (1966); *Chem. Abstr.* **66**, 37449 (1967).

P115. C. Y. Wu and H. E. Swift (B. F. Goodrich Co.), U.S. Patent 3,723,553 (1973); *Chem. Abstr.* **78**, 159063 (1973).

P116. H. Hashimoto and Y. Inoue, Jap. Patent 73 30,244 (1973); *Chem. Abstr.* **80**, 3057 (1974).

P117. H. Singer (Henkel & Cie.), Ger. Offen. 2,142,444 (1973); *Chem. Abstr.* **78**, 135688 (1973).

P118. H. Morikawa (Mitsubishi Petrochem. Co.), Jap. Patent 73 16,915 (1973); *Chem. Abstr.* **79**, 78228 (1973).

P119. M. Nishino and A. Miyake (Toray Ind. Inc.), Jap. Patent 73 19,305 (1973); *Chem. Abstr.* **79**, 78227 (1973).

P120. S. Tokizane and A. Miyake (Toray Ind. Inc.), Jap. Patent 73 03,834 (1973); *Chem. Abstr.* **79**, 18219 (1973).

P121. K. Nishio and A. Miyake (Toray Ind. Inc.), Jap. Patent 73 03,836 (1973); *Chem. Abstr.* **79**, 18218 (1973).

P122. Y. Nishino and A. Miyake (Toray Ind. Inc.), Jap. Patent 73 04,034 (1973); *Chem. Abstr.* **78**, 147448 (1973).

P123. Y. Nishino and A. Miyake (Toray Ind. Inc.), Jap. Patent 73 04,035 (1973); *Chem. Abstr.* **78**, 147449 (1973).

P124. E. A. Zuech (Phillips Petrol. Co.), U.S. Patent 3,352,031 (1967); *Chem. Abstr.* **68**, 12559 (1968).

P125. P. Arthur and B. C. Pratt (E.I. du Pont), U.S. Patent 2,571,099 (1951); *Chem. Abstr.* **46**, 3068 (1952).

P126. P. Albanese, L. Benzoni, G. Carnisio, and A. Crivelli (Montecatini Edison), Ital. Patent 869,900 (1970); *Chem. Abstr.* **78**, 135701 (1973).

P127. D. Rose (Henkel & Cie.), Ger. Offen. 2,161,750 (1973); *Chem. Abstr.* **79**, 52777 (1973).

P128. J. Itakura, H. Tanaka, and H. Ito (Toa Gosei Chem. Ind.), Jap. Patent 73 26,724 (1973); *Chem. Abstr.* **79**, 136460 (1973).

P129. H. Seibt and N. von Kutepow (B.A.S.F.), Belg. Patent 635,483 (1964); *Chem. Abstr.* **61**, 11891 (1964).

P130. G. Longoni, P. Chini, and F. Canziani (Snam Progetti S.p.A.), Ger. Offen. 2,148,925 (1972); *Chem. Abstr.* **77**, 114570 (1972).

P131. Studiengesellschaft Kohle mbH., Brit. Patent 1,197,500 (1970); see *Chem. Abstr.* **71**, 123910 (1969).

P132. C. Duscheck, H. Fuellbier, and J. Beger, Ger. (East) Patent 98,502 (1973); *Chem. Abstr.* **80**, 59416 (1974).

P133. Y.-T. Chia, W. C. Drinkard, and E. N. Square (E.I. du Pont), U.S. Patent 3,766,237 (1973); *Chem. Abstr.* **80**, 70373 (1974).

P134. S. Miyamoto, T. Yoshimura, and R. Tomita (Mitsui Petrochemical Ind. Ltd.), Jap. Kokai 74 13,101 (1974); *Chem. Abstr.* **80**, 120199 (1974).

P135. Ciba-Geigy A. G. Neth. Appl. 73 03,381 (1973).

P136. S. Nishino and A. Miyake (Toray Ind. Inc.) Jap. Patent 74 07,152 (1974); *Chem. Abstr.* **81**, 13182 (1974).
P137. S. Nishino and A. Miyake (Toray Ind. Inc.) Jap. Patent 74 07,151 (1974); *Chem. Abstr.* **81**, 13183 (1974).
P138. J. Furukawa, J. Kiji, and K. Masui (Research Inst. Prod. Develop.) Jap. Patent 74 06,901 (1974); *Chem. Abstr.* **81**, 37652 (1974).

Reviews

A selection of general review articles relevant to the material discussed in this chapter is listed below.

R. Baker, π-Allylmetal derivatives in organic synthesis. *Chem. Rev.* **73**, 487 (1973).
C. W. Bird, "Transition Metal Intermediates in Organic Synthesis," Chapter 2. Academic Press, New York.
P. Heimbach, Cyclooligomerisation an Übergangsmetall-Katalysatoren. *Angew. Chem.* **85**, 1035 (1973).
P. Heimbach, Organic synthesis by combination of catalytic and classical reactions (Japan). *Yuki Gosei Kagaku Kyokai Shi* **31**, 299 (1973).
P. Heimbach, Nickel catalyzed syntheses of methyl-substituted cyclic olefins. *Advan. Homogen Catal.* **2** (in press).
P. Heimbach, P. W. Jolly, and G. Wilke, π-Allylnickel intermediates in organic synthesis. *Advan. Organometal. Chem.* **8**, 29 (1970).
W. Keim, π-Allyl systems in catalysis. *In* "Transition Metals in Homogeneous Catalysis" (G. N. Schrauzer, ed.), p. 60. Dekker, New York, 1971.
A. Miyake, H. Kondo, M. Nishino, and S. Tokizane, Catalytic formation of macrocyclic polyenes from butadiene. *Int. Congr. Pure Appl. Chem., 23rd, 1971* Vol. 6, p. 201 (1971).
K. Nishimura, Linear telomerization of butadiene (Jap.). *Sikiyu Gakhai Shi* **16**, 548 (1973).
J. Tsuji, Addition reactions of butadiene catalyzed by palladium complexes. *Accounts Chem. Res.* **6**, 8 (1973).
G. Wilke, Über nickelorganische Verbindungen. *Pure Appl. Chem.* **17**, 179 (1968).
G. Wilke *et al.*, Cyclooligomerisation von Butadien und Übergangsmetall-π-Komplexe. *Angew. Chem.* **75**, 10 (1963).

CHAPTER IV

The Polymerization of Butadiene and Related Reactions

I. Introduction

The commercial interest in polybutadiene has insured that a great many chemists have been involved in the search for catalysts active for the polymerization reaction. Although a large number of nickel-based catalysts are known, some of which are able to polymerize butadiene under surprisingly mild conditions, only two of them have been used to produce polybutadiene on a large scale. A ternary catalyst, $Ni(naphth)_2–Al(C_2H_5)_3–BF_3 \cdot O(C_2H_5)_2$, developed by the Bridgestone Tire Company, is used in the production of a polymer having a cis-1,4 content of around 97% (5, 53, 67, 71, 128, 133,137). On a more modest scale the preparation of a low molecular weight liquid polymer having a cis-1,4 content of 70–80% has been developed by the Chemische Werke Hüls using the catalyst $Ni(acac)_2–(C_2H_5)_3Al_2Cl_3$ with a regulator (109).

The principal investigators of the polymerization of butadiene with nickel-containing catalysts have regularly summarized their work in the review articles listed as refs. 31, 32, 46, 47, 65, 135, 162, 164, 165, and 202.

II. The Polymerization of Butadiene (Table IV-1)

A. *The Catalyst*

The wide variety of nickel catalysts capable of polymerizing butadiene can be conveniently divided into those based on π-allylnickel complexes, bis(cyclooctadiene)nickel, π-cyclopentadienylnickel complexes, nickel tetra-carbonyl, tetrakisligand nickel complexes, various metal salts, and metallic nickel.

π-Allylnickel complexes are active alone (Table IV-1, row 1); or in combination with organic and inorganic acids (Table IV-1, row 2); substituted quinones (Table IV-1, row 3); metal halides (Table IV-1, row 4); metal salts of organic acids (Table IV-1, row 5); iodine, water, oxygen and peroxides (Table IV-1, rows 6 and 7); and various ligands, e.g., amines, phosphines, and olefins (Table IV-1, row 8). Active catalysts have also been prepared by depositing bis(π-allyl)nickel or π-allylnickel halides on silica or alumina carriers (181, 215). In addition catalysts formed by reduction with aluminum-alkyls (198, P64) and by reaction with the π-allyl complexes of chromium and titanium have been reported (91, 118, 191). Various catalyst systems are based on bis(cyclooctadiene)nickel (25, 26, 28, 35, 38–40, 117, P28, P33, P40, P76, P80–P83), and it is probable that in these cases an α,ω-bis(π-allyl)-C_{12}-nickel complex that is known to be produced on reaction of $(COD)_2Ni$ with buta-diene is involved (150). The π-olefin complexes $(COD)NiX$ $(X = Br, I)$ and methyl maleate- and methyl fumarate-nickel bromide (34) also polymerize butadiene (Table IV-1, row 9); as do the related compounds formed during the electrolysis of nickel iodide in the presence of COD and maleic acid or maleic anhydride (131, 134). Systems based on acrylonitrile–nickel with added aluminum tribromide or trifluoracetic acid are also active (203).

π-Cyclopentadienylnickel complexes are active when combined with metal halides, aluminum-alkyls, organic acids, or chloranil (Table IV-1, row 10).

Binary catalysts have been prepared by reacting nickel tetracarbonyl with a metal halide or a reducing agent, and ternary catalysts by combining the tetracarbonyl and reducing agent with a metal halide or boron trifluoride (Table IV-1, row 11). The use of phosphine- and phosphite-substituted nickel carbonyl complexes in combination with a nickel halide has also been patented (Table IV-1, row 12). Tetrakisligand nickel complexes combined with an inorganic acid or a metal halide, with or without a reducing agent, are also active (Table IV-1, row 13).

The majority of the patents covering the nickel-catalyzed polymerization of butadiene are concerned with variations on the theme nickel salt–metal halide–reducing agent. Although catalysts have been prepared by activating the nickel salt by irradiation, heat treatment, or addition of a metal halide, e.g., $AlCl_3$ (Table IV-1, row 14), greater activity is obtained by treating the nickel salt with a reducing agent, e.g., $(C_2H_5)_2AlCl$, or lithium-alkyl (Table IV-1, row 15). Three component catalysts are also popular and are formed by treating the above catalyst with a metal salt (frequently a halide), boron trifluoride or a related compound [e.g., $Ni(BF_4)_2$], or an organofluorine compound (Table IV-1, row 16). Active systems have also been formed by reducing the nickel salt in the presence of a donor ligand, e.g., amine, phosphine, quinone, cyclooctadiene, or maleic anhydride (Table IV-1, row 17). It is also claimed that it is advantageous to use a catalyst formed by the reduction of a mixture of cobalt and nickel salts (Table IV-1, row 18).

The combination nickel salt–reducing agent can be replaced by Raney nickel or reduced nickel. The metal may be activated by adding a second metal halide, an organic halide, or an acid (Table IV-1, row 19); or by forming a three-component system with a reducing agent (Table IV-1, row 20).

B. The Product

The nature of the polybutadiene produced using nickel-containing catalysts depends to some extent on the catalyst composition and polymers have been isolated consisting mainly of cis-1,4 units (1), trans-1,4 units (2), and 1,2-vinyl units (3). In addition, polymers containing varying proportions of all three units are possible.

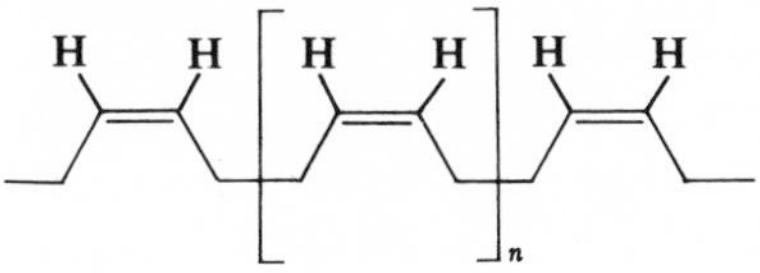

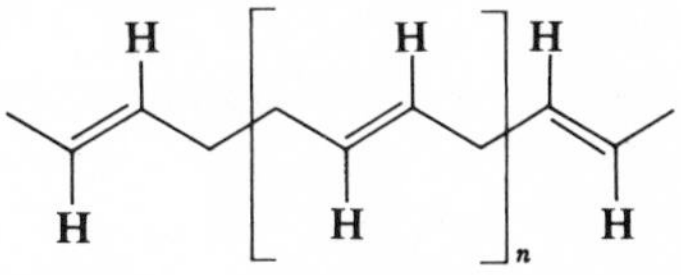

In the case of the polymerization catalyzed by $Ni(octanoate)_2–Al(C_2H_5)_3–BF_3 \cdot O(C_2H_5)_2$, evidence has been published indicating that the structure of the polymer is a function of the molecular weight; low molecular weight polymer has a trans-1,4 content of as high as 40% and this decreases to ca. 4% for high molecular weight polymer (27). Branching within a stereoregular cis-polybutadiene has also been reported (154). Particular attention has been focused on what was originally called "equibinary cis,trans-polybutadiene" and was believed to contain alternating cis and trans units. It has been suggested, however, on the basis of high-resolution NMR studies, that this may well be a block copolymer consisting of cis-1,4 sequences separated by trans-1,4 sequences (167, 168; see also 169–173, 209–211). See, however, 212.

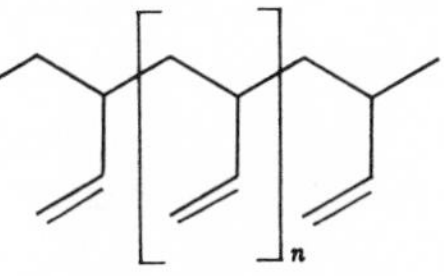

The polymerization of butadiene using nickel-containing catalysts is believed to involve a π-allylNiX system (see Section C), and it is therefore to be expected that the course of polymerization is influenced by the variation of

X and by the introduction of coordinating solvents and ligands. The effect of varying the group X is seen most clearly in the reaction involving the dimeric π-allylnickel halides: the chloride produces *cis*-1,4-polybutadiene, the iodide produces *trans*-1,4-polybutadiene, whereas the polymer produced using the bromide consists of a mixture of the two microstructures (Table IV-1, row 1). The effect of solvent variation and addition of donor ligands has been most thoroughly studied for the catalysts based on $(\pi\text{-allylNiOCOCF}_3)_2$. The product using heptane as solvent is *cis*-1,4-polybutadiene but addition of benzene or *o*-dichlorobenzene causes the trans-1,4 content to increase to a maximum of 50%. Addition of an excess of trifluoracetic acid to the catalyst has the same effect. Ethanol, when added to the solvent, causes formation of *trans*-1,4-polybutadiene, although at a much decreased rate, and this is also the product using the triphenyl phosphite adduct to $\pi\text{-C}_3\text{H}_5\text{NiOCOCF}_3$ as catalyst (23, 35, 38, 85, 97, P169). In addition to the effect of the group X, solvent, and added ligands, it has been observed that the nature of the original π-allyl group attached to the nickel has a minor influence on the microstructure of the product (23).

Of the various cocatalysts that have been investigated, Lewis acids, e.g., AlBr_3, and transition metal halides, e.g., TiCl_4 and MoCl_5, are particularly effective, although it is not always clear at which metal atom the polymerization is occurring. Enhanced activity has also been claimed for oxygen, peroxides, and iodine (80, 82, 83, 86, 87) but this has been contested (129). Particularly effective additives are substituted quinones and in some cases catalysts result that are capable of polymerizing butadiene rapidly at $-15°$ (2, 72, 87, 110, P161). [The chemistry of the reaction between π-allylnickel complexes and quinones (140–145, 207) is discussed in Volume I, p. 351].

The various effects mentioned above for the systems based on the π-allylnickel complexes are mirrored in the more complicated binary and ternary catalysts and strongly suggest that common intermediates are involved. For example, mainly *trans*-1,4-polybutadiene is obtained using the following iodine-containing catalysts: (COD)NiI, $(\pi\text{-C}_5\text{H}_5)_2\text{Ni-TiI}_4$, and Ni(octanoate)$_2$–Al(C$_2H_5$)$_3$–HI, as well as with the triphenyl phosphite-modified catalysts $[(\text{C}_6\text{H}_5\text{O})_3\text{P}]_4\text{Ni-ZnCl}_2$ and $\pi\text{-C}_3\text{H}_5\text{NiOCOCF}_3[\text{P}(\text{OC}_6\text{H}_5)_3]$.

Polybutadiene having a high 1,2-vinyl content is not normally observed with nickel catalysts. The ternary catalysts Ni(stearate)$_2$–Al(C$_2$H$_5$)$_3$–MoCl$_5$ and Ni(octanoate)$_2$–Al(C$_2$H$_5$)$_3$–WOCl$_4$ are exceptions, yielding polymers with a vinyl-1,2 content as high as 80% (P55, P123, P124).

It is convenient to mention here that the addition of nickel acetylacetonate to the anionic telomerization of butadiene with toluene, which is catalyzed by butyllithium–potassium butoxide, causes an increase in the cis-1,4 content at the expense of the 1,2-vinyl and trans-1,4 content (P179).

TABLE IV-1

THE POLYMERIZATION OF BUTADIENE (BD)

Catalyst	Ni:Cocat.	Solvent	Temp. (°C)	% BD conv. (hr)	Polymer microstructure (%)			Ref.^{a}
					1,4-cis	1,4-trans	1,2	
1a $(\pi\text{-}CH_3CHCHCH_2NiCl)_2$	—	Benzene	30	40 (40)	93.5	5	1.5	*14*, 2, 12, 13, 17, 23, 29–32, 35, 38, 42, 57–59, 62, 63, 65, 68, 70, 73–77, 80, 82, 83, 85, 86, 88–90, 97–99, 110, 111, 113, 116, 124, 132, 136, 149, 187, 194, P1, P79, P168
1b $(\pi\text{-}CH_3CHCHCH_2NiBr)_2$	—	Benzene	50	25 (18)	45	53	2	68
1c $(\pi\text{-}CH_3CHCHCH_2NiI)_2$	—	Benzene	50	100 (15)	4	93	3	68
1d $(\pi\text{-}CH_3CHCHCH_2NiSCN)_2$	—	Benzene	50	93.5 (17)	70.5	27	2.5	116
1e $(\pi\text{-}CH_3CHCHCH_2\text{-}NiOCOCH_3)_2$	—	Benzene	50	90 (18)	91	6	3	116
1f $(\pi\text{-}CH_3CHCHCH_2\text{-}NiOCOCF_3)_2$	—	Benzene	50	18.1 (18)	87	9	4	116
2a $\pi\text{-}C_{12}H_{18}Ni\text{-}HCl$	1:1	Heptane	55	13 (3)	84	13	3	35, 28, 38–40, 42, 55, 60, 63, 83, 86, 89, 110–112, 116, 117, 125, 132, 166, 185, P1, P78, P80, P82, P169
2b $\pi\text{-}C_{12}H_{18}Ni\text{-}CCl_3CO_2H$	1:1	Heptane	55	30 (3)	91	6	3	35
2c $\pi\text{-}C_{12}H_{18}Ni\text{-}CF_3CO_2H$	1:1	Heptane	55	90 (3)	91	4	5	35, P22

(continued)

TABLE IV-1 (*continued*)

Catalyst	Ni:Cocat.	Solvent	Temp. (°C)	% Bd conv. (hr)	1,4-cis	1,4-trans	1,2	Ref.^{a}
					Polymer microstructure (%)			
3a (π-CH₃CHCHCH₂NiCl)₂–chloranil	1:1	Benzene	25	83 (48)	94	3	3	72, 2–4, 35, 72, 87, 110, 116, 132, 140, 145, P1, P84, P85, P131
3b (π-CH₃CHCHCH₂NiCl)₂–fluoranil	1:1	Benzene	−15	26.5 (0.2)	98	1	1	3
4a (π-CH₃CHCHCH₂NiCl)₂–TiCl₄	1:1	Benzene	30	100 (17)	90	8	2	14, 2, 12, 13, 17, 18, 25, 26, 56, 63–66, 68, 72, 83, 88, 90, 97, 103, 104, 106, 110, 113, 115, 118, 139, 145, 146, 148, 193, 198, 199, 200, 208, 213, P1, P64, P81, P83
4b (π-CH₃CHCHCH₂NiCl)₂–VCl₄	1:1	Benzene	20	40 (0.5)	91	7	2	68
4c (π-CH₃CHCHCH₂NiCl)₂–AlCl₃	1:1	Benzene	50	70 (0.5)	90	8	2	68
5 (π-CH₃CHCHCH₂NiCl)₂–Zn(OCOCCl₃)₂	1:3	Benzene	−15	90 (0.3)	92.5	6.0	2.0	89, 83, 86, 125, 126, 132, P1
6 (π-CH₃CHCHCH₂NiCl)₂–I₂	2:1	Benzene	−15	61 (21)	96	2.5	1.5	110, 35, 40, 113
7 (π-CH₃CHCHCH₂NiCl)₂–O₂	1:1	Benzene	40	42 (3)	94	3	3	80, 82, 83, 86, 87, 129, P78, P91
8 π-C₁₂H₁₈Ni–CF₃CO₂H–P(OC₆H₅)₃	1:1:1	Heptane	55	75 (48)	0	96	4	35, 23, 65, 83, 86, 87, 89, 90, 114, 146
9a (COD)NiBr	—	Benzene	—	—	60.5	37.5	2	105, 131, 134
9b (COD)NiI	—	Benzene	—	—	0	99	1	105, 100
10a (π-C₅H₅)₂Ni–TiCl₄	1:1	Benzene	50	76 (6)	93	4	3	11, 15, 19, 39, 56, 110, 118, 120, 184, P100, P152–P154, P190

		Solvent	Temp	Yield	cis	trans	1,2	References
10b $(\pi\text{-}C_5H_5)_2Ni\text{-}TiBr_4$	1:1	Benzene	50	—	37	60	3	19
10c $(\pi\text{-}C_5H_5)_2Ni\text{-}TiI_4$	1:1	Benzene	50	33 (20)	0	96	4	19
10d $(\pi\text{-}C_5H_5NiCO)_2\text{-}VOCl_4$	1:1	Benzene	50	7 (17)	94	4	2	15
11a $Ni(CO)_4\text{-}TiCl_4$	1:1	Benzene	50	50 (17)	87	10	3	15, 5, 16, 19, 22, 95, 155, P21, P31, P32, P35, P36, P39, P100, P132, P134–P136, P158
11b $Ni(CO)_4\text{-}Al(C_2H_5)_3\text{-}BF_3\cdot O(C_2H_5)_2$	1:1:1	Benzene	25	79.7 (1)	96	3.5	0.5	5, P17, P34
12 $(C_6H_5)_3PNi(CO)_3\text{-}BF_3$	0.4:1.8	Benzene	50	95 (3)	73	26	1	P29, P52
13a $[(C_6H_5O)_3P]_4Ni\text{-}ZnCl_2$	1:6	Benzene	120	—(7)	0	71	29	P52, 61, P89
13b $(Cl_3P)_4Ni\text{-}LiC_4H_9\text{-}AlBr_3$	0.1:3:2	Hexane	25	100	76	22	2	48
13c $[(C_2H_5O)_3P]_4Ni\text{-}H_2SO_4$	1:1	Iso-propanol	80	100 (24–48)	—	100	—	175
14a $NiCl_2$	—	Benzene	50	63.5 (16)	93	5	2	94, 21, 96, 138, 188, 190, 214, P150, P164, P165
14b NiI_2	—	Benzene	50	33 (17.5)	17	75	8	94
14c $Ni(OCOCF_3)_2$	—	Heptane	50	70 (16)	93	4	3	33, 192, P77, P177
14d $CF_3OCONiCl$	—	Chloro-benzene	25	79 (0.5)	93	2	5	33, P77, P168, P176
14e $NiCl_2\text{-}WCl_6$	—	Benzene	—	—	79	17	14	16, 51, 69, 130, 190, P38, P98, P113
15 $Ni(acac)_2\text{-}Al(C_2H_5)_2Cl$	1:4	Benzene	30	89 (6)	83	16	1	109, 24, 45, 53, 78, 79, 114, 119, 130, 153, P25–P27, P44, P48–P51, P60, P62, P63, P72, P75, P102, P103, P105, P107, P109, P110, P114–P117, P136–P138, P142–P144, P146–P149, P151, P155, P170, P175, P180, P181, P184, P191, P193

(continued)

TABLE IV-1 (*continued*)

Catalyst	Ni:Cocat.	Solvent	Temp. (°C)	% BD conv. (hr)	Polymer microstructure (%)			Ref.^{a}
					1,4-cis	1,4-trans	1,2	
16a Ni(naphthenate)$_2$– Al(C$_2$H$_5$)$_3$–TiCl$_4$	0.9:1 (Al:Ti)	Benzene	50	64.5 (17)	95.5	3.4	1.1	*P14*, 79, 84, 128, 137, P55, P58, P66, P67, P73, P88, P104, P118, P119, P121, P123, P124, P139–P141, P145, P178
16b Ni(naphthenate)$_2$– Al(C$_2$H$_5$)$_3$– BF$_3$·O(C$_2$H$_5$)$_2$	1.2:1 (Al:B)	Benzene	40	91 (0.75)	97.4	2.4	0.2	*P14*, 5, 27, 37, 41, 50, 53, 67, 71, 84, 102, 108, 121, 122, 128, 133, 137, 151, 152, 195, P3, P6, P12, P13, P15–P17, P23, P56, P57, P61– P63, P68–P70, P86, P87, P90, P93–P97, P101, P110, P111, P119, P120, P122, P174, P186–P188, P192, P194
16c Ni(octanoate)$_2$– Al(C$_2$H$_5$)$_3$–(CF$_3$)$_2$CHOH	0.025:0.3: 0.2	Benzene	50	78 (18)	96.4	—	—	*P65*, P59, P125, P126, P128, P162, P182, P183
16d Ni(octanoate)$_2$– Al(C$_2$H$_5$)$_3$–HF	0.05:0.7: 1.5	Benzene	50	81 (4)	97	2	1	*123*, P71, P120, P138, P172, P173, P195
16e Ni(octanoate)$_2$– Al(C$_2$H$_5$)$_3$–HI	0.05:0.7: 1.5	Benzene	50	70 (18)	13	81	6	123

								References
17a Ni(acac)$_2$–Al(C$_2$H$_5$)$_2$Cl P(C$_6$H$_5$)$_3$	1:10:4	Benzene	RT	90 (20)	49	11	40	78, 5, 71, P45, P99, P125, P127–P130, P157, P159, P161, P163, P182
17b Ni(acac)$_2$–Al(C$_2$H$_5$)$_3$–*p*-chloranil	0.2:1:1	Toluene	40	95 (4)	94.1	4.1	5.2	205, P161
17c NiI$_2$–electrolysis–maleic anhydride	1:1	Ethanol	RT	32 (4 days)	0	99	1	134, 131
18 Ni(acac)$_2$–Co(acac)$_3$–Al$_2$(C$_2$H$_5$)$_3$Cl$_3$ (Ni:Co)	1:1	Benzene	25	90 (1)	90	—	—	109, P41–P43, P74, P92, P112
19a Ni(Raney)–TiCl$_4$	—	Benzene	40	70.9 (3)	93.2	4.9	1.9	137, 93, 114, 128, P30, P108, P167
19b Ni(red)–CCl$_4$	0.5:1	Benzene	60	26.2 (24)	79.0	16.3	4.7	1, 93
19c Ni–CF$_3$CO$_2$H	—	Benzene	—	55 (20)	83	14	3	39, 128, 137
19d Ni(Raney)–BF$_3$·O(C$_2$H$_5$)$_2$	1:1	Benzene	40	59.8 (20)	91.2	6.1	2.7	128, 137
20 Ni(red)–Al(C$_2$H$_5$)$_3$–BF$_3$·O(C$_2$H$_5$)$_2$ (Al:B)	1:1	Benzene	40	70.2 (5)	98.2	1.7	0.1	128, 137, P11, P15, P40, P160, P166

^{a} The reference cited in the table is shown in italics. The remaining references refer not only to the specific example quoted but also to related systems.

C. Mechanistic Considerations

Insight into the mechanism of the polymerization reaction has been obtained from kinetic and NMR studies as well as from investigations using [^{14}C]-labeled organic ligands.

The kinetics (see Table IV-2) of the reaction have been studied for systems involving π-allylnickel halides (32, 57–59, 62, 75, 76, 136, 149, 156, 157, 164) and π-allylnickel haloacetates (23, 29, 32, 35, 85, 158); the binary systems (π-$C_4H_7NiCl)_2$–$Ni(OCOCCl_3)_2$ (125), (π-$C_4H_7NiCl)_2$–CCl_3CO_2H (55, 125, 185), $Ni(CO)_4$–CCl_3CO_2H (22), (π-$C_4H_7NiI)_2$–oxidant (e.g., O_2 and I_2) (129, 156), (π-allylNiCl)$_2$–substituted quinone (3, 4), (π-$C_3H_5)_2Ni$–$TiCl_4$ (64), (π-$C_3H_5NiCl)_2$–$TiCl_4$ (66, 163, 200), $Ni(acac)_2$–R_2AlCl (24, 153), and Ni_2O_3–$AlCl_3$ (69); and the ternary systems $Ni(naphth)_2$–$BF_3 \cdot O(C_2H_5)_2$–$Al(C_2H_5)_3$ (128, 133) and (π-$C_3H_5NiCl)_2$–$TiCl_4$–additive, e.g., C_6H_5OH (66).

The reaction catalyzed by a π-allylnickel iodide has been studied in greatest detail (57, 59, 62, 75, 76, 136, 156, 157, 164) and is briefly discussed here. At low monomer concentration the rate law for the reaction is

$$-\frac{d[C_4H_6]}{dt} = v = k[(\pi\text{-allylNiI})_2]^{1/2}[C_4H_6]$$

i.e., the reaction is first order in monomer concentration and one-half order in catalyst concentration. At high monomer concentrations the order in monomer falls to around 0.5. The half-order dependence on catalyst concentration is interpreted as indicating that the π-allylnickel iodide dimer itself is inactive and is monomerized by the diene in an initial fast reaction. This is then followed by a slower step involving insertion of the coordinated diene into the allyl–nickel bond. Further information has been obtained by studying the effect of conversion, catalyst concentration, and temperature on the molecular weight of and molecular weight distribution in the resulting polymer. At low conversion the product has a binodal molecular weight distribution (Fig. IV-1). With increasing conversion, the average molecular weight increases, the higher molecular weight component of the binodal distribution decreases, and the breadth of the distribution increases. Such behavior is typical for a polymerization reaction proceeding by a slow stepwise propagation with occasional transfer.

Three types of polymer chain in the product have been distinguished (156; see also 186).

1. Living polymers, i.e., growing polymer chains that started growing at the beginning of the reaction and have never transferred. The molecular weight of the living polymer is proportional to the conversion and is characterized by a narrow distribution. The fraction of living polymer present in the total decreases with increasing conversion.

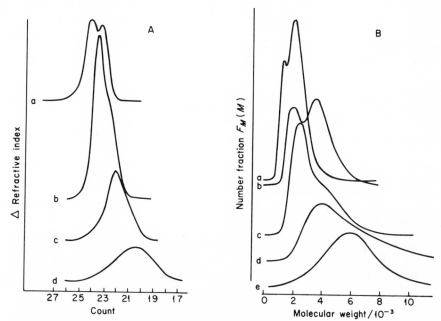

Fig. IV-1. The polymerization of butadiene catalyzed by $(\pi\text{-}C_4H_7NiI)_2$ (156). A, Gel permeation chromatograms produced at different conversions. $[C_4H_6] = 1.5$ M, [catalyst] $= 0.004$ M, T $= 50°$. Conversions: (a) 0.25 M; (b) 0.30 M; (c) 0.75 M; (d) 1.92 M. B, Molecular weight distributions for polymers produced at different catalyst concentrations. $[C_4H_6] = 1.5$ M, conversion $= 0.36$ M, T $= 50°$. Catalyst concentrations: (a) 0.0134 M; (b) 0.0086 M; (c) 0.0065 M; (d) 0.0036 M; (e) 0.0020 M.

2. Dead polymer chains that result from random transfer. The fraction present in the total increases with increasing conversion.

3. Reinitiated polymer chains. These are actively growing chains that appear at the same rate as the initial living polymer chains disappear and in the early stages of the polymerization have a lower molecular weight than the other two types of polymer.

A similar rate law to that found for the iodide is observed using π-allylnickel chloride or bromide. In contrast, the rate of polymerization with $(\pi\text{-allylNiOCOCF}_3)_2$ is found to be first order in both monomer and catalyst concentration and this relationship is observed for a variety of other systems (see Table IV-2). The rate of reaction using the catalyst $Ni(acac)_2\text{-}R_2AlCl$ is found to be first order in both monomer and nickel component and second order or $1\frac{1}{2}$ order in aluminum-alkyl where R is $iso\text{-}C_4H_9$ or C_2H_5, respectively. The reaction involving $Ni_2O_3\text{-}AlCl_3$ is first order in catalyst concentration and second order in monomer concentration. Typical rate equations and activation energies are shown in Table IV-2.

TABLE IV-2

KINETIC DATA FOR THE NICKEL-CATALYZED POLYMERIZATION OF BUTADIENE

Catalyst	Rate law	Activation energy (kcal/mole)	Ref.
$(\pi\text{-}C_3H_5NiCl)_2$	$v = k[\text{cat}]^{0.5}[C_4H_6]$	16.0	75, 76
$(\pi\text{-}C_3H_5NiBr)_2$	$v = k[\text{cat}]^{0.5}[C_4H_6]$	15.2	75, 76
$(\pi\text{-}C_4H_7NiI)_2$	$v = k[\text{cat}]^{0.5}[C_4H_6]$	14.5	57, 76
$(\pi\text{-}C_3H_5NiOCOCCl_3)_2$	$v = k[\text{cat}][C_4H_6]^{1.6}$	11	158
$(\pi\text{-}C_4H_7NiOCOCF_3)_2$	$v = k[\text{cat}][C_4H_6]$	10.2	23
$(\pi\text{-}C_4H_7NiCl)_2\text{-}CCl_3CO_2H$	$v = k[\text{cat}][C_4H_6]^2$	12	125, 185
$(\pi\text{-}C_4H_7NiCl)_2\text{-}2 Ni(OCOCCl_3)_2$	$v = k[\text{cat}][C_4H_6]$	6.4	125
$(\pi\text{-}C_4H_7NiCl)_2\text{-benzoquinone}$	—	12.9	3, 4
$(\pi\text{-}C_4H_7NiCl)_2\text{-}2 \text{ chloranil}$	$v = k[\text{cat}][C_4H_6]$	6.3	3, 4
$Ni(acac)_2\text{-}(C_2H_5)_2AlCl$	$v = k[\text{ni}][\text{al}]^{0.5}[C_4H_6]$	8.9	24
$Ni(acac)_2\text{-}(iso\text{-}C_4H_9)_2AlCl$	$v = k[\text{ni}][\text{al}]^2[C_4H_6]$	8.0	24
$Ni_2O_3\text{-}AlCl_3$	$v = k[\text{cat}][C_4H_6]^2$	—	69
$Ni(naphth)\text{-}BF_3 \cdot O(C_2H_5)_2\text{-}$ $Al(C_2H_5)_3$	$v = k[\text{cat}][C_4H_6]$	~ 10	128, 133

A binodal molecular weight distribution similar to that observed in the reaction catalyzed by π-allylnickel iodide has also been observed using the catalyst $(\pi\text{-}C_3H_5NiCl)_2\text{-}TiCl_4$ (103, 104, 106). In contrast, a uninodal distribution is observed in the reaction catalyzed by $(\pi\text{-}C_4H_7NiCl)_2\text{-}2Ni$-$(OCOCCl_3)_2$, suggesting that here the living polymer mechanism is predominant (126). An interesting molecular weight jump has been observed in the reaction catalyzed by $Ni(naphth)_2\text{-}Al(C_2H_5)_3\text{-}BF_3 \cdot O(C_2H_5)_2$ if triethylaluminum or triethylamine are added at the end of the reaction; this is thought to be the result of a coupling reaction between living polymer ends (67, 71). A similar effect may be responsible for the lack of incorporation of radioactive terminators in the reaction catalyzed by $(\pi\text{-}C_4H_7NiCl)_2\text{-}CCl_3CO_2H$ and terminated by labeled water or methanol (182).

The ternary catalyst $Ni(naphth)_2\text{-}Al(C_2H_5)_3\text{-}BF_3 \cdot O(C_2H_5)_2$ has been investigated by conductivity, ESR, and IR measurements. Because many experiments were carried out in the absence of butadiene, the relevance of the results is questionable. It could be shown, however, that fluorination of both the nickel and aluminum components occurs (102, 121, 122, 195).

The fate of the original π-allyl group in the polymerization reactions catalyzed by π-allylnickel complexes has been studied using $[^{14}C]$-labeled allyl groups: with $(\pi\text{-}C_4H_7NiX)_2$ as catalyst, approximately two crotyl groups are incorporated per macromolecule which, it is suggested, indicates that dimerization of the living polymer occurs. Experiments with the catalyst

$(\pi\text{-}C_4H_7NiX)_2\text{-}TiCl_4$ show that only one labeled crotyl group is incorporated per macromolecule (13, 17). The incorporation of only 10% of the labeled crotyl group using the catalyst $(\pi\text{-}C_4H_7NiCl)_2\text{-}CCl_3CO_2H$ is probably the result of protonolysis of the crotyl group before the start of the catalytic reaction (54, 55, 60; see also 166, 182).

NMR techniques have also been used to follow the course of the polymerization (23, 70, 73, 74, 77, 85, 174, 187, 189, 194, 196, 206) and a study of the reaction of π-crotylnickel iodide with butadiene has been particularly valuable. Using both perdeuterobutadiene and a perdeuterated π-crotylnickel complex it has been shown that the original crotyl group attached to the nickel is incorporated terminally in the living polymer chain, and also that the propagating end of the chain forms a π-allyl complex with the metal; insertion occurring predominantly into the unsubstituted side of the π-crotyl group (73, 74, 187, 206).

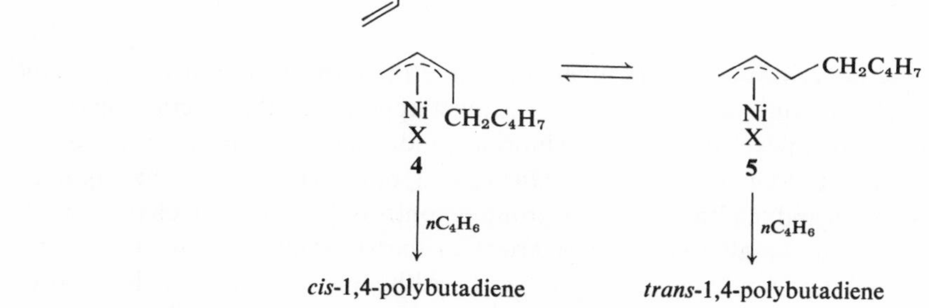

Many authors have speculated on the mechanism of the polymerization process (43, 44, 49, 57, 77–79, 87, 123, 135, 179, 180). The results of the physical studies described above, combined with the known chemistry of the π-allylnickel system, enable a mechanism to be formulated for at least the early stages of the process, in which living polymer formation predominates. The initial step is the formation of a π-allylnickel system. Further reaction probably involves the insertion of a complexed single-cis butadiene molecule into a σ-allyl group to form a new, anti-substituted π-allylnickel system (4) from which *cis*-1,4-polybutadiene is derived. *trans*-1,4-Polybutadiene is produced if 4 rearranges, before further reaction, to the thermodynamically preferred syn-substituted π-allylnickel system 5. These processes are illustrated below for a system derived from $(\pi\text{-crotylNiX})_2$. That only minor amounts of polymer having a 1,2 microstructure are formed indicates that the insertion of only one double bond of the butadiene into the π-allyl system is of little significance. The termination step for the reaction is probably the

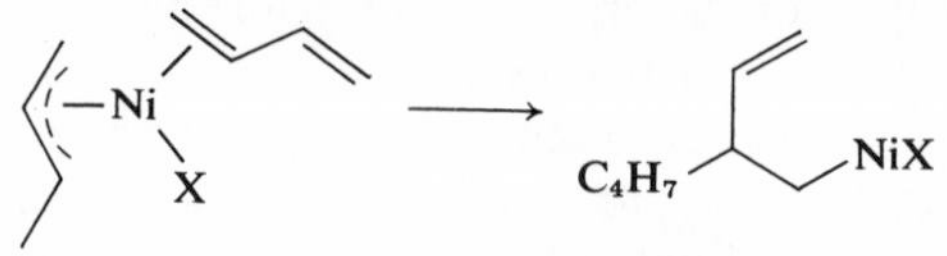

β elimination of a nickel-hydride, which is then able to react further with butadiene. The reverse of this process, i.e., the addition of a nickel-hydride

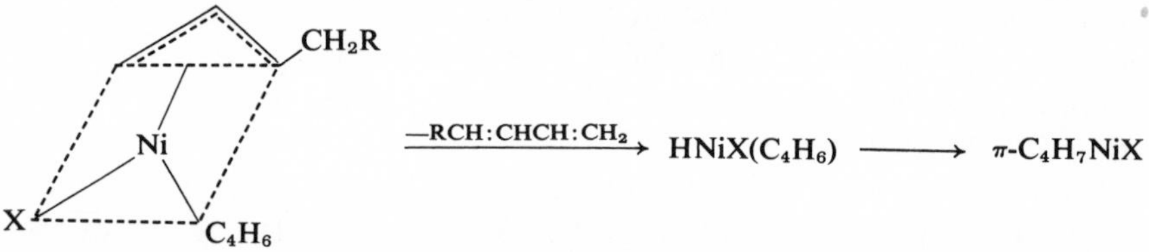

to a 1,3-diene, has been studied using $[\text{HNiLig}_4]^+$; the *anti*-π-allylnickel complex is first formed and then isomerizes to the thermodynamically more stable syn form, suggesting that the initial intermediate contains single-cis complexed butadiene (160, 161). Various mechanisms for the syn–anti isomerization of π-allyl complexes have been discussed in Volume I, p. 360.

There is some evidence that the penultimate double bond in the growing chain interacts with the nickel atom (187; see also 179) and in certain palladium complexes, e.g., **6**, such an interaction has been confirmed spectroscopically (197).

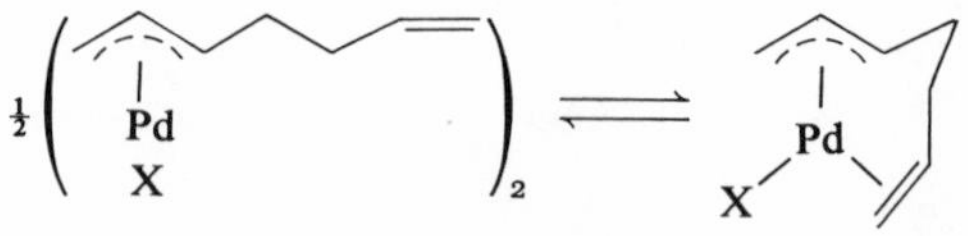

6

A hitherto unexplained phenomenon is the effect of the group X on the microstructure of the polymer. It is most apparent in the reaction involving the π-allylnickel halides, the chloride producing *cis*-1,4-polybutadiene and the iodide *trans*-1,4-butadiene. One can suppose that this difference is steric in origin and that the size of the group X controls the rate at which the initially formed *anti*-π-allyl system isomerizes to the *syn*-π-allyl system from which *trans*-1,4-polybutadiene is produced. Although mechanisms have been suggested that account for the formation of equibinary *cis,trans*-polybutadiene (132), recent spectroscopic results suggest that they are superfluous (see p. 215).

Other catalyst systems that do not contain a π-allylnickel complex probably proceed by an initial reaction with butadiene in which a π-allyl group is generated. For example, the addition of a nickel-hydride to butadiene generates a π-crotyl species that is also known to be formed on reaction of nickelocene with butadiene (159). In the case of $NiCl_2$ activated by heat treatment it has been suggested that an interaction of the butadiene with a surface subhalide occurs to give a π-allyl ion radical that reacts further. Supporting ESR evidence has been published (96).

III. The Polymerization of Isoprene and Related Dienes (Table IV-3)

Most of the catalysts capable of polymerizing butadiene are active with other 1,3-dienes. For a complete account it is necessary to consult the patent literature listed at the end of the chapter, because many of the patents covering polybutadiene have been "stretched" to include other 1,3-dienes. We have confined our attention to publications in which the polymerization of isoprene or other 1,3-dienes is described in detail.

Isoprene, having two nonidentical double bonds, can form 12 simple polymers; of these, the most interesting are *cis*-1,4-polyisoprene (**7**, natural rubber) and *trans*-1,4-polyisoprene (**8**, gutta percha). A selection of nickel-

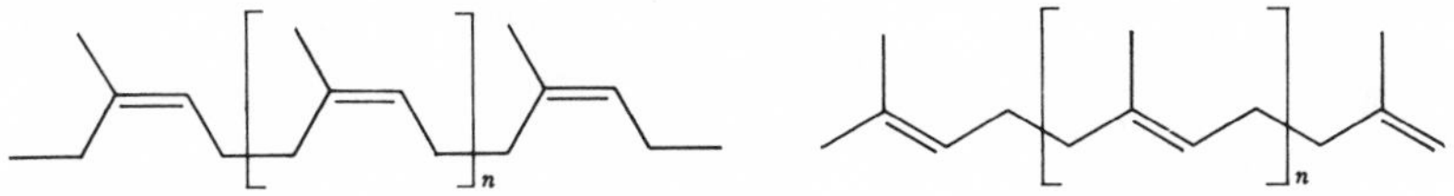

7

8

containing catalysts active in the polymerization of isoprene is shown in Table IV-3. Of particular interest is the formation of *trans*-1,4-polyisoprene using $(\pi\text{-}C_3H_5NiI)_2$ as catalyst and the formation of the so-called equibinary *cis,trans*-polyisoprene from the reaction catalyzed by $(\pi\text{-}C_3H_5NiOCOCF_3)_2$ in *o*-dichlorobenzene. The absence of a 1,2 bond in the polymer in all cases indicates that the π-allyl intermediates involved have structures **9** and **10**.

9

10

The kinetics of the polymerization have been studied for the reactions

TABLE IV-3

THE POLYMERIZATION OF ISOPRENE

| Catalyst | Solvent | Temp. | % BD conv. (hr.) | Polymer microstructure (%) | | | Ref.^{a} |
				cis-1,4	trans-1,4	3,4	
$(\pi\text{-}C_3H_5NiOCOCF_3)_2$	n-Heptane	60	12 (20)	34	60	6	36, 196
	$o\text{-}C_6H_4Cl_2$	55	32 (20)	55	45	0	36
$(\pi\text{-}C_4H_7NiCl)_2\text{-}TiCl_4$	Benzene	-15	20.6 (44)	50	38	12	81, 196
$(\pi\text{-}C_4H_7NiCl_2)$–chloranil	Benzene	50	28 (13)	46	39	15	81
$(\pi\text{-}C_3H_5NiI)_2$	Benzene	—	—	0	95	5	127, 194, 204
$Ni(CO)_4\text{-}AlBr_3\text{-}Al(C_2H_5)_3$	Benzene	14	60 (20)	78	9	13	5, 9, P20, P22, P37

^{a} The references refer not only to the specific example quoted in the table but also to related systems.

catalyzed by $(\pi\text{-}C_3H_5NiI)_2$: the reaction is found to be one-half order in catalyst concentration and to have an activation energy of 18 ± 1 kcal/mole (204).

Other 1,3-dienes that have been polymerized with nickel-containing catalysts include 2-ethyl-, 2-isopropyl-, and 2-*tert*-butyl-1,3-butadiene [with $(\pi\text{-}C_4H_7NiX)_2$ (196)], *trans*-piperylene [with $Ni(acac)_2\text{-}C_2H_5AlCl_2$ (P106) and $(\pi\text{-}C_3H_5NiI)_2$ (204)]; 2-chloro-1,3-butadiene [with $Ni(stearate)_2\text{-}PR_3$–$AlR_3$ (P53)]; 2,3-dimethylbutadiene [with $(\pi\text{-}C_3H_5NiI)_2$ (204) and $(\pi\text{-}C_4H_7NiCl)_2$–chloranil (176, 178)]; cyclopentadiene [with $NiCl_2\text{-}AlCl_3$ (201)]; and cyclohexadiene [with $(\pi\text{-}C_3H_5NiI)_2$ (127, 178, 204)]. The rate of the reaction catalyzed by $(\pi\text{-}C_3H_5NiI)_2$ has been shown (204) to decrease in the order butadiene > isoprene > piperylene $\geq$ cyclohexadiene > 2,3-dimethylbutadiene. The polymer obtained by treating 2,3-dimethylbutadiene with $(\pi\text{-}C_4H_7NiCl)_2$–chloranil is unusual in that it is reported to have an almost exclusive trans-1,4 microstructure (176).

IV. Copolymerization

The copolymerization of two different 1,3-dienes has been studied for the following combinations: butadiene with isoprene (52, 135, 147, 180, P18, P19, P24, P47), with 2,3-dimethylbutadiene (176, 178, 180, P5, P8, P18), with *trans*-piperylene (P5, P8, P18), with methyl-2,4-pentadienoate (20, 180), and with 1,3-cyclohexadiene (177, 178, 180, P171) and isoprene with 1,3-cyclohexadiene (127, 178, 180, P165).

Of the copolymers formed by butadiene with olefins that with styrene is the most important and has been obtained using as catalyst $(\pi\text{-allylNiX})_2$ (59, 92, 180), $(\pi\text{-}C_4H_7NiCl)_2\text{-}CCl_3CO_2H$ or *p*-chloranil (92, 96, 180), $Ni(CO)_4\text{-}TiCl_4$ or $AlCl_3$ (P31, P32, P35, P133), thermally activated nickel dihalides (96, P165), $Ni(acac)_2\text{-}(C_2H_5)_3Al_2Cl_3$ (P45, P46, P49, P50), and $Ni(naphth)_2\text{-}BF_3\cdot O(C_2H_5)_3\text{-}Al(C_2H_5)_3$ (10, P7, P9, P10, P185, P189). This last mentioned ternary catalyst (and related systems) has also been used to prepare graft polymers from preformed polybutadiene (6–8, 10, P2–P4, P6). A feature of these copolymers is that the styrene has no effect on the configuration of the incorporated butadiene molecules; i.e., using $(\pi\text{-}C_3H_5NiI)_2$ as catalyst the copolymer contains trans-1,4 units, whereas in the presence of $(\pi\text{-}C_4H_7NiCl)_2$ cis-1,4 units are incorporated. The ready dimerization and oligomerization of styrene (59, 101, 107), which is also catalyzed by nickel-containing complexes, is indicative of the facility with which β elimination occurs. As a result a considerable reduction in the molecular weight of the copolymer is observed. A careful analysis of the copolymer obtained using $(\pi\text{-}C_3H_5NiI)_2$ as catalyst has shown that the incorporation of a styrene molecule is not automatically followed by termination of the polymer growth as a result of β elimination but that incorporation into the polymer chain can also occur. This last observation suggests that stabilization of the intermediate through a substituted π-benzylnickel species (e.g., **11**) may be involved.

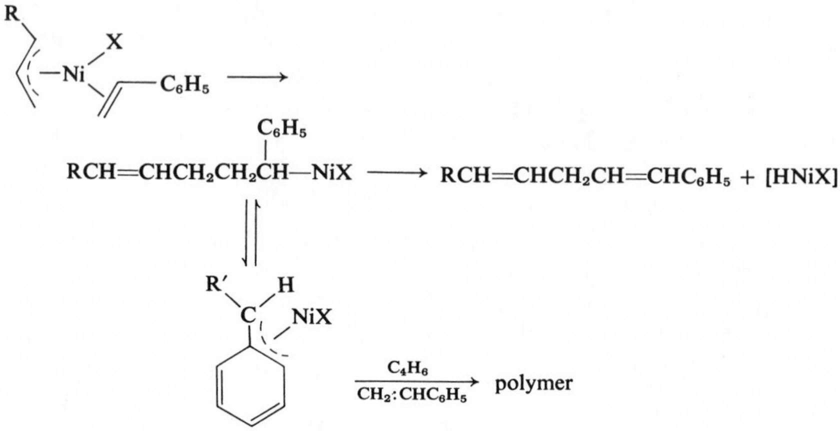

$$RCH{=}CHCH_2CH_2\overset{\overset{\displaystyle C_6H_5}{|}}{C}H{-}NiX \longrightarrow RCH{=}CHCH_2CH{=}CHC_6H_5 + [HNiX]$$

11

It is convenient to mention here that a random copolymer has been produced by copolymerizing butadiene and acetylene in the presence of a nickel naphthenate–triethylaluminum system (183). Moreover, the copolymerization of olefin oxides (e.g., propylene oxide) with 1,3-dienes (e.g., butadiene) has been reported in the patent literature to be catalyzed by bisligand nickel dihalides with a reducing agent, e.g. $(C_4H_9)_2Zn$ (P54, P156).

References

1. S. Aoki, S. Kubota, and T. Otsu, *J. Polym. Sci., Part A-1* **7**, 1567 (1969).
2. A. G. Azizov, O. K. Sharaev, E. I. Tinyakova, and B. A. Dolgoplosk, *Vysokomol. Soedin., Ser. B* **11**, 746 (1969); *Chem. Abstr.* **72**, 32935 (1970).
3. A. G. Azizov, O. K. Sharaev, E. I. Tinyakova, and B. A. Dolgoplosk, *Proc. Acad. Sci. USSR* **190**, 65 (1970).
4. A. G. Azizov, O. K. Sharaev, E. I. Tinyakova, and B. A. Dolgoplosk, *Proc. Acad. Sci. USSR* **197**, 268 (1971).
5. S. Anzai, A. Onishi, T. Saegusa, and J. Furukawa, *Kogyo Kagaku Zasshi* **72**, 2058 (1969).
6. S. Anzai, K. Irako, A. Onishi, and J. Furukawa, *Kogyo Kagaku Zasshi* **72**, 2090 (1969).
7. S. Anzai, Y. Hayakawa, K. Irako, A. Onishi, and J. Furukawa, *Kogyo Kagaku Zasshi* **72**, 2101 (1969).
8. S. Anzai, Y. Hayakawa, K. Irako, A. Onishi, and J. Furukawa, *Kogyo Kagaku Zasshi* **72**, 2113 (1969).
9. S. Anzai, Y. Ishizuka, A. Onishi, and J. Furukawa, *Kogyo Kagaku Zasshi* **72**, 2068 (1969).

10. S. Anzai, K. Irako, A. Onishi, and J. Furukawa, *Kogyo Kagaku Zasshi* **72**, 2081 (1969).
11. B. D. Babitskii, T. G. Golenko, B. A. Dolgoplosk, V. A. Kormer, V. I. Skoblikova, and E. I. Tinyakova, *Proc. Acad. Sci. USSR* **161**, 325 (1965).
12. B. D. Babitskii, B. A. Dolgoplosk, V. A. Kormer, M. I. Lobach, E. I. Tinyakova, and V. A. Yakovlev, *Bull. Acad. Sci. USSR* p. 1478 (1965).
13. B. D. Babitskii, V. A. Kormer, M. I. Lobach, I. Y. Poddubnyi, and V. N. Sokolov, *Vysokomol. Soedin., Ser. B* **9**, 6 (1967); *Chem. Abstr.* **66**, 116416 (1967).
14. B. D. Babitskii, B. A. Dolgoplosk, V. A. Kormer, M. I. Lobach, E. I. Tinyakova, and V. A. Yakovlev, *Proc. Acad. Sci. USSR* **161**, 283 (1965).
15. B. D. Babitskii, V. A. Kormer, and I. M. Lapuk, *Proc. Acad. Sci. USSR* **165**, 1053 (1965).
16. B. D. Babitskii, V. A. Kormer, M. I. Lobach, and N. N. Chesnikova, *Proc. Acad. Sci. USSR* **160**, 83 (1965).
17. B. D. Babitskii, V. A. Kormer, M. I. Lobach, I. Y. Poddubnyi, and V. N. Sokolov, *Proc. Acad. Sci. USSR* **180**, 80 (1968).
18. B. D. Babitskii, B. A. Dolgoplosk, V. A. Kormer, M. I. Lobach, E. I. Tinyakova, N. N. Chesnokova, and V. A. Yakovlev, *Vysokomol. Soedin.* **6**, 2202 (1964); *Chem. Abstr.* **62**, 7972 (1965).
19. B. D. Babitskii, V. A. Kormer, I. M. Lapuk, and V. I. Skoblikova, *J. Polym. Sci., Part C* **16**, 3219 (1968).
20. T. I. Bevza, N. A. Pokatilo, M. P. Teterina, and B. A. Dolgoplosk, *Vysokomol. Soedin., Ser. A* **10**, 207 (1968); *Chem. Abstr.* **68**, 69402 (1968).
21. T. I. Bevza, S. I. Beilin, Y. V. Korshak, and B. A. Dolgoplosk, *Vysokomol. Soedin., Ser. B* **10**, 865 (1968); *Chem. Abstr.* **70**, 38627 (1969).
22. K. Bouchal and F. Hrabák, *Collect Czech. Chem. Commun.* **34**, 1113 (1969).
23. P. Bourdauducq and F. Dawans, *J. Polym. Sci., Part A-1* **10**, 2527 (1972).
24. Chih-Ch'uan Ch'en, Lien-Sheng Chiang, Hsing-Ya Li, Chung-Chi Chung, and Chun Ch'u Yang, *Ko Fen Tzu T'ung Hsun* **7**, 323 (1965); *Chem. Abstr.* **64**, 16097 (1966).
25. F. Dawans and P. Teyssie, *J. Polym. Sci., Part B* **3**, 1045 (1965).
26. F. Dawans and P. Teyssie, *C.R. Acad. Sci.* **261**, 4097 (1965).
27. W. M. Saltman and L. J. Kuzma, *Rubber Chem. Technol.* **46**, 1055 (1973).
28. F. Dawans and P. Teyssie, *C.R. Acad. Sci., Ser. C* **263**, 1512 (1966).
29. F. Dawans and P. Teyssie, *J. Polym. Sci., Part B* **7**, 111 (1969).
30. F. Dawans, J. C. Marechal, and P. Teyssie, *J. Organometal. Chem.* **21**, 259 (1970).
31. F. Dawans, J. P. Marchand, and P. Teyssie, *Int. Syn. Rubber Symp. Lect., 4th 1969* p. 20 (1969).
32. F. Dawans and P. Teyssie, *Amer. Chem. Soc., Div. Org. Coatings Plast. Chem., Pap.* **30**, 208 (1970).
33. F. Dawans, J. P. Durand, and P. Teyssie, *J. Polym. Sci., Part B* **10**, 493 (1972).
34. M. Dubini and F. Montino, *Chem. Commun.* p. 749 (1966).
35. J. P. Durand, F. Dawans, and P. Teyssie, *J. Polym. Sci., Part A-1* **8**, 979 (1970).
36. J. P. Durand and F. Dawans, *J. Polym. Sci., Part B* **8**, 743 (1970).
37. E. W. Duck, D. P. Grieve, and M. N. Thornber, *Int. Syn. Rubber Symp. Lect., 4th, 1969* p. 88 (1969).
38. J. P. Durand and P. Teyssie, *J. Polym. Sci., Part B* **6**, 299 (1968).
39. J. P. Durand, F. Dawans, and P. Teyssie, *J. Polym. Sci., Part B* **6**, 757 (1968).
40. J. P. Durand, F. Dawans, and P. Teyssie, *J. Polym. Sci., Part B* **5**, 785 (1967).
41. E. W. Duck, D. K. Jenkins, D. P. Grieve, and M. N. Thornber, *Eur. Polym. J.* **7**, 55 (1971).

42. B. A. Dolgoplosk, I. A. Oreshkin, E. I. Tinyakova, and V. A. Yakovlev, *Bull. Acad. Sci. USSR* p. 2059 (1967).
43. B. A. Dolgoplosk, B. D. Babitskii, V. A. Kormer, M. I. Lobach, and E. I. Tinyakova, *Proc. Acad. Sci. USSR* **164**, 1013 (1965).
44. B. A. Dolgoplosk, I. I. Moiseev, and E. I. Tinyakova, *Proc. Acad. Sci. USSR* **173**, 340 (1967).
45. B. A. Dolgoplosk, E. N. Kropacheva, E. K. Khrennikova, E. I. Kuznetsova, and K. G. Golodova, *Proc. Acad. Sci. USSR* **135**, 1357 (1960).
46. B. A. Dolgoplosk and E. I. Tinyakova, *Mekh. Kinet. Slezhnykh Katal. React., Lektsii, Simp., 1st 1968* p. 142 (1970).
47. B. A. Dolgoplosk and E. I. Tinyakova, *Bull. Acad. Sci. USSR* p. 291 (1970).
48. C. Dixon, E. W. Duck, and D. K. Jenkins, *Organometal. Chem. Syn.* **1**, 77 (1970/ 1971).
49. B. A. Dolgoplosk, E. I. Tinyakova, P. A. Vinogradov, O. P. Parenago, and B. S. Turov, *J. Polym. Sci., Part C* **16**, 3685 (1968).
50. C. Dixon, E. W. Duck, and D. K. Jenkins, *Eur. Polym. J.* **8**, 13 (1972).
51. V. S. Feldblyum and A. I. Leshcheva, *J. Org. Chem. USSR* **8**, 1115 (1972).
52. J. Furukawa, T. Saegusa, T. Naruyama, and S. Kurahashi, *Kogyo Kagaku Zasshi* **65**, 2082 (1962); *Chem. Abstr.* **58**, 14263 (1963).
53. J. Furukawa, T. Tsuruta, T. Miki, A. Onishi, A. Kawasaki, T. Fueno, S.-T. Chen, N. Yamamoto, and T. Matsumoto, *Kogyo Kagaku Zasshi* **61**, 1046 (1958); *Chem. Abstr.* **55**, 18166 (1961).
54. V. M. Frolov, A. V. Volkov, O. P. Parenago, and B. A. Dolgoplosk, *Proc. Acad. Sci. USSR* **177**, 1200 (1967).
55. V. M. Frolov, G. V. Isagulyants, V. M. Gorelik, A. P. Klimov, O. P. Parenago, Y. I. Derbentsev, and B. A. Dolgoplosk, *Proc. Acad. Sci. USSR* **195**, 883 (1970).
56. T. G. Golenko, K. L. Makovetskii, V. A. Libina, A. N. Karasev, E. I. Tinyakova, and B. A. Dolgoplosk, *Bull. Acad. Sci. USSR* p. 2147 (1968).
57. J. F. Harrod and L. R. Wallace, *Macromolecules* **2**, 449 (1969).
58. J. F. Harrod and L. R. Wallace, *Amer. Chem. Soc., Div. Org. Coatings Plast. Chem. Pap.* **30**, 219 (1970).
59. G. Henrici-Olivé, S. Olivé, and E. Schmidt, *J. Organometal. Chem.* **39**, 201 (1972).
60. G. V. Isagulyants, V. M. Frolov, A. P. Klimov, V. M. Gorelik, O. P. Parenago, and B. A. Dolgoplosk, *Proc. Acad. Sci. USSR* **198**, 524 (1971).
61. D. K. Jenkins, D. G. Timms, and E. W. Duck, *Polym. J.* **7**, 419 (1966).
62. A. I. Kadantseva, K. G. Ogorodnikova, V. A. Vashkevich, and S. S. Medvedev, *Uch. Zap. Mosk. Inst. Tokoi Khim. Tekhnol.* **1**, 100 (1970); *Chem. Abstr.* **76**, 127489 (1972).
63. V. A. Kormer, A. S. Lyashch, B. D. Babitskii, G. V. Lvova, M. I. Lobach, and T. Y. Chepurnaya, *Proc. Acad. Sci. USSR* **180**, 88 (1968).
64. E. V. Kristalnyi, E. V. Zablotskaya, and S. S. Medvedev, *Vysokomol. Soedin., Ser. B* **10**, 318 (1968); *Chem. Abstr.* **69**, 27899 (2968).
65. V. A. Kormer, B. D. Babitskiy, and M. I. Lobach, *Advan. Chem. Ser.* **91**, 306 (1969).
66. E. V. Kristalnyi, N. V. Orekhova, E. V. Zabolotskaya, A. R. Gantmakher, and S. S. Medvedev, *Polym. Sci. USSR A* **12**, 947 (1970).
67. K. Komatsu, E. Okuya, A. Sakaguchi, S. Shoji, H. Yasunaga, and J. Furukawa, *Kogyo Kagaku Zasshi* **74**, 2173 (1971).
68. V. A. Kormer, B. D. Babitskii, M. I. Lobach, and N. N. Chesnokova, *J. Polym. Sci., Part C* **16**, 4351 (1969).

69. K. Komatsu, J. Hirota, H. Yasunaga, and J. Furukawa, *Kogyo Kagaku Zasshi* **74**, 2377 (1971).
70. V. A. Kormer and M. I. Lobach, *J. Polym. Sci., Part B* **10**, 177 (1972).
71. K. Komatsu, E. Okuya, S. Nishiyama, J. Hirota, H. Yasunaga, and J. Furukawa, *Kogyo Kagaku Zasshi* **74**, 2529 (1971).
72. G. Lugli, W. Marconi, A. Mazzei, and N. Palladino, *Inorg. Chim. Acta* **3**, 151 (1969).
73. M. I. Lobach, V. A. Kormer, I. Y. Tsereteli, G. P. Kondratenkov, B. D. Babitskii, and V. I. Klepikova, *J. Polym. Sci., Part B* **9**, 71 (1971).
74. M. I. Lobach, V. A. Kormer, I. Y. Tsereteli, G. P. Kondratenkov, B. D. Babitskii, and V. I. Klepikova, *Proc. Acad. Sci. USSR* **196**, 22 (1971).
75. A. M. Lazutkin, V. A. Vashkevich, S. S. Medvedev, and N. N. Vasileva, *Proc. Acad. Sci. USSR* **175**, 583 (1967).
76. A. M. Lazutkin, A. I. Kadantseva, V. A. Vashkevich, and S. S. Medvedev, *Vysokomol. Soedin., Ser. B* **12**, 635 (1970); *Chem. Abstr.* **73**, 110200 (1970).
77. T. Matsumoto and J. Furukawa, *J. Polym. Sci., Part B* **5**, 935 (1967).
78. K. Matsuzaki and T. Yasukawa, *J. Polym. Sci., Part A-1* **5**, 521 (1967).
79. K. Matsuzaki and T. Yasukawa, *J. Polym. Sci., Part A-1* **5**, 511 (1967).
80. T. Matsumoto, J. Furukawa, and H. Morimura, *J. Polym. Sci., Part A-1* **9**, 1971 (1971).
81. I. N. Markevich, A. E. Tiger, E. I. Tinyakova, N. N. Stefanovskaya, and B. A. Dolgoplosk, *Vysokomol. Soedin., Ser. B* **11**, 185 (1969); *Chem. Abstr.* **70**, 116023 (1969).
82. T. Matsumoto, J. Furukawa, and H. Morimura, *J. Polym. Sci., Part B* **7**, 541 (1969).
83. T. Matsumoto, J. Furukawa, and H. Morimura, *J. Polym. Sci., Part B* **6**, 869 (1968).
84. T. Matsumoto and A. Ohnishi, *Kogyo Kagaku Zasshi* **71**, 2059 (1968).
85. J. C. Marechal, F. Dawans, and P. Teyssie, *J. Polym. Sci., Part A-1* **8**, 1993 (1970).
86. T. Matsumoto, J. Furukawa, and H. Morimura, *J. Polym. Sci., Part A-1* **9**, 875 (1971).
87. T. Matsumoto and J. Furukawa, *J. Macromol. Sci., Chem.* **6**, 281 (1972).
88. E. A. Mushina, T. K. Vydrina, E. V. Sakharova, E. I. Tinyakova, and B. A. Dolgoplosk, *Proc. Acad. Sci. USSR* **170**, 881 (1966).
89. E. A. Mushina, T. K. Vydrina, E. V. Sakharova, E. I. Tinyakova, and B. A. Dolgoplosk, *Proc. Acad. Sci. USSR* **177**, 1035 (1967).
90. E. V. Mushina, T. K. Vydrina, E. V. Sakharova, V. A. Yakovlev, E. I. Tinyakova, and B. A. Dolgoplosk, *Vysokomol. Soedin., Ser. B* **9**, 784 (1967); *Chem. Abstr.* **68**, 13827 (1968).
91. I. A. Oreshkin, G. M. Chernenko, E. I. Tinyakova, and B. A. Dolgoplosk, *Dokl. Akad. Nauk SSSR* **169**, 1102 (1966); *Chem. Abstr.* **65**, 15630 (1966).
92. I. Y. Ostrovskaya, K. L. Makovetskii, E. I. Tinyakova, and B. A. Dolgoplosk, *Proc. Acad. Sci. USSR* **181**, 701 (1968).
93. T. Otsu and M. Yamaguchi, *J. Polym. Sci., Part A-1* **7**, 387 (1969).
94. I. Y. Ostrovskaya, K. L. Makovetskii, B. A. Dolgoplosk, and E. I. Tinyakova, *Izv. Akad. Nauk SSSR, Ser. Khim.* p. 1632 (1967); *Chem. Abstr.* **68**, 13828 (1968).
95. S. Otsuka and M. Kawakami, *Kogyo Kagaku Zasshi* **68**, 874 (1965).
96. I. Y. Ostrovskaya, K. L. Makovetskii, G. P. Karpacheva, E. I. Tinyakova, and B. A. Dolgoplosk, *Proc. Acad. Sci. USSR* **197**, 350 (1971).
97. L. Porri, G. Natta, and M. C. Gallazzi, *J. Polym. Sci., Part C* **16**, 2525 (1967).
98. L. Porri, G. Natta, and M. C. Gallazzi, *Chim. Ind. (Milan)* **46**, 428 (1964).
99. L. Porri, *Corsi Semin. Chim.* p. 86 (1968).

100. L. Porri and G. Allegra, *Corsi Semin. Chim.* p. 83 (1968).
101. F. Dawans, *Tetrahedron Lett.* p. 1943 (1971).
102. R. Prikryl, A. Tkac, and A. Stasko, *Collect. Czech. Chem. Commun.* 37, 1295 (1972).
103. N. I. Pakuro, E. V. Zabolotskaya, and S. S. Medvedev, *Vysokomol. Soedin., Ser. B* 10, 3 (1968); *Chem. Abstr.* 68, 69914 (1968).
104. N. I. Pakuro, E. V. Zabolotskaya, N. A. Pravikova, and S. S. Medvedev, *Proc. Acad. Sci. USSR* 190, 58 (1970).
105. L. Porri, M. C. Gallazzi, and G. Vitulli, *J. Polym. Sci., Part B* 5, 629 (1967).
106. N. I. Pakuro, N. A. Pravikova, E. V. Zabolotskaya, and S. S. Medvedev, *Dokl. Akad. Nauk SSSR* 194, 617 (1970); *Chem. Abstr.* 74, 32513 (1971).
107. L. I. Redkina, K. L. Makovetskii, E. I. Tinyakova, and B. A. Dolgoplosk, *Proc. Acad. Sci. USSR* 186, 105 (1969).
108. R. Sakata, J. Hosono, A. Onishi, and K. Ueda, *Makromol. Chem.* 139, 73 (1970).
109. B. Schleimer and H. Weber, *Angew. Makromol. Chem.* 16/17, 253 (1971).
110. O. K. Sharaev, A. V. Alferov, E. I. Tinyakova, B. A. Dolgoplosk, V. A. Kormer, and B. D. Babitskii, *Proc. Acad. Sci. USSR* 177, 1003 (1967).
111. O. K. Sharaev, A. V. Alferov, E. I. Tinyakova, and B. A. Dolgoplosk, *Bull. Acad. Sci. USSR* p. 2469 (1967).
112. O. K. Sharaev, A. V. Alferov, E. I. Tinyakova, and B. A. Dolgoplosk, *Bull. Acad. Sci. USSR* p. 2468 (1967).
113. O. K. Sharaev. A. V. Alferov, E. I. Tinyakova, and B. A. Dolgoplosk, *Bull. Acad. Sci. USSR* p. 1141 (1967).
114. O. K. Sharaev, A. V. Alferov, B. A. Dolgoplosk, and E. I. Tinyakova, *Proc. Acad. Sci. USSR* 164, 854 (1965).
115. O. K. Sharayev, A. V. Alferov, E. I. Tinyakova, and B. A. Dolgoplosk, *Polym. Sci. USSR A9*, 709 (1967).
116. N. P. Simanova, M. I. Lobach, B. D. Babitskii, and V. A. Kormer, *Vysokomol. Soedin., Ser. B* 10, 588 (1968); *Chem. Abstr.* 70, 5059 (1969).
117. P. Teyssie, F. Dawans, and J. P. Durand, *J. Polym. Sci., Part C* 22, 221 (1968).
118. E. I. Tinyakova, A. V. Alferov, T. G. Golenko, B. A. Dolgoplosk, I. A. Oreshkin, O. K. Sharaev, G. N. Chernenko, and V. A. Yakovlev, *J. Polym. Sci., Part C* 16, 2625 (1967).
119. M. Tsutsui and J. Ariyoshi, *J. Polym. Sci., Part A* 3, 1729 (1965).
120. Y. Tajima and E. Kunioka, *Bull. Chem. Soc. Jap.* 40, 697 (1967).
121. A. Tkac and A. Stasko, *Collect. Czech. Chem. Commun.* 37, 1006 (1972).
122. A. Tkac and A. Stasko, *Collect. Czech. Chem. Commun.* 37, 573 (1972).
123. M. C. Throckmorton and F. S. Farson, *Rubber Chem. Technol.* 45, 268 (1972).
124. P. Teyssie, F. Dawans, J. P. Durand, and J. C. Marechal, *Colloq. Int. Cent. Nat. Rech. Sci.* p. 351 (1970).
125. T. K. Vydrina, E. I. Tinyakova, and B. A. Dolgoplosk, *Proc. Acad. Sci. USSR* 183, 1013 (1968).
126. T. K. Vydrina, S. V. Novikov, E. I. Tinyakova, and B. A. Dolgoplosk, *Proc. Acad. Sci. USSR* 194, 623 (1970).
127. L. M. Vardanyan, N. S. Quang, Y. V. Korshak, and B. A. Dolgoplosk, *Vysokomol. Soedin., Ser. B* 13, 19 (1971); *Chem. Abstr.* 74, 126152 (1971).
128. K.-I. Ueda, A. Oonishi, T. Yoshimoto, J.-I. Hosono, K. Maeda, and T. Matsumoto, *Amer. Chem. Soc., Div. Petrol. Chem., Prepr.* 9, A47 (1964).
129. L. R. Wallace and J. F. Harrod, *Macromolecules* 4, 656 (1971).
130. J. Zachoval, J. Kalal, B. Veruovic, and L. Stefka, *Collect. Czech. Chem. Commun.* 30, 1326 (1965).

131. N. Yamazaki, S. Murai, K. Ebina, T. Ohta, and S. Nakahama, *Kinet. Mech. Polyreact.* **2**, 411 (1969).

132. V. A. Yakovlev, B. A. Dolgoplosk, K. L. Makovetskii, and E. I. Tinyakova, *Proc. Acad. Sci. USSR* **187**, 557 (1969).

133. T. Yoshimoto, K. Komatsu, R. Sakata, K. Yamamoto, Y. Takeuchi, A. Onishi, and K. Ueda, *Makromol. Chem.* **139**, 61 (1970).

134. N. Yamazaki, M. Kase, and T. Ohta, *Polym. J.* **2**, 364 (1971).

135. B. A. Dolgoplosk, *Polym. Sci. USSR A* **13**, 367 (1971).

136. A. I. Lazutkina, L. Y. Alt, T. L. Matveeva, A. M. Lazutkin, and Y. I. Ermakov, *Kinet. Catal. (USSR)* **12**, 1034 (1971).

137. K.-I. Ueda, A. Oonishi, T. Yoshimoto, J.-I. Hosono, K. Maeda, and T. Matsumoto, *Kogyo Kagaku Zasshi* **66**, 1103 (1963).

138. W. S. Anderson, *J. Polym. Sci., Part A-1* **5**, 429 (1967).

139. V. I. Skoblikova, S. S. Passova, B. D. Babitskii, and V. A. Kormer, *Vysokomol. Soedin., Ser. B* **10**, 590 (1968); *Chem. Abstr.* **70**, 5060 (1969).

140. A. G. Azizov, T. K. Vydrina, O. K. Sharaev, E. I. Tinyakova, and B. A. Dolgoplosk, *Proc. Acad. Sci. USSR* **195**, 797 (1970).

141. E. N. Zavadovskaya, M. P. Teterina, O. K. Sharaev, A. G. Azizov, T. K. Vydrina, and E. I. Tinyakova, *Zh. Prikl. Spektrosk.* **13**, 851 (1970); *Chem. Abstr.* **74**, 69720 (1971).

142. E. N. Zavadovskaya, M. P. Teterina, O. K. Sharaev, A. G. Azizov, T. K. Vydrina, E. I. Tinyakova, and B. A. Dolgoplosk, *Proc. Acad. Sci. USSR* **188**, 798 (1969).

143. A. G. Azizov, O. K. Sharaev, E. I. Tinyakova, G. N. Bondarenko, M. P. Teterina, N. A. Shimanko, and B. A. Dolgoplosk, *Proc. Acad. Sci. USSR* **197**, 299 (1971).

144. M. R. Galding and N. A. Buzina, *Proc. Acad. Sci. USSR* **197**, 238 (1971).

145. V. I. Skoblikova, Z. D. Stepanova, V. D. Babitskii, and V. A. Kormer, *J. Gen. Chem. USSR* **39**, 207 (1969).

146. G. Wilke *et al.*, *Angew. Chem.* **78**, 157 (1966).

147. I. N. Smirnova, B. A. Dolgoplosk, and V. A. Krol, *Kinet. Mech. Polyreact.* **2**, 291 (1969).

148. N. I. Pakuro and E. V. Zabolotskaya, *Vysokomol. Soedin, Ser. B* **15**, 796 (1973); *Chem. Abstr.* **80**, 134493 (1974).

149. V. M. Breitman and A. V. Zak, *Prom. Sin. Kauch Nauch.-Tekh. Sb.* p. 10 (1971); *Chem. Abstr.* **78**, 4603 (1973).

150. B. Bogdanović, P. Heimbach, M. Kröner, G. Wilke, E. G. Hoffmann, and J. Brandt, *Justus Liebigs Ann. Chem.* **727**, 143 (1969).

151. C. Dixon, E. W. Duck, D. P. Grieve, D. K. Jenkins, and M. N. Thornber, *Eur. Polym. J.* **6**, 1359 (1970).

152. T. Yoshimoto, K. Komatsu, H. Matsui, R. Sakata, K. Yamamoto, Y. Takeuchi, A. Oonishi, and K. Ueda, *Kogyo Kagaku Zasshi* **70**, 1993 (1967).

153. F.-S. Wang, C.-C. Chang, F.-S. Yu, and Y.-C. Liao, *Ko Fen Tzu T'ung Hsun* **6**, 332 (1964); *Chem. Abstr.* **64**, 815 (1966).

154. L. A. Nedoinova, V. A. Grechanovskii, I. Y. Poddubnyi, and L. R. Avdeeva, *Vysokomol. Soedin. Ser. A* **16**, 183 (1974); *Chem. Abstr.* **80**, 134525 (1974).

155. S. Otsuka and M. Kawakami, *Angew. Chem.* **75**, 858 (1963).

156. J. F. Harrod and L. R. Wallace, *Macromolecules* **5**, 682 (1972).

157. J. F. Harrod and L. R. Wallace, *Macromolecules* **5**, 685 (1972).

158. V. A. Yakovlev, B. A. Dolgoplosk, Y. I. Tinyakova, and O. N. Yakovleva, *Polym. Sci. USSR A* **11**, 1869 (1969).

159. D. W. McBride, E. Dudek, and F. G. A. Stone, *J. Chem. Soc., London* p. 1752 (1964).
160. C. A. Tolman, *J. Amer. Chem. Soc.* 92, 6777 (1970).
161. C. A. Tolman, *J. Amer. Chem. Soc.* 92, 6785 (1970).
162. F. Dawans and P. Teyssie, *Ind. Eng. Chem., Prod. Res. Develop.* 10, 261 (1971).
163. N. I. Pakuro, E. V. Zabolotskaya, and S. S. Medvedev, *Dokl. Akad. Nauk SSSR* 185, 1323 (1969); *Chem. Abstr.* 71, 30727 (1969).
164. S. S. Medvedev, *Russ. Chem. Rev.* 37, 842 (1968).
165. B. A. Dolgoplosk and Y. Osada, *Kobunshi* 20, 835 (1971).
166. A. V. Volkov, O. P. Parenago, V. M. Frolov, and B. A. Dolgoplosk, *Proc. Acad. Sci. USSR* 187, 563 (1969).
167. J. Furukawa, E. Kobayashi, T. Kawagoe, N. Katsuki, and M. Imanari, *J. Polym. Sci., Polym. Lett. Ed.* 11, 239 (1973).
168. J. M. Thomassin, E. Walckiers, R. Warin, and P. Teyssié, *J. Polym. Sci., Polym. Lett. Ed.* 11, 229 (1973).
169. V. D. Mochel, *J. Polym. Sci., Part A-1* 10, 1009 (1972).
170. K. F. Elgert, B. Stützel, P. Frenzel, H. J. Cantow, and R. Streck, *Makromol. Chem.* 170, 257 (1973).
171. E. R. Santee, R. Chang, and M. Morton, *J. Polym. Sci., Polym. Lett. Ed.* 11, 449 (1973).
172. E. R. Santee, V. D. Mochel, and M. Morton, *J. Polym. Sci., Polym. Lett. Ed.* 11, 453 (1973).
173. A. D. H. Clague, J. A. M. van Broekhoven, and J. W. de Haan, *J. Polym. Sci. Polym. Lett. Ed.* 11, 299 (1973).
174. R. Warin, P. Teyssié, P. Bourdauducq, and F. Dawans, *J. Polym. Sci., Polym. Lett. Ed.* 11, 177 (1973).
175. J. Furukawa, J. Kiji, H. Konishi, K. Yamamoto, S. I. Mitani, and S. Yoshikawa, *Makromol. Chem.* 174, 65 (1973).
176. N. S. Kueng, Y. V. Korshak, G. N. Bondarenko, M. P. Teterina, and B. A. Dolgoplosk, *Vysokomol. Soedin., Ser. B* 15, 375 (1973); *Chem. Abstr.* 79, 66863 (1973).
177. G. M. Chernenko, S. I. Beilin, B. A. Dolgoplosk, M. P. Teterina, and V. M. Frolov, *Proc. Acad. Sci. USSR* 203, 225 (1972).
178. B. A. Dolgoplosk, S. I. Beilin, Y. V. Korshak, G. M. Chernenko, L. M. Vardanyan and M. P. Teterina, *Eur. Polym. J.* 9, 895 (1973).
179. B. A. Dolgoplosk, K. L. Makovetskii, L. I. Redkina, T. V. Soboleva, E. I. Tinyakova, and V. A. Yakovlev, *Proc. Acad. Sci. USSR* 205, 602 (1972).
180. B. A. Dolgoplosk, S. I. Beilin, Y. V. Korshak, K. L. Makovetsky, and E. I. Tinyakova, *J. Polym. Sci., Polym. Chem. Ed.* 11, 2569 (1973).
181. Y. I. Ermakov, A. M. Lazutkin, E. A. Demin, Y. P. Grabovskii, and V. A. Zakharov, *Kinet. Catal.* 13, 1269 (1972).
182. V. M. Frolov, G. V. Isagulyants, V. M. Gorelik, A. P. Klimov, O. P. Parenago, and B. A. Dolgoplosk, *Kinet. Catal.* 13, 446 (1972).
183. J. Furukawa, E. Kobayashi, and T. Kawagoe, *J. Polym. Sci., Polym. Lett. Ed.* 11, 573 (1973).
184. T. G. Golenko, G. N. Bondarenko, K. L. Makovetskii, E. I. Tinyakova, M. P. Teterina, and B. A. Dolgoplosk, *Bull. Acad. Sci. USSR* p. 2134 (1972).
185. V. M. Gorelik, E. S. Novikova, O. P. Parenago, V. M. Frolov, and B. A. Dolgoplosk, *Kinet. Catal.* 13, 1274 (1972).
186. V. A. Grechanovskii, L. A. Nedoinova, I. Y. Poddubnyi, and L. R. Avdeeva, *Proc. Acad. Sci. USSR* 206, 820 (1972).

187. V. I. Klepikova, G. P. Kondratenkov, V. A. Kormer, M. I. Lobach, and L. A. Churlyaeva, *J. Polym. Sci., Polym. Lett. Ed.* **11**, 193 (1973).
188. K. L. Makovetsky, I. Y. Ostrovskaya, and G. P. Karpacheva, *Nuova Chim.* **49**, 75 (1973).
189. M. I. Lobach, *Nuova Chim.* **49**, 43 (1973).
190. K. G. Miesserov, I. L. Kershenbaum, R. E. Lobach, and B. A. Dolgoplosk, *Vysokomol. Soedin., Ser. B* **10**, 673 (1968); *Chem. Abstr.* **70**, 12392 (1969).
191. I. A. Oreshkin, Y. I. Tinyakova, and B. A. Dolgoplosk, *J. Polym. Sci. USSR* **A11**, 2096 (1969).
192. I. Y. Ostrovskaya and K. L. Makovetskii, *Proc. Acad. Sci. USSR* **208**, 126 (1973).
193. N. I. Pakuro and E. V. Zabolotskaya, *Vysokomol. Soedin., Ser. B* **15**, 185 (1973); *Chem. Abstr.* **79**, 42904 (1973).
194. V. A. Vasilev, V. I. Klepikova, G. P. Kondratenkov, V. A. Kormer, and M. I. Lobach, *Proc. Acad. Sci. USSR* **206**, 719 (1972).
195. A. Tkac and V. Adamcik, *Collect. Czech. Chem. Commun.* **38**, 1346 (1973).
196. V. A. Vasiliev, N. A. Kalinicheva, V. A. Kormer, M. I. Lobach, and V. I. Klepikova, *J. Polym. Sci., Polym. Chem. Ed.* **11**, 2489 (1973).
197. R. P. Hughes, T. Jack, and J. Powell, *J. Organometal. Chem.* **63**, 451 (1973).
198. A. I. Kadantseva, V. S. Byrikhin, and S. S. Medvedev, *Vysokomol. Soedin., Ser. A* **14**, 2668 (1972); *Chem. Abstr.* **78**, 98681 (1973).
199. E. V. Kristalnyi, E. V. Zabolotskaya, and A. R. Gantmakker, *Vysokomol. Soedin., Ser. B* **14**, 569 (1972); *Chem. Abstr.* **78**, 5140 (1973).
200. N. I. Pakuro, E. V. Zabolotskaya, and B. A. Dolgoplosk, *Dokl. Akad. Nauk SSSR* **209**, 631 (1973); *Chem. Abstr.* **79**, 53866 (1973).
201. I. I. Yukelson and Z. A. Okladnikova, *Izv. Vyssh. Ucheb. Zaved., Khim. Khim. Tekhnol.* **16**, 1416 (1973); *Chem. Abstr.* **80**, 27536 (1974).
202. B. A. Dolgoplosk, K. L. Makovetskii, E. I. Tinyakova, and O. K. Sharaev, "The Polymerization of Dienes with π-Allyl Complexes" (in Russian). Nauka Press, Moscow, 1968.
203. S. Miyamoto and T. Arakawa, *Kogyo Kagaku Zasshi* **74**, 1394 (1971).
204. T. V. Soboleva, V. A. Yakovlev, E. I. Tinyakova, and B. A. Dolgoplosk, *Proc. Acad. Sci. USSR* **212**, 810 (1973).
205. J. Beger, H. Füllbier, and W. Gaube, *J. Prakt. Chem.* **316**, 173 (1974).
206. V. I. Klepikova, V. A. Vasilev, G. P. Kondratenkov, M. I. Lobach, and V. A. Kormer, *Proc. Acad. Sci. USSR* **211**, 641 (1973).
207. A. N. Nesmeyanov, L. S. Isaeva, L. N. Lorens, A. M. Vainberg, and Y. S. Nekrasov, *Proc. Acad. Sci. USSR* **205**, 678 (1972).
208. E. V. Kristalnyi, E. V. Zabolotskaya, and A. R. Gantmakher, *Proc. Acad. Sci. USSR* **204**, 528 (1972).
209. F. Conti, A. Segre, P. Pini, and L. Porri, *Polymer* **15**, 5 (1974).
210. K. Hatada, Y. Terawaki, H. Okuda, Y. Tanaka, and H. Sato, *J. Polym. Sci., Polym. Lett. Ed.* **12**, 305 (1974).
211. Y. Tanaka, H. Sato, M. Ogawa, K. Hatada, and Y. Terawaki, *J. Polym. Sci. Polym. Let. Ed.* **12**, 369 (1974).
212. M. Julémont, E. Walckiers, R. Warin, and P. Teyssié, *Makromol. Chem.* **175**, 1673 (1974).
213. E. V. Kristalnyi, N. I. Galkina, E. V. Zabolotskaya, and A. Gantmakher, *Vysokomol. Soedin. Ser. B* **16**, 64 (1974).
214. S. A. Radzinskii, A. P. Sheinker, F. S. Yakushin, and A. D. Abkin, *Dokl. Akad. Nauk SSSR* **213**, 887 (1973).
215. S. Malinowski and W. Skupinski, *Rocz. Chem.* **48**, 359 (1974).

Patents

P1. A. V. Alferov *et al.*, U.S. Patent 3,468,866 (1969); *Chem. Abstr.* **71**, 125722 (1969).

P2. A. Onishi, S. Anzai, and M. Ishii (Bridgestone Tire Co. Ltd.), U.S. Patent 3,639,520 (1972); *Chem. Abstr.* **76**, 142115 (1972).

P3. S. Anzai, K. Irako, and A. Ohnishi (Bridgestone Tire Co., Ltd.), Jap. Patent 71 21,738 (1971); *Chem. Abstr.* **76**, 15574 (1972).

P4. A. Onishi, S. Anzai, T. Ishikawa, A. Koga, K. Irako, and M. Ishii (Bridgestone Tire Co., Ltd.), U.S. Patent 3,591,658 (1971); *Chem. Abstr.* **75**, 130650 (1971).

P5. K. Irako, S. Anzai, Y. Ishizuka, H. Yoshii, T. Ishikawa, and A. Oonishi (Bridgestone Tire Co., Ltd.), Jap. Patent 71 10,872 (1971); *Chem. Abstr.* **75**, 130642 (1971).

P6. S. Anzai, K. Irako, H. Kawamoto, Y. Hayakawa, and A. Onishi (Bridgestone Tire Co., Ltd.), Jap. Patent 70 38,069 (1970); *Chem. Abstr.* **74**, 65265 (1971).

P7. S. Anzai, K. Irako, Y. Hayakawa, T. Wakamatsu, S. Miyamoto, and A. Onishi (Bridgestone Tire Co., Ltd.), Jap. Patent 70 24,588 (1970); *Chem. Abstr.* **74**, 13986 (1971).

P8. T. Ishikawa, Y. Ishizuka, K. Irako, S. Anzai, and A. Onishi (Bridgestone Tire Co., Ltd.), Jap. Patent 70 13,590 (1970); *Chem. Abstr.* **73**, 78319 (1970).

P9. A. Onishi, S. Anzai, T. Ishikawa, A. Koga, K. Irako, and M. Ishii (Bridgestone Tire Co., Ltd.), Brit. Patent 1,077,160 (1967); *Chem. Abstr.* **67**, 82827 (1967).

P10. Bridgestone Tire Co., Ltd., Fr. Patent 1,458,322 (1966); *Chem. Abstr.* **67**, 22718 (1967).

P11. J. Hosono, A. Onishi, and K. Ueda (Bridgestone Tire Co., Ltd.), Jap. Patent 8191 ('62) (1962); *Chem. Anstr.* **58**, 11556 (1963).

P12. K. Maeda, A. Onishi, and K. Ueda (Bridgestone Tire Co., Ltd.), Jap. Patent 8193('62) (1962); *Chem. Abstr.* **58**, 11556 (1963).

P13. K. Komatsu, Y. Takeuchi, S. Nishiyama, A. Onishi, and K. Ueda (Bridgestone Tire Co., Ltd.), Jap. Patent 14,459('67) (1967); *Chem. Abstr.* **68**, 30872 (1968).

P14. Bridgestone Tire Co., Ltd., Brit. Patent 905,099 (1962); *Chem. Abstr.* **57**, 16830 (1962).

P15. K. Ueda, A. Onishi, T. Yoshimoto, J. Hosono, and K. Maeda (Bridgestone Tire Co., Ltd.), U.S. Patent 3,170,904 (1965); see *Chem. Abstr.* **57**, 6090 (1962).

P16. K. Ueda, A. Onishi, T. Yoshimoto, J. Hosono, and K. Maeda (Bridgestone Tire Co., Ltd.), U.S. Patent 3,170,907 (1965); *Chem. Abstr.* **62**, 16486 (1965).

P17. K. Ueda, A. Onishi, T. Yoshimoto, J. Hosono, and K. Maeda (Bridgestone Tire Co., Ltd.), U.S. Patent 3,170,905 (1965); *Chem. Abstr.* **62**, 16487 (1965).

P18. J. Furukawa, T. Saegusa, A. Onishi, T. Ishikawa, S. Anazi, K. Irako, T. Narumiya, and Y. Ishizuka (Bridgestone Tire Co., Ltd.), Ger. Aus. 1,570,293 (1970); see *Chem. Abstr.* **66**, 19556 (1967).

P19. J. Furukawa, T. Saegusa, K. Ueda, A. Onishi, K. Irako, and T. Ebata (Bridgestone Tire Co., Ltd.), Jap. Patent 21,995('65) (1965); *Chem. Abstr.* **64**, 3812 (1966).

P20. S. Anzai, A. Onishi, K. Ueda, and T. Saegusa (Bridgestone Tire Co., Ltd.), Jap. Patent 18,546('63) (1963); *Chem. Abstr.* **60**, 3178 (1964).

P21. J. Furukawa, T. Miki, S. Anzai, and T. Narumiya (Bridgestone Tire Co., Ltd.), Jap. Patent 17,687('62) (1962); *Chem. Abstr.* **59**, 8897 (1963).

P22. S. Anzai, A. Onishi, and K. Ueda (Bridgestone Tire Co., Ltd.), Japan. Patent 15,662('65) (1965); *Chem. Abstr.* **63**, 18422 (1965).

P23. Y. Takeuchi, K. Komatsu, S. Kato, A. Onishi, and K. Ueda (Bridgestone Tire Co., Ltd.), Jap. Patent 12,996('65) (1965); *Chem. Abstr.* **63**, 18426 (1965).

P24. J. Furukawa, T. Saegusa, T. Narumiya, and S. Anzai (Bridgestone Tire Co., Ltd.), Jap. Patent 20,196('65) (1965); *Chem. Abstr.* **65**, 12388 (1966).

P25. Bataafse Petroleum Maatsch. N. V., Brit. Patent 906,052 (1962); *Chem. Abstr.* **58**, 646 (1963).

P26. Bataafse Petroleum Maatsch. N. V., Brit. Patent 906,056 (1962); *Chem. Abstr.* **58**, 3582 (1963).

P27. N. V. de Bataafsche Petroleum Maatsch., Brit. Patent 838,368 (1960); *Chem. Abstr.* **54**, 23401 (1960).

P28. H. Naarmann, H. P. Hofmann, and E.-G. Kastning (B.A.S.F.), Brit. Patent 1,089,465 (1967); see *Chem. Abstr.* **64**, 14308 (1966).

P29. B.A.S.F., Neth. Patent Appl. 6,410,331 (1965); *Chem. Abstr.* **63**, 4494 (1965).

P30. H. Müller (B.A.S.F.), Brit. Patent 946,058 (1964); Conc. Belg. 617,188 (1962); *Chem. Abstr.* **58**, 9310 (1963).

P31. H. Müller, H. Seibt, and H. Overwien (B.A.S.F.), Ger. Patent 1,292,847 (1969); *Chem. Abstr.* **70**, 116074 (1969).

P32. H. Müller (B.A.S.F.), Ger. Patent 1,301,490 (1969); *Chem. Abstr.* **71**, 92458 (1969).

P33. H. Müller (B.A.S.F.), Ger. Aus. 1,174,071 (1964); *Chem. Abstr.* **61**, 12175 (1964).

P34. H. Lautenschlager (B.A.S.F.), Ger. Aus. 1,156,986 (1963); *Chem. Abstr.* **60**, 3178 (1964).

P35. H. Müller (B.A.S.F.), Ger. Patent 1,225,391 (1966); *Chem. Abstr.* **66**, 11629 (1967).

P36. H. Müller (B.A.S.F.), Ger. Aus. 1,209,752 (1960); *Chem. Abstr.* **64**, 12929 (1966).

P37. H. Kröper and H. W. Weitz (B.A.S.F.), Ger. Patent 1,026,959 (1958); *Chem. Abstr.* **55**, 1078 (1961).

P38. H. Müller and H. Lautenschlager (B.A.S.F.), Ger. Patent 1,268,397 (1968); *Chem. Abstr.* **69**, 20205 (1968).

P39. H. Müller (B.A.S.F.), Ger. Patent 1,269,811 (1968); *Chem. Abstr.* **69**, 28440 (1968).

P40. H. Müller and D. Wittenberg (B.A.S.F.), Ger. Aus. 1,174,507 (1964); *Chem. Abstr.* **61**, 10853 (1964).

P41. B. Schleimer and H. Weber (Chemische Werke Hüls A.G.), S. Afr. Patent 67 01,959 (1968); *Chem. Abstr.* **70**, 38661 (1969).

P42. Chemische Werke Hüls A.G., Fr. Patent 1,518,385 (1968); *Chem. Abstr.* **70**, 97795 (1969).

P43. H. Newby (Chemische Werke Hüls A.G.), Brit. Patent 1,140,018 (1969); *Chem. Abstr.* **70**, 69155 (1969).

P44. B. Schleimer and H. Weber (Chemische Werke Hüls A.G.), U.S. Patent 3,329,734 (1967); *Chem. Abstr.* **67**, 91231 (1967).

P45. Chemische Werke Hüls A.G., Fr. Patent 1,492,544 (1967); *Chem. Abstr.* **69**, 10937 (1968).

P46. B. Schleimer and H. Weber (Chemische Werke Hüls A.G.), U.S. Patent 3,321,541 (1967); *Chem. Abstr.* **67**, 44378 (1967).

P47. Chemische Werke Hüls A.G., Fr. Patent 1,472,291 (1967); *Chem. Abstr.* **68**, 40772 (1968).

P48. B. Schleimer (Chemische Werke Hüls A.G.), Ger. Aus. 1,241,119 (1967); *Chem. Abstr.* **67**, 109473 (1967).

P49. Chemische Werke Hüls A.G., Fr. Patent 1,447,507 (1966); *Chem. Abstr.* **66**, 46773 (1967).

P50. B. Schleimer and H. Weber (Chemische Werke Hüls A.G.), Ger. Aus. 1,212,302 (1966); *Chem. Abstr.* **64**, 17837 (1966).

P51. Chemische Werke Hüls A.G., Belg. Patent 630,428 (1963); *Chem. Abstr.* **61**, 808 (1964).

P52. J. J. Hawkins, C. D. Storrs, and S. D. Zimmerman (Columbian Carbon Co.), U.S. Patent 3,661,882 (1972); *Chem. Abstr.* **77**, 49813 (1972).

P53. G. Pampus and N. Schön (Farb. Bayer A. G.), Ger. Offen. 1,595,702 (1970).

P54. W. Oberkirch, H. Herlinger, and P. Gunther (Farb. Bayer A.G.), Brit. Patent 1,164,577 (1969); *Chem. Abstr.* **71**, 102439 (1969).

P55. W. A. Judy and M. C. Throckmorton (Goodyear Tire and Rubber Co.), U.S. Patent 3,652,529 (1972); *Chem. Abstr.* **77**, 21196 (1972).

P56. M. C. Throckmorton (Goodyear Tire and Rubber Co.), U.S. Patent 3,624,000 (1971); *Chem. Abstr.* **76**, 142117 (1972).

P57. M. C. Throckmorton (Goodyear Tire and Rubber Co.), Ger. Offen. 2,128,260 (1972); *Chem. Abstr.* **76**, 128460 (1972).

P58. M. C. Throckmorton and W. M. Saltman (Goodyear Tire and Rubber Co.), Ger. Offen. 2,118,915 (1971); *Chem. Abstr.* **76**, 73480 (1972).

P59. M. C. Throckmorton (Goodyear Tire and Rubber Co.), Ger. Offen. 2,113,527 (1971); *Chem. Abstr.* **76**, 60719 (1972).

P60. M. C. Throckmorton (Goodyear Tire and Rubber Co.), Ger. Offen. 2,019,751 (1970); *Chem. Abstr.* **74**, 43301 (1971).

P61. K. J. Frech, F. H. Hoppstock, and J. A. Goodwin (Goodyear Tire and Rubber Co.), Ger. Offen. 2,043,746 (1971); *Chem. Abstr.* **75**, 7113 (1971).

P62. M. C. Throckmorton and W. M. Saltman (Goodyear Tire and Rubber Co.), U.S. Patent 3,528,957 (1970); see *Chem. Abstr.* **70**, 107325 (1969).

P63. M. C. Throckmorton (Goodyear Tire and Rubber Co.), Ger. Offen. 2,028,283 (1971); *Chem. Abstr.* **74**, 100517 (1971).

P64. W. M. Saltman and M. L. Wise (Goodyear Tire and Rubber Co.), Ger. Offen. 2,016, 987 (1970); *Chem. Abstr.* **74**, 65261 (1971).

P65. M. C. Throckmorton and J. Lal (Goodyear Tire and Rubber Co.), Ger. Offen. 2,017,006 (1970); *Chem. Abstr.* **74**, 32525 (1971).

P66. M. C. Throckmorton (Goodyear Tire and Rubber Co.), Fr. Patent 1,584,923 (1970); *Chem. Abstr.* **73**, 110699 (1970).

P67. M. C. Throckmorton and W. M. Saltman (Goodyear Tire and Rubber Co.), S. Afr. Patent 68 01,063 (1968); *Chem. Abstr.* **70**, 48429 (1969).

P68. M. C. Throckmorton (Goodyear Tire and Rubber Co.), S. Afr. Patent 68 00,447 (1968); *Chem. Abstr.* **70**, 48428 (1969).

P69. W. M. Saltman and M. C. Throckmorton (Goodyear Tire and Rubber Co.), Ger. Offen. 1,901,952 (1969); *Chem. Abstr.* **71**, 102913 (1969).

P70. Goodyear Tire and Rubber Co., Ger. Offen. 1,809,495 (1969); *Chem. Abstr.* **71**, 82399 (1969).

P71. M. C. Throckmorton (Goodyear Tire and Rubber Co.), S. Afr. Patent 68 00,446 (1968); *Chem. Abstr.* **70**, 69158 (1969).

P72. Goodyear Tire and Rubber Co., Brit. Patent 859,698 (1961); *Chem. Abstr.* **55**, 16018 (1961).

P73. Goodyear Tire and Rubber Co., Brit. Patent 1,009,285 (1965); *Chem. Abstr.* **64**, 3811 (1966).

P74. Goodyear Tire and Rubber Co., Neth. Appl. 6,414,376 (1965); *Chem. Abstr.* **64**, 2261 (1966).

P75. C. E. Brockway and A. F. Ekar (Goodrich-Gulf Chemicals Inc.), U.S. Patent 2,977,349 (1961); *Chem. Abstr.* **55**, 16012 (1961).

P76. F. Dawans (Institut Français du Pétrole), Fr. Patent 1,583,616 (1969); *Chem. Abstr.* **73**, 56978 (1970).

P77. G. Codet, F. de Charentenay, F. Dawans, and P. Teyssié (Institut Français du Pétrole), Ger. Offen. 1,949,242 (1970); *Chem. Abstr.* 73, 36295 (1970).

P78. F. Dawans, J. P. Durand, and P. Teyssié (Institut Français du Pétrole), Fr. Addn. 94,311 (1969); *Chem. Abstr.* 72, 112515 (1970).

P79. F. Dawans (Institut Français du Pétrole), Ger. Offen, 1,936,807 (1970); *Chem. Abstr.* 72, 67986 (1970).

P80. F. Dawans, J. P. Durand, and P. Teyssié (Institut Français du Pétrole), Fr. Patent 1,556,962 (1969); *Chem. Abstr.* 71, 82389 (1969).

P81. F. Dawans, P. Teyssié, and E. Goldenberg (Institut Français du Pétrole), Fr. Addn. 90,152 (1967); *Chem. Abstr.* 69, 59947 (1968).

P82. Institut Français du Pétrole, Neth. Appl. 6,608,304 (1966); *Chem. Abstr.* 67, 3319 (1967).

P83. Institut Français du Pétrole, Neth. Appl. 6,601,988 (1966); *Chem. Abstr.* 66, 38689 (1967).

P84. B. A. Dolgoplosk, E. I. Tinyakova, S. I. Beilin, and K. L. Makovetskii (Topchiev, A.V., Institute of Petrochem. Syn.), Fr. Patent 2,071,644 (1971); *Chem. Abstr.* 77, 6948 (1972).

P85. B. A. Dolgoplosk, E. I. Tinyakova, S. I. Beilin, K. L. Makovetskii, G. M. Chernenko, I. Y. Ostrovskaya, V. A. Krol, and E. K. Khrennikova (Topchiev, A.V., Institute of Petrochem. Syn.), Ger. Offen. 2,038,721 (1971); *Chem. Abstr.* 75, 141758 (1971).

P86. D. K. Jenkins and C. G. P. Dixon (Int. Syn. Rubber Co. Ltd.), Brit. Patent 1,134,739 (1968); *Chem. Abstr.* 70, 29887 (1969).

P87. C. G. P. Dixon and D. K. Jenkins (Int. Syn. Rubber Co. Ltd.), Brit. Patent 1,120,060 (1968); *Chem. Abstr.* 69, 52844 (1968).

P88. International Synthetic Rubber Co. Ltd., Neth. Appl. 6,607,552 (1966); *Chem. Abstr.* 66, 105743 (1967).

P89. International Synthetic Rubber Co. Ltd., Neth. Appl. 6,615,309 (1966); see *Chem. Abstr.* 70, 29553 (1969).

P90. International Synthetic Rubber Co. Ltd., Neth. Appl. 6,515,817 (1966); *Chem. Abstr.* 65, 17177 (1966).

P91. T. Matsumoto (Jap. Synthetic Rubber Co., Ltd.), Jap. Patent 71 34,985 (1971); *Chem. Abstr.* 76, 60727 (1972).

P92. H. Mori, K. Ikeda, I. Nagaoka, S. Hirayanagi, S. Saka, T. Kikuchi, and A. Kihi (Jap. Synthetic Rubber Co., Ltd.), Jap. Patent 70 35,316 (1970); *Chem. Abstr.* 74, 65285 (1971).

P93. A. Kiso, H. Mori, Y. Tanaka, and K. Matsuzaka (Jap. Synthetic Rubber Co., Ltd.), Jap. Patent 69 27,824 (1969); *Chem. Abstr.* 74, 32526 (1971).

P94. H. Yasunaga, K. Komatsu, Y. Ninomiya, J. Hirota, and K. Maeda (Jap. Synthetic Rubber Co., Ltd.), Jap. Patent 70 08,427 (1970); *Chem. Abstr.* 73, 99829 (1970).

P95. E. Okuya, T. Ito, A. Sakaguchi, K. Komatsu, and H. Yasunaga (Jap. Synthetic Rubber Co., Ltd.), Jap. Patent 70 02,386 (1970); *Chem. Abstr.* 73, 4737 (1970).

P96. H. Yasunaga, K. Komatsu, E. Okutani, I. Ito, and A. Sakaguchi (Jap. Synthetic Rubber Co., Ltd.), Jap. Patent 70 02,387 (1970); *Chem. Abstr.* 72, 112511 (1970).

P97. Y. Takeda, K. Komatsu, M. Yoshizawa, H. Yasunaga, and K. Ueda (Jap. Synthetic Rubber Co., Ltd.), Jap. Patent 68 20,306 (1968); *Chem. Abstr.* 70, 58774 (1969).

P98. K. Komatsu, S. Nishiyama, H. Yasunaga, and K. Ueda (Jap. Synthetic Rubber Co., Ltd.), Jap. Patent 68 08,826 (1968); *Chem. Abstr.* 70, 5107 (1969).

P99. S. Otsuka, H. Mori, T. Kikuchi, S. Hashimoto, Y. Tanaka, F. Imaizumi, M. Abe, S. Muranishi, and I. Nagaoka (Jap. Synthetic Rubber Co., Ltd.), Jap. Patent 4984('67) (1967); *Chem. Abstr.* **67**, 65235 (1967).

P100. S. Otsuka, M. Kawakami, T. Kikuchi, Y. Tanaka, M. Endo, and S. Yokoi (Jap. Synthetic Rubber Co., Ltd.), Jap. Patent 19,833('66) (1966); *Chem. Abstr.* **66**, 76758 (1967).

P101. Japan. Synthetic Rubber Co., Ltd., Fr. Patent 1,440,731 (1966); *Chem. Abstr.* **66**, 76484 (1967).

P102. K. Ukita, T. Sadamori, and Y. Tokuyama (Japanese Geon Co., Ltd.), Ger. Offen. 2,053,484 (1972); *Chem. Abstr.* **77**, 62503 (1972).

P103. Mitsui Chemical Industry Co., Ltd., Belg. Patent 655,015 (1965); *Chem. Abstr.* **64**, 19945 (1966).

P104. Montecatini Edison S.p.A., Ital. Patent 846,007 (1969); *Chem. Abstr.* **75**, 110909 (1971).

P105. Montecatini Soc. Gen., Ital. Patent 588,825 (1957); *Chem. Abstr.* **56**, 2544 (1962).

P106. Montecatini Soc. Gen., Belg. Patent 617,545 (1962); *Chem. Abstr.* **58**, 6942 (1963).

P107. Montecatini Soc. Gen., Ital. Patent 594,618 (1959); *Chem. Abstr.* **55**, 12936 (1961).

P108. Montecatini Soc. Gen., Ital. Patent 594,665 (1959); *Chem. Abstr.* **55**, 17091 (1961).

P109. T. Yamawaki, T. Suzuki, and S. Hinjo (Mitsubishi Chem. Ind. Co., Ltd.), Ger. Offen. 2,141,844 (1972); *Chem. Abstr.* **77**, 21193 (1972).

P110. T. Yamawaki, M. Usami, T. Suzuki, and T. Uematsu (Mitsubishi Chem. Ind. Co., Ltd.), Jap. Patent 72 06,411 (1972); *Chem. Abstr.* **77**, 21199 (1972).

P111. Mitsubishi Chem. Ind. Co., Ltd., Fr. Patent 2,039,808 (1971); *Chem. Abstr.* **75**, 152760 (1971).

P112. S. Kambara, M. Sano, A. Takahashi, and T. Ajiro (Mitsui Toatsu Chemicals Co., Ltd.), Jap. Patent 70 01,630 (1970); *Chem. Abstr.* **72**, 112544 (1970).

P113. T. Arakawa (Mitsui Petrochemical Ind. Ltd.), Jap. Patent 69 29,465 (1969); *Chem. Abstr.* **72**, 67983 (1970).

P114. T. Yamawaki and M. Usami (Mitsubishi Chem. Ind. Co., Ltd.), Jap. Patent 71 17,134 (1971); *Chem. Abstr.* **75**, 110905 (1971).

P115. A. Kawasaki, M. Taniguchi, and T. Nishiyama (Maruzen Petrochemical Co., Ltd.), Jap. Patent 71 09,593 (1971); *Chem. Abstr.* **75**, 37621 (1971).

P116. Mitsubishi Chem. Ind. Co., Ltd., Fr. Patent 1,559,871 (1969); *Chem. Abstr.* **71**, 82398 (1969).

P117. S. Tanaka, A. Nakamura, E. Kubo, and K. Tanaka (Mitsubishi Petrochem Co., Ltd.), Jap. Patent 15,661('65) (1965); *Chem. Abstr.* **63**, 18422 (1965).

P118. R. H. Gaeth (Phillips Petroleum Co.), U.S. Patent 3,557,075 (1971); *Chem. Abstr.* **74**, 76794 (1971).

P119. R. H. Gaeth (Phillips Petroleum Co.), U.S. Patent 3,594,360 (1971); *Chem. Abstr.* **75**, 141763 (1971).

P120. R. H. Smith and D. D. Norwood (Phillips Petroleum Co.), U.S. Patent 3,513,149 (1970); *Chem. Abstr.* **74**, 65260 (1971).

P121. R. P. Zellinski and R. H. Gaeth (Phillips Petroleum Co.), U.S. Patent 3,560,405 (1971); *Chem. Abstr.* **74**, 65262 (1971).

P122. R. H. Gaeth (Phillips Petroleum Co.), U.S. Patent 3,525,729 (1970); *Chem. Abstr.* **73**, 88861 (1970).

P123. R. H. Gaeth (Phillips Petroleum Co.), U.S. Patent 3,492,282 (1970); *Chem. Abstr.* **72**, 67987 (1970).

P124. R. H. Gaeth and F. E. Naylor (Phillips Petroleum Co.), Fr. Patent 1,512,894 (1968); *Chem. Abstr.* **70**, 69160 (1969).

P125. Y. Yagi, H. Sato, S. Narisawa, S. Yasui, A. Kobayashi, M. Hino, and K. Hata (Sumitomo Chemical Co., Ltd.), Ger. Offen. 2,164,515 (1972); *Chem. Abstr.* **77**, 127717 (1972).

P126. Y. Yagi, S. Narisawa, T. Oshima, and K. Hata (Sumitomo Chemical Co., Ltd.), Brit. Patent 2,147,851 (1972); *Chem. Abstr.* **77**, 36004 (1972).

P127. Solvay et Cie., Fr. Addn. 2,030,561 (1970); see *Chem. Abstr.* **76**, 127751 (1972).

P128. Y. Yagi, A. Kobayashi, and H. Sato (Sumitomo Chemical Co., Ltd.), Ger. Offen. 2,127,621 (1971); *Chem. Abstr.* **76**, 100930 (1972).

P129. Y. Yagi, A. Kobayashi, H. Hirata, and H. Sato (Sumitomo Chemical Co., Ltd.), Ger. Offen. 2,024,345 (1970); *Chem. Abstr.* **74**, 54878 (1971).

P130. Y. Yagi, I. Hirata, and A. Kobayashi (Sumitomo Chemical Co., Ltd.), Ger. Offen. 2,017,316 (1970); *Chem. Abstr.* **74**, 4454 (1971).

P131. SNAM Progetti S.p.A., Brit. Patent 1,132,425 (1968); *Chem. Abstr.* **70**, 20846 (1969).

P132. Stamicarbon N.V., Neth. Patent Appl. 6,508,841 (1967); *Chem. Abstr.* **67**, 12350 (1967).

P133. Stamicarbon N.V., Neth. Patent Appl. 6,700,205 (1968); *Chem. Abstr.* **69**, 78284 (1968).

P134. R. J. L. Graf and A. Marchal (Stamicarbon N.V.), Ger. Offen. 2,109,281 (1971); *Chem. Abstr.* **76**, 15571 (1972).

P135. C. E. P. V. van den Berg (Stamicarbon N.V.), Neth. Patent Appl. 70 02,967 (1971); *Chem. Abstr.* **76**, 47129 (1972).

P136. G. J. van Amerongen and K. Nozaki (Shell Internationale Research Maatsch. N.V.), Ger. Aus. 1,139,647 (1962); see *Chem. Abstr.* **57**, 1066 (1962).

P137. Shell Internationale Research Maatsch. N.V., Belg. Patent 620,290 (1963); *Chem. Abstr.* **59**, 4061 (1963).

P138. Shell Internationale Research Maatsch. N.V., Brit. Patent 908,335 (1962); *Chem. Abstr.* **58**, 3582 (1963).

P139. E. W. Duck and J. A. Waterman (Shell Internationale Research Maatsch. N.V.), Ger. Offen. 1,139,277 (1962); *Chem. Abstr.* **58**, 14266 (1963).

P140. Shell Internationale Research Maatsch. N.V., Belg. Patent 614,967 (1962); *Chem. Abstr.* **58**, 11556 (1963).

P141. Shell Internationale Research Maatsch. N.V., Brit. Patent 904,404 (1962); *Chem. Abstr.* **58**, 11556 (1963).

P142. Shell Internationale Research Maatsch. N.V., Brit. Patent 884,071 (1961); *Chem. Abstr.* **56**, 11769 (1962).

P143. Shell Internationale Research Maatsch. N.V., Brit. Patent 890,139 (1962); *Chem. Abstr.* **57**, 1066 (1963).

P144. Shell Internationale Research Maatsch. N.V., Brit. Patent 890,140 (1962); *Chem. Abstr.* **57**, 1066 (1962).

P145. Shell Internationale Research Maatsch. N.V., Belg. Patent 629,605 (1963); *Chem. Abstr.* **61**, 3281 (1964).

P146. E. A. Youngman (Shell Oil Co.), U.S. Patent 3,005,811 (1958); *Chem. Abstr.* **56**, 3613 (1962).

P147. T. L. Higgins and C. H. Wilcoxen (Shell Oil Co.), U.S. Patent 3,068,217 (1962); *Chem. Abstr.* **58**, 5800 (1963).

P148. W. S. Anderson and L. M. Porter (Shell Oil Co.), U.S. Patent 2,965,625 (1960); *Chem. Abstr.* **55**, 7921 (1961).

P149. G. J. van Amerongen and K. Nozaki (Shell Oil Co.), U.S. Patent 3,068,180 (1962); *Chem. Abstr.* **58**, 9246 (1963).

P150. W. S. Anderson (Shell Oil Co.), U.S. Patent 2,943,987 (1960); *Chem. Abstr.* **54**, 23400 (1960).

P151. E. A. Youngman (Shell Oil Co.), U.S. Patent 3,066,128 (1962); *Chem. Abstr.* **59**, 8955 (1963).

P152. K. Tani and S. Yuguchi (Toyo Rayon Co.), Jap. Patent 797('67) (1967); *Chem. Abstr.* **66**, 116488 (1967).

P153. T. Yoshida and S. Yuguchi (Toyo Rayon Co.), Jap. Patent 25,077 ('65) (1965); *Chem. Abstr.* **64**, 8451 (1966).

P154. K. Tani and S. Yaguchi (Toyo Rayon Co.), Jap. Patent 798('67) (1967); *Chem. Abstr.* **67**, 3517 (1967).

P155. S. Sugiura, F. Tasaka, H. Ueno, M. Kono, N. Katagiri, and N. Sakinaga (Ube Industries Ltd.), Ger. Patent 1,267,849 (1968); *Chem. Abstr.* **69**, 20209 (1968).

P156. United States Rubber Co., Neth. Patent Appl. 6,507, 223 (1965); *Chem. Abstr.* **64**, 16013 (1966).

P157. C. W. Childers (United States Rubber Co.), Fr. Patent 1,377,675 (1964); *Chem. Abstr.* **62**, 13357 (1965).

P158. C. W. Childers (United States Rubber Co.), Fr. Patent 1,377,676 (1964); *Chem. Abstr.* **62**, 11998 (1965).

P159. C. W. Childers (United States Rubber Co.), Belg. Patent 634,251 (1963); *Chem. Abstr.* **61**, 739 (1964).

P160. S. Sugiura, H. Ueno, S. Nakatomi, H. Ishikawa, and T. Inoue (Ube Industries Ltd.), Ger. Aus. 1,268,849 (1968); *Chem. Abstr.* **69**, 20215 (1968).

P161. Y. Yagi, A. Kobayashi, and I. Hirata (Sumitomo Chemical Co. Ltd.), U.S. Patent 3,684,789 (1972); see *Chem. Abstr.* **74**, 4454 (1971).

P162. Y. Yagi, H. Sato, S. Narisawa, S. Yasui, A. Kobayashi, M. Hino, and K. Hata (Sumitomo Chemical Co. Ltd.), Ger. Offen. 2,164,514 (1972); *Chem. Abstr.* **78**, 17346 (1973).

P163. M. Iwamoto and J. Matsui (Toray Industries Inc.), Jap. Patent 72 25,866 (1972) *Chem. Abstr.* **78**, 16803 (1973).

P164. T. Yamawaki and M. Usami (Mitsubishi Chemical Industries Co. Ltd.), Jap. Patent 71 17,133 (1971); *Chem. Abstr.* **75**, 130651 (1971).

P165. B. A. Dolgoplosk *et al.* (Topchiev A.V., Institute of Petrochemical Synthesis), Ger. Offen. 2,064,767 (1971); *Chem. Abstr.* **75**, 130652 (1971).

P166. T. Yoshimoto, A. Onishi, and K. Ueda (Bridgestone Tire Co. Ltd.), Jap. Patent 8192('62) (1962); *Chem. Abstr.* **58**, 11556 (1963).

P167. T. Otsu, S. Aoki, and R. Imoto (Toyo Soda Manufg. Co., Ltd.), Jap. Patent 70 38,068 (1970); *Chem. Abstr.* **74**, 65263 (1971).

P168. E. Goldenberg, F. Dawans, J. P. Durand, and G. Martino (Institut Français du Petrole), Ger. Offen. 2,232,767 (1973); *Chem. Abstr.* **78**, 148787 (1973).

P169. J. P. Durand, F. Dawans, and P. Teyssié (Institut Français du Pétrole), Fr. Patent 1,590,083 (1970); *Chem. Abstr.* **78**, 59551 (1973).

P170. K. D. Hesse and H. von Portatius (Chemische Werke Hüls A.G.), Ger. Offen. 2,122,956 (1972); *Chem. Abstr.* **78**, 59008 (1973).

P171. B. A. Dolgoplosk *et al.*, U.S. Patent 3,740,382 (1973); *Chem. Abstr.* **79**, 54640 (1973).

P172. M. C. Throckmorton and W. M. Saltman (Goodyear Tire and Rubber Co.), Ger. Offen. 2,257,137 (1973).

P173. M. C. Throckmorton, C. E. Traina, and R. A. Mournighan (Goodyear Tire and Rubber Co.), Ger. Offen. 2,256,340 (1973).

P174. M. C. Throckmorton (Goodyear Tire and Rubber Co.), U.S. Patent 3,732,195 (1973).

P175. N. H. Phung and G. Lefebvre (Institut Français du Pétrole), Fr. Addn. 95,693 (1971); *Chem. Abstr.* **78**, 137090 (1973).

P176. J. P. Durand and F. Dawans (Institut Français du Pétrole), Fr. Patent 2,153,787 (1973); *Chem. Abstr.* **79**, 79476 (1973).

P177. P. Teyssie and F. Dawans (Institut Français du Pétrole), Fr. Patent 1,602,281 (1970); *Chem. Abstr.* **79**, 6529 (1973).

P178. M. Ichikawa, Y. Takeuchi, Y. Harita, and T. Miura (Jap. Synthetic Rubber Co.), Jap. Kokai 73 11,385 (1973); *Chem. Abstr.* **78**, 160738 (1973).

P179. J. H. Merkley (Lithium Corp.), Ger. Offen. 2,148,148 (1973); *Chem. Abstr.* **78**, 160360 (1973).

P180. K. Ukita and T. Sadamori (Japanese Geon Co. Ltd.), U.S. Patent 3,725,492 (1973); *Chem. Abstr.* **79**, 19433 (1973).

P181. H. Hasegawa and T. Goh (Nippon Zeon Co. Ltd.), Jap. Kokai 73 01,077 (1973); *Chem. Abstr.* **78**, 125672 (1973).

P182. Y. Yagi, A. Kobayashi, I. Hirata, and H. Sato (Sumitomo Chem. Co. Ltd.), Jap. Patent 72 11,811 (1972); *Chem. Abstr.* **77**, 89759 (1972).

P183. Y. Yagi, S. Narisawa, T. Oshima, and K. Hata (Sumitomo Chemical Co. Ltd.), Ger. Offen. 2,147,851 (1972); *Chem. Abstr.* **78**, 125665 (1973).

P184. T. Yamada, T. Watanabe, R. Miyazaki, and Y. Yagi (Sumitomo Chemical Co. Ltd.), Jap. Kokai 73 17,887 (1973); *Chem. Abstr.* **79**, 5924 (1973).

P185. T. Yamawaki, T. Suzuki, and S. Hino (Mitsubishi Chem. Ind. Co.), Jap. Patent 73 24,833 (1973); *Chem. Abstr.* **80**, 96614 (1974).

P186. W. H. Saltman and M. C. Throckmorton (Goodyear Tire and Rubber Co.), U.S. Patent 3,769,270 (1973).

P187. K. Komatsu and A. Sakguchi (Jap. Synthetic Rubber Co. Ltd,), Jap. Patent 73 06,940 (1973); *Chem. Abstr.* **79**, 137825 (1973).

P188. T. Yamawaki, T. Suzuki, and S. Hino (Mitsubishi Chemical Ind. Co. Ltd.), Jap. Patent 73 06,184 (1973); *Chem. Abstr.* **79**, 137829 (1973).

P189. S. Anzai, K. Irako, Y. Hayakawa, M. Kojima, and A. Oonishi (Bridgestone Tire Co. Ltd.), Jap. Patent 71 20,058 (1971); *Chem. Abstr.* **75**, 152759 (1971).

P190. H. Naarmann (B.A.S.F.), Belg. Patent 670,029 (1966); *Chem. Abstr.* **66** 66007 (1967).

P191. T. Yamawaki, M. Usami, T. Suzuki, T. Uematsu, and S. Hino (Mitsubishi Chem. Ind. Co., Ltd.), Jap. Patent 72 50,228 (1972); *Chem. Abstr.* **80**, 28176 (1974).

P192. T. Yamawaki, M. Usami, T. Suzuki, and T. Uematsu (Mitsubishi Chem. Ind. Co., Ltd.), Jap. Patent 73 12,194 (1973); *Chem. Abstr.* **80**, 71871 (1974).

P193. S. Yasui, Y. Shinohara, and Y. Ibaraki (Sumitomo Chem. Co. Ltd.), Ger. Offen. 2,310,276 (1973); *Chem. Abstr.* **80**, 71874 (1974).

P194. S. Yasui, Y. Shinohara, and Y. Yagi (Sumitomo Chem. Co., Ltd.), Jap. Kokai 73 43,084 (1973); *Chem. Abstr.* **80**, 48561 (1974).

P195. S. Yasui, T. Noguchi, M. Yamamoto, and Y. Yagi (Sumitomo Chem. Co., Ltd.), Ger. Offen. 2,334, 267 (1974).

CHAPTER V

Coupling Reactions and Related Processes

I. Introduction

This chapter is basically concerned with two reactions: the coupling of organic halides in the presence of stoichiometric amounts of zerovalent nickel complexes, and the nickel-catalyzed cross-coupling reaction between organic halides and Grignard reagents. These reactions are mechanistically related and both have found use in synthetic organic chemistry. Reviews on various aspects of these reactions are to be found in references 41, 59, 132, 133, and 138.

II. The Coupling of Organic Halides and Related Reactions

A. *The Coupling of Allyl Halides* (*Table V-1*)

The first report of the reaction of an allyl halide with an organonickel complex to give a coupling product is contained in a patent issued in 1943, in which the reaction of methallyl chloride with nickel tetracarbonyl to give 2,5-dimethyl-1,5-hexadiene is described (P7). Almost a quarter of a century

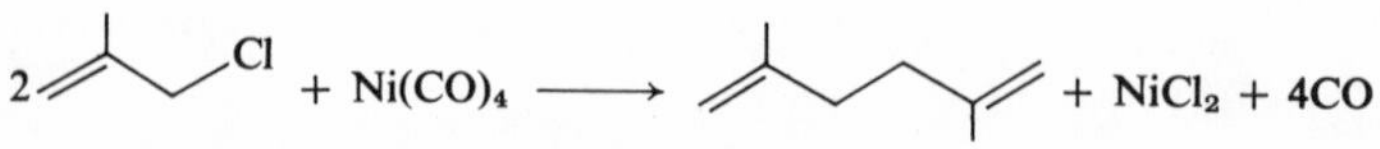

$$2 \text{ (methallyl chloride)} + Ni(CO)_4 \longrightarrow \text{(2,5-dimethyl-1,5-hexadiene)} + NiCl_2 + 4CO$$

was to pass before it was recognized that a π-allylnickel intermediate was involved.

Numerous examples of the coupling of allyl groups have now been reported and, although the majority involve the halides, examples are known for acetate and tosylate derivatives. The organonickel reagent is frequently nickel tetracarbonyl; other complexes that are active include triphenylphosphine nickel tricarbonyl, bis(cyclooctadiene)nickel, bis(π-allyl)nickel, and π-allylnickel halides. The carbonyl-free reagents are preferred because the undesired insertion of CO, which is sometimes observed with nickel tetracarbonyl, is thereby avoided.

The coupling of substituted allylic compounds generally occurs at the least substituted end of the allyl group, e.g.,

$$-\text{NiBr}/2 + \text{C}_2\text{H}_5\text{OCH}_2\text{CH}=\text{C}(\text{CH}_3)\text{CH}_2\text{Br} \xrightarrow{-\text{NiBr}_2}$$
$$\text{C}_2\text{H}_5\text{OCH}_2\text{CH}=\text{C}(\text{CH}_3)\text{CH}_2\text{CH}_2\text{CH}=\text{C}(\text{CH}_3)_2$$

whereas the reaction between compounds able to exist in cis or trans forms is generally stereochemically nonspecific. However, both the direction of coupling and the geometry of the product can be influenced by varying the solvent or by adding additional ligands or quaternary ammonium salts. For example, the principal product of the reaction between allyl chloride and $(\pi\text{-CH}_2\text{CHCHCN–NiCl})_2$ is the cis isomer **1** in toluene or ether and the trans isomer **2** in acetonitrile with added 2,6-lutidine, whereas the addition of a quaternary ammonium salt to this last combination causes the 3-substituted derivative **3** to be formed (15).

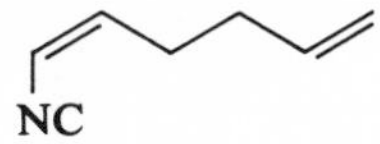

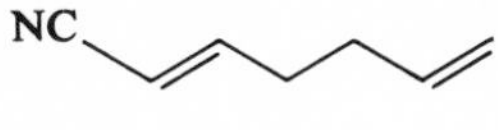

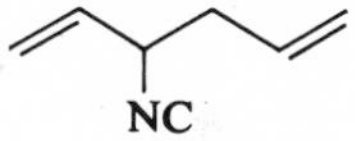

Bis(π-allyl)nickel complexes react with allyl halides in a manner similar to the π-allylnickel halides. For example, the α,ω-bis(π-allyl)-C_{12}-nickel species **4** reacts with allyl bromide to produce a mixture of pentaenes in which **5** predominates (91).

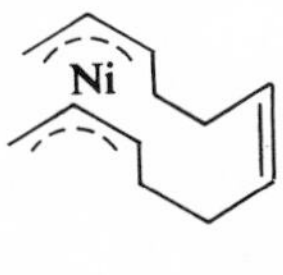

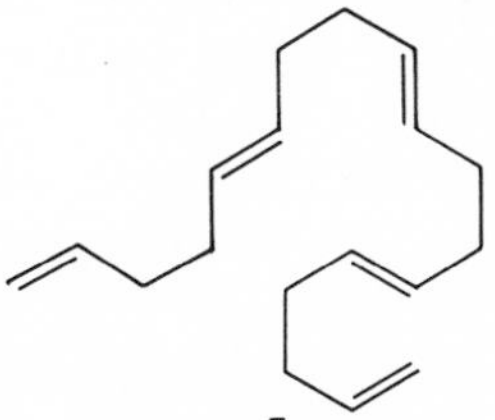

Among the most interesting synthetic possibilities is the formation of medium sized rings by the intramolecular coupling of long-chain diallylic halides. In this way it has been possible to synthesize the nine-membered ring

1,4,7-trimethylenecyclononane (6), the 4,5-cis isomer (7) of the 11-membered sesquiterpene humulene, the diene macrolide 8, the 18-membered cyclic diene 9, and a precursor of the 14-membered cyclic diterpene cembrene (146).

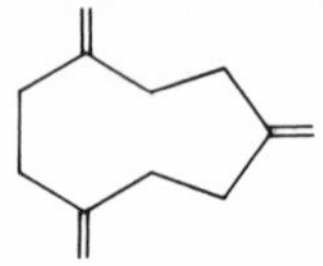

6 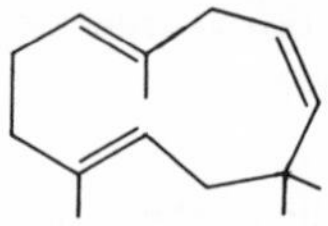   7 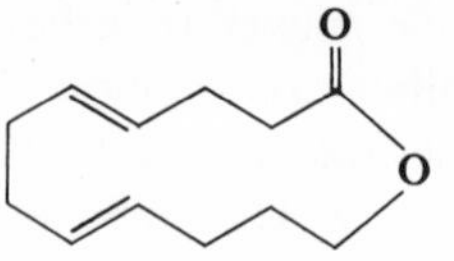   8

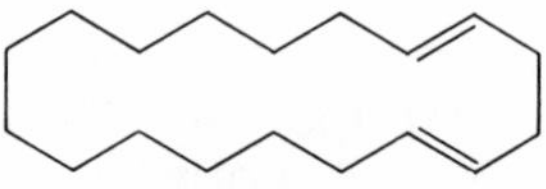

9

Eight- or ten-membered rings cannot, however, be synthesized by this procedure, the principal product obtained from the reaction of the C_8 and C_{10} dihalides $BrCH_2CH:CH(CH_2)_nCH:CHCH_2Br$ ($n = 2$ or 4) being vinylcyclohexene and divinylcyclohexane, respectively. Even this observation has found application in a synthesis of the sesquiterpenic alcohol *dl*-elemol (10) from the corresponding dibromide (4).

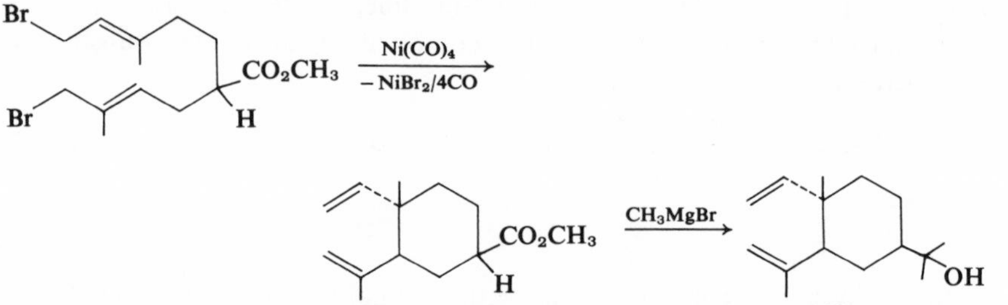

Olefins with two substituents allylic to the double bond can react either by coupling or by elimination of both substituents to give a 1,3-diene. The former behavior has been observed for $C_2H_5OCH_2CH:CHCH_2Br$ (29) and the latter in reactions involving $ClCH_2CH:CHCH_2Cl$ (1, 66–68), $HOCH_2CH:CHCH_2OH$ (66), $CH_3OCOCH_2CH:CHCH_2Br$ (66), and $CH_3OCOCH_2C(CH_3):CHCH_2Cl$ (66). A nickel-containing polymer, formulated as 11, has been isolated from the reaction of 1,4-dichloro-2-butene with nickel tetracarbonyl in benzene (1). A similar polymer has been isolated from the reaction with 1,4-di(chloromethyl)benzene.

11

TABLE V-1
The Coupling of Allylic Compounds

Product	Reactant		Ref.
	Organic component	Nickel component	
$(CH_2:CHCH_2)_2$	$CH_2:CHCH_2Br$	$(\pi\text{-}CH_2CHCH_2NiBr)_2$	10, 46
	$CH_2:CHCH_2Br$	$Ni(CO)_4$	14, 62, 84
	$CH_2:CHCH_2Br$	$(COD)_2Ni$	54
	$CH_2:CHCH_2OCOCH_3$	$Ni(CO)_4$	14, 51
	$CH_2:CHCH_2OSO_2C_6H_4\text{-}p\text{-}CH_3$	$Ni(CO)_4$	14
	$CH_2:CHCH_2Br$		91
$CH_2:CH[(CH_2)_2CH:CH]_3(CH_2)_2CH:CH_2$, $CH_2:CHCH_2[(CH_2)_2CH:CH]_2(CH_2)_2CH:CH_2$, $CH_3CH:CH[(CH_2)_2CH:CH]_2(CH_2)_2CH:CH_2$			
$[CH_2:C(CH_3)CH_2]_2$	$CH_2:C(CH_3)CH_2Cl$	$Ni(CO)_4$	P7
$(CH_2:CHCH_2)_2$, $[CH_2:C(CH_3)CH_2]_2$, $CH_2:C(CH_3)CH_2CH_2CH:CH_2$	$CH_2:C(CH_3)CH_2Br$	$(\pi\text{-}CH_2CHCH_2NiBr)_2$	10, 50
	$CH_2:C(CH_3)CH_2Br$	$[\pi\text{-}CH_2C(CH_3)CH_2NiBr]_2$	10, 50
	$CH_2:CHCH_2OSO_2C_6H_4\text{-}p\text{-}CH_3$	$[\pi\text{-}CH_2C(CH_3)CH_2NiBr]_2$	50
$(CH_3CH:CHCH_2)_2$,	$CH_3CH:CHCH_2Cl$	$Ni(CO)_4$	34, P4
$CH_2:CHCH(CH_3)CH_2CH:CHCH_3$	$CH_2:CHCHClCH_3$	$Ni(CO)_4$	34, P4
$[(CH_3)_2C:CHCH_2]_2$,	$(CH_3)_2C:CHCH_2Cl$	$Ni(CO)_4$	34, P4
$CH_2:CHC(CH_3)_2CH_2CH_2CH:C(CH_3)_2$	$CH_2:CHC(CH_3)_2Cl$	$Ni(CO)_4$	34, P4
$cis\text{-}$ and $trans\text{-}(CH_3)_2C:CH(CH_2)_2$ $C(CH_3):CH(CH_2)_2C:CH_2)CH:CH_2$	$CH_2:CHC(:CH_2)CH_2Br$	$[\pi\text{-}CH_2CHC(CH_3)(CH_2)_2CH:C(CH_3)_2NiBr]_2$	50
			10, 41
		$Ni(CO)_4$	10

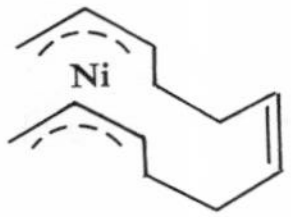

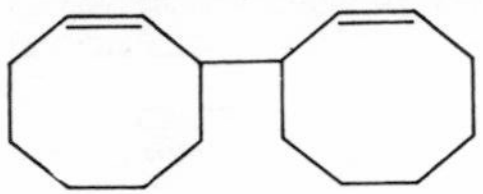

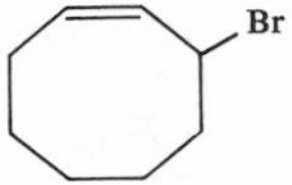

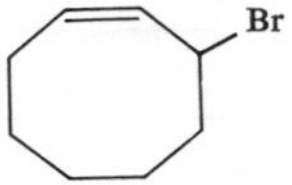

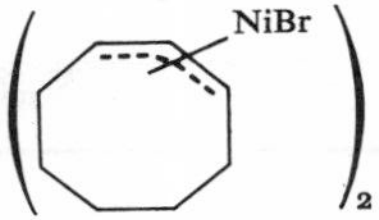

(continued)

TABLE V-1 (*continued*)

Product	Reactant — Organic component	Reactant — Nickel component	Ref.
cis- and trans-CH₂:CH(CH₂)₂CH:CHCN, CH₂:CHCHCNCH₂CH:CH₂	cis-NCCH:CHCH₂Br	(π-CH₂CHCH₂NiBr)₂	16
	trans-NCCH:CHCH₂Br	(π-CH₂CHCH₂NiBr)₂	16
	CH₂:CHCH₂Br	(π-CH₂CHCHCNNiCl)₂	15, 16
(NCCH₂CH:CHCH₂)₂	NCCH₂CH:CHCH₂Cl	Ni(CO)₄	P2
cis,cis-; cis,trans-; and trans,trans-(NCCH:CHCH₂)₂	cis- and trans-NCCH:CHCH₂Cl	Ni(CO)₄	34, P4
(CH₃OCH₂CH₂CH:CHCH₂)₂	CH₃OCH₂CH₂CH:CHCH₂Cl	Ni(CO)₄	34, P4
	CH₂:CHCHClCH₂CH₂OCH₃	Ni(CO)₄	26, 28
trans-(CH₃)₂C:CH(CH₂)₂C(CH₃):CHCH₂OC₂H₅	trans-C₂H₅OCH₂CH:C(CH₃)CH₂Br	[π-(CH₃)₂CCHCH₂NiBr]₂	26, 28
trans,trans- and trans,cis-[C₂H₅OCH₂CH:C(CH₃)CH₂]₂	trans-C₂H₅OCH₂CH:C(CH₃)CH₂Br	Ni(CO)₄	26, 28
[ClCH₂CH(OC₂H₅)CH₂CH:CHCH₂]₂	ClCH₂CH(OC₂H₅)CH₂CH:CHCH₂Cl	Ni(CO)₄	P3
cis- and trans-CH₂:CH(CH₂)₂CH:CHCO₂CH₃	trans-CH₃CO₂CH:CHCH₂Br	(π-CH₂CHCH₂NiBr)₂	16
	CH₂:CHCH₂Br	(π-CH₂CHCH₂CO₂CH₃NiBr)₂	16
cis- and trans-CH₂:CH(CH₂)₂C(CH₃):CHCO₂R, CH₂:CHCH₂C(CO₂R)C(CH₃):CH₂	CH₂:CHCH₂Br	[π-CH₂C(CH₃)CHCO₂RNiBr]₂	15
cis- and trans-(CH₃)₂C:CH(CH₂)₂-C(CH₃):CHCO₂C₂H₅	cis- and trans-C₂H₅OCOCH:C(CH₃)CH₂Br	[π-(CH₃)₂CCHCH₂NiBr]₂	26, 28
cis- and trans-(CH₃)₂C:CH(CH₂)₂-C(CH₃):CHCH₂OCOCH₃	(CH₃)₂C:CHCH₂Cl	[π-CH₂C(CH₃)CHCH₂OCOCH₃NiCl]₂	15, 16
trans,trans- and cis,trans-(CH₃OCOCH:CHCH₂)₂	cis- and trans-CH₃OCOCH₂CH:C(CH₃)CH₂Br	[π-(CH₃)₂CCHCH₂NiBr]₂	26, 28
trans,trans- and cis,trans-(C₂H₅OCOCH:CHCH₂)₂	cis- and trans-CH₃OCOCH:CHCH₂Br	Ni(CO)₄	52
[CH₂:C(CO₂C₂H₅)CH₂]₂	trans-C₂H₅OCOCH:CHCH₂Br	Ni(CO)₄	5, 26, 28
cis,trans-[C₂H₅OCOCH:C(CH₃)CH₂]₂	CH₂:C(CO₂C₂H₅)CH₂Br	Ni(CO)₄	26, 28
[CH₂:C(CO₂C₂H₅)(CH₂)₂]₂C:CH₂	cis- and trans-C₂H₅OCOCH:C(CH₃)CH₂Br	[π-CH₂C(CO₂C₂H₅)CH₂NiBr]₂	5, 50
trans,trans-(C₆H₅CH:CHCH₂)₂	CH₂:C(CH₂Cl)₂	K₄[Ni₂(CN)₆]	20, 22
	trans-C₆H₅CH:CHCH₂Br	Ni(CO)₄	51
	trans-C₆H₅CH:CHCH₂OCOCH₃	Ni(CO)₄	31
(Cl₂C:CHCCl₂)₂	Cl₂C:CHCCl₃	Ni(CO)₄	1
⁺CH₂CH:CHCH₂⁾ₙ(NiCl₂)ₙ	ClCH₂CH:CHCH₂Cl		
	BrCH₂CH:CH(CH₂)₂CH:CHCH₂Br	Ni(CO)₄	42
	BrCH₂CH:CH(CH₂)₂CH:CHCH₂Br	(C₆H₅)₃PNi(CO)₃	42

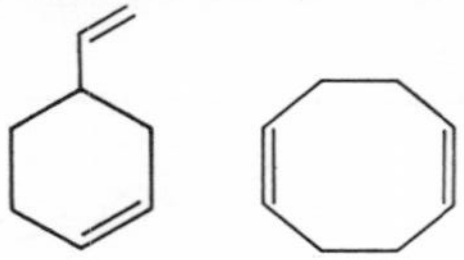

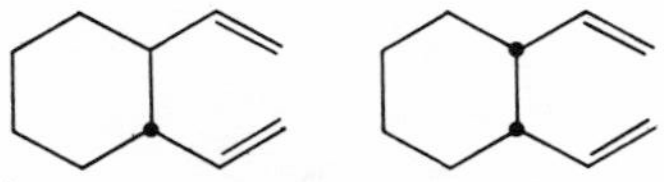

42

4, 41

5
5

5
5
5

7

Ni(CO)$_4$ or (C$_6$H$_5$)$_3$PNi(CO)$_3$

Ni(CO)$_4$

(C$_6$H$_5$)$_3$PNi(CO)$_3$
(C$_6$H$_5$)$_3$PNi(CO)$_3$

Ni(CO)$_4$
Ni(CO)$_4$
Ni(CO)$_4$

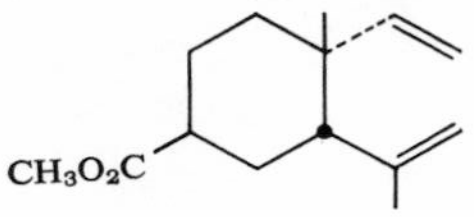

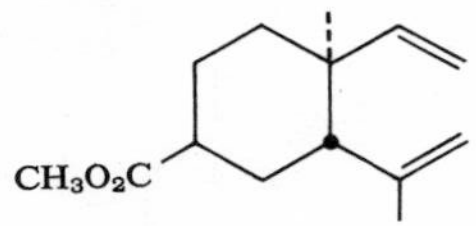

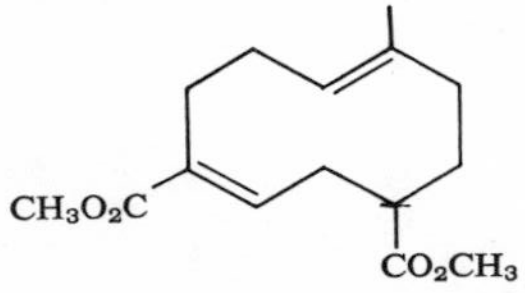

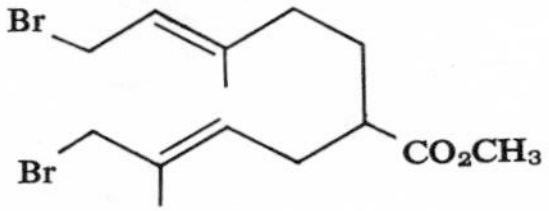

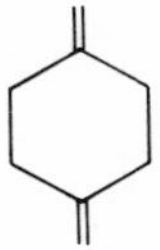

Ni(CO)$_4$

cis,cis- or *trans,trans-*
BrCH$_2$CH:CH(CH$_2$)$_4$CH:CHCH$_2$Br

CH$_2$:C(CH$_2$Cl)$_2$
[ClCH$_2$C(:CH$_2$)CH$_2$]$_2$

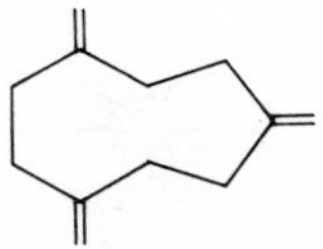

CH$_2$:C(CH$_2$Cl)$_2$
CH$_2$:C[CH$_2$CH$_2$C(:CH$_2$)CH$_2$Cl]$_2$
CH$_2$:C(CH$_2$Cl)$_2$–[ClCH$_2$C(:CH$_2$)CH$_2$]$_2$

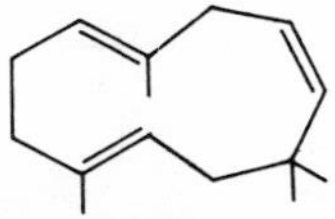

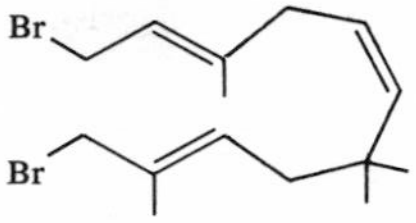

(continued)

TABLE V-1 (*continued*)

Product	Reactant		Ref.
	Organic component	Nickel component	
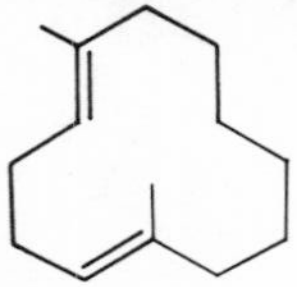	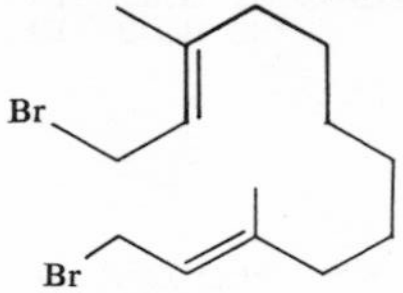	$Ni(CO)_4$	41, 53
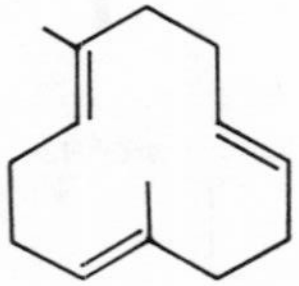	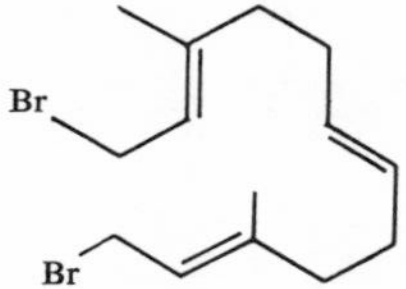	$Ni(CO)_4$	9
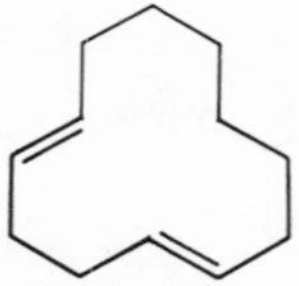	*cis,cis-* or *trans,trans-* $BrCH_2CH:CH(CH_2)_6CH:CHCH_2Br$	$Ni(CO)_4$	41, 42
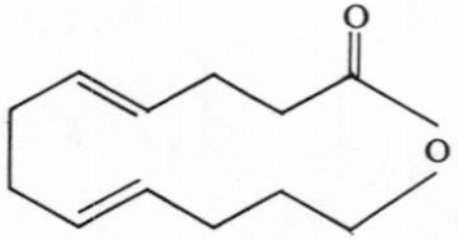	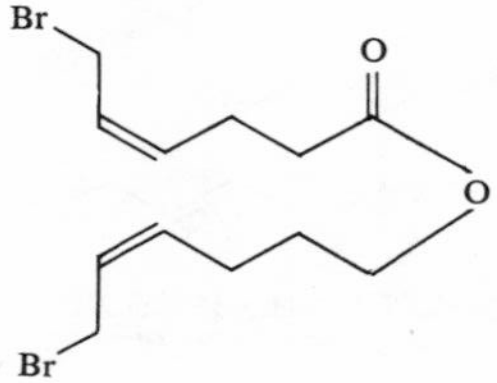	$Ni(CO)_4$	74

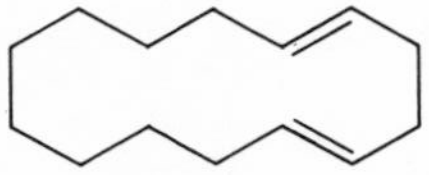

cis,cis- or *trans,trans-* BrCH₂CH:CH(CH₂)₈CH:CHCH₂Br	Ni(CO)₄	41, 42
trans,trans- BrCH₂CH:C(CH₃)(CH₂)₈C(CH₃):CHCH₂Br	Ni(CO)₄	146
	Ni(CO)₄	146
cis,cis- or *trans,trans-* BrCH₂CH:CH(CH₂)₁₂CH:CHCH₂Br	Ni(CO)₄	41, 42
CH₂:CHCH₂Br–HC:CH, CH₂:C(CH₃)CH₂Br–HC:CH	(π-CH₂CHCH₂NiBr)₂, [π-CH₂C(CH₃)CH₂NiBr]₂	17, 17
CH₂:CHCH₂Br–HC:CH	(π-Cyclohexenyl NiBr)₂	17

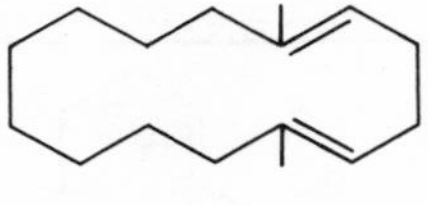

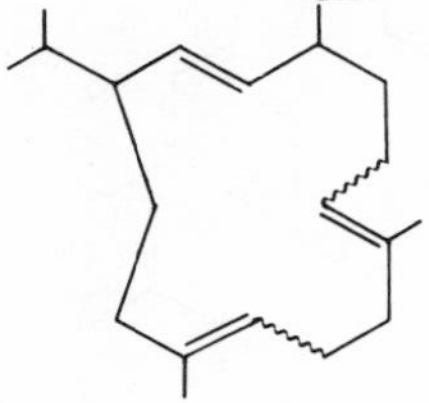

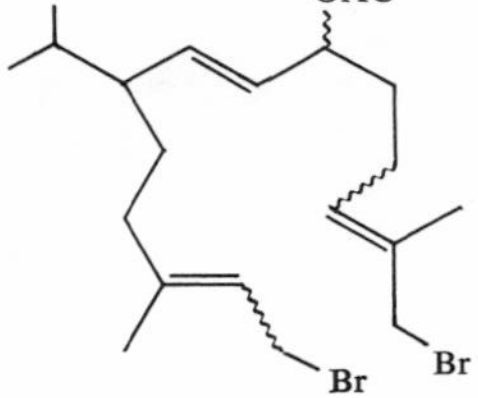

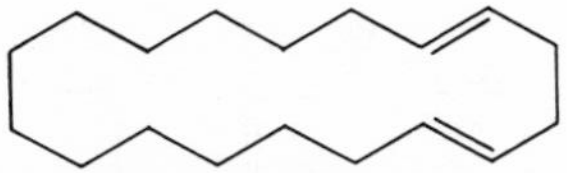

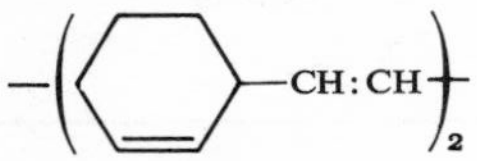

cis,cis-(CH₂:CHCH₂CH:CH)₂
cis,cis-[CH₂:C(CH₃)CH₂CH:CH]₂

cis,cis-(⟨cyclohexene⟩–CH:CH)₂

(continued)

TABLE V-1 (*continued*)

Product	Reactant		Ref.
	Organic component	Nickel component	
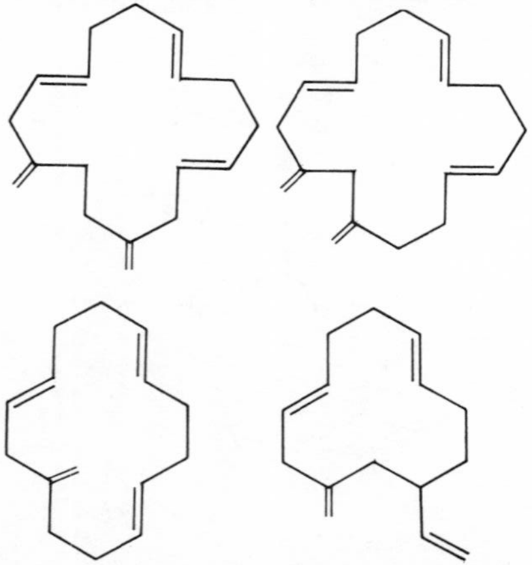	$CH_2:C:CH_2$	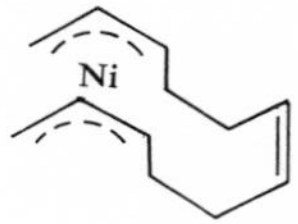	92

An interesting extension of the coupling reaction is the generation of 1,4-diallyl-*cis*,*cis*-1,3-butadiene by the reaction of π-allylnickel bromide with allyl bromide and acetylene: the coupling reaction is preceded by the insertion of two molecules of acetylene (17). A similar reaction has been observed

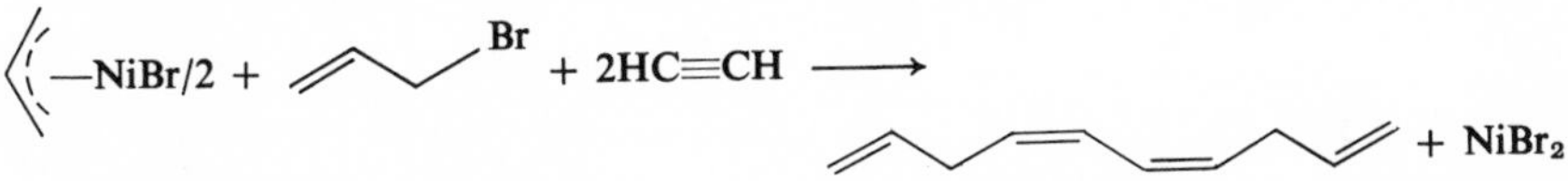

$$\text{—NiBr/2} + \quad \overset{\text{Br}}{\diagup\!\!\diagdown} + \text{2HC}\equiv\text{CH} \longrightarrow \qquad + \text{NiBr}_2$$

between allene and the α,ω-bis(π-allyl)-C_{12}-nickel species **4**: the product is a complex mixture that includes **12** and **13** and results from the insertion of one or two allene molecules (92; see also 137).

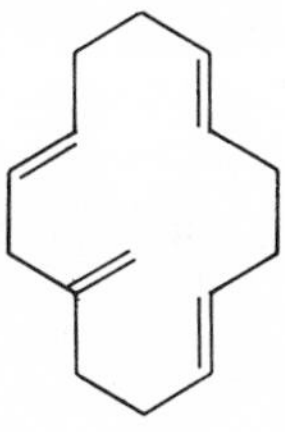

| 12 | 13 |

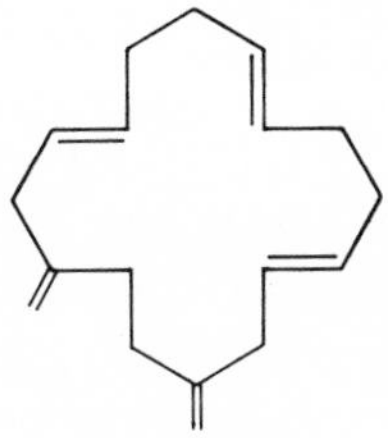

B. The Coupling of Allyl Halides with Non-allylic Halides (*Tables V-2 and V-3*)

The scope of the coupling reaction was greatly increased by the discovery that the π-allylnickel halides react with a wide variety of non-allylic halides,

$$\text{—NiBr/2} + \text{R}'\text{Br} \longrightarrow \qquad \overset{\text{R}'}{} + \text{NiBr}_2$$

and this has introduced a useful method for the selective combination of unlike groups. The preferred coupling with the least substituted end of the π-allyl group that is a characteristic of this reaction has found use in a direct synthesis of α-santalene (**14**), epi-β-santalene (**15**), and vitamin $K_{2(5)}$ (**16**).

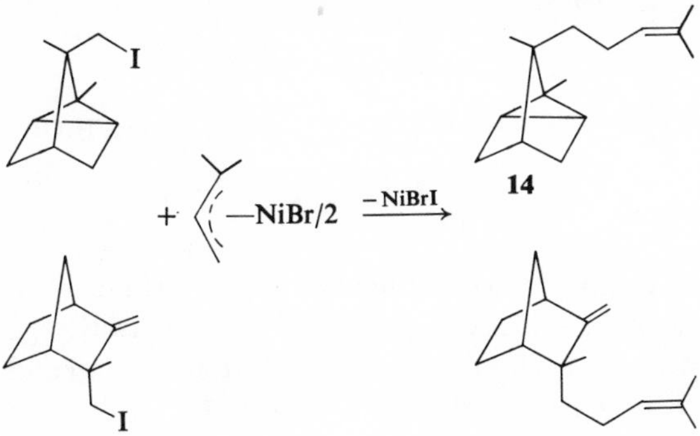

$$+ \quad \text{—NiBr/2} \xrightarrow{\;-\text{NiBrI}\;}$$

14

15

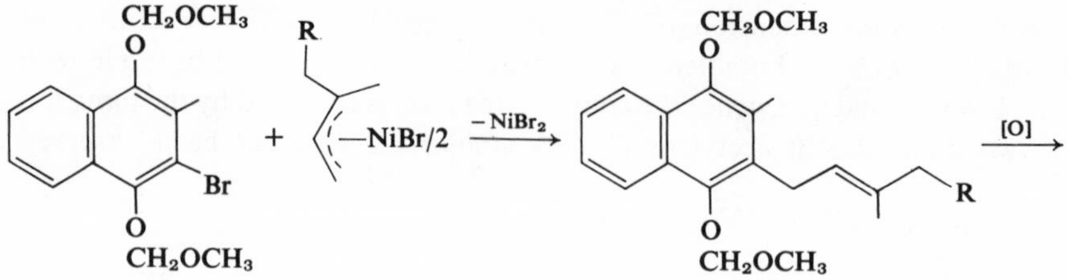

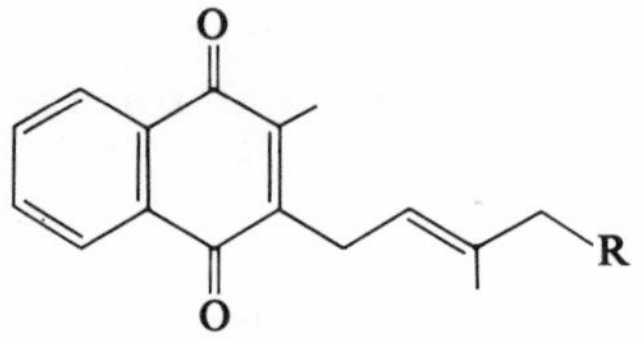

$$R = \{CH_2CH_2CH(CH_3)CH_2\}_3 H$$

16

Active oxygen compounds (e.g., aldehydes and epoxides) also react, yielding unsaturated alcohols.

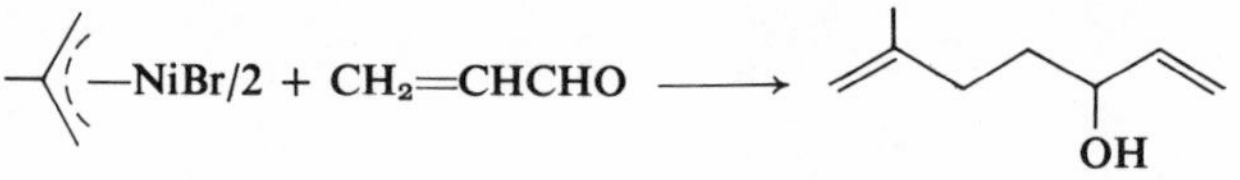

The analogous reaction with π-2-methoxyallylnickel bromide has been used to introduce the acetonyl functional group (139). Bis(π-allyl)nickel complexes

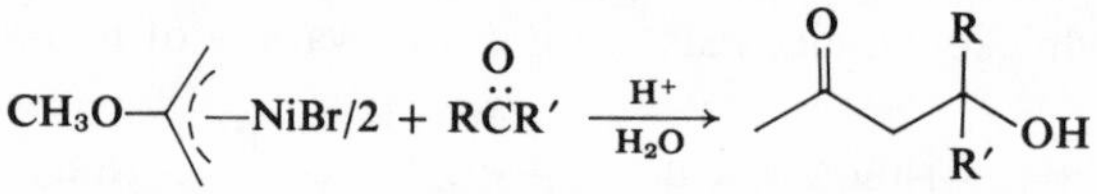

react similarly and in the reaction with benzaldehyde an intermediate alkoxide (**17**) has been isolated and shown to hydrolyze to 4-phenyl-1-butene-4-ol (58, P8). This reaction has been extended to that of the α,ω-bis(π-

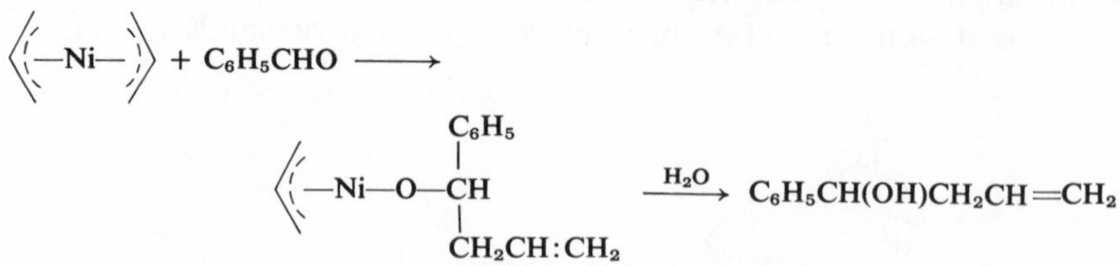

17

allyl)-C_{12}-nickel complex **4** with acetaldehyde, acrolein, and acetyl chloride. The reaction with acetaldehyde produces, after hydrolysis, a mixture of alcohols in which **18** predominates. The addition of a second molecule of acetaldehyde to give **19** has also been observed, and one of the products of the reaction of **4** with a mixture of acetaldehyde and allyl bromide is **20** (91).

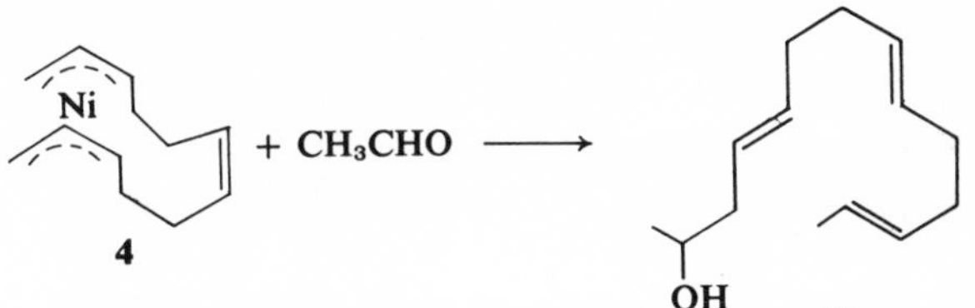

The reaction of butadiene and acetone to give 2-methyl-4,7,9-decatriene-2-ol, which is catalyzed by $\text{Ni(naphth)}_2\text{–RMgX–P(OR)}_3$, is probably related (P18).

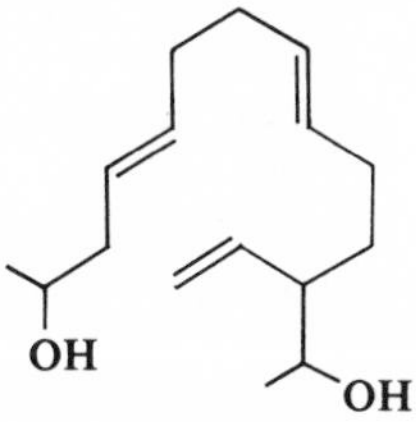

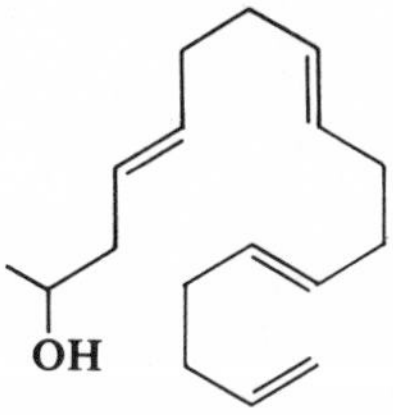

Other reactions involving π-allyl complexes include that of π-allylnickel bromide with ethyldiazoacetate, which produces *trans*-vinylacrylate; an intermediate carbene is believed to be generated by loss of nitrogen (24). A

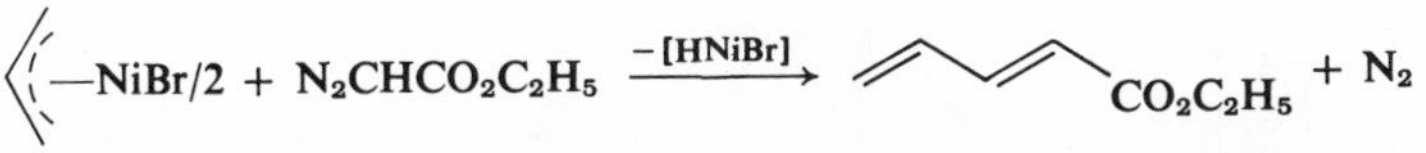

further example is probably to be found in the reaction of allyl alcohols with morpholine, which is catalyzed by $[(C_4H_9)_3P]_2\text{NiBr}_2\text{–KO}tert\text{-}C_4H_9$ and in which allylamines are formed (90). Related to these reactions is the irreversible

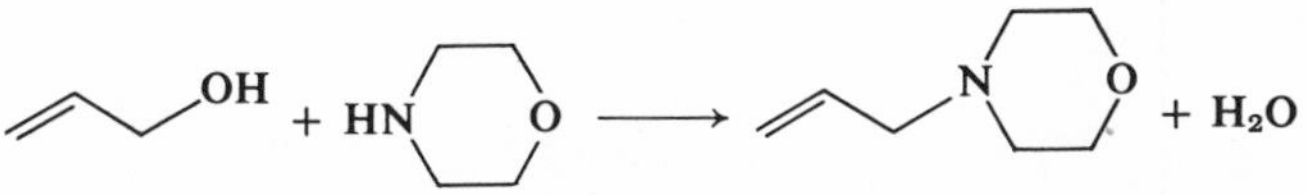

transfer of a halogen atom to a π-allyl group that has been observed on treating π-allylnickel bromide with a fluorophosphine (93). A convenient method

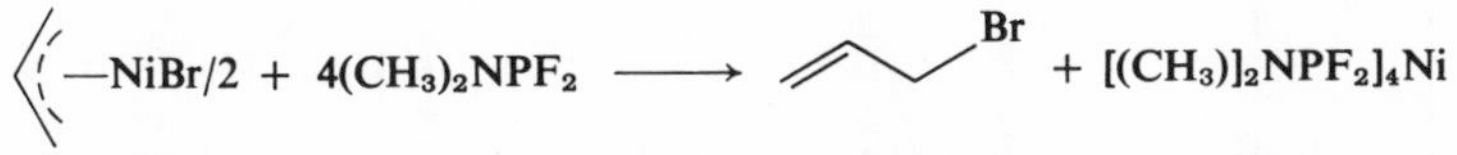

for regenerating protected hydroxyl or amino functions is based on the cleavage of the allyloxycarbonyl group from oxygen or nitrogen by treatment with nickel tetracarbonyl (94).

$$\text{ROH} + \text{CH}_2{=}\text{CHCH}_2\text{OCOCl} \xrightarrow{-\text{HCl}} \text{CH}_2{=}\text{CHCH}_2\text{OCOOR} \xrightarrow{\text{Ni(CO)}_4} \text{ROH}$$

$$(cyclo\text{-}C_6H_{11})_2\text{NCOOCH}_2\text{CH}{=}\text{CH}_2 \xrightarrow{\text{Ni(CO)}_4} (cyclo\text{-}C_6H_{11})_2\text{NH}$$

TABLE V-2
THE COUPLING OF ALLYLIC COMPOUNDS WITH ORGANIC HALIDES

Product	Reactant		Ref.
	Organic component	Nickel component	
$CH_2:C(CH_3)CH_2CH_3$	CH_3I	$[\pi\text{-}CH_2C(CH_3)CH_2NiBr]_2$	6
$(CH_3)_2C:CHCH_2CH_3$	CH_3Br	$[\pi\text{-}CH_2C(CH_3)CH_2NiBr]_2$	6
$CH_2:C(CH_3)CH_2C(CH_3)_3$	CH_3I	$[\pi\text{-}CH_2CHC(CH_3)_2NiBr]_2$	6
	$(CH_3)_3Cl$	$[\pi\text{-}CH_2C(CH_3)CH_2NiBr]_2$	6
$CH_2:C(CH_3)CH_2CH_2CH:CH_2$	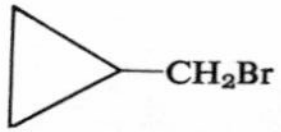 CH_2Br (cyclopropane)	$[\pi\text{-}CH_2C(CH_3)CH_2NiBr]_2$	50
$CH_2:CHCH_2CH_2CH_2CH(CH_3)_2$	$CH_2:CHCH_2Br\text{-}(CH_3)_2CHCH_2CH_2I$	$Ni(CO)_4$	26, 28
$CH_2:C(CO_2C_2H_5)CH_2CH_2CH_2CH_2Cl$	$ICH_2CH_2CH_2Cl$	$[\pi\text{-}CH_2C(CO_2C_2H_5)CH_2NiBr]_2$	6
$CH_2:C(CH_3)CH_2$—(cyclohexyl, I) 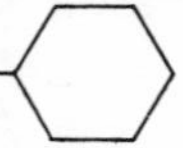	 (cyclohexyl-I)	$[\pi\text{-}CH_2C(CH_3)CH_2NiBr]_2$	6
cis- and trans-$(CH_3)_2C:CHCH_2$- (cyclohexyl)			
$CH_2C(CH_3):CHCH_2$—(cyclohexyl)	(cyclohexyl-I)	$[\pi\text{-}CH_2CHC(CH_3)CH_2CH_2CH_2CH:C(CH_3)_2NiBr]_2$	6, 41
cis- and trans-$(CH_3)_2C:CH(CH_2)_2$- (cyclohexyl)			
$C(CH_4):CHCH_2$—(cyclohexyl)	(cyclohexyl-I)	$[\pi\text{-}CH_2CHC(CH_3)CH_2CH_2CH:C(CH_3)_2NiBr]_2$	50
CH_3COCH_2—(cyclohexyl)			
$CH_2:C(CH_3)CH_2$—(cyclohexyl-OH)	(cyclohexyl, HO, I)	$[\pi\text{-}CH_2C(OCH_3)CH_2NiBr]_2$	139
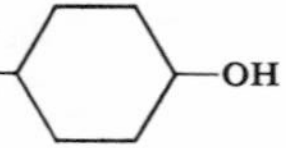	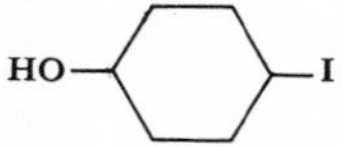	$[\pi\text{-}CH_2C(CH_3)CH_2NiBr]_2$	6

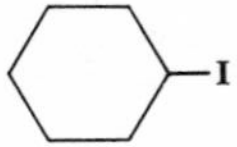

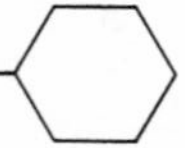

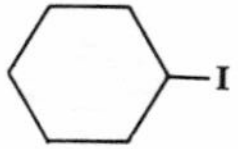

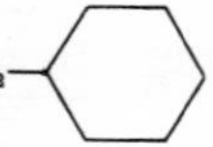

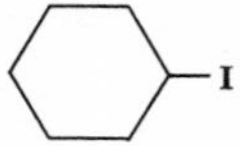

258

CH$_3$COCH$_2$CH(CH$_3$)(CH$_2$)$_5$CH$_3$	CH$_3$(CH$_2$)$_5$CH(CH$_3$)I	[π-CH$_2$C(OCH$_3$)CH(CH$_3$)CH$_2$NiBr]$_2$	139
Geranylacetone	Geranyl bromide	[π-CH$_2$C(OCH$_3$)CH$_3$)CH$_2$NiBr]$_2$	139
CH$_2$:C(CH$_3$)CH$_2$COCH$_3$	CH$_3$C(OCH$_3$)CH$_2$Cl	[π-CH$_2$C(CH$_3$)CH$_2$NiBr]$_2$	6
CH$_2$:C(CH$_3$)CH$_2$COC$_6$H$_4$-p-Br	p-Br-C$_6$H$_4$COCH$_2$Br	[π-CH$_2$C(CH$_3$)CH$_2$NiBr]$_2$	50
CH$_3$SCH$_2$Cl	CH$_3$SCH$_2$Cl	[π-CH$_2$(CH$_3$)CH$_2$NiBr]$_2$	6
CH$_2$:C(CH$_3$)CH$_2$CH$_2$SCH$_3$	C$_6$H$_5$OCH$_2$Cl	[π-CH$_2$C(CH$_3$)CH$_2$NiBr]$_2$	50
CH$_2$:C(CH$_3$)CH$_2$CH$_2$OC$_6$H$_5$	C$_6$H$_5$SCH$_2$Cl	[π-CH$_2$C(CH$_3$)CH$_2$NiBr]$_2$	6
CH$_2$:C(CH$_3$)CH$_2$CH$_2$SC$_6$H$_5$	C$_6$H$_5$CH$_2$Br	[π-CH$_2$C(OCH$_3$)CH$_2$NiBr]$_2$	139
CH$_3$COCH$_2$CH$_2$C$_6$H$_5$	C$_6$H$_5$CH$_2$Br	[π-CH$_2$C(OCH$_3$)CH$_2$NiBr]$_2$	6
CH$_2$:C(CH$_3$)CH$_2$CH$_2$C$_6$H$_5$	CH$_2$:C(CH$_3$)CH$_2$C$_6$H$_5$	[π-CH$_2$C(CH$_3$)CH$_2$NiBr]$_2$	139
CH$_2$:C(CH$_3$)CH$_2$C(CO$_2$C$_2$H$_5$):CH$_2$	CH$_2$:C(CO$_2$C$_2$H$_5$)CH$_2$Br	[π-CH$_2$C(OCH$_3$)CH$_2$NiBr]$_2$	6
CH$_3$:C(CH$_3$)CH$_2$CH:CH$_2$	CH$_2$:CHBr	[π-CH$_2$C(CH$_3$)CH$_2$NiBr]$_2$	139
CH$_3$COCH$_2$CH:CHCH$_3$, CH$_3$COCH:CHCH$_2$CH$_3$	CH$_3$CH:CHBr	[π-CH$_2$C(OCH$_3$)CH$_2$NiBr]$_2$	139
CH$_3$COCH$_2$CH:CHC$_6$H$_5$	C$_6$H$_5$CH:CHBr	[π-CH$_2$C(OCH$_3$)CH$_2$NiBr]$_2$	139
CH$_2$:CHCH$_2$C$_6$H$_5$	C$_6$H$_5$I-CH$_2$:CHCH$_2$Br	Ni(CO)$_4$	26, 28
CH$_2$:C(CH$_3$)CH$_2$C$_6$H$_5$	C$_6$H$_5$I	[π-CH$_2$C(OCH$_3$)CH$_2$NiBr]$_2$	6
	C$_6$H$_5$Br	Ni(CO)$_4$	39
	C$_6$H$_5$I-CH$_2$:C(CH$_3$)CH$_2$Br	[π-CH$_2$C(CH$_3$)CH$_2$NiBr]$_2$	26, 28
CH$_3$CH:CHCH$_2$C$_6$H$_5$	C$_6$H$_5$I-CH$_3$CH:CHCH$_2$Br	Ni(CO)$_4$	26, 28
(CH$_3$)$_2$C:CHCH$_2$C$_6$H$_5$	C$_6$H$_5$I-(CH$_3$)$_2$C:CHCH$_2$Br	Ni(CO)$_4$	26, 28
CH$_3$COCH$_2$C$_6$H$_5$	C$_6$H$_5$I	[π-CH$_2$C(OCH$_3$)CH$_2$NiBr]$_2$	139
CH$_2$:C(CH$_3$)CH$_2$ ⟨C$_6$H$_4$⟩ CH$_2$C(CH$_3$):CH$_2$	(Br⟨C$_6$H$_4$⟩Br)	[π-CH$_2$C(CH$_3$)CH$_2$NiBr]$_2$	16
	(I⟨C$_6$H$_4$⟩I)	[π-CH$_2$C(CH$_3$)CH$_2$NiBr]$_2$	16
6′-Acetonylpapaverine	6′-Bromopapaverine	[π-CH$_2$C(OCH$_3$)CH$_2$NiBr]$_2$	139
(bicyclic structure)	(bicyclic iodide structure)	[π-(CH$_3$)$_2$CCHCH$_2$NiBr]$_2$	6
(bicyclic methylene structure)	(bicyclic methylene iodide structure)	[π-(CH$_3$)$_2$CCHCH$_2$NiBr]$_2$	6

259

(continued)

TABLE V-2 (*continued*)

Product	Reactant		Ref.
	Organic component	Nickel component	

Reactant:
Organic component — steroid with OCOCH₃ group; —CH₃I; Ni(CO)₄ — Ref. 21

Organic component — aromatic ring with OCOCH₃, Br, OCOCH₃ substituents; [π-(CH₃)₂CCHCH₂NiBr]₂ — Ref. P17

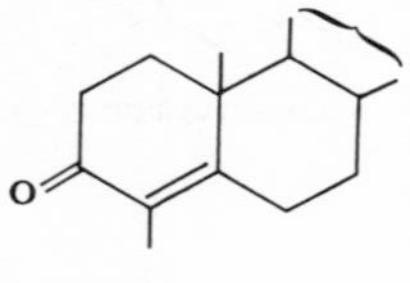

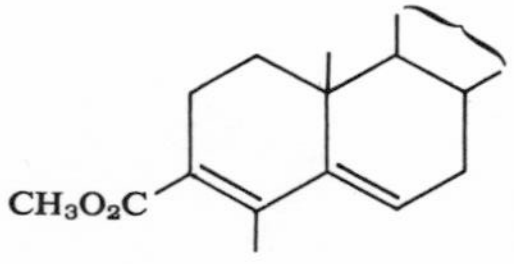

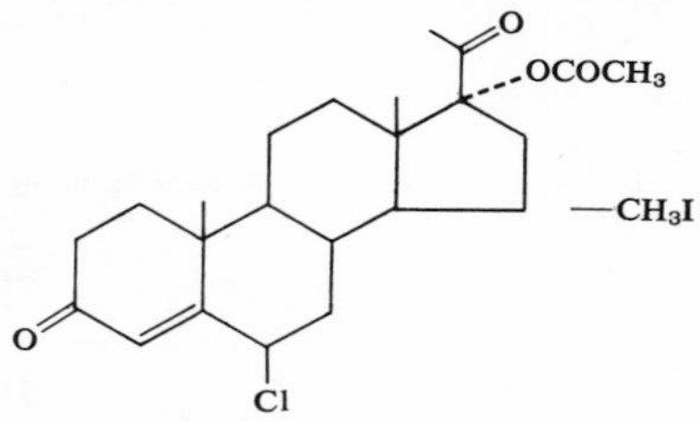

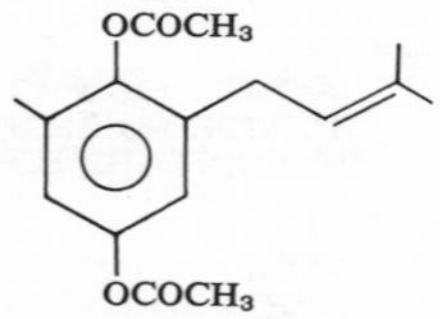

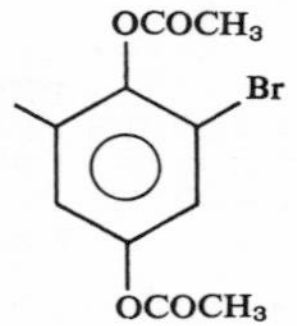

Organic component — naphthalene with OCH₂OCH₃, Br, OCH₂OCH₃ substituents; [π-(CH₃)₂CCHCH₂NiBr]₂ — Ref. 27, 49

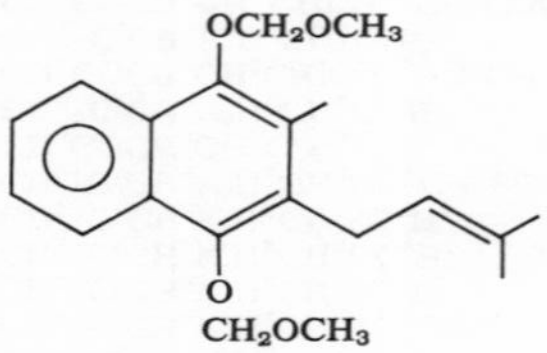

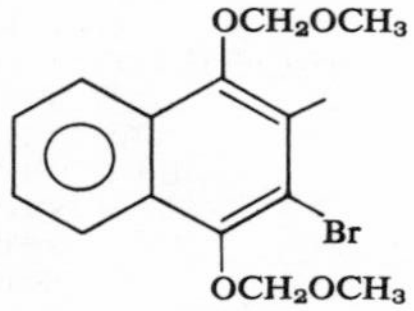

260

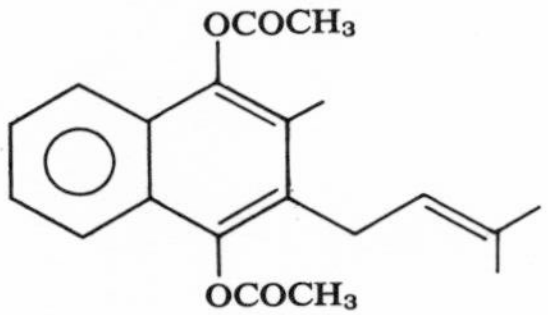

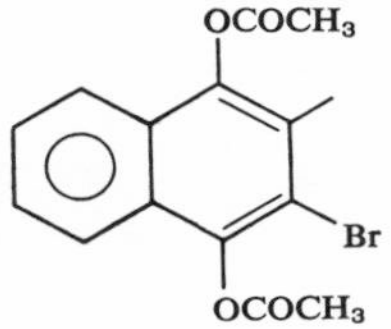

261

$[\pi\text{-}(CH_3)_2CCHCH_2NiBr]_2$ — 27, 49

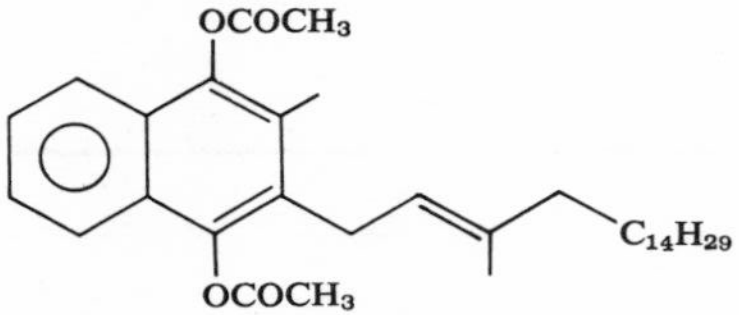

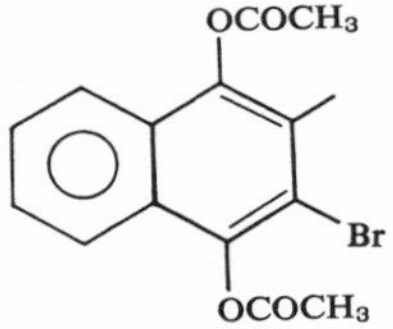

$[\pi\text{-}CH_3C(C_{15}H_{31})CHCH_2NiBr]_2$ — 27, 49

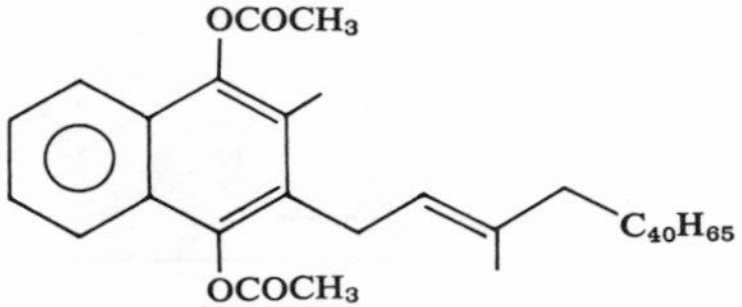

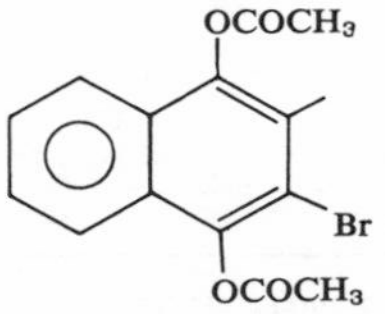

$[\pi\text{-}CH_3C(C_{41}H_{67})CHCH_2NiBr]_2$ — 49

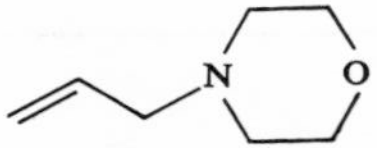

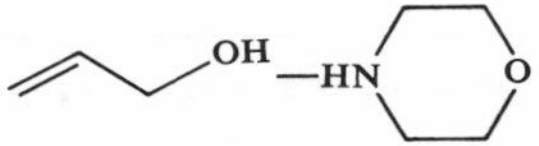

$[(C_4H_9)_3P]_2NiBr_2\text{-}KOtert\text{-}C_4H_9$ — 90

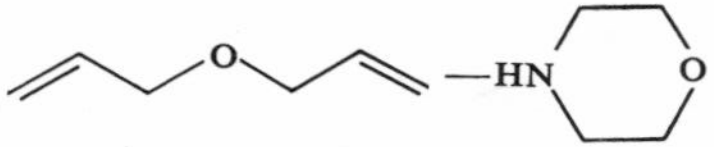

$[(C_4H_9)_3P]_2NiBr_2\text{-}KOtert\text{-}C_4H_9$ — 90

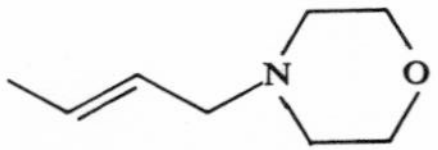

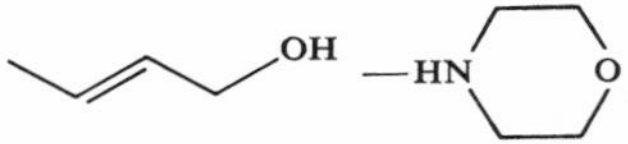

$[(C_4H_9)_3P]_2NiBr_2\text{-}KOtert\text{-}C_4H_9$ — 90

(continued)

TABLE V-2 (*continued*)

Product	Reactant		Ref.
	Organic component	Nickel component	
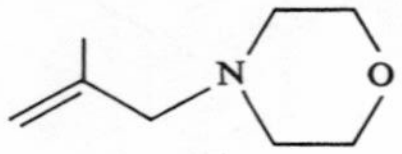 (morpholine N-methallyl structure)	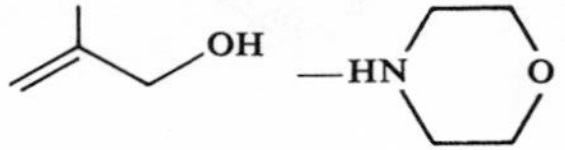 (2-methylallyl alcohol with morpholine, —HN O)	$[(C_4H_9)_3P]_2NiBr_2$–KO$tert$-C_4H_9	90
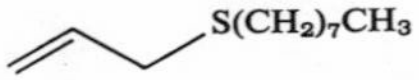 S(CH₂)₇CH₃	 OH —CH₃(CH₂)₇SH	$[(C_4H_9)_3P]_2NiBr_2$–KO$tert$-C_4H_9	90
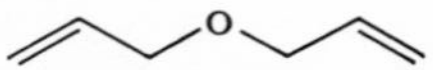 (diallyl ether structure)	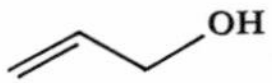 OH	Ni(acac)₂–P(C₄H₉)₃–NaBH₄	90

262

TABLE V-3

THE REACTION OF π-ALLYLNICKEL COMPLEXES WITH ACTIVE OXYGEN COMPOUNDS AND ACTIVATED OLEFINS

Product	Reactant		Ref.
	Organic component	Nickel component	
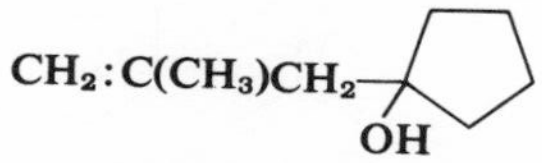 $CH_2{:}C(CH_3)CH_2$— (cyclopentanol)	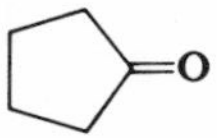 (cyclopentanone)	$[\pi\text{-}CH_2C(CH_3)CH_2NiBr]_2$	6
$CH_2{:}C(CH_3)CH_2CH(OH)CH{:}CH_2$	$CH_2{:}CHCHO$	$[\pi\text{-}CH_2C(CH_3)CH_2NiBr]_2$	6
$CH_3COCH_2CH(OH)(CH_2)_5CH_3$	$CH_3(CH_2)_5CHO$	$[\pi\text{-}CH_2C(OCH_3)CH_2NiBr]_2$	139
$CH_2{:}CHCH_2CH(OH)C_6H_5$	$CH_2{:}CHCH_2Br\text{–}C_6H_5CHO$	Raney Ni	38
$CH_2{:}C(CH_3)CH_2CH(OH)C_6H_5$	C_6H_5CHO	$[\pi\text{-}CH_2C(CH_3)CH_2NiBr]_2$	6, 41
$CH_3COCH_2CH(OH)C_6H_5$	C_6H_5CHO	$[\pi\text{-}CH_2C(OCH_3)CH_2NiBr]_2$	139
3-Acetonyl-5-d-cholestan-3-ol	5-α-Cholestan-3-ol	$[\pi\text{-}CH_2C(OCH_3)CH_2NiBr]_2$	139
$CH_2{:}C(CH_3)CH_2CH(C_6H_5)CH_2OH$	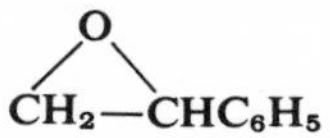 $CH_2{-}CHC_6H_5$ (epoxide)	$[\pi\text{-}CH_2C(CH_3)CH_2NiBr]_2$	6
$CH_3CH(OH)CH_2[CH{:}CH(CH_2)_2]_3CH{:}CH_2$	$CH_3CHO\text{–}CH_2{:}CHCH_2Br$	 Ni (diene complex)	91
$CH_3CH(OH)CH_2[CH{:}CH(CH_2)_2]_2CH{:}CHCH_3$, $CH_3CH(OH)CH_2[CH{:}CH(CH_2)_2]_2CH_2CH{:}CH_2$	CH_3CHO	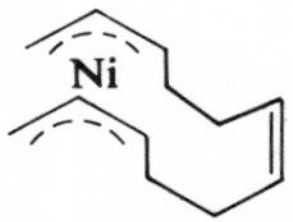 Ni (diene complex)	91

(continued)

TABLE V-3 (continued)

Product	Reactant		Ref.
	Organic component	Nickel component	
$CH_2:CH(OH)CH_2[CH:CH(CH_2)_2]_2CH_2CH:CH_2$	$CH_2:CHCHO$	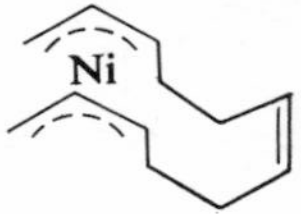	91
$CH_3CH(OH)CH_2[CH:CH(CH_2)_2]_2CH_2CH:CH_2$	CH_3COCl	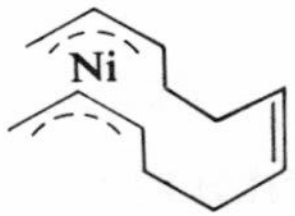	91
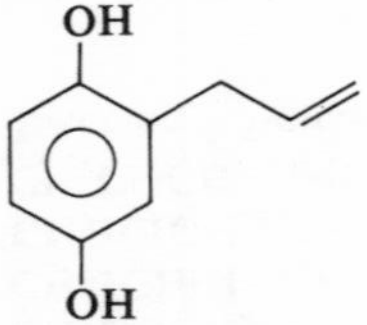		$(\pi\text{-}CH_2CHCH_2NiBr)_2$	18
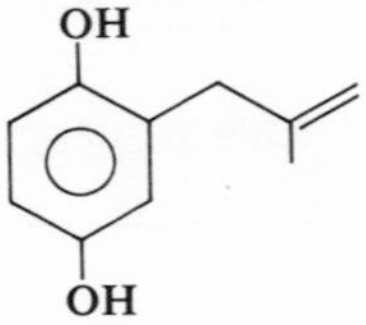	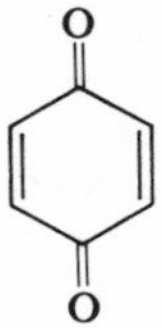	$[\pi\text{-}CH_2C(CH_3)CH_2NiBr]_2$	18
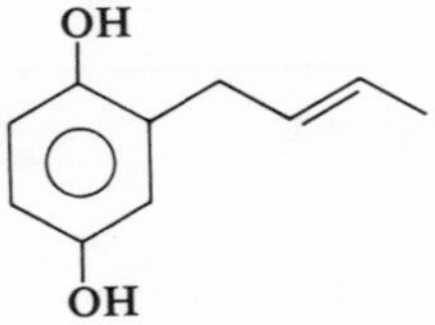		$(\pi\text{-}CH_3CHCHCH_2NiBr)_2$	18

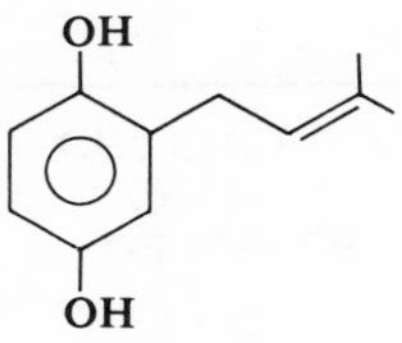

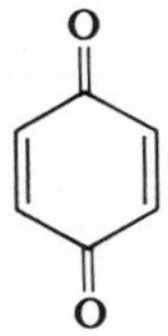

$[\pi\text{-}(CH_3)_2CCHCH_2NiBr]_2$ 18

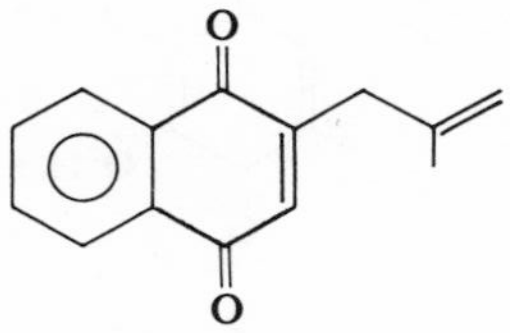

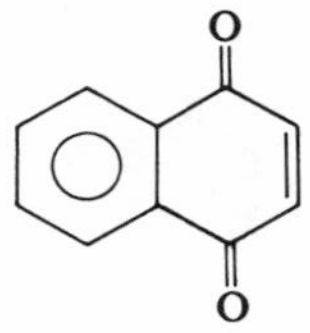

$[\pi\text{-}CH_2C(CH_3)CH_2NiBr]_2$ 18

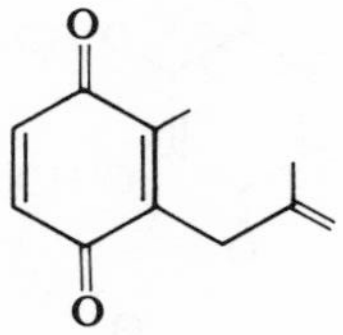

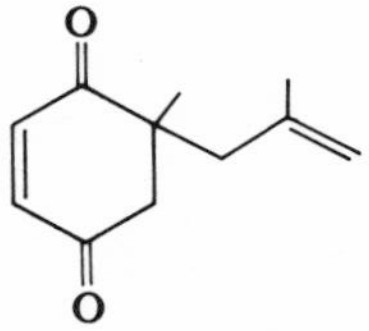

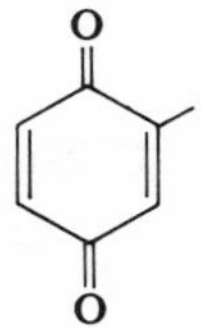

$[\pi\text{-}CH_2C(CH_3)CH_2NiBr]_2$ 18

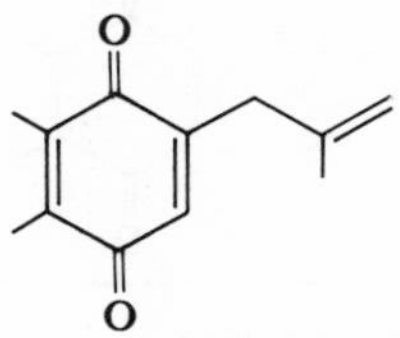

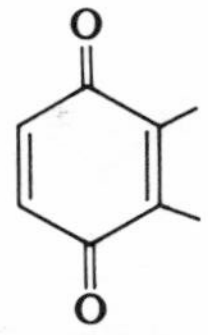

$[\pi\text{-}CH_2C(CH_3)CH_2NiBr]_2$ 18

(continued)

TABLE V-3 (*continued*)

Product	Reactant		Ref.
	Organic component	Nickel component	
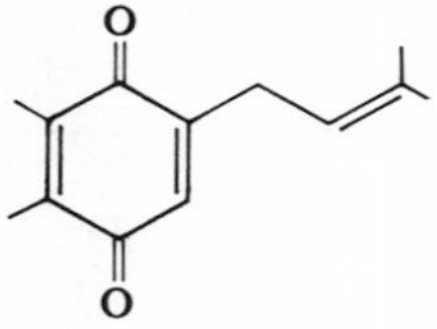	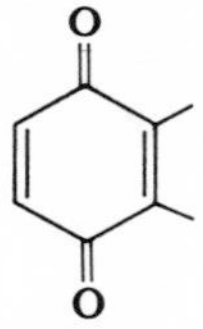	$[\pi\text{-}(CH_3)_2CCHCH_2NiBr]_2$	18
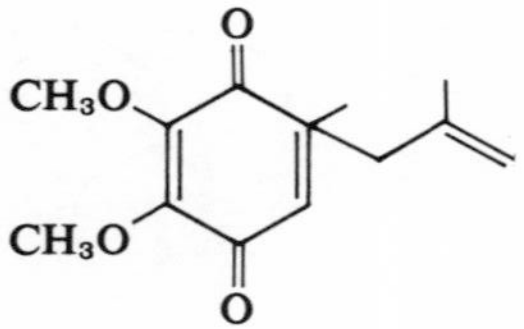	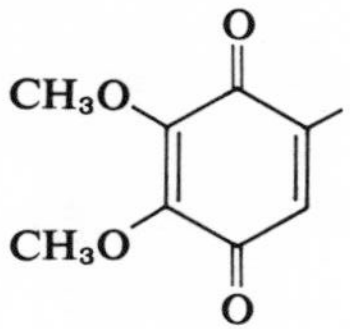	$[\pi\text{-}CH_2C(CH_3)CH_2NiBr]_2$	18
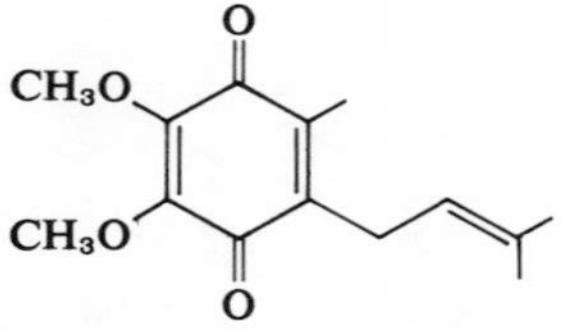	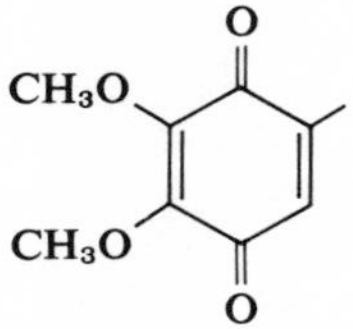	$[\pi(CH_3)_2CCHCH_2NiBr]_2$	18
cis- and $trans$-CH_2:CHCH:CHCO$_2$C$_2$H$_5$	CHN$_2$CO$_2$C$_2$H$_5$	$(\pi\text{-}CH_2CHCH_2NiBr)_2$	24
CH$_2$:CHCH$_2$CH:CHCO$_2$CH$_3$,	CH$_2$:CHCO$_2$CH$_3$	$(\pi\text{-}CH_2CHCH_2NiBr)_2$	55
CH$_2$:CHCH$_2$CH$_2$CH$_2$CO$_2$CH$_3$			

Insertion of activated olefins or alkynes into the π-allylnickel bond has been observed for methylacrylate, acrylonitrile (55, 89), and methyl propiolate (99).

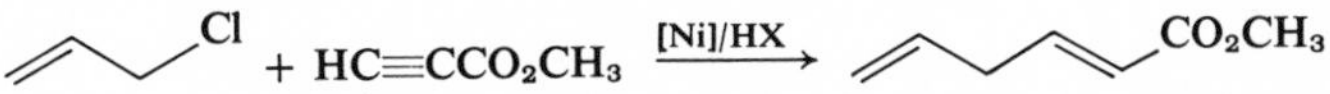

$$\text{allyl-Cl} + HC{\equiv}CCO_2CH_3 \xrightarrow{\text{[Ni]/HX}} \text{chain-}CO_2CH_3$$

Nucleophilic attack of an allyl group is observed in the reactions of π-allyl-nickel halides with quinones (18). This reaction has been used in a one-step synthesis of the coenzyme Q_1 (**21**). Related to these are the reactions of

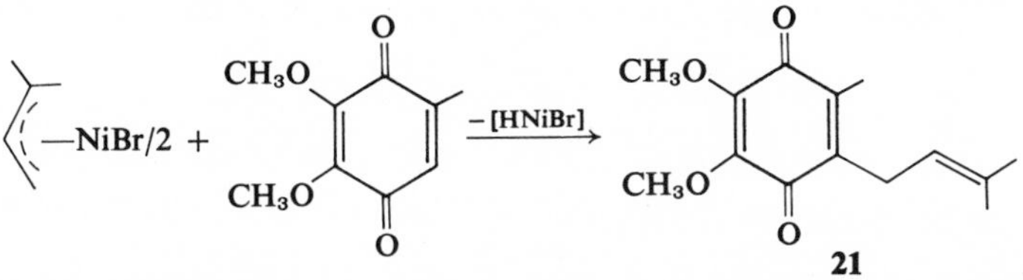

$$\text{—NiBr/2} + \text{(}CH_3O\text{-quinone}CH_3O\text{)} \xrightarrow{-\text{[HNiBr]}} \text{(21)}$$

21

π-allylnickel complexes with acrylonitrile (98) and bicycloheptene (82), from which the organonickel species **22** and **23** have been isolated.

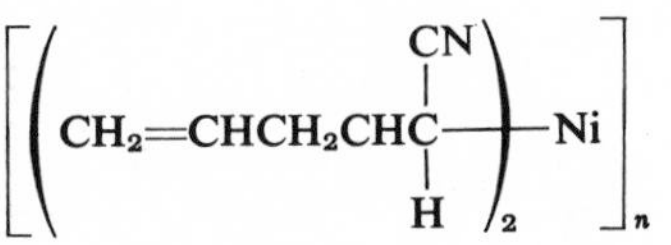

$$\left[\left(CH_2{=}CHCH_2CH\underset{\underset{H}{|}}{\overset{\overset{CN}{|}}{C}}{-}Ni\right)_2\right]_n$$

22

$$\left(\text{bicycloheptene}{-}NiOCOCH_3\right)_2$$

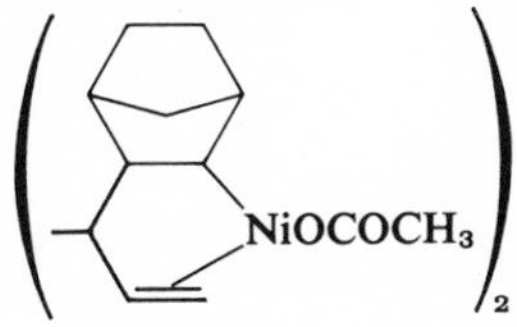

23

C. Mechanistic Considerations

The postulated intermediacy of π-allylnickel species during the reactions involving allyl halides is supported by the fact that π-allylnickel complexes are regularly prepared by reacting allyl halides with zerovalent nickel complexes. Moreover, reaction of these with excess allyl halide produces diallyls. The detailed mechanism of these reactions is not completely understood, however. It is possible that the product is the result of several simultaneously occurring reactions and probable that mechanistic variations exist depending on the ligands (e.g., solvent, CO, COD, etc.) present in solution. Two basic mechanisms can be considered, the first involving π-allylnickel halide intermediates and the second a bis(π-allyl)nickel species. It may be supposed that coupling in both cases is preceded by ligand-induced conversion of a π-allyl group to a σ-allyl group.

Examples from the chemistry of the π-allylnickel complexes can be cited in support of both mechanisms. For instance, complexes (e.g., **26** and **27**) related to **24** have been isolated from the reaction of methallyl bromide with nickel tetracarbonyl (10) or of ammonia with π-allylnickel bromide (64), whereas an olefin complex (**28**) related to **25** is formed on reacting π-methallyl-nickel chloride with bicycloheptene (82). The possibility that ionic inter-

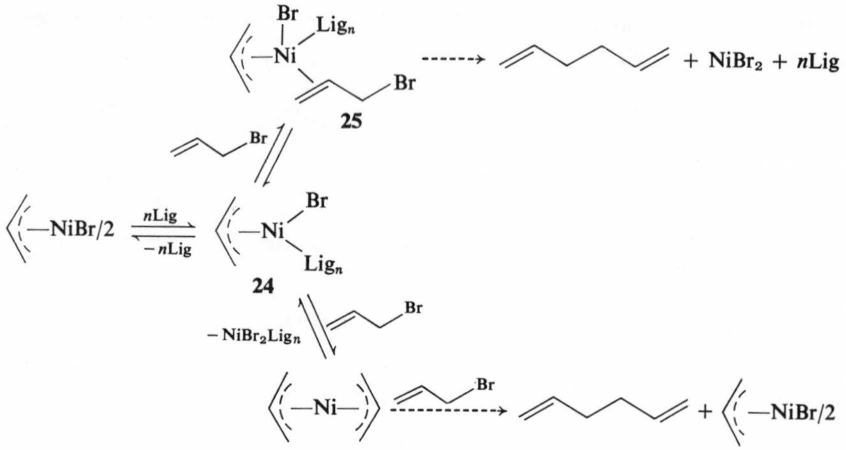

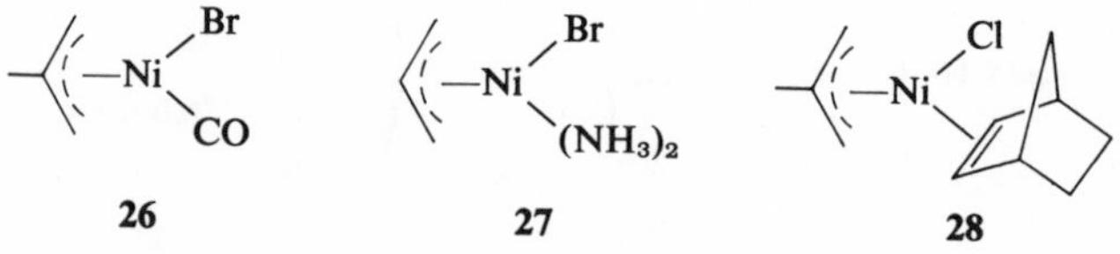

mediates stabilized by polar solvent are involved, e.g., $[\pi\text{-}C_3H_5Ni(Lig)_n]^+X^-$, should also be mentioned. The existence of one such system, $[\pi\text{-}C_3H_5Ni\text{-}(thiourea)_2]^+Cl^-$, has been confirmed by an x-ray structural determination (72, 73). Support for the second mechanism is found in the observed dispro-portionation of π-allylnickel halides in coordinating solvents (e.g., DMF, NH_3) (11, 64, P1): the corresponding tosylates or alkoxides disproportionate spontaneously at room temperature or slightly above (79, 80). The $\pi\text{--}\sigma$ conversion of the allyl group that precedes coupling has been well established and can be instigated either by donor ligands or by coordinating solvents (a detailed discussion is to be found in Volume I, pages 342 and 361). The lack of stereospecificity observed in many of the coupling reactions indicates a facile syn–anti isomerization of the π-allyl group, and various mechanisms for this process are also discussed in Volume I (p. 360).

The observed formation of symmetrical products, as well as cross-coupling products, from the reaction of a π-allylnickel halide or a bis(π-allyl)nickel complex with a second, different, allyl halide indicates a further complication; namely, that a facile exchange reaction takes place (10). The first of these exchange reactions accounts for the observed scrambling of the labeled C atom in $CH_2:CH^{14}CH_2Br$ after contact with nickel tetracarbonyl (14, 62).

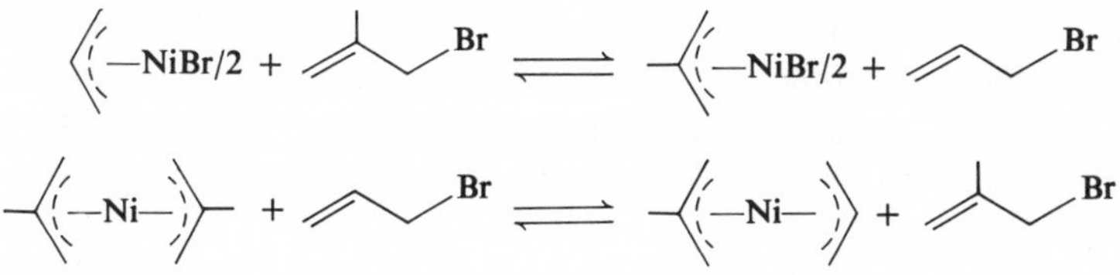

A clear distinction between the two basic mechanisms is seldom possible. In the case of cyclooctenyl bromide, a reaction proceeding exclusively through a bis(π-allyl)nickel complex can be excluded because it has been shown that bis(π-cyclooctenyl)nickel in DMF reacts with CO to give a mixture of the ketone **29** and bis(cyclooctenyl) (**30**), whereas reaction of π-cyclooctenyl-nickel bromide with CO in the same solvent or of cyclooctenyl bromide with nickel tetracarbonyl produces only bis(cyclooctenyl) (10, 65, 81). The reaction

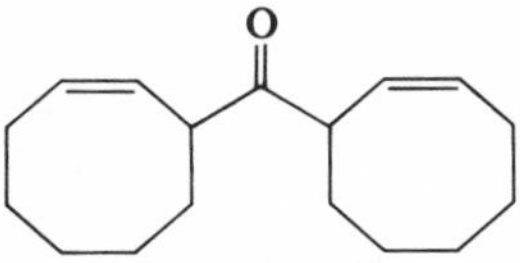

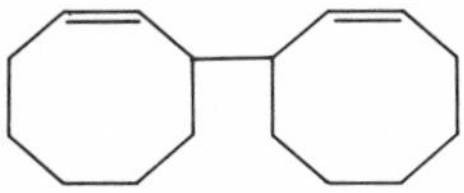

of the C_8 dibromide, $BrCH_2CH:CH(CH_2)_2CH:CHCH_2Br$, with nickel tetracarbonyl in DMF suggests that the bis(π-allyl)nickel mechanism can be of importance; the product of the coupling reaction is not cyclooctadiene; instead, vinylcyclohexene is formed. This is also the principal dimer produced on reacting the bis(π-allyl)-C_8-nickel ligand complex **31** with CO (83).

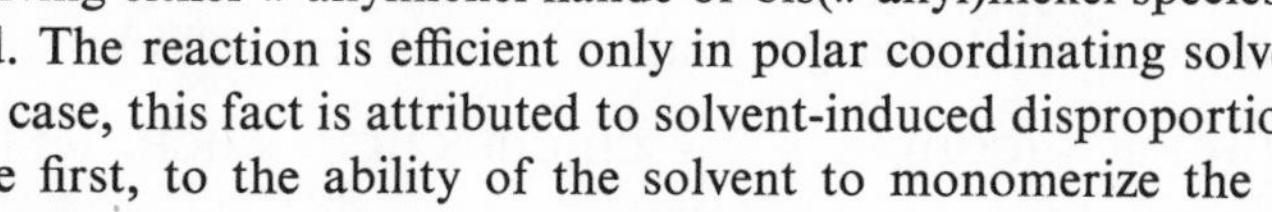

The mechanism of the coupling of an allyl halide with a second, non-allylic, organic halide is even more speculative. As in the previous reaction, mechanisms involving either π-allylnickel halide or bis(π-allyl)nickel species can be formulated. The reaction is efficient only in polar coordinating solvents. In the second case, this fact is attributed to solvent-induced disproportionation, and, in the first, to the ability of the solvent to monomerize the dimeric π-allylnickel halide, to convert a π-allyl group to a σ-allyl group, and to induce coupling between the two allyl fragments.

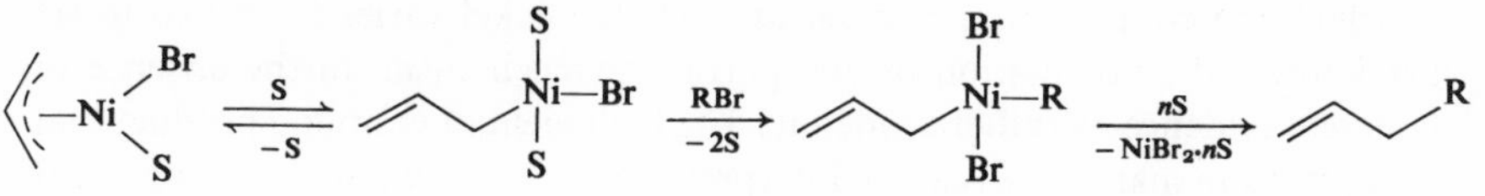

It has been convincingly argued that the reaction of π-allylnickel halides with aldehydes or ketones in DMF to give unsaturated alcohols involves bis(π-allyl)nickel species: these have been shown to take part in the same type of reaction and in one case an intermediate alkoxide (**17**) (see page 256) has been isolated (58, P8).

The alkylation of quinones by π-allylnickel halides is known to involve the initial formation of an unstable adduct and this is presumably followed by 1,4 addition of the π-allylnickel halide to the quinone, with ensuing elimination of an [HNiBr] species (18).

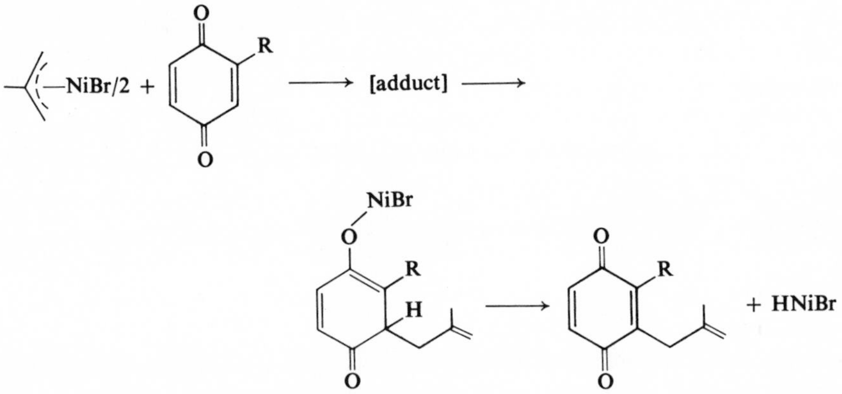

D. The Coupling of Non-allylic Organic Halides and Related Reactions (*Table V-4*)

The coupling reaction is not confined to systems involving allyl halides and has been observed for a number of activated non-allylic halides, e.g., benzyl halides and alkenyl halides. The most commonly used nickel reagents are $(COD)_2Ni$ or $K_4[Ni_2(CN)_6]$. Polar solvents are necessary and the presence of a donor ligand is, in some cases, advantageous. The reaction is limited to the synthesis of symmetrical products and is perhaps of particular interest as a method for synthesizing 1,3-dienes from alkenyl halides.

$$2CH_3CH{=}CHBr + (COD)_2Ni \longrightarrow CH_3CH{=}CHCH{=}CHCH_3 + NiBr_2 + 2COD$$

The reaction probably involves the oxidative addition of the organic halide to the nickel atom. However, because examples of bis(alkyl)nickel complexes in which the coordination number around the nickel varies from two to six are known, the postulation of any particular mechanism, in the absence of concrete evidence, is rather academic. The high yield of coupling product that is observed in many cases can be interpreted as evidence for a cis arrangement of the two σ-bonded organic groups at the nickel atom, which suggests that, where $(COD)_2Ni$ is employed, the COD is acting as a chelating ligand.

Intramolecular coupling to give cyclic products is limited to the reaction of *cis,cis*-1,6-diiodo-1,5-hexadiene with bis(cyclooctadiene)nickel from which 1,3-cyclohexadiene has been isolated (40).

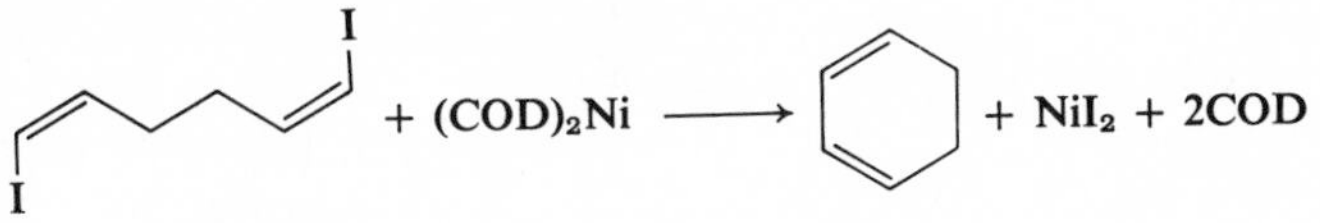

The reaction of phenacyl halides with nickel tetracarbonyl is solvent dependent. In THF the expected coupling product, e.g., **32**, is obtained, whereas in DMF cyclization occurs to give a disubstituted furan, e.g., **33** (35).

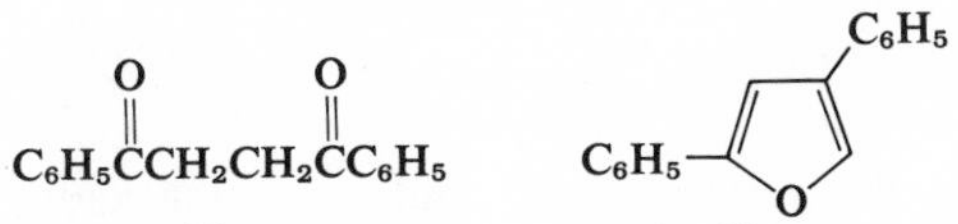

A plausible route for the cyclization reaction is suggested by the observation that α-bromoketones react with nickel tetracarbonyl in DMF to give β-epoxy ketones, which rearrange to furan derivatives on heating (36).

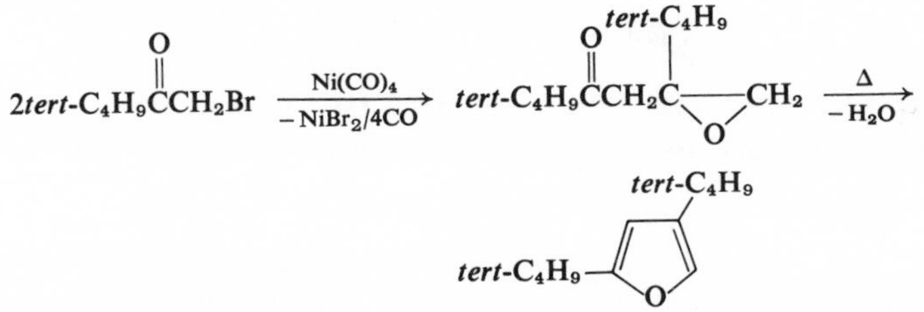

The reaction of benzoyl chloride with nickel tetracarbonyl is complicated by the subsequent reaction of the product (benzil) with benzoyl chloride to give the enediol **34** in THF (2, 61), or the benzoate of benzoin in moist acetone (60). There is some evidence that the benzil coordinates to the nickel

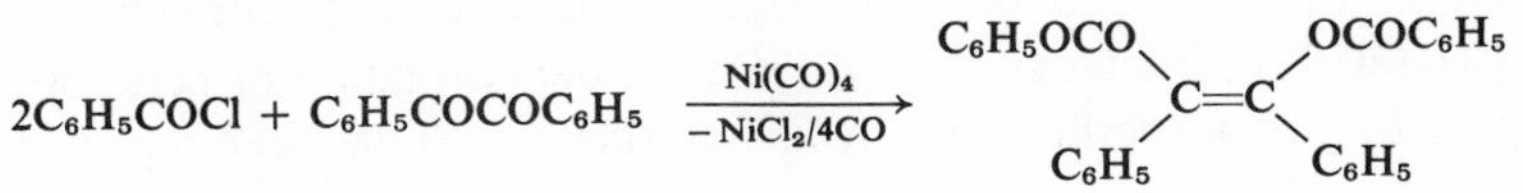

in the second stage of the reaction. Similar behavior has been reported for perfluoroacyl halides (12). Vinylacetyl chloride, however, is decarbonylated to give diallyl. Presumably the intermediate nickel-acyl complex involved in this last reaction is unstable and decomposes liberating CO (76).

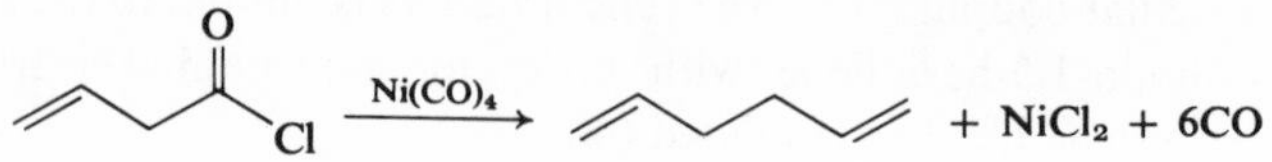

The dehalogenation of 1,2-dihalo systems as a method of introducing unsaturation into a molecule has apparently only been used to synthesize cyclobutene derivatives (e.g., **35**) from the corresponding 1,2-dichlorocyclobutane (69, 70). The reaction takes a different course when carried out with alkenyl-

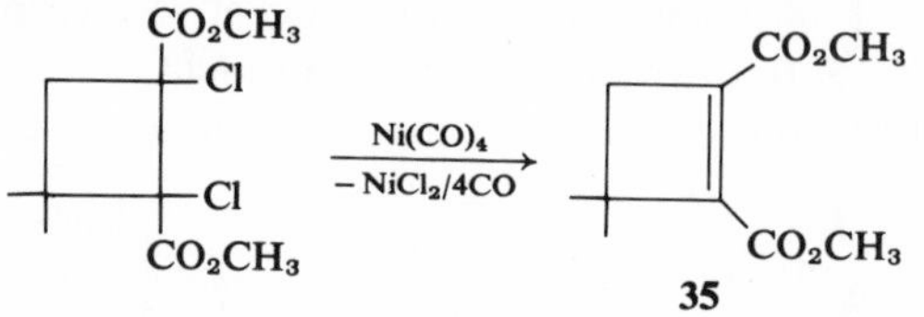

substituted cyclobutane derivatives, and a cyclohexadiene (e.g., **36**) is produced as the result of ring expansion (29). The best known example of

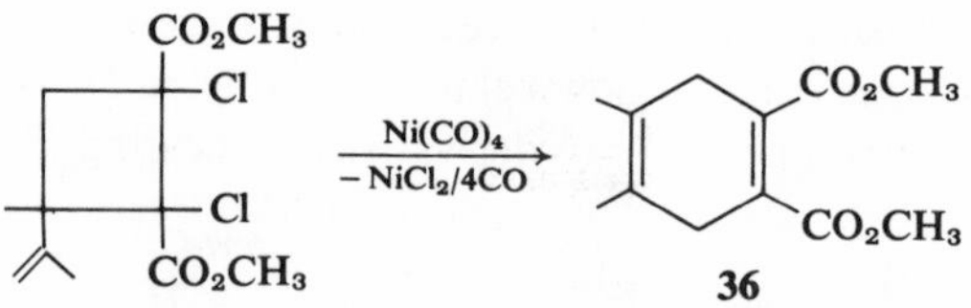

this type of reaction, the synthesis of π-tetramethylcyclobutadienenickel dichloride (71), is discussed with related reactions in Volume I, Chapter VII.

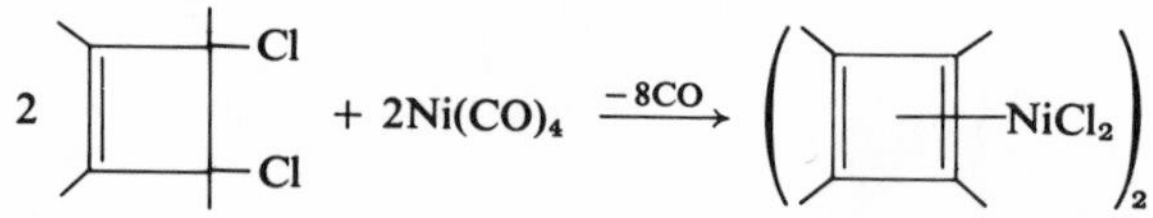

A trivial but historically interesting example of this type of reaction is the formation of carbon monoxide by treating oxalyl chloride with nickel tetracarbonyl (85).

Reactions perhaps related to those discussed in this section include the formation of diphenylbutadiyne from the treatment of lithium phenylacetylene with nickel tetracarbonyl (96); the defluorination of perfluoroazopropane to give bis(perfluoropropylidene)azine, which has been reported in a

$$C_3F_7N{=}NC_3F_7 + Ni(CO)_4 \longrightarrow C_2F_5CF{=}NN{=}CFC_2F_5 + NiF_2 + 4CO$$

patent (P9), and the formation of tetrafluoroethylene by treating CF_2Cl_2 or CF_2Br_2 with nickel tetracarbonyl at room temperature or slightly above (P15).

Active oxygen compounds react with alkyl halides, in a manner analogous to that discussed for the allyl halides, to form alcohols. For example, β-hydroxyphenylpropionitrile is formed on reacting bromo- or iodoacetonitrile

$$\text{BrCH}_2\text{CN} + \text{C}_6\text{H}_5\text{CHO} \xrightarrow{\text{Ni(CO)}_4} \text{C}_6\text{H}_5\text{CHOHCH}_2\text{CN}$$

with benzaldehyde (48). A novel and little explored variation of this reaction is the cyclization of certain haloketones using a reagent prepared by treating nickel tetraphenylporphine (NiTPP) with lithium naphthalide. The reaction can be conducted under extremely mild conditions, allowing, for example, the synthesis of 1-methylcyclopentanol from 6-iodohexanone to be carried out at $-50°$ (75).

An insertion reaction has been observed on treating *trans*-bromostyrene with $K_4[\text{Ni}_2(\text{CN})_6]$ or Ni(CO)_4 and acrylonitrile or ethyl acrylate (19, 20).

$$\text{C}_6\text{H}_5\text{CH}{=}\text{CHBr} + \text{CH}_2{=}\text{CHCN} \xrightarrow{[\text{Ni}_2(\text{CN})_6]^{4-}} \text{C}_6\text{H}_5\text{CH}{=}\text{CHCH}_2\text{CH}_2\text{CN}$$

E. Miscellaneous Reactions

A number of unrelated reactions involving the systems discussed in the previous sections that do not fit readily into the classification used in this chapter should be mentioned. These include a nickel-catalyzed Arbuzov reaction in which trialkyl phosphites are converted, in the presence of nickel salts, into phosphonates. This reaction is catalyzed by nickel halides and

$$\text{RX} + \text{P(OC}_2\text{H}_5)_3 \xrightarrow[\text{ca. 150°}]{\text{NiX}_2} \text{RPO(OC}_2\text{H}_5)_2 + \text{C}_2\text{H}_5\text{X}$$

nickel acetylacetonate and has been reported for allyl halides (P13), vinyl halides (136, P16), and aryl halides (88, P16). Aryl halides have also been reported to react with triphenylphosphine, in the presence of nickel catalysts, to give phosphonium salts (142).

The coupling reactions discussed in Sections II,A, II,B, and II,D are occasionally accompanied by reduction of the organic halide. This is the main reaction observed on reacting allyl halides of the type $\text{RCH:CHCH}_2\text{X}$ (where R is CO_2H, CO_2CH_3, or CN) with nickel tetracarbonyl in aqueous ethanol and is accompanied by an allylic rearrangement (68, 77, 87, 100).

$$\text{HO}_2\text{CCH}{=}\text{CHCH}_2\text{Cl} \xrightarrow{\text{Ni(CO)}_4/\text{H}_2\text{O}} \text{HO}_2\text{CCH}_2\text{CH}{=}\text{CH}_2 + 4\text{CO} + \text{ClNiOH}$$

TABLE V-4

COUPLING OF NON-ALLYLIC ORGANIC HALIDES

Product	Reactant		Ref.
	Organic component	Nickel component	
NC(CH₂)₄CN	NCCH₂CH₂Br	Ni(CO)₄	P6
	NCCH₂CH₂X (X = Cl, Br)	[(C₆H₅)₃P]₂Ni(CO)₂	P5
NCCH₂Si(CH₃)₃	NCCH₂Br–(CH₃)₃SiCl	Ni(CO)₄	78
C₆H₅CH₂CH₂C₆H₅	C₆H₅CH₂X (X = Br, I)	Ni(CO)₄	37, 47
$\left(\text{CH}_3\text{—C}_6\text{H}_4\text{—CH}_2\right)_2$	C₆H₅CH₂Br	K₄[Ni₂(CN)₆]	19, 20, 22
$\left(\text{CH}_3\text{O—C}_6\text{H}_4\text{—CH}_2\right)_2$	CH₃—C₆H₄—CH₂Br	K₄[Ni₂(CN)₆]	20
$\left(\text{NC—C}_6\text{H}_4\text{—CH}_2\right)_2$	CH₃O—C₆H₄—CH₂Br	K₄[Ni₂(CN)₆]	20
$\left(\text{—CH}_2\text{—C}_6\text{H}_4\text{—CH}_2\text{—}\right)_n$ (NiCl₂)ₙ	NC—C₆H₄—CH₂Br	K₄[Ni₂(CN)₆]	20
(C₆H₅COCH₂)₂	ClCH₂—C₆H₄—CH₂Cl	Ni(CO)₄	1
	C₆H₅COCH₂Br	Ni(CO)₄	35
	C₆H₅COCH₂Br	K₄[Ni₂(CN)₆]	20
$\left(\text{CH}_3\text{—C}_6\text{H}_4\text{—COCH}_2\right)_2$	CH₃—C₆H₄—COCH₂Br	Ni(CO)₄	35

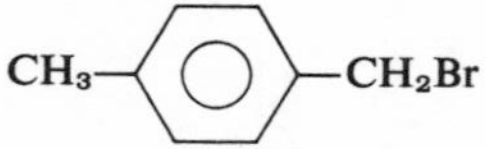

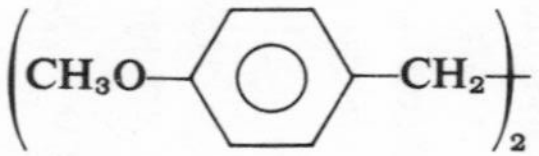

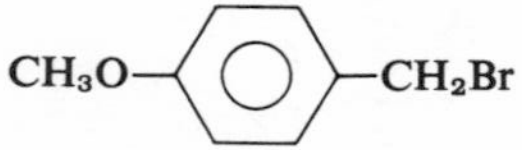

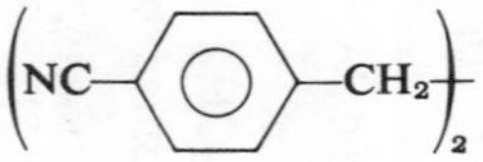

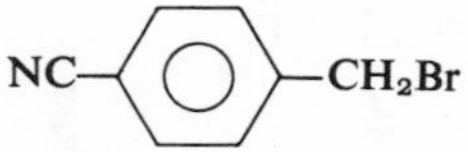

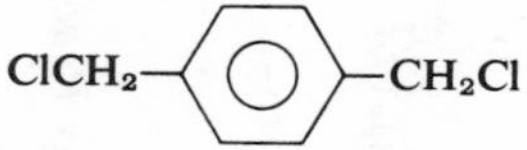

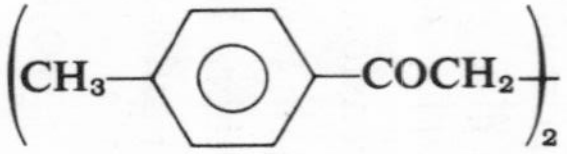

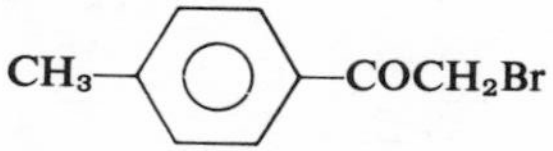

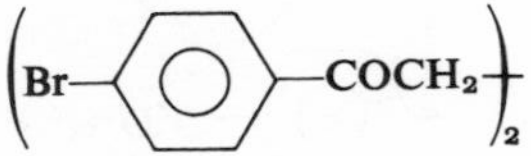

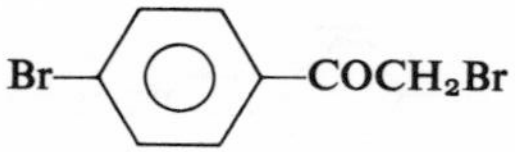

Halide	Product	Catalyst	Reference
(Br—⬡—COCH₂—)₂	Br—⬡—COCH₂Br	K₄[Ni₂(CN)₆]	20
C₆H₅CCl₂CCl₂C₆H₅	C₆H₅CCl₃	Ni(CO)₄	35
(C₆H₅(CH₂)₂CHBrCH₂CH₂CBr₂—)₂	C₆H₅(CH₂)₂CHBrCH₂CBr₃	Ni(CO)₄	23
(C₆H₅)₂C:C(C₆H₅)₂	(C₆H₅)₂CCl₂	Ni(CO)₄	23
trans-C₆H₅CH:CHCH₃	trans-C₆H₅CH:CHBr	Ni(CO)₄	33
		NiI₂—CH₃Li	8
⬡ (cyclohexadiene)	cis,cis-ICH:CH(CH(CH₂)₂CH:CHI	(COD)₂Ni	40
trans,trans- and cis,trans-CH₃CH:CHCH:CHCH₃	trans-CH₃CH:CHBr	(COD)₂Ni	40
	cis-CH₃CH:CHBr	(COD)₂Ni	40
trans,trans-C₆H₅CH:CHCH:CHC₆H₅	trans-C₆H₅CH:CHBr	(COD)₂Ni	40
	trans-C₆H₅CH:CHBr	K₄[Ni₂(CN)₆]	19, 20
CH₃CO₂C(:CH₂)C(:CH₂)CO₂CH₃	CH₂:CBrCO₂CH₃	(COD)₂Ni	40
	CH₂:CClCO₂CH₃	(COD)₂Ni	40
cis,cis-CH₃CO₂CH:CHCH:CHCO₂CH₃	cis-BrCH:CHCO₂CH₃	(COD)₂Ni	40
trans,trans-CH₃CO₂CH:CHCH:CHCO₂CH₃	trans-BrCH:CHCO₂CH₃	(COD)₂Ni	40
trans,trans-C₆H₅CH₂CO₂CH:CHCH:CHCO₂CH₂C₆H₅	trans-BrCH:CHCO₂CH₂C₆H₅	(COD)₂Ni	40
C₆H₅—C₆H₅	C₆H₅X (X = Cl, Br, I)	(COD)₂Ni	30, 56
CH₃—⬡—⬡—CH₃	CH₃—⬡—I	(COD)₂Ni	30
CH₃,CH₃ biphenyl structure	⬡(Br)(CH₃)	(COD)₂Ni	30

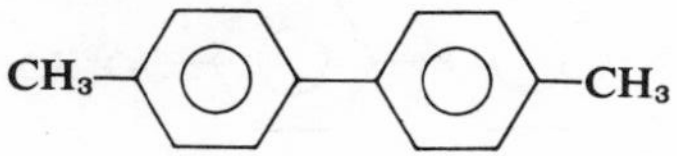

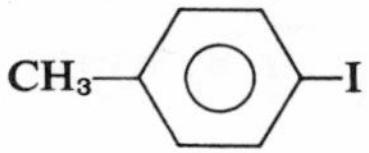

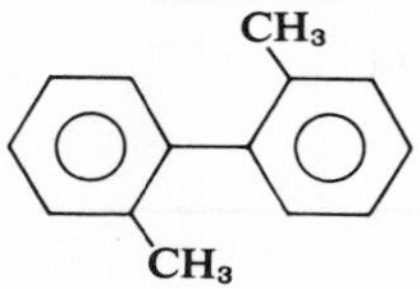

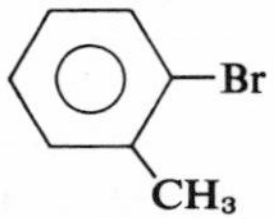

(continued)

TABLE V-4 (*continued*)

Product	Reactant		Ref.
	Organic component	Nickel component	
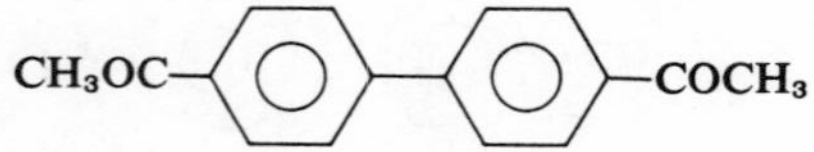 CH_3OC—⬡—⬡—$COCH_3$	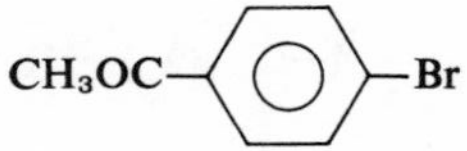 CH_3OC—⬡—Br	$(COD)_2Ni$	30
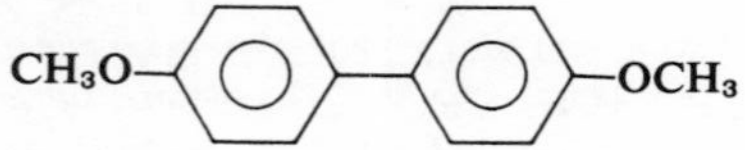 CH_3O—⬡—⬡—OCH_3	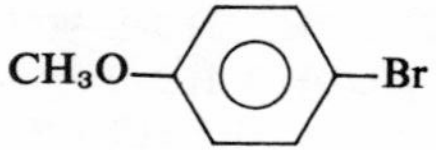 CH_3O—⬡—Br	$(COD)_2Ni$	30
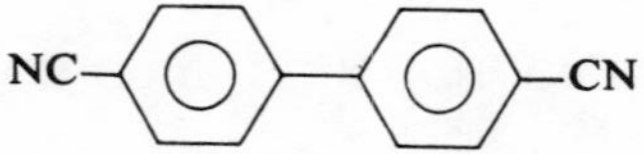 NC—⬡—⬡—CN	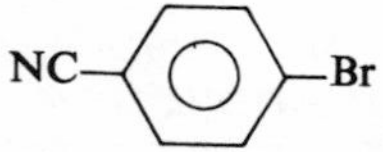 NC—⬡—Br	$(COD)_2Ni$	30
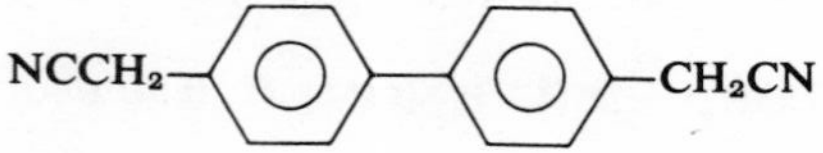 $NCCH_2$—⬡—⬡—CH_2CN	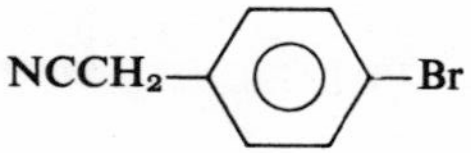 $NCCH_2$—⬡—Br	$(COD)_2Ni$	30
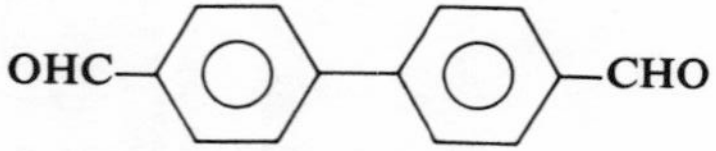 OHC—⬡—⬡—CHO	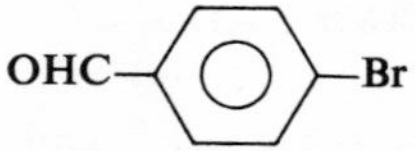 OHC—⬡—Br	$(COD)_2Ni$	30
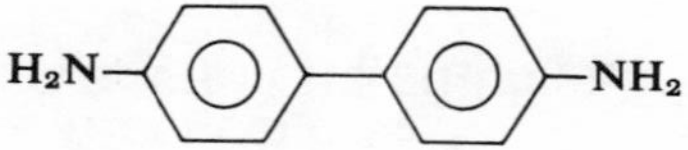 H_2N—⬡—⬡—NH_2	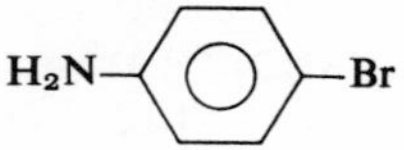 H_2N—⬡—Br	$(COD)_2Ni$	30
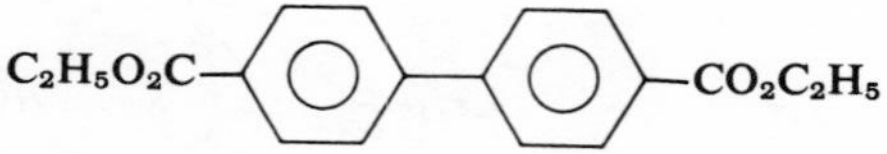 $C_2H_5O_2C$—⬡—⬡—$CO_2C_2H_5$	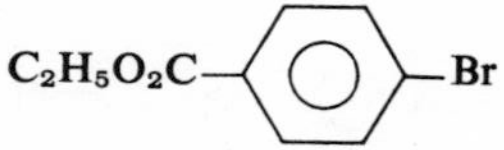 $C_2H_5O_2C$—⬡—Br	$(COD)_2Ni$	30

Product	Organic halide	Catalyst	Reference
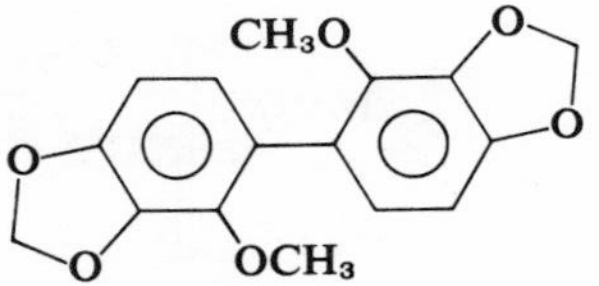 [CH₃O / OCH₃ bis(methylenedioxy)biphenyl structure]	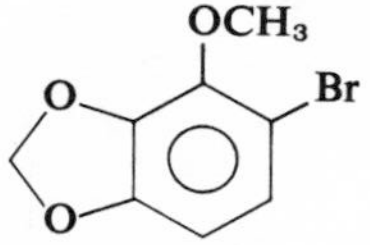 [OCH₃, Br methylenedioxybenzene structure]	$(COD)_2Ni$	30
$C_6F_5-C_6F_5$	C_6F_5X (X = Br, I)	$Ni(CO)_4$	3, 56
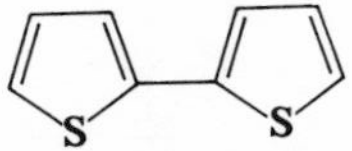 [bithienyl structure]	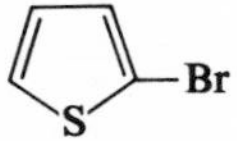 [2-bromothiophene]	$(COD)_2Ni$	30
$CH_3COCOCH_3$	CH_3COCl	$Ni(CO)_4$	25
$C_6H_5OCOC(C_6H_5):C(C_6H_5)OCOC_6H_5$	C_6H_5COX (X = Cl, Br)	$Ni(CO)_4$	2, 61
$C_6H_5OCOC(C_6H_5):C(OH)C_6H_5$	C_6H_5COCl	$Ni(CO)_4$	60
$C_3F_7CO_2C(C_3F_7):C(C_3F_7)CO_2C_3F_7$	C_3F_7COCl	$Ni(CO)_4$	12
$C_4F_8HCO_2C(C_4F_8H):C(C_4F_8H)CO_2C_4F_8H$	C_4F_8HCOCl	$Ni(CO)_4$	12
$CH_3COC(CH_3)C(CH_3)\!-\!CHCH_3$ (epoxide)	$CH_3COCH(CH_3)Br$	$Ni(CO)_4$	36
$C_2H_5COCH_2C(C_2H_5)\!-\!CH_2$ (epoxide)	$C_2H_5COCH_2Br$	$Ni(CO)_4$	36
$C_2H_5COC(CH_3)C(C_2H_5)CHCH_3$ (epoxide)	$C_2H_5COCH(CH_3)Br$	$Ni(CO)_4$	36
$tert\text{-}C_4H_9COCH_2C(tert\text{-}C_4H_9)CH_2$ (epoxide)	$tert\text{-}C_4H_9COCH_2Br$	$Ni(CO)_4$	36
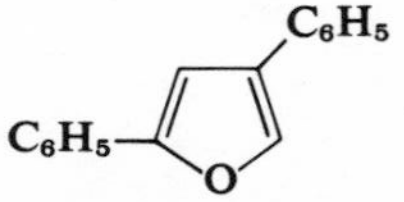 [furan structure, C_6H_5 substituents]	$C_6H_5COCH_2Br$	$Ni(CO)_4$	35

(continued)

TABLE V-4 *(continued)*

	Reactant		
Product	Organic component	Nickel component	Ref.
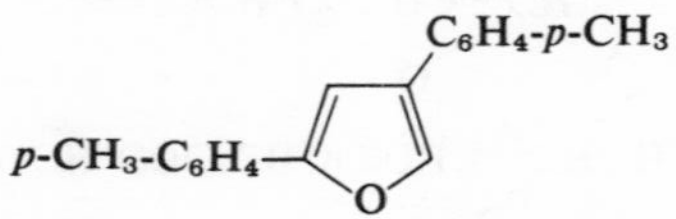	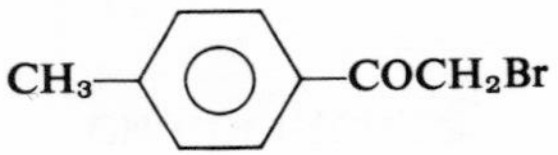	$Ni(CO)_4$	35
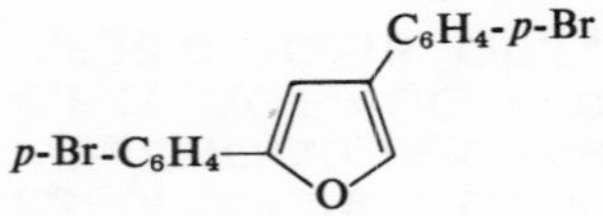	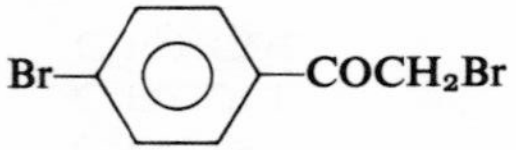	$Ni(CO)_4$	35
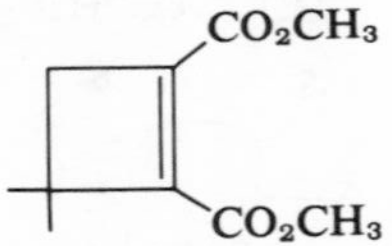	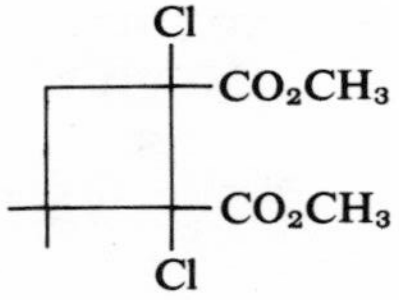	$Ni(CO)_4$	69
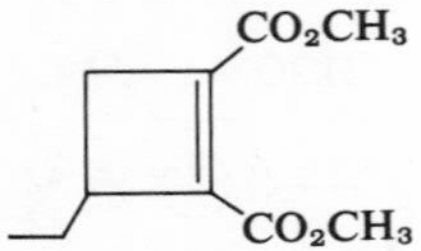	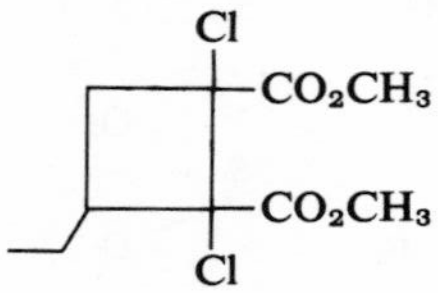	$Ni(CO)_4$	70

Product	Reactant	Reagent	Ref.
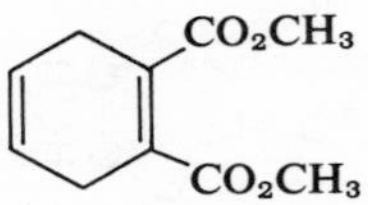 (dimethyl ester cyclohexene)	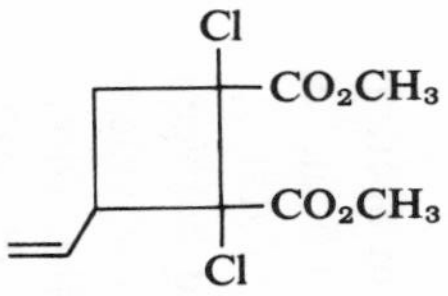 (dichloro diester vinyl)	$Ni(CO)_4$	29
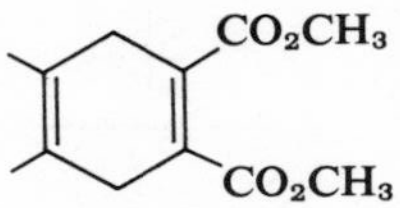 (dimethyl ester methylcyclohexene)	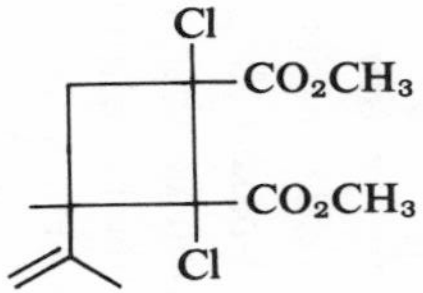 (dichloro diester isopropenyl)	$Ni(CO)_4$	29
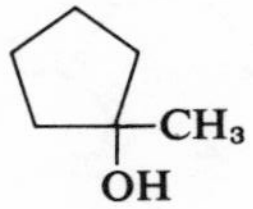 (1-methylcyclopentanol)	$I(CH_2)_4COCH_3$	$NiTPP^a$–LiNaphth	75
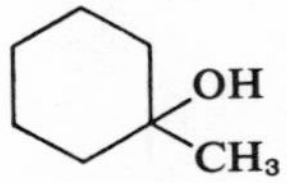 (1-methylcyclohexanol)	$I(CH_2)_5COCH_3$	$NiTPP^a$–LiNaphth	75
$C_6H_5CH(OH)CH_2CN$	$BrCH_2CN$–C_6H_5CHO	$Ni(CO)_4$	78

^{a} NiTPP = nickel tetraphenylporphine.

Reduction has also been observed in reactions involving $ArCH_2Br$ (20, 23, 37), $(C_6H_5)_3CCl$ (43), $C_6H_5CCl_3$ (44), C_6F_5X (3), 1-iodoalkynes (86), and mercaptoles (95). Frequently an undesirable side reaction, it has found synthetic application in the reduction of some perhaloorganic compounds, e.g., DDT, and in the synthesis of 5-bromolyxose and 5-bromoxylose (37) (23, 32).

$$(p\text{-}ClC_6H_4)_2CHCCl_3 \xrightarrow{\text{Ni(CO)}_4} (p\text{-}ClC_6H_4)_2CHCHCl_2$$

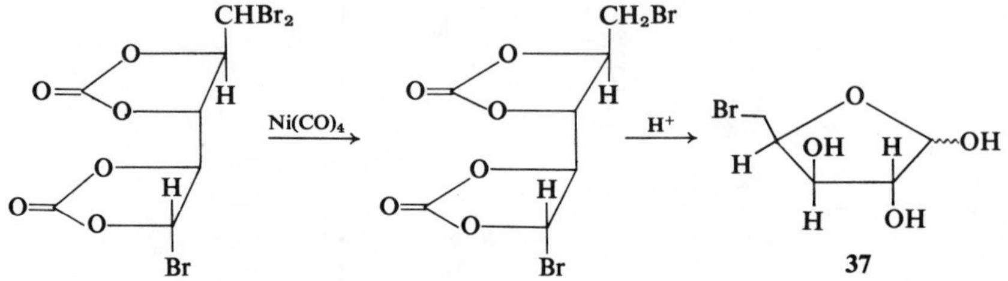

The reaction of allyl halides with epoxides has been mentioned in Section II,B. In the absence of an organic halide, styrene oxide reacts with nickel tetracarbonyl to form phenylacetaldehyde (97) [nickel-based Ziegler catalysts polymerize epoxides (P14)]. A rather unusual reaction is observed between epoxides and carbon dioxide in the presence of bisphosphine zerovalent nickel complexes or bis(cyclooctadiene)nickel. Insertion into the three-membered ring occurs to give alkylene carbonates in good yields (48, P12). The rate of reaction depends on the structure of the catalyst and epoxide and decreases in the order

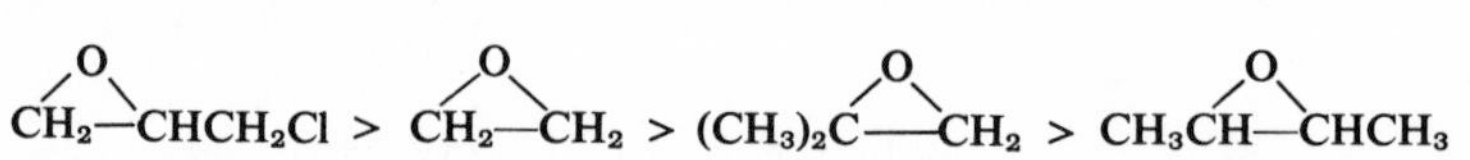

Practically no reaction is observed in the case of 2,3-epoxybutane. A mechanism involving opening of the three-membered ring, insertion of CO_2, and reductive elimination has been suggested, although evidence is lacking.

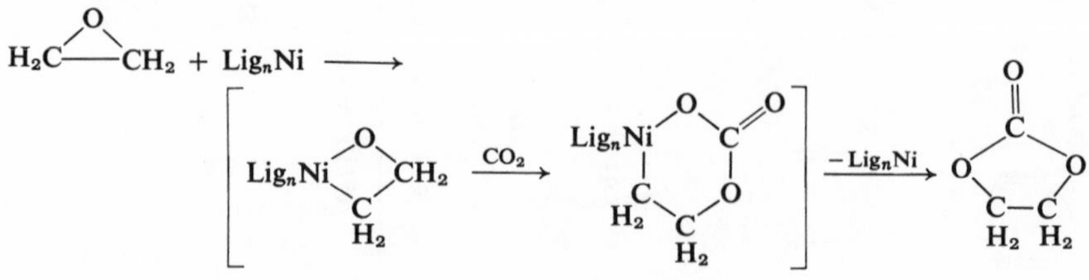

III. The Reaction of Organic Halides with Grignard Reagents (Table V-5)

The coupling of the organic groups that is observed during the stoichiometric reaction of organo-main group element compounds with transition metal salts (a reaction used to prepare zerovalent transition metal complexes) is hardly of interest as an organic synthetic method (examples involving nickel have been mentioned in Volume I, p. 157, and in refs. 124, 126–129, and 135). However, the extension in the case of the organomagnesium compounds to cross-coupling with organic halides (124) is of more significance and in recent years the reactions catalyzed by nickel have attracted particular attention.

The reaction occurs readily and, under suitable conditions, a high yield of coupling product is obtained. The reaction is general for all classes of

$$RX + R'MgY \xrightarrow{[Ni]} R-R' + MgXY$$

Grignard reagent but is apparently confined to three types of organic halide, viz., alkenyl, aryl, and allyl halides, the chlorides being particularly reactive. The reaction is catalyzed by both "ligand-free" and phosphine-modified nickel catalysts. Side reactions (see below) can be suppressed to a large extent by using catalysts modified by chelating ligands.

The coupling reaction is of particular interest for the synthesis of alkyl-substituted aromatic compounds, e.g., **38**, derived from *m*-dichlorobenzene.

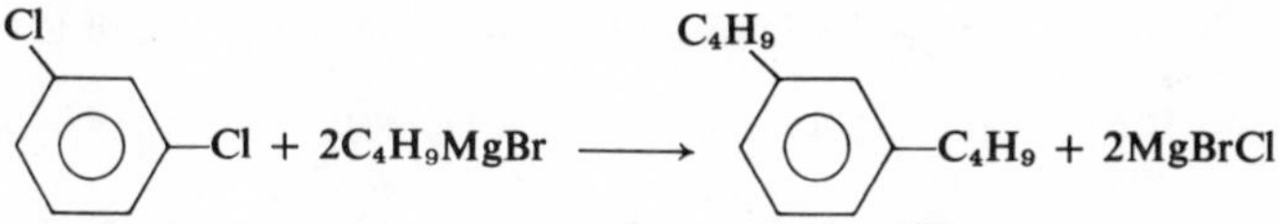

By using a catalyst modified by the optically active bidentate phosphine, 2,3-*O*-isopropylydene-2,3-dihydroxy-1,4-bis(diphenylphosphino)butane (diop) **39**, an asymmetric cross-coupling between a racemic secondary Grignard reagent and an organic halide has been achieved (101, 107). The optical induction, however, is modest; for example, **40** is formed with an optical purity of 17%. The reaction with monohaloolefins occurs stereospecifically,

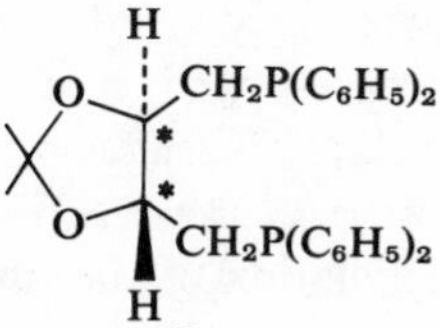

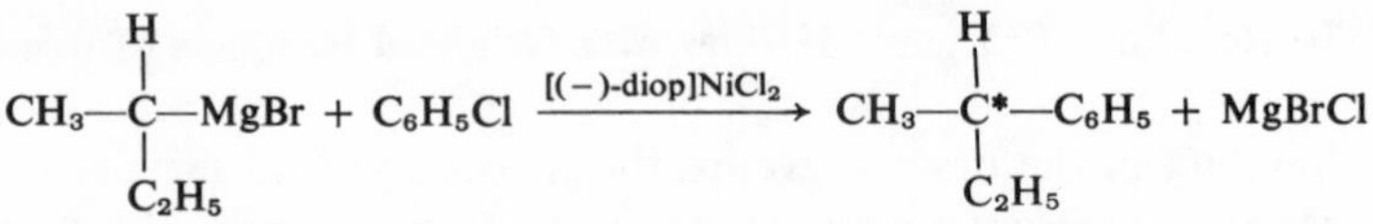

$$\underset{\overset{|}{C_2H_5}}{\overset{\overset{H}{|}}{CH_3-C-MgBr}} + C_6H_5Cl \xrightarrow{[(-)\text{-diop}]NiCl_2} \underset{\overset{|}{C_2H_5}}{\overset{\overset{H}{|}}{CH_3-C^*-C_6H_5}} + MgBrCl$$

40

whereas that with 1,2-dihaloethylene produces the cis disubstituted olefin preferentially (123). This last reaction is accompanied by the evolution of acetylene and in the case of polyhaloolefins alkyne formation is the main reaction observed (106).

$$Cl_2C{=}CCl_2 + nC_6H_5MgBr \longrightarrow C_6H_5C{\equiv}CC_6H_5$$

An interesting variation of the cross-coupling reaction is that of Grignard reagents with trisubstituted silanes or germanes, which has been shown to be catalyzed by a variety of nickel complexes, including $NiCl_2$, $[(C_6H_5)_3P]_2$-$NiCl_2$, π-$C_5H_5NiX[P(C_6H_5)_3]$, $[(cyclo\text{-}C_6H_{11})_3P]_2NiCl(H)$, and $[(C_6H_5)_3P]_2$-$Ni(CH_2{:}CH_2)$ (45, 104, 109, 140, 141).

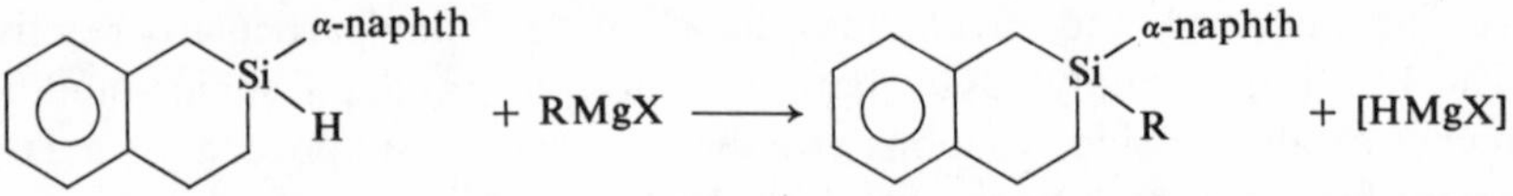

A number of nickel-catalyzed side reactions have been observed that involve the Grignard reagent and in unfavorable cases these can predominate. Grignard reagents that have β hydrogen atoms can eliminate an olefin molecule and the resulting species is able to reduce the organic halide (117–121,

$$CH_3CH_2MgBr + RX \longrightarrow CH_2{=}CH_2 + RH + MgBrX$$

123; see also 131). This reaction may be suppressed by using catalysts modified by chelating phosphines. A second side reaction is the isomerization of the alkyl group bonded to the magnesium, which results in the formation of a mixture of isomeric coupling products (103, 112, 115, 116, 118, 125). The

$$iso\text{-}C_3H_7MgX + RX \longrightarrow \begin{array}{c} iso\text{-}C_3H_7R \\ \text{and} \\ n\text{-}C_3H_7R \end{array} + MgX_2$$

extent of isomerization apparently depends on the nature of the ligand present in the catalyst and increases with increasing basicity of the phosphine. The nature of the organic halide also affects the extent of isomerization. Anomalous behavior is reported in the reaction of perfluorobenzene with ethylmagnesium bromide; metallation of the ring occurs, giving C_6F_5MgBr (122).

The β-hydrogen transfer mentioned above has found use in a highly stereo-specific synthesis of trisubstituted silanes from the corresponding methoxide, fluoride, or chloride (45, 108, 134).

$$\alpha\text{-naphth}\overset{C_2H_5}{\underset{C_6H_5}{\text{Si}}}\text{OCH}_3 + C_2H_5MgX \xrightarrow{-\text{MgXOCH}_3} \alpha\text{-naphth}\overset{C_2H_5}{\underset{C_6H_5}{\text{Si}}}\text{H} + CH_2{=}CH_2$$

Related to the reaction of organic halides with Grignard reagents are those of aryl halides with lithium enolates (57) and alkali metal cyanides (13, 19, 20, 22, 63). The first are catalyzed (inefficiently) by zerovalent nickel complexes, e.g., $[(C_6H_5)_3P]_4Ni$ and $(COD)_2Ni$, and have been used to synthesize phenylbenzyl ketone and cephalotaxinone (**41**).

$$C_6H_5Br + LiCH_2COC_6H_5 \xrightarrow{[Ni]} C_6H_5CH_2COC_6H_5 + LiBr$$

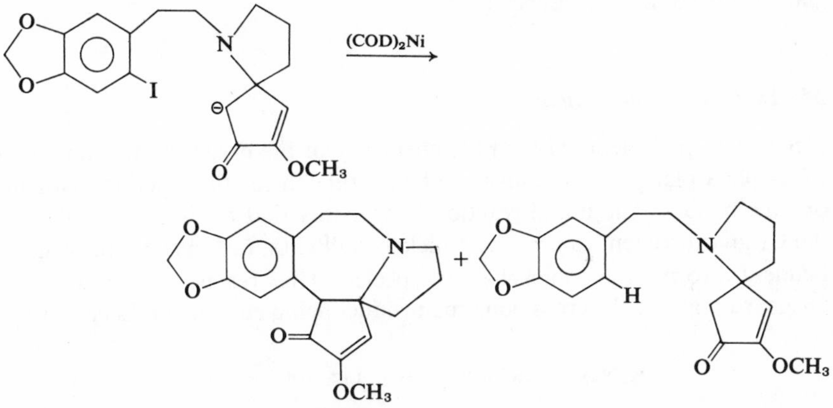

41

This last reaction is accompanied by considerable reduction and it has been suggested that the THF which is used as solvent is the source of hydrogen. The cyanation reaction has been observed for $trans\text{-}C_6H_5CH{:}CHBr$ and $trans\text{-}C_6H_5CH{:}CHCH_2Br$ in the presence of $K_4[Ni_2(CN)_6]$ and free cyano ion. It is probably related to the formation of aryl cyanides by reaction of aryl halides with sodium cyanide, which is catalyzed by tetrakisphosphine nickel complexes (63).

$$trans\text{-}C_6H_5CH{=}CHBr + KCN \xrightarrow[-KBr]{[Ni]} trans\text{-}C_6H_5CH{=}CHCN$$

$$Ar\text{--}X + NaCN \xrightarrow{Lig_4Ni} Ar\text{--}CN + NaX$$

It is convenient to mention here that disilanes may be alkylated by reaction with organic halides in the presence of $[(C_4H_9)_3P]_2NiBr_2$ (P11),

$$Cl_3Si-SiCl_3 + CH_2=CHCH_2Cl \xrightarrow{[Ni]} Cl_3SiCH_2CH=CH_2 + SiCl_4$$

while amides, imides, and amines may be silylated by reaction with trisubstituted silanes in the presence of $NiCl_2-(C_2H_5)_2S$ (143).

$$RCONH_2 \xrightarrow{HSiR_3} RCONH(SiR_3) \xrightarrow{HSiR_3} RCON(SiR_3)_2$$

More than 30 years ago it was shown that the reaction between Grignard reagents and organic carbonyl compounds is also influenced by the presence of transition metal halides: the reaction of methylmagnesium bromide and isophorone produces a mixture of 1,3,5,5-tetramethyl-1-cyclohexene-3-ol and 1,3,5,5-tetramethyl-1,3-cyclohexadiene in the absence of nickel chloride but mainly pinacol in its presence (130).

Mechanistic Considerations

Sufficient precedent exists in the chemistry of the organonickel complexes to enable a plausible mechanism to be postulated for the reaction involving organic halides. The initial reaction is probably that of the nickel salt with the Grignard reagent to produce a nickel–dialkyl that reacts with the organic halide to form a halonickel–alkyl species. This reacts further with the Grignard reagent, the cross-coupling product being eliminated. Each of these

$$Lig_2NiX_2 + 2RMgY \longrightarrow Lig_2NiR_2 + 2MgYX$$

$$Lig_2NiR_2 + R'X \longrightarrow Lig_2Ni\overset{X}{\underset{R'}{\diagdown}} + R-R$$

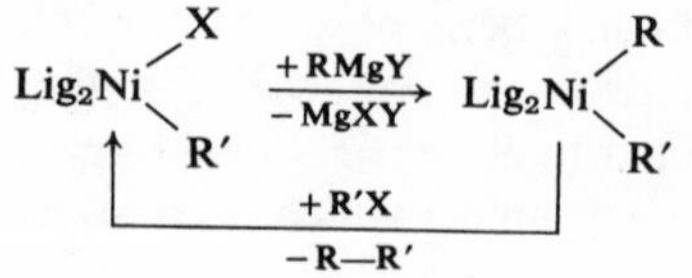

$$Lig_2Ni\overset{X}{\underset{R'}{\diagdown}} \xrightarrow[-MgXY]{+RMgY} Lig_2Ni\overset{R}{\underset{R'}{\diagdown}} \quad \xrightarrow[-R-R']{+R'X}$$

reactions is well documented (see Volume I, Chapter IV). However, this mechanism is probably an oversimplification, neglecting, as it does, interaction between the intermediate nickel–alkyl systems and the Grignard reagent.

The intermediacy of a π-allylnickel species in the reactions involving allyl halides and allyl alcohol is suggested by the isolation, for example, of 1-phenyl-1-butene from the reaction of methylmagnesium bromide with both cinnamyl alcohol and α-phenylallyl alcohol (103, 110, 113, 118). Further sup-

port comes from the stoichiometric reactions of π-allylnickel halides with Grignard reagents (144).

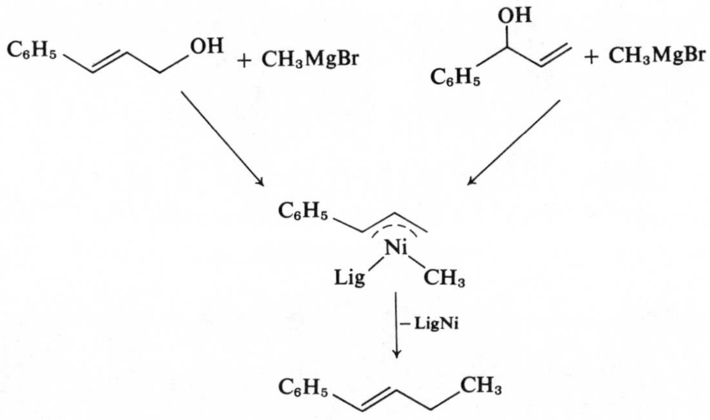

The elimination of an olefin from the Grignard reagent and the isomerization of the alkyl group originally bonded to magnesium are probably related reactions and involve a nickel-hydride species.

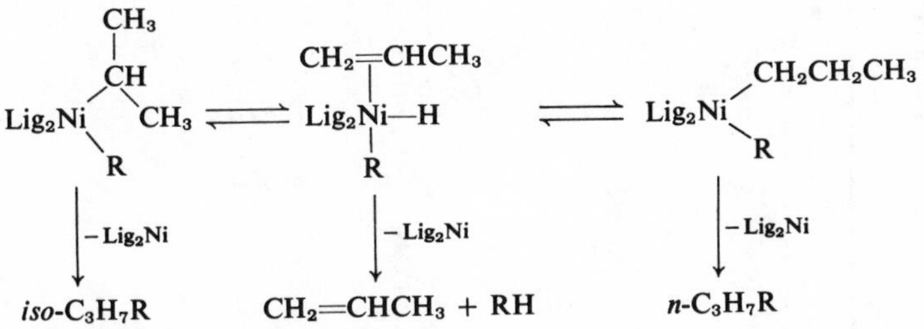

The proposal of a meaningful mechanism for the reaction of silanes with Grignard reagents is prevented by a lack of knowledge of the modes of reaction of silanes with organonickel complexes. It has been suggested that oxidative addition of the Grignard reagent to a zerovalent nickel species may be involved (45, see also 145).

TABLE V-5

Cross-coupling of Organic Halides with Grignard Reagents

$$RX + R'MgY \rightarrow R-R' + MgXY$$

Product (R–R')	RX	R'MgY	Catalyst	Ref.
$C_2H_5CHCH_3C_6H_5$, $C_2H_5CH_2CH_2C_6H_5$	C_6H_5X (X = F, Cl, Br)	$C_2H_5CHCH_3MgCl$	$[(-)\text{-diop}]NiCl_2{}^a$	101, 107
$C_2H_5C_6H_5$	C_6H_5X (X = F, Cl, Br)	C_2H_5MgBr	$[(C_6H_5)_2PC_2H_4P(C_6H_5)_2]NiCl_2$	102
$iso\text{-}C_3H_7C_6H_5$, $C_3H_7C_6H_5$	C_6H_5X (X = F, Cl, Br)	$iso\text{-}C_3H_7MgCl$	$[(CH_3)_2PC_2H_4P(CH_3)_2]NiCl_2$	115, 125
$C_4H_9C_6H_5$	C_6H_5X (X = F, Cl, Br)	C_4H_9MgBr	$NiCl_2$	102
$C_6H_5C_6H_5$	C_6H_5X (X = F, Cl, Br)	C_6H_5MgBr	$[(CH_3)_2PC_2H_4P(CH_3)_2]NiCl_2$	124
$p\text{-}CH_3C_6H_4iso\text{-}C_3H_7$, $p\text{-}CH_3C_6H_4C_3H_7$	$p\text{-}CH_3C_6H_4Cl$	$iso\text{-}C_3H_7MgCl$	$[(CH_3)_2PC_2H_4P(CH_3)_2]NiCl_2$	115
$p\text{-}CH_3OC_6H_4iso\text{-}C_3H_7$, $p\text{-}CH_3OC_6H_4C_3H_7$	$p\text{-}CH_3OC_6H_4Cl$	$iso\text{-}C_3H_7MgCl$	$[(CH_3)_2PC_2H_4P(CH_3)_2]NiCl_2$	115
$m\text{-}CF_3C_6H_4iso\text{-}C_3H_7$, $m\text{-}CF_3C_6H_4C_3H_7$	$m\text{-}CF_3C_6H_4Cl$	$iso\text{-}C_3H_7MgCl$	$[(CH_3)_2PC_2H_4P(CH_3)_2]NiCl_2$	115
$p\text{-}CF_3C_6H_4iso\text{-}C_3H_7$, $p\text{-}CF_3C_6H_4C_3H_7$	$p\text{-}CF_3C_6H_4Cl$	$iso\text{-}C_3H_7MgCl$	$[(CH_3)_2PC_2H_4P(CH_3)_2]NiCl_2$	115
$o\text{-}CF_3C_6H_4iso\text{-}C_3H_7$, $o\text{-}CF_3C_6H_4C_3H_7$	$o\text{-}CF_3C_6H_4Cl$	$iso\text{-}C_3H_7MgCl$	$[(CH_3)_2PC_2H_4P(CH_3)_2]NiCl_2$	115
$o\text{-}C_4H_9C_6H_4C_4H_9$	$o\text{-}ClC_6H_4Cl$	C_4H_9MgBr	$[(C_6H_5)_2PC_2H_4P(C_6H_5)_2]NiCl_2$	102
$m\text{-}C_4H_9C_6H_4C_4H_9$	$m\text{-}ClC_6H_4Cl$	C_4H_9MgBr	$[(C_6H_5)_2PC_2H_4P(C_6H_5)_2]NiCl_2$	102
$p\text{-}C_4H_9C_6H_4C_4H_9$	$p\text{-}ClC_6H_4Cl$	C_4H_9MgBr	$[(C_6H_5)_2PC_2H_4P(C_6H_5)_2]NiCl_2$	102
$p\text{-}C_6H_5C_6H_4C_6H_5$	$p\text{-}BrC_6H_4Br$	C_6H_5MgBr	$Ni(acac)_2$	105
$p\text{-}(m\text{-}CH_3C_6H_4)C_6H_4\text{-}m\text{-}CH_3C_6H_4$,	$p\text{-}BrC_6H_5Br$	$m\text{-}CH_3C_6H_5MgBr$	$Ni(acac)_2$	105
$\alpha\text{-}Naphthyl\text{-}iso\text{-}C_3H_7$, $\alpha\text{-}Naphthyl\text{-}C_3H_7$	$\alpha\text{-}Naphthyl\text{-}Br$	$iso\text{-}C_3H_7MgCl$	$[(CH_3)_2PC_2H_4P(CH_3)_2]NiCl_2$	115
$C_2H_5CHCH_3\text{-}\alpha\text{-}Naphthyl$	$\alpha\text{-}Naphthyl\text{-}Br$	$C_2H_5CHCH_3MgCl$	$[(-)\text{-diop}]NiCl_2{}^a$	101
$C_2H_5CHCH_3CH:CH_2$,	$CH_2:CHCl$	$C_2H_5CHCH_3MgBr$	$[(-)\text{-diop}]NiCl_2{}^a$	107
$C_6H_5CHCH_3CH_2CH:CH_2$				
$C_6H_5CHCH_3CH:CH_2$,	$CH_2:CHCl$	$C_6H_5CHCH_3MgBr$	$[(-)\text{-diop}]NiCl_2{}^a$	101, 107
$C_6H_5CH:CH_2$				
$\alpha\text{-}Naphthyl\text{-}CH:CH_2$	$CH_2:CHCl$	C_6H_5MgBr	$[(C_6H_5)_2PC_2H_4P(C_6H_5)_2]NiCl_2$	102, P10
$C_8H_{17}CH:CH_2$	$CH_2:CHCl$	$\alpha\text{-}Naphthyl\text{-}MgBr$	$[(C_6H_5)_2PC_2H_4P(C_6H_5)_2]NiCl_2$	102
$C_3H_7CH(CH_3)CH:CH_2$	$CH_2:CHCl$	$C_8H_{17}MgCl$	$[(C_6H_5)_2PC_2H_4P(C_6H_5)_2]NiCl_2$	102
$(C_6H_5)_2C:CH_2$	$CH_2:CCl_2$	$C_3H_7CH(CH_3)MgCl$	$[(CH_3)_2PC_2H_4P(CH_3)_2]NiCl_2$	115
$cis\text{- and }trans\text{-}C_6H_5CH:CHC_6H_5$	$ClCH:CHCl$	C_6H_5MgBr	$[(C_6H_5)_2PC_2H_4P(C_6H_5)_2]NiCl_2$	102, 123
$m\text{-}CH_3C_6H_4CH:CHC_6H_4\text{-}m\text{-}CH_3$	$ClCH:CHCl$	C_6H_5MgBr	$[(C_6H_5)_2PC_2H_4P(C_6H_5)_2]NiCl_2$	102
$p\text{-}CH_3C_6H_4CH:CHC_6H_4\text{-}p\text{-}CH_3$	$ClCH:CHCl$	$m\text{-}CH_3C_6H_4MgX$	$Ni(acac)_2$	105
$trans\text{-}C_6H_5CH:CHC_6H_5$	$ClCH:CHCl$	$p\text{-}CH_3C_6H_4MgX$	$Ni(acac)_2$	105
$trans\text{-}C_6H_5CH:CHBr$	C_6H_5MgBr	$[(C_6H_5)_2PC_6H_6P(C_6H_5)_2]NiCl_2$	123	
$p\text{-}CH_3OC_6H_4CH:CHC_6H_4\text{-}p\text{-}OCH_3$	$trans\text{-}C_6H_5CH:CHBr$	$p\text{-}CH_3OC_6H_4MgBr$	$Ni(acac)_2$	105
$p\text{-}CH_3C_6H_4CH:CHC_6H_4\text{-}p\text{-}CH_3$	$trans\text{-}C_6H_5CH:CHBr$	$p\text{-}CH_3C_6H_4MgBr$	$Ni(acac)_2$	105
$m\text{-}CH_3C_6H_4CH:CHC_6H_4\text{-}m\text{-}CH_3$	$trans\text{-}C_6H_5CH:CHBr$	$m\text{-}CH_3C_6H_4MgBr$	$Ni(acac)_2$	105
$p\text{-}BrC_6H_4CH:CHC_6H_4\text{-}p\text{-}Br$	$trans\text{-}C_6H_5CH:CHBr$	$p\text{-}BrC_6H_4MgBr$	$Ni(acac)_2$	105
$2,4\text{-}(CH_3)_2C_6H_3CH:CHC_6H_3\text{-}2,4\text{-}(CH_3)_2$	$trans\text{-}C_6H_5CH:CHBr$	$2,4\text{-}(CH_3)_2C_6H_3MgBr$	$Ni(acac)_2$	105

Product	Grignard reagent	Catalyst	Reference
α-Naphthyl-CH:CH-C$_6$H$_5$	α-Naphthyl-MgBr	Ni(acac)$_2$	105
α-Thienyl-CH:CH-C$_6$H$_5$	α-Thienyl-MgBr	Ni(acac)$_2$	105
(C$_6$H$_5$)$_2$CH$_2$	C$_6$H$_5$CH$_2$Br	NiCl$_2$	126
CH$_2$:CHCH$_2$CH$_3$	CH$_3$MgI	NiCl$_2$	118
CH$_2$:CHCH$_2$C$_4$H$_9$ (traces)	C$_4$H$_9$MgBr	NiCl$_2$	111, 117, 118
CH$_2$:CHCH$_2$CH$_3$	CH$_3$MgBr	[(C$_6$H$_5$)$_3$P]$_2$NiCl$_2$	103
CH$_2$:CHCH$_2$CH$_2$C$_6$H$_5$	C$_6$H$_5$CH$_2$MgBr	[(C$_6$H$_5$)$_3$P]$_2$NiCl$_2$	103
CH$_2$:CHCH$_2$C$_6$H$_5$	C$_6$H$_5$MgBr	[(C$_6$H$_5$)$_3$P]$_2$NiCl$_2$	103
(CH$_3$)$_2$CHCH:CH$_2$, CH$_3$CH:CHCH$_2$CH$_3$	CH$_3$MgBr	[(C$_6$H$_5$)$_3$P]$_2$NiCl$_2$	103, 113
CH$_3$CH:CHCH$_2$C$_6$H$_5$, CH$_3$CH(C$_6$H$_5$)CH:CH$_2$	C$_6$H$_5$MgBr	[(C$_6$H$_5$)$_3$P]$_2$NiCl$_2$	113
(CH$_3$)$_2$CHCH:CH$_2$, CH$_3$CH:CHCH$_2$CH$_3$	CH$_3$MgBr	[(C$_6$H$_5$)$_3$P]$_2$NiCl$_2$	103, 113, 114
C$_6$H$_5$C(CH$_3$)HCH:CH$_2$, C$_6$H$_5$CH:CHCH$_2$CH$_3$	C$_6$H$_5$MgBr	[(C$_6$H$_5$)$_3$P]$_2$NiCl$_2$	103
[cyclohexane, 1-vinyl, 1-OH substituted structure]	CH$_3$MgBr	[(C$_6$H$_5$)$_3$P]$_2$NiCl$_2$	103
[cyclohexene/propenylidene structure]	CH$_3$MgBr	[(C$_6$H$_5$)$_3$P]$_2$NiCl$_2$ (?)	110
[cyclohexane, vinyl, OH structure]	CH$_3$MgBr	[(C$_6$H$_5$)$_3$P]$_2$NiCl$_2$	103
C$_6$H$_5$CH:CHCH$_2$CH$_3$	C$_3$H$_7$MgBr	[(C$_6$H$_5$)$_2$PC$_3$H$_6$P(C$_6$H$_5$)$_2$]NiCl$_2$	110
C$_6$H$_5$CH:CHCH$_2$C$_3$H$_7$	CH$_3$MgBr	[(C$_6$H$_5$)$_3$P]$_2$NiCl$_2$	103
C$_6$H$_5$CH(OH)CH:CH$_2$	CH$_3$MgBr	[(C$_6$H$_5$)$_3$P]$_2$NiCl$_2$	103
C$_6$H$_5$Si(CH$_3$)$_3$	CH$_2$:CHCH$_2$MgBr	[(C$_6$H$_5$)$_3$P]$_2$NiCl$_2$	45, 109
C$_6$H$_5$(CH$_3$)$_2$SiCH$_2$CH:CH$_2$	C$_6$H$_5$CH$_2$MgBr	[(C$_6$H$_5$)$_3$P]$_2$NiCl$_2$	45, 109
C$_6$H$_5$(CH$_3$)$_2$SiCH$_2$C$_6$H$_5$	CH$_3$MgBr	[(C$_6$H$_5$)$_3$P]$_2$NiCl$_2$	109
(C$_6$H$_5$)$_2$Si(CH$_3$)$_2$	CH$_2$:CHCH$_2$MgBr	[(C$_6$H$_5$)$_3$P]$_2$NiCl$_2$	109
CH$_3$(C$_6$H$_5$)$_2$SiCH$_2$CH:CH$_2$	C$_6$H$_5$CH$_2$MgBr	[(C$_6$H$_5$)$_3$P]$_2$NiCl$_2$	109
CH$_3$(C$_6$H$_5$)$_2$SiCH$_2$C$_6$H$_5$	CH$_3$MgBr	[(C$_6$H$_5$)$_3$P]$_2$NiCl$_2$	109
(C$_6$H$_5$)$_3$SiCH$_3$	CH$_3$MgBr	[(C$_6$H$_5$)$_3$P]$_2$Ni(CH$_2$:CH$_2$)	45
(C$_6$H$_5$)$_3$SiCH$_2$CH:CH$_2$	CH$_2$:CHCH$_2$MgBr	[(C$_6$H$_5$)$_3$P]$_2$NiCl$_2$	109
[silyl-α-naphth CH$_3$ cyclohexane structure]	CH$_3$MgBr	[(C$_6$H$_5$)$_3$P]$_2$NiCl$_2$	104, 109

(continued)

TABLE V-5 (*continued*)

Product (R–R′)	RX	R′MgY	Catalyst	Ref.
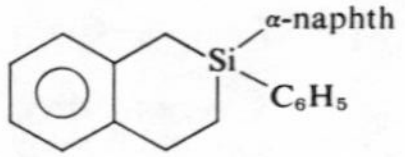	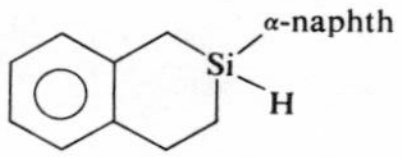	C_6H_5MgBr	$[(C_6H_5)_3P]_2NiCl_2$	104, 109
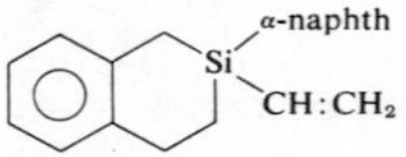	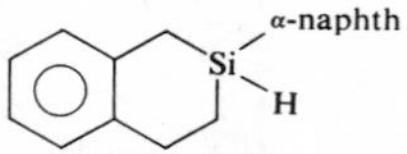	$CH_2{:}CHMgBr$	$[(C_6H_5)_3P]_2NiCl_2$	104, 109
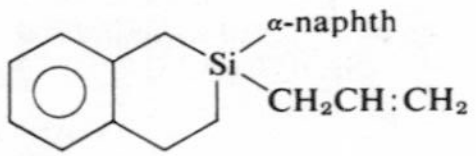	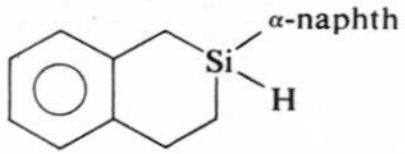	$CH_2{:}CHCH_2MgBr$	$[(C_6H_5)_3P]_2NiCl_2$	104, 109
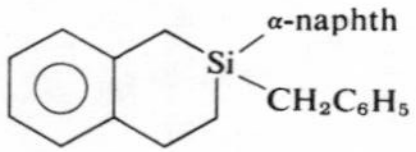	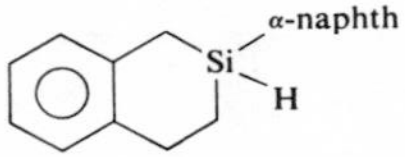	$C_6H_5CH_2MgCl$	$[(C_6H_5)_3P]_2NiCl_2$	104, 109
Linear product	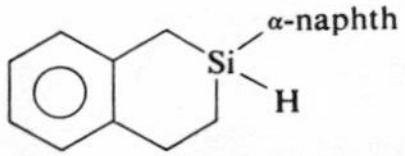	$CH_2{:}C(CH_3)CH_2MgBr$	$[(C_6H_5)_3P]_2NiCl_2$	104, 109
C_6H_5(α-Naphthyl)$Si(OCH_3)C_2H_5$	C_6H_5(α-Naphthyl)$Si(OCH_3)CH{:}CH_2$	C_2H_5MgBr	$[(C_6H_5)_3P]_2NiCl_2$	108
C_6H_5(α-Naphthyl)$SiH(C_2H_5)$	C_6H_5(α-Naphthyl)$SiH(CH{:}CH_2)$	C_2H_5MgBr	$[(C_6H_5)_3P]_2NiCl_2$	108
C_6H_5(α-Naphthyl)$SiH(C_2H_5)$	C_6H_5(α-Naphthyl)$Si(OCH_3)C_2H_5$	C_2H_5MgBr	$[(C_6H_5)_3P]_2Ni(CH_2{:}CH_2)$	45, 108
CH_3(α-Naphthyl)$GeC_2H_5(C_6H_5)$	CH_3(α-Naphthyl)$GeH(C_6H_5)$	C_2H_5MgBr	$NiCl_2$	141
CH_3(α-Naphthyl)$GeCH_2C_6H_5(C_6H_5)$	CH_3(α-Naphthyl)$GeH(C_6H_5)$	$C_6H_5CH_2MgBr$	$NiCl_2$	140
CH_3(α-Naphthyl)$Ge(C_6H_5)_2$	CH_3(α-Naphthyl)$GeH(C_6H_5)$	C_6H_5MgBr	$NiCl_2$	141
CH_3(α-Naphthyl)$GeCH_2CH{:}CH_2(C_6H_5)$	CH_3(α-Naphthyl)$GeH(C_6H_5)$	$CH_2{:}CHCH_2MgBr$	$NiCl_2$	141
iso-C_3H_7(α-Naphthyl)$GeCH_3(C_6H_5)$	*iso*-C_3H_7(α-Naphthyl)$GeH(C_6H_5)$	CH_3MgBr	$NiCl_2$	141
iso-C_3H_7(α-Naphthyl)$GeC_2H_5(C_6H_5)$	*iso*-C_3H_7(α-Naphthyl)$GeH(C_6H_5)$	C_2H_5MgBr	$NiCl_2$	141
iso-C_3H_7(α-Naphthyl)$GeCH_2CH{:}CH_2(C_6H_5)$	*iso*-C_3H_7(α-Naphthyl)$GeH(C_6H_5)$	$CH_2{:}CHCH_2MgBr$	$NiCl_2$	141
iso-C_3H_7(α-Naphthyl)$GeCH_2C_6H_5(C_6H_5)$	*iso*-C_3H_7(α-Naphthyl)$GeH(C_6H_5)$	$C_6H_5CH_2MgBr$	$NiCl_2$	140
$(C_6H_5)_3GeCH_3$	$(C_6H_5)_3GeH$	CH_3MgBr	$NiCl_2$	141

[a] diop = 2,3-*O*-isopropylydene-2,3-dihydroxy-1,4-bis(diphenylphosphino)butane; see ref. 39.

References

1. M. Asfazadourian and M. Prillieux, *J. Polym. Sci.*, *Part C* 22, 267 (1968).
2. N. L. Bauld, *Tetrahedron Lett.* p. 1841 (1963).
3. W. F. Beckert and J. U. Lowe, *J. Org. Chem.* 32, 1215 (1967).
4. E. J. Corey and E. A. Broger, *Tetrahedron Lett.* p. 1779 (1969).
5. E. J. Corey and M. F. Semmelhack, *Tetrahedron Lett.* p. 6237 (1966).
6. E. J. Corey and M. F. Semmelhack, *J. Amer. Chem. Soc.* 89, 2755 (1967).
7. E. J. Corey and E. Hamanaka, *J. Amer. Chem. Soc.* 89, 2758 (1967).
8. E. J. Corey and G. H. Posner, *Tetrahedron Lett.* p. 315 (1970).
9. E. J. Corey and E. Hamanaka, *J. Amer. Chem. Soc.* 86, 1641 (1964).
10. E. J. Corey, M. F. Semmelhack, and L. S. Hegedus, *J. Amer. Chem. Soc.* 90, 2416 (1968).
11. E. J. Corey, L. S. Hegedus, and M. F. Semmelhack, *J. Amer. Chem. Soc.* 90, 2417 (1968).
12. J. J. Drysdale and D. D. Coffman, *J. Amer. Chem. Soc.* 82, 5111 (1960).
13. E. J. Corey and L. S. Hegedus, *J. Amer. Chem. Soc.* 92, 1233 (1969).
14. M. Dubini, G. P. Chiusoli, and F. Montino, *Tetrahedron Lett.* p. 1591 (1963).
15. F. Guerrieri and G. P. Chiusoli, *Proc. Int. Symp. Reactiv. Bonding Transition Organometal. Compounds, 3rd, 1970* C5 (1970).
16. F. Guerrieri and G. P. Chiusoli, *Chim. Ind. (Milan)* 51, 1252 (1969).
17. F. Guerrieri and G. P. Chiusoli, *J. Organometal. Chem.* 19, 453 (1969).
18. L. S. Hegedus, E. L. Waterman, and J. Catlin, *J. Amer. Chem. Soc.* 94, 7155 (1972).
19. I. Hashimoto, M. Ryang, and S. Tsutsumi, *Tetrahedron Lett.* p. 4567 (1970).
20. I. Hashimoto, N. Tsuruta, M. Ryang, and S. Tsutsumi, *J. Org. Chem.* 35, 3748 (1970).
21. I. T. Harrison, E. Kimura, E. Bohme, and J. H. Fried, *Tetrahedron Lett.* p. 1589 (1969).
22. I. Hashimoto, M. Ryang, and S. Tsutsumi, *Tetrahedron Lett.* p. 3291 (1969).
23. T. Kunieda, T. Tamura, and T. Takizawa, *Chem. Commun.* p. 885 (1972).
24. I. Moritani, Y. Yamamoto, and H. Konishi, *Chem. Commun.* p. 1457 (1969).
25. H. Reihlen, A. Gruhl, and E. von Hessling, *Justus Liebigs Ann. Chem.* 472, 268 (1929).
26. K. Sato and S. Inoue, *Azahu Garasu Kogyo Gijutsu Shoreikai Kenku Hokuku* 19, 259 (1971).
27. K. Sato, S. Inoue, and K. Saito, *Chem. Commun.* p. 953 (1972).
28. K. Sato, S. Inoue, S. Ota, and Y. Fujita, *J. Org. Chem.* 37, 462 (1972).
29. H.-D. Scharf and F. Korte, *Chem. Ber.* 99, 3925 (1966).
30. M. F. Semmelhack, P. M. Helquist, and L. D. Jones, *J. Amer. Chem. Soc.* 93, 5908 (1971).
31. J. S. Swanson and G. D. Stucky, *Organometal. Chem. Syn.* 1, 467 (1972).
32. T. Tamura, T. Kunieda, and T. Takizawa, *Tetrahedron Lett.* p. 2219 (1972).
33. C. E. Coffey, *J. Amer. Chem. Soc.* 83, 1623 (1961).
34. I. D. Webb and G. T. Borcherdt, *J. Amer. Chem. Soc.* 73, 2654 (1951).
35. E. Yoshisato and S. Tsutsumi, *Chem. Commun.* p. 33 (1968).
36. E. Yoshisato and S. Tsutsumi, *J. Amer. Chem. Soc.* 90, 4488 (1968).
37. E. Yoshisato and S. Tsutsumi, *J. Org. Chem.* 33, 869 (1968).
38. G. Agnes, G. P. Chiusoli, and A. Marraccini, *J. Organometal. Chem.* 49, 239 (1973).

39. M. F. Semmelhack and P. M. Helquist, *Org. Syn.* **52**, 115 (1972).
40. M. F. Semmelhack, P. M. Helquist, and J. D. Gorzynski, *J. Amer. Chem. Soc.* **94**, 9234 (1972).
41. M. F. Semmelhack, *Org. React.* **19**, 115 (1972).
42. E. J. Corey and E. K. W. Wat, *J. Amer. Chem. Soc.* **89**, 2757 (1967).
43. W. Schlenk, L. Mair, and C. Bornhardt, *Chem. Ber.* **44**, 1169 (1911).
44. J. B. Lee and B. Cubberley, *Tetrahedron Lett.* p. 1061 (1969).
45. E. Colomer, R. J. P. Corriu, and B. Meunier, *J. Organometal. Chem.* **71**, 197 (1974).
46. M. Dubini and F. Montino, *Chim. Ind. (Milan)* **49**, 1283 (1967).
47. K. Jacob, I. Wiswedel, T. Zeine, and K. H. Thiele, *Z. Anorg. Allg. Chem.* **402**, 193 (1973).
48. R. J. de Pasquale, *Chem. Commun.* p. 157 (1973).
49. K. Sato, S. Inoue, and K. Saito, *J. Chem. Soc., Perkin Trans. 1* p. 2289 (1973).
50. M. F. Semmelhack, Ph.D. Thesis, Harvard University, Cambridge, Massachusetts (1967).
51. N. L. Bauld *Tetrahedron Lett.* p. 859 (1962).
52. G. P. Chiusoli and G. Cometti, *Chim. Ind. (Milan)* **45**, 401 (1963).
53. E. Hamanaka, Ph.D. Thesis, Harvard University, Cambridge, Massachusetts (1967).
54. G. Wilke *et al.*, unpublished results (ca. 1964).
55. M. Dubini, F. Montino, and G. P. Chiusoli, *Chim. Ind. (Milan)* **47**, 839 (1965).
56. K. J. Klabunde and J. Y. F. Low, *J. Organometal. Chem.* **51**, C33 (1973).
57. M. F. Semmelhack, R. D. Stauffer, and T. D. Rogerson, *Tetrahedron Lett.* p. 4519 (1973).
58. G. Wilke, *Pure Appl. Chem.* **17**, 179 (1968).
59. P. Heimbach, P. W. Jolly, and G. Wilke, *Advan. Organometal. Chem.* **8**, 29 (1970).
60. L. Cassar and G. P. Chiusoli, *Tetrahedron Lett.* p. 2805 (1966).
61. N. L. Bauld, *J. Amer. Chem. Soc.* **84**, 4345 (1962).
62. M. Dubini, G. P. Chiusoli, and F. Montino, *Chim. Ind. (Milan)* **45**, 1237 (1963).
63. L. Cassar, *J. Organometal. Chem.* **54**, C57 (1973).
64. U. Birkenstock, Ph.D. Thesis, Technische Hochschule, Aachen (1966).
65. W. Keim, Ph.D. Thesis, Technische Hochschule Aachen (1963).
66. J. B. Mettalia and E. H. Specht, *J. Org. Chem.* **32**, 3941 (1967).
67. G. P. Chiusoli, *Chim. Ind. (Milan)* **41**, 506 (1959).
68. G. P. Chiusoli and L. Cassar, *Angew. Chem.* **79**, 177 (1967).
69. H.-D. Scharf and F. Korte, *Chem. Ber.* **98**, 3672 (1965).
70. H.-D. Scharf and F. Korte, *Chem. Ber.* **99**, 1299 (1966).
71. R. Criegee and G. Schröder, *Justus Liebigs Ann. Chem.* **623**, 1 (1959).
72. A. Sirigu, *Inorg. Chem.* **9**, 2245 (1970).
73. F. Guerrieri, *Chem. Commun.* p. 983 (1968).
74. E. J. Corey and H. A. Kirst, *J. Amer. Chem. Soc.* **94**, 667 (1972).
75. E. J. Corey and I. Kuwajima, *J. Amer. Chem. Soc.* **92**, 395 (1970).
76. G. P. Chiusoli, S. Merzoni, and G. Mondelli, *Tetrahedron Lett.* p. 2777 (1964).
77. G. P. Chiusoli, G. Bottaccio, and A. Cameroni, *Chim. Ind. (Milan)* **44**, 131 (1962).
78. E. Yoshisato, T. Abe, S. Murai, N. Sonoda, and S. Tsutsumi, unpublished results, quoted by M. Ryang, *Organometal Chem. Rev. Sect. A* **5**, 67 (1970).
79. V. A. Yakovlev, E. I. Tinyakova, and B. A. Dolgoplosk, *Bull. Acad. Sci. USSR* p. 1350 (1968).
80. H. Bönnemann, Ph.D. Thesis, Technische Hochschule Aachen (1967).

81. G. Wilke *et al.*, *Angew. Chem.* **78**, 157 (1966).
82. M. C. Gallazi, T. L. Hanlon, G. Vitulli, and L. Porri, *J. Organometal. Chem.* **33**, C45 (1971).
83. B. Büssemeier, Ph.D. Thesis, University of Bochum (1973).
84. W. T. Dent, R. Long, and G. H. Whitfield, *J. Chem. Soc., London* p. 1588 (1964).
85. H. O. Jones and H. S. Tasker, *J. Chem. Soc., London* p. 1904 (1909).
86. E. R. H. Jones, T. Y. Shen, and M. C. Whiting, *J. Chem. Soc., London* p. 763 (1951).
87. G. P. Chiusoli and G. Bottaccio, *Chim. Ind. (Milan)* **43**, 1022 (1961).
88. P. Tavs, *Chem. Ber.* **103**, 2428 (1970).
89. G. P. Chiusoli, *Chim. Ind. (Milan)* **43**, 365 (1961).
90. J. Furukawa, J. Kiji, K. Yamamoto, and T. Tojo, *Tetrahedron* **29**, 3149 (1973).
91. R. Baker, B. N. Blackett, R. C. Cookson, R. C. Cross, and D. P. Madden, *Chem. Commun.* p. 343 (1972).
92. R. Baker, B. N. Blackett, and R. C. Cookson, *Chem. Commun.* p. 802 (1972).
93. H. T. Dodd and J. F. Nixon, *J. Organometal Chem.* **32**, C67 (1971).
94. E. J. Corey and J. W. Suggs, *J. Org. Chem.* **38**, 3223 (1973).
95. W. E. Truce and F. E. Roberts, *J. Org. Chem.* **28**, 961 (1963).
96. I. Rhee, M. Ryang, and S. Tsutsumi, *Tetrahedron Lett.* p. 4593 (1969).
97. S. Fukuoka, M. Ryang, and S. Tsutsumi, *J. Org. Chem.* **35**, 3184 (1970).
98. M. Dubini and F. Montino, *J. Organometal. Chem.* **6**, 188 (1966).
99. G. P. Chiusoli, C. Venturello, and S. Merzoni, unpublished results, quoted in G. P. Chiusoli, *Aspects Homogen. Catal.* **1**, 85 (1970).
100. L. Cassar and G. P. Chiusoli, *Tetrahedron Lett.* p. 3295 (1965).
101. Y. Kiso, K. Tamao, N. Miyake, K. Yamamoto, and M. Kumada, *Tetrahedron Lett.* p. 3 (1974).
102. K. Tamao, K. Sumitani, and M. Kumada, *J. Amer. Chem. Soc.* **94**, 4374 (1972).
103. C. Chuit, H. Felkin, C. Frajerman, G. Roussi, and G. Swierczewski, *Chem. Commun.* p. 1604 (1968).
104. R. J. P. Corriu and J. P. Massé, *Chem. Commun.* p. 213 (1970).
105. R. J. P. Corriu and J. P. Massé, *Chem. Commun.* p. 144 (1972).
106. A. Collet and J. Jacques, *Synthesis* p. 38 (1972).
107. G. Consiglio and C. Botteghi, *Helv. Chim. Acta* **56**, 460 (1973).
108. R. J. P. Corriu and B. Meunier, *Chem. Commun.* p. 164 (1973).
109. R. J. P. Corriu, J. P. R. Massé, and B. Meunier, *J. Organometal. Chem.* **55**, 73 (1973).
110. H. Felkin, *Abstr. Int. Conf. Organometal. Chem. 6th*, 1973 P13 (1973).
111. Y. Ohbe, K. Doi, and T. Matsudo, *Nippon Kagaku Kaishi*, p. 193 (1974).
112. H. Felkin and G. Swierczewski, *C.R. Acad. Sci., Ser. C* **266**, 1611 (1968).
113. H. Felkin and G. Swierczewski, *Tetrahedron Lett.* p. 1433 (1972).
114. M. L. H. Green, M. J. Smith, H. Felkin, and G. Swierczewski, *Chem. Commun.* p. 158 (1971).
115. Y. Kiso, K. Tamao, and M. Kumada, *J. Organometal. Chem.* **50**, C12 (1973).
116. Y. Ohbe and T. Matsuda, *Nippon Kagaku Zasshi* **89**, 298 (1968).
117. Y. Ohbe and T. Matsuda, *Bull. Chem. Soc. Jap.* **45**, 2947 (1972).
118. Y. Ohbe and T. Matsuda, *Tetrahedron* **29**, 2989 (1973).
119. V. D. Parker and C. R. Noller, *Tetrahedron Lett.* p. 1737 (1963).
120. V. D. Parker, L. H. Piette, R. M. Salinger, and C. R. Noller, *J. Amer. Chem. Soc.* **86**, 1110 (1964).
121. V. D. Parker and C. R. Noller, *J. Amer. Chem. Soc.* **86**, 1112 (1964).

122. W. L. Respess and C. Tamborski, *J. Organometal. Chem.* **18**, 263 (1969).
123. K. Tamao, M. Zembayashi, Y. Kiso, and M. Kumada, *J. Organometal. Chem.* **55**, C91 (1973).
124. M. S. Kharasch and E. K. Fields, *J. Amer. Chem. Soc.* **63**, 2316 (1941).
125. K. Tamao, Y. Kiso, K. Sumitani, and M. Kumada, *J. Amer. Chem. Soc.* **94**, 9268 (1972).
126. G. Vavon, C. Chaminade, and G. Quesnel, *C.R. Acad. Sci.* **220**, 850 (1945).
127. H. Gilman and M. Lichtenwalter, *J. Amer. Chem. Soc.* **61**, 957 (1939).
128. H. Gilman, R. G. Jones, and L. A. Woods, *J. Amer. Chem. Soc.* **76**, 3615 (1954).
129. L. Farády and L. Markó, *J. Organometal. Chem.* **43**, 51 (1972).
130. M. S. Kharasch and P. O. Tawney, *J. Amer. Chem. Soc.* **63**, 2308 (1941).
131. L. I. Zakharkin, *Bull. Acad. Sci. USSR* p. 932 (1967).
132. R. Baker, *Chem. Rev.* **73**, 487 (1973).
133. W. Carruthers, *Chem. Ind. (London)* p. 931 (1973).
134. A. J. P. Corriu and B. Meunier, *J. Organometal. Chem.* **60**, 31 (1973).
135. I. L. Kershenbaum, K. L. Makovetskii, and B. A. Dolgoplosk, *Bull. Acad. Sci. USSR* p. 1135 (1973).
136. P. Tavs and H. Weitkamp, *Tetrahedron* **26**, 5529 (1970).
137. R. Baker, R. C. Cookson, and J. R. Vinsen, *Chem. Commun.* p. 515 (1974).
138. P. W. Jolly and G. Wilke, *Kontakte (Merck)* No. 2, p. 14 (1974).
139. L. S. Hegedus and R. K. Stiverson, *J. Amer. Chem. Soc.* **96**, 3250 (1974).
140. F. Carre and R. Corriu, *J. Organometal. Chem.* **73**, C49 (1974).
141. F. H. Carre and R. J. P. Corriu, *J. Organometal. Chem.* **74**, 49 (1974).
142. L. Cassar and M. Foa, *J. Organometal. Chem.* **74**, 75 (1974).
143. M. Paul and E. Frainnet, *C.R. Acad. Sci.* **279C**, 213 (1974).
144. Y. Ohbe, M. Takagi, and T. Matsuda, *Tetrahedron.* **30**, 2669 (1974).
145. Y. Kiso, K. Tamao, and M. Kumada, *J. Organometal. Chem.* **76**, 95 (1974).
146. W. G. Dauben, G. H. Beasley, M. D. Broadhurst, B. Muller, D. J. Peppard, P. Pesnelle, and C. Suter, *J. Amer. Chem. Soc.* **96**, 4724 (1974).

Patents

P1. G. Wilke, B. Bogdanović, U. Birkenstock, and H. Pauling (Studiengesellschaft Kohle m.b.H.), D.D.R. Patent 68,506 (1969); see *Chem. Abstr.* **72**, 43881 (1970).
P2. W. W. Prichard and G. M. Whitman (E. I. du Pont), U.S. Patent 2,524,833 (1950); *Chem. Abstr.* **45**, 1618 (1951).
P3. G. E. Tabet (E. I. du Pont), U.S. Patent 2,570,887 (1951); *Chem. Abstr.* **46**, 3071 (1952).
P4. I. D. Webb (E. I. du Pont), U.S. Patent 2,654,787 (1953); *Chem. Abstr.* **48**, 10048 (1954).
P5. O. T. Onsager (Halcon Internat. Inc.), Ger. Offen. 2,008,569 (1970); *Chem. Abstr.* **73**, 120124 (1970).
P6. O. T. Onsager (Halcon Internat. Inc.), Ger. Offen. 2,008,568 (1971); *Chem. Abstr.* **75**, 48480 (1971).
P7. I. G. Farbenindustrie A.G., Belg. Patent 448,884 (1943); *Chem. Abstr.* **41**, 6576 (1947).
P8. Studiengesellschaft Kohle m.b.H., Brit. Patent 1,197,500 (1970); see *Chem. Abstr.* **71**, 123910 (1969).

P9. W. J. Chambers (E. I. du Pont), U.S. Patent 3,117,996 (1964); *Chem. Abstr.* **60**, 6745 (1964).

P10. M. Kumada and K. Tamao, Jap. Kokai 73 86, 801 (1973); *Chem. Abstr.* **80**, 95450 (1974).

P11. W. H. Atwell and G. N. Bokerman (Dow Corning Corp.), Ger. Offen. 2,260,282 (1973); *Chem. Abstr.* **79**, 66560 (1973).

P12. R. J. de Pasquale (Shell Oil Co.), U.S. Patent 3,748,345 (1973); *Chem. Abstr.* **79**, 105234 (1973).

P13. K. Kimura and H. Ito (Toa Gosei Chem. Ind. Co., Ltd.), Jap. Kokai 73 75, 528 (1973); *Chem. Abstr.* **80**, 70962 (1974).

P14. See for example, K. Hashimoto, N. Mogi, and M. Shindo (Showa Denko K.K.), Ger. Patent 1,645,326 (1973); *Chem. Abstr.* **79**, 79528 (1973).

P15. H. S. Block (Universal Oil Products Co.), U.S. Patent 3,799,996 (1974); *Chem. Abstr.* **80**, 132769 (1974).

P16. Shell Internationale Research, Neth. Patent Appl. 68 17,815 (1968); *Chem. Abstr.* **73**, 88024 (1974).

P17. K. Sato and S. Inoue (Eisai Co. Ltd.), Jap. Kokai 74 00,267 (1974); *Chem. Abstr.* **80**, 108376 (1974).

P18. K. Yamamoto, K. Ueda, S. Akutagawa, and A. Komatsu (Takasago Perfumery Co. Ltd.), Jap. Patent 73 34,572 (1973); *Chem. Abstr.* **80**, 108019 (1974).

CHAPTER VI

Carbonylation and Related Reactions

I. Introduction

The discovery that, in the presence of nickel, carbon monoxide can be directly incorporated into organic systems is principally the result of the investigations carried out by W. Reppe and his co-workers during the years 1938–1945. At the end of the second world war these results were made generally available, initially in the form of summaries published by various interested organizations (11, 24, 40, 77, 106) or by Reppe and his colleagues (88–92, 110, 111) and instigated a period of intense activity in laboratories all over the world. The early work is concerned principally with the carbonylation of alkynes, olefins, alcohols and ethers: in recent years attention has shifted to the carbonylation of organic halides. Although all of these reactions have much in common, our lack of knowledge of the detailed mechanism of carbonylation prevents a general discussion of the whole area and we have chosen to divide the topic into two: the first section is concerned with the classical Reppe-carbonylation reaction while the second is devoted to reactions involving organic halides.

Nickel is, of course, not the only transition metal that catalyzes the carbonylation reaction and although it is a preferred metal for the carbonylation of alkynes, olefins, and organic halides, other metals are of greater importance for the carbonylation of alcohols, ethers, etc., and cobalt catalysts are preeminent for the hydroformylation of olefins. More balanced treatments of the use of transition metals in carbonylation reactions are to be found in the list of reviews included at the end of the chapter.

II. The Carbonylation of Alkynes, Olefins, Alcohols, and Related Compounds

The nickel used in the Reppe carbonylation reaction can be present as a stoichiometric reagent, in the form of an aqueous acidic solution of nickel tetracarbonyl, or as a catalyst, if the reaction is conducted in an atmosphere of carbon monoxide. Both possibilities have been explored in detail. The advantage of the stoichiometric reaction is that it may be carried out at atmospheric pressure and moderate temperatures (particularly important for the carbonylation of acetylene) and its viability rests on an efficient reconversion of the resulting nickel salt into nickel tetracarbonyl (see, for example, 243, P56, P158, P159, and Volume I, p. 3). A semicatalytic process has been developed for the carbonylation of acetylene that enables the reaction to be carried out at moderate temperature and pressure: the reaction is started stoichiometrically using nickel tetracarbonyl and HCl and is continued catalytically by feeding CO and acetylene into the reaction mixture. The presence of a halide (particularly iodide) greatly facilitates the carbonylation reactions.

A. The Carbonylation of Alkynes (Table VI-1)

Alkynes react with carbon monoxide in the presence of nickel compounds to give unsaturated carboxylic acids or their derivatives. The reaction is

$$RC{\equiv}CR' + CO + H_2O \xrightarrow{[Ni]} \begin{array}{c} R \\ \diagdown \\ H \end{array} C{=}C \begin{array}{c} R' \\ \diagup \\ CO_2H \end{array}$$

normally carried out in alcoholic media and an ester is produced. Thioesters (48, P65), amides (48, P39, P68, P70, P95–P97), or anhydrides (48, 100, P37, P54) may be synthesized by carrying the reactions out in the presence of thiols, amines, or carboxylic acids: the acid is first formed and then reacts further.

$$HC{\equiv}CH + CO + ROH \longrightarrow CH_2{=}CHCO_2R$$
$$HC{\equiv}CH + CO + RSH \longrightarrow CH_2{=}CHCOSR$$
$$HC{\equiv}CH + CO + R_2NH \longrightarrow CH_2{=}CHCONR_2$$
$$HC{\equiv}CH + CO + RCO_2H \longrightarrow CH_2{=}CHCOOCOR$$

In the presence of butadiene a Diels-Alder reaction with the unsaturated acid occurs to give cyclohexene-3-carboxylic acid (P20).

Although both mono- and disubstituted alkynes may be carbonylated, the reaction involving acetylene has been studied in greatest detail. [Dialkynes cannot, apparently, be carbonylated (79).] The carbonylation reaction may be carried out in both a stoichiometric and a catalytic manner.

The stoichiometric reaction involves nickel tetracarbonyl in an aqueous acid (e.g., HCl or CH_3CO_2H) and takes place at temperatures ranging from

30° to 85°. The reaction can be performed in a variety of solvents (81), although water or an alcohol are commonly used. The presence of water is

$$4HC{\equiv}CH + Ni(CO)_4 + 2HCl + 4H_2O \longrightarrow 4CH_2{=}CHCO_2H + NiCl_2 + [H_2]$$

essential and its absence results either in a complete lack of reaction (94) or in extensive hydrogenation of the product (48). The reaction is characterized by an induction period that may, in certain cases, be shortened by the addition of pyridine (80) or by irradiation with ultraviolet light (74, 94). The predicted evolution of hydrogen is not observed and attempts have been made to determine its fate in reactions involving acetylene (46, 48, 56, 57) and propyne (55): practically no gaseous hydrogen is detected and it is probable that the principal reactions leading to its consumption are the nickel-catalyzed homogeneous hydrogenation of the reactant and the products. Although the stoichiometric reaction has been used to carbonylate acetylene, the main application is to be found in the carbonylation of substituted alkynes on a laboratory scale. One example of the carbonylation of an alkyne under basic conditions has been reported: diphenylacetylene reacts with nickel tetracarbonyl in methanolic sodium hydroxide to give α-phenyl-*trans*-cinnamic acid as well as tetraphenylbutadiene (58). However, the suggestion that $[Ni_3(CO)_8]^{2-}$ is the active species should be treated with caution (see Volume I, p. 20).

Considerable attention has been given to the catalytic carbonylation. Catalysts have been reported based on nickel tetracarbonyl, either alone (39, 53, P16, P48, P65, P70) or in the presence of tetrasubstituted ammonium or phosphonium salts (P13, P66, P76) or a promoter, e.g., $CuCl_2$ (P15, P87). A catalyst may also be prepared from a nickel salt, either alone (15, 18, 19, 48, 55, 61, 62, 65, P2, P38, P39, P72, P79, P84, P95) or in the presence of a promoter, e.g., $CuCl_2$ or Hg_2Cl_2 (P3, P14, P20, P62, P73, P75, P77, P88), selenium, or reduced nickel (61, P5, P89, P93). Modification of the catalyst by reacting the nickel salt in the presence of a suitable ligand is the subject of many patents; suitable ligands include triphenylphosphine (48, 54, 55, 63, 65, P22, P74, P92), trithiophosphite (P81), tetrasubstituted phosphonium salts (P90), phosphorus pentasulfide (P59), thioacetamide (P11), aminobenzenethiol (P58), 2-thiopyrimidinetrione (P80), imidazoline-2-thione (P58), diethylmercaptoethane (P15), mercaptobenzoic acid (P60), N-methylpyrrollidone (P12, P21), N,N-trimethylethylenediamine (P15), ethanolamine (P61), α-picoline (P85), 1-nitroso-2-naphthol (P77), acetylacetone (P15), dimethylglyoxime (P58, P62), and benzoxazole (P61). The effect of varying the ligand has been systematically studied only in the case of the carbonylation of 1-hexyne using $[(p\text{-}XC_6H_4)_3P]_2NiBr_2$ as catalyst (54): the rate of reaction is found to increase with increasing donor strength of the group X. The catalytic carbonylation reaction is typically carried out at 150–200° and with a pressure of 30 atm.

A semicatalytic process has also been developed whereby the reaction is started stoichiometrically using $Ni(CO)_4$ and HCl and is continued catalytically by feeding CO and acetylene into the reaction. This procedure has the advantage of allowing the reaction to be carried out at essentially atmospheric pressure and with moderate temperatures (32, 55, 92, 96, 97, 107, P53, P63–P67, P70). As found for the stoichiometric reaction, the nature of the solvent is of secondary importance (32).

The majority of the publications mentioned in the discussion given above are concerned with the conversion of acetylene into acrylic acid or its derivatives. Side reactions are only observed if carbonylation is carried out in the absence of water [hydrogenation of the product to give propionic acid occurs (48)] or as the result of further reaction of the acrylic acid with either acetylene to give vinylacrylate, 2,4,6-heptatriene-1-carboxylate, and higher homologs (8, 75, P13, P78), or with further CO to give succinic acid (P98). Unsaturated dicarboxylic acids have also been isolated (P6, P13, P78). The reaction with further CO is included in the following section and that with acetylene is mentioned in Chapter II, Section III.

A study of the products of the carbonylation of mono- and disubstituted alkynes indicates that predominantly cis Markownikoff addition of the elements of formic acid (H—CO_2H) occurs. The most detailed study involves monosubstituted diphenylacetylene (21). In the case in which the substituent

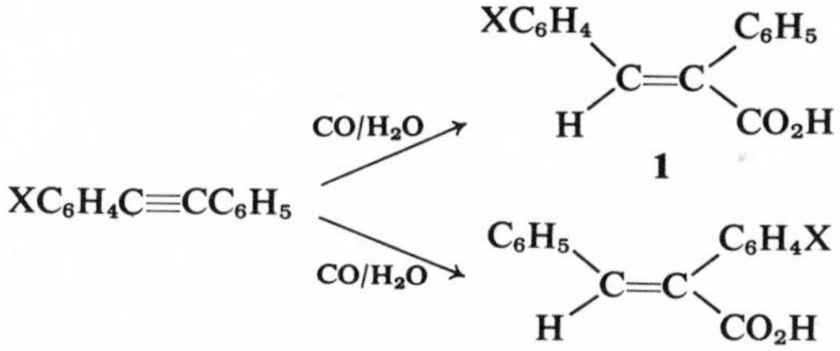

X occupies a meta or para position, a correlation is found between the distribution of the products and the Hammett σ parameter for X. However, the presence of ortho substituents (irrespective of their electronic properties) results in the preferential formation of **2**, the carboxyl group adding to the more sterically crowded alkyne C atom.

o-Nitrodiphenylacetylene reacts anomalously: 3-benzylidene oxindole (**3**) in traces is the only characterisable product (21, 105).

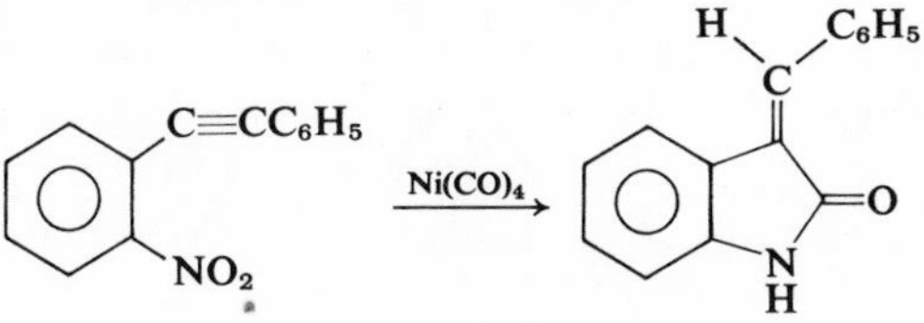

Incorporation of a second molecule of a disubstituted alkyne has been observed in the reaction of nickel tetracarbonyl with diphenylacetylene in dioxan: tetraphenylcyclopentenone (4) is formed (87). It is not clear whether 4 is the product of the hydrogenation of tetraphenylcyclopentadienone or whether it is the result of an acid-catalyzed rearrangement of an intermediate divinyl ketone. Both reactions have precedence: the formation of a cyclo-

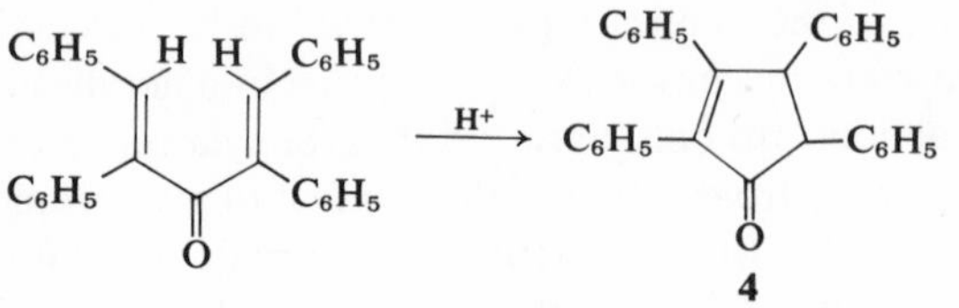

pentadienone derivative is believed to occur on treating cyclooctyne with nickel tetracarbonyl in ether (60) and the formation of divinylketones has been mentioned briefly (59).

Vinyl-substituted alkynes can also be carbonylated. The product from the reaction is a cyclic dimer. The primary product, a 1,3-dienoic acid, probably reacts further in a Diels-Alder reaction. The product from the reaction involving vinylacetylene, 5, is known as "mikanecic acid" (28, 48, 67, 68).

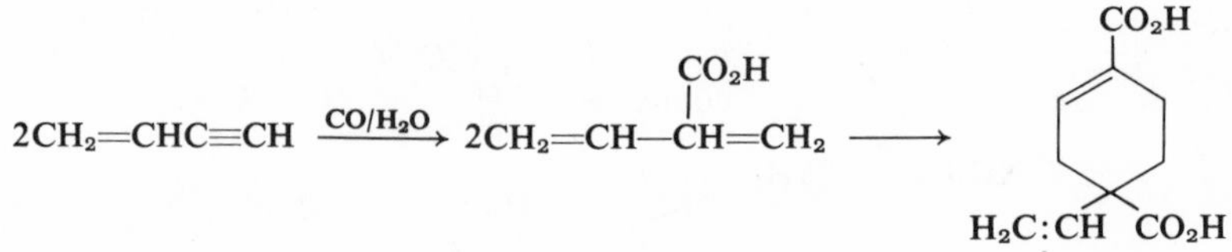

By analogy, the compounds formed by carbonylation of 3-pentene-1-yne (80), 3-methyl-3-butene-1-yne (70), and 1-cyclohexenylacetylene (80) have been formulated as 6, 7, and 8 (93).

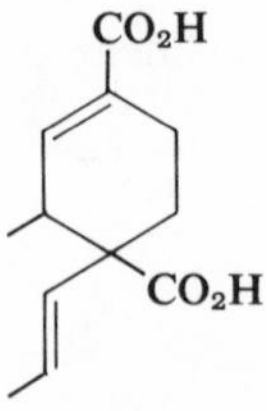

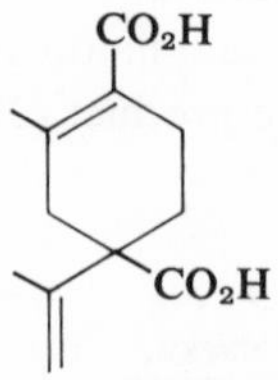

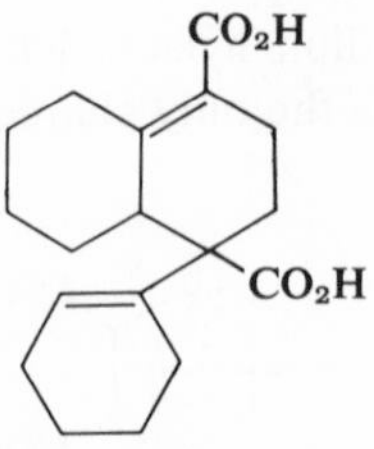

RELATED REACTIONS

Allene is not as reactive as propyne but it may also be carbonylated to give methyl methacrylate (38, 45, P23, P41, P42). The reaction of 1,2-hexadiene leads mainly to polymeric material (38, 45).

Diphenylacetylene and diphenylketene have been reported to react together with nickel tetracarbonyl in DMF or THF to give a cyclopentene-1,3-dione derivative (9) in addition to the thermal reaction product, 2,3,4-triphenyl-1-naphthyl diphenylacetate (103). The intermediacy of a nickelacyclohexene-1,3-dione system has been suggested (34).

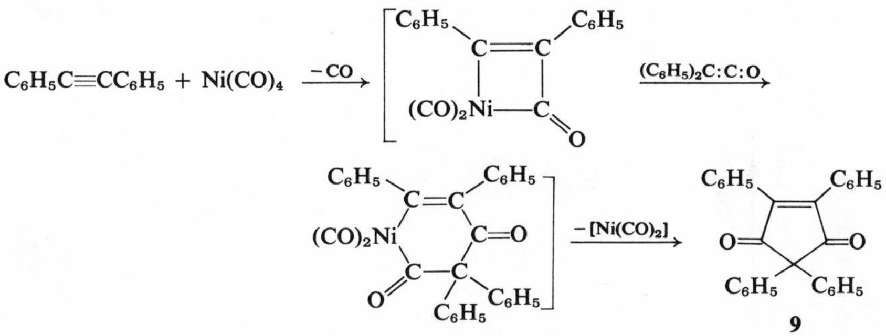

An investigation of the reaction of cyclopropenone derivatives with nickel tetracarbonyl was prompted by the early suggestion that these are involved as intermediates in the carbonylation of alkynes (48, 98). Diphenylcyclopropenone, dibutylcyclopropenone, and cycloheptacyclopropenone react with aqueous acidic nickel tetracarbonyl to form unsaturated carboxylic acids, e.g., *trans*-α-phenylcinnamic acid derived from diphenylcyclopropenone (10, 17). The possibility that this reaction involves initial decarbonylation to

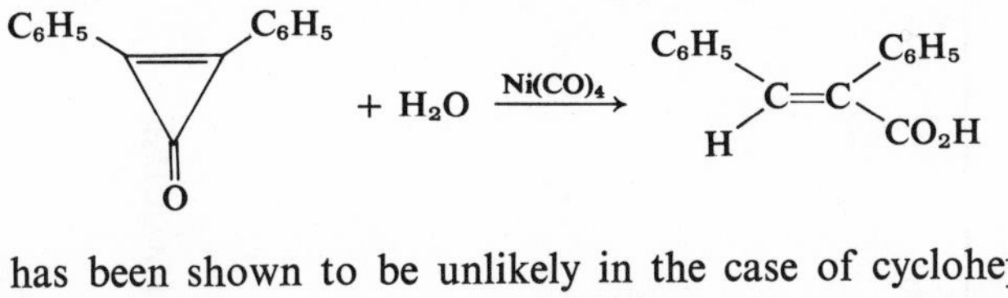

the alkyne has been shown to be unlikely in the case of cycloheptacyclopropenone: in anhydrous benzene trimerization without decarbonylation occurs (244; see, however, ref. 60). [The same trimer is also produced by reacting 2,3,4,5-bis(pentamethylene)benzoquinone with cycloheptacyclopropenone in the presence of catalytic amounts of bis(cyclooctadiene)nickel (244).] The reaction of diphenylcyclopropenone with nickel tetracarbonyl in anhydrous benzene, however, is accompanied by considerable decarbonylation and a mixture of diphenylacetylene, tetraphenylcyclopentadienone, and

TABLE VI-1

THE CARBONYLATION OF ALKYNES

Product	Alkyne	Nickel component	Type^a	Ref.
$CH_2:CHCO_2H$	$HC:CH$	$Ni(CO)_4$	S	31, 32, 41, 46–48, 56, 57, 77, 85, 86, 87, P6, P24, P25, P37, P44, P45, P71, P96, P97
		$Ni(CO)_4$	C	P16, P64, P65, P76, P87
		NiX_2	C	15, 18, 19, 23, 48, 61, P2, P3, P5, P14, P39, P72, P73, P75, P79, P83, P84, P86, P88, P89, P91, P93, P95
		$NiX_2–Lig$	C	48, 61–65, P11, P12, P15, P21, P22, P58–P62, P74, P77, P80, P81, P85, P90, P92
		$Ni(CO)_4$	SC	32, 92, 96, 97, P53, P54, P63, P66–P68
$CH_2:C(CO_2H)CH_3$,	$HC:CCH_3$	$Ni(CO)_4$	S	55, 68, P49
($trans$-$CH_3CH:CHCO_2CH_3$)		NiX_2	C	55, P38
		$Ni(CO)_4$	SC	39, 45, 55, P47, P48, P50
$CH_2:C(CO_2H)C_4H_9$,	$HC:CC_4H_9$	$Ni(CO)_4$	S	21, 45, 54, 68, 78, 81
($C_4H_9CH:CHCO_2H$)				
$CH_2:C(CO_2H)C_6H_{13}$	$HC:CC_6H_{13}$	$NiX_2–Lig$	C	54, P38
$CH_2:C(CO_2H)CH_2OH$,	$HC:CCH_2OH$	$Ni(CO)_4$	S	48, 94, 95
($HOCH_2CH:CHCO_2R$)		$Ni(CO)_4$	S/C	53

Product	Reactant	Catalyst		Ref.
$CH_2{:}C(CO_2H)CH_2OCOCH_3$	$HC{:}CCH_2OCOCH_3$	$Ni(CO)_4$	S	53, 78
$CH_2{:}C(CO_2H)CH_2CO_2C_2H_5$	$HC{:}CCH_2CO_2C_2H_5$	$Ni(CO)_4$	S	82
$CH_2{:}C(CO_2H)CH(OCOCH_3)C_6H_5$	$HC{:}CCH(OCOCH_3)C_6H_5$	$Ni(CO)_4$	S	78
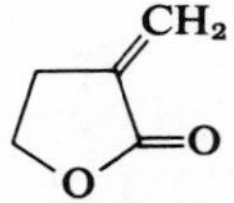 (α-methylene-γ-butyrolactone structure)	$HC{:}CCH_2CH_2OH$	$Ni(CO)_4$	S	78, P43
$CH_2{:}C(CO_2H)CH_2CH_2OCOCH_3$	$HC{:}CCH_2CH_2OCOCH_3$	$Ni(CO)_4$	S	78
$CH_2{:}C(CO_2H)CH_2CH_2CH_2O$ (tetrahydropyran) 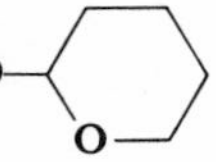	$HC{:}CCH_2CH_2O$ 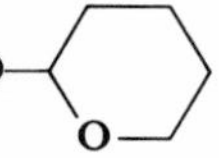 (tetrahydropyran)	$Ni(CO)_4$	S	78
$CH_2{:}C(CO_2H)CH_2CH_2CO_2C_2H_5$	$HC{:}CCH_2CH_2CO_2C_2H_5$	$Ni(CO)_4$	S	82
$CH_2{:}C(CO_2H)CH(OH)CH_3$	$HC{:}CCH(OH)CH_3$	$Ni(CO)_4$	S	48
$CH_2{:}C(CO_2H)CH(OCOCH_3)CH_3$	$HC{:}CCH(OCOCH_3)CH_3$	$Ni(CO)_4$	S	29
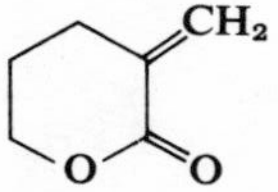 (α-methylene-δ-valerolactone structure)	$HC{:}C(CH_2)_3OH$	$Ni(CO)_4$	S	78
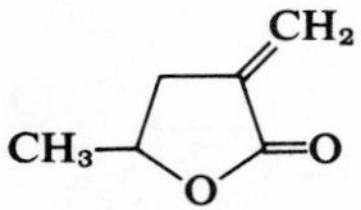 (methylene lactone with CH_3 structure)	$HC{:}CCH_2CH(OH)CH_3$	$Ni(CO)_4$	S	78, P43
$CH_2{:}C(CO_2H)(CH_2)_3CN$	$HC{:}CC(CH_2)_3CN$	$Ni(CO)_4$	S	82
$CH_2{:}C(CO_2H)(CH_2)_3CO_2C_2H_5$	$HC{:}CC(CH_2)_3CO_2C_2H_5$	$Ni(CO)_4$	S	82
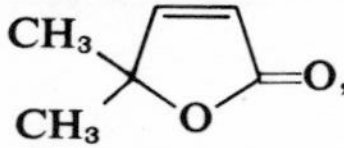 (dimethyl furanone structure, CH_3, CH_3)	$HC{:}CCOH(CH_3)_2$	$Ni(CO)_4$	S	70, 83
$(CH_3)_2C{:}C{:}CHCO_2C_4H_9$				

(continued)

TABLE VI-1 (*continued*)

Product	Alkyne	Nickel component	Type[a]	Ref.
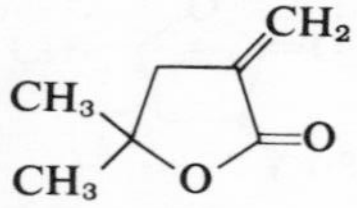 (structure: 4-methylene-5,5-dimethyl-dihydrofuran-2-one)	$HC:CCH_2COH(CH_3)_2$	$Ni(CO)_4$	S	78
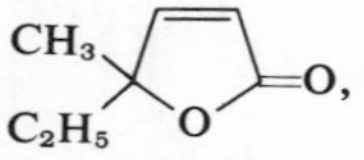 (structure: 5-methyl-5-ethyl-2(5H)-furanone)	$HC:CC(CH_3)OHCH_2CH_3$	$Ni(CO)_4$	S	83
$CH_3CH_2C(CH_3):C:CHCO_2H$ $CH_2:C(CO_2H)CH(OCOCH_3)C_3H_7$ $CH_2:C(CO_2H)(CH_2)_4CO_2C_2H_5$	$HC:CCH(OCOCH_3)C_3H_7$ $HC:C(CH_2)_4CO_2C_2H_5$	$Ni(CO)_4$ $Ni(CO)_4$	S S	78 82
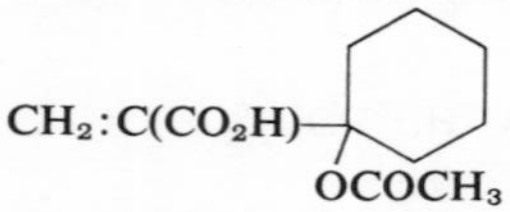 $CH_2:C(CO_2H)$ (cyclohexane ring with $OCOCH_3$)	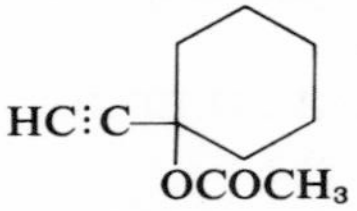 $HC:C$ (cyclohexane ring with $OCOCH_3$)	$Ni(CO)_4$	S	78
$CH_2:C(CO_2H)C_6H_5,$ $(C_6H_5CH:CHCO_2H)$	$HC:CC_6H_5$	$Ni(CO)_4$	S	21, 48, 78, 81
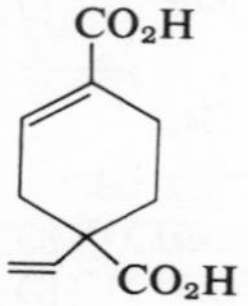 (cyclohexene ring with CO_2H and CO_2H and vinyl group)	$HC:CCH:CH_2$	$Ni(CO)_4$	S	24, 48, 67, 68

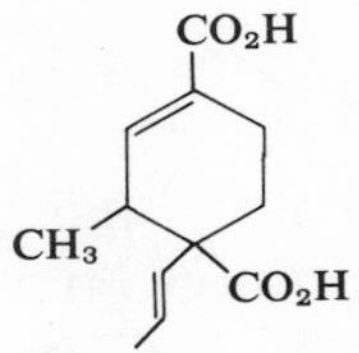

Substrate	Catalyst		Ref.	Product
$HC{:}CCH{:}CHCH_3$	$Ni(CO)_4$	S	80	(cyclohexene with CO_2H, CO_2H, CH_3 substituents)
$HC{:}CC(CH_3){:}CH_2$	$Ni(CO)_4$	S	70	(cyclohexene with CO_2H, CO_2H, CH_3 substituents)
$HC{:}C$ (cyclohexenyl)	$Ni(CO)_4$	S	80	(fused bicyclic with CO_2H, CO_2H substituents)
$CH_3C{:}CCOCH_3$	$Ni(CO)_4$	S	79	$CH_3C(CO_2H){:}CHCOCH_3$
$CH_3C{:}CCH(OH)CH_3$	$Ni(CO)_4$	S	79	$CH_3C(CO_2H){:}CHCH(OH)CH_3$
$CH_3C{:}CC_5H_{11}$	$Ni(CO)_4$	S	109	$CH_3C(CO_2H){:}CHC_5H_{11}$
$CH_3C{:}CC_6H_{13}$	$Ni(CO)_4$	S	48	$CH_3C(CO_2H){:}CHC_6H_{13}$, $C_6H_{13}C(CO_2H){:}CHCH_3$
$CH_3C{:}CC_6H_5$	$Ni(CO)_4$	S	21, 48, P19	$CH_3C(CO_2H){:}CHC_6H_5$ $C_6H_5C(CO_2H){:}CHCH_3$

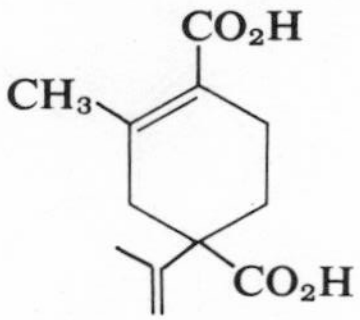

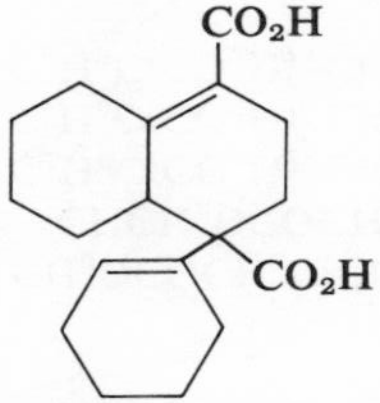

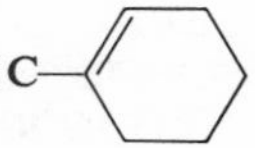

(continued)

TABLE VI-1 (*continued*)

Product	Alkyne	Nickel component	Type^a	Ref.
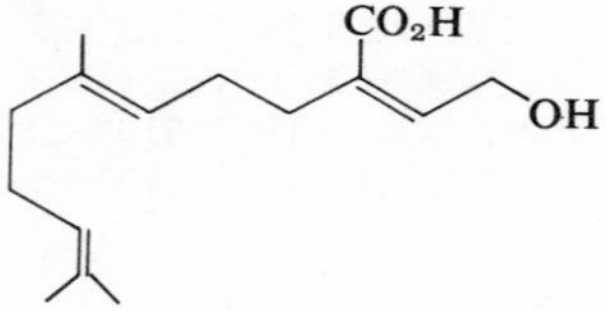	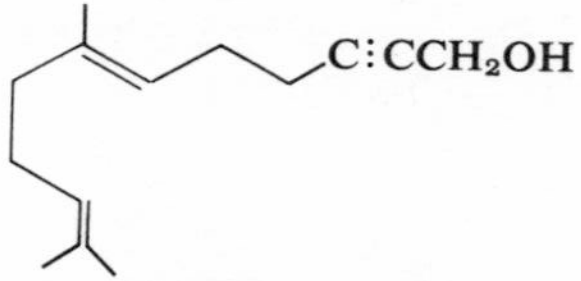 $C:CCH_2OH$	$Ni(CO)_4$	S	27
$CH_3OCOCH_2C(CO_2H):$ $CHCH_2OCOCH_3$	$CH_3OCOCH_2C:$ CCH_2OCOCH_3	$Ni(CO)_4$	S	79
$C_4H_9C(CO_2H):CHCOCH_3$	$C_4H_9C:CCOCH_3$	$Ni(CO)_4$	S	79
$C_4H_9C(CO_2H):CHCO_2H$	$C_4H_9C:CCO_2H$	$Ni(CO)_4$	S	79
$C_4H_9C(CO_2H):CHC_4H_9$	$C_4H_9C:CC_4H_9$	$Ni(CO)_4$	S	79
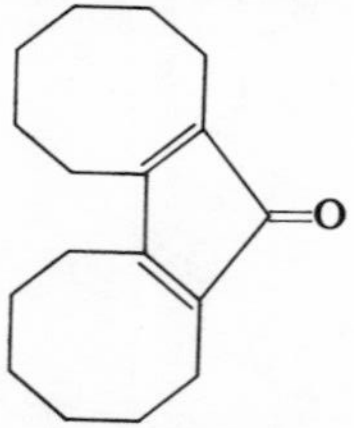	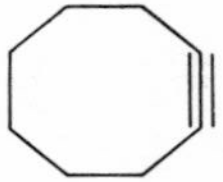	$Ni(CO)_4$	S	60
$C_6H_5C(CO_2H):CHC_6H_5$	$C_6H_5C:CC_6H_5$	$Ni(CO)_4$	S	34, 48, 58, 79, P19, P31
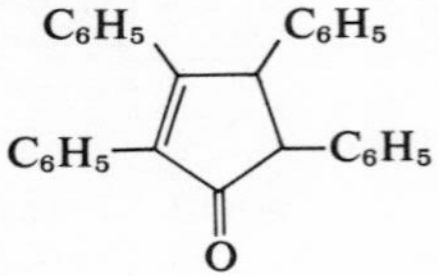	$C_6H_5C:CC_6H_5$	$Ni(CO)_4$	S	87

p-CH$_3$OC$_6$H$_4$C(CO$_2$H):CHC$_6$H$_5$, C$_6$H$_5$C(CO$_2$H):CHC$_6$H$_4$-p-OCH$_3$	p-CH$_3$OC$_6$H$_4$C:CC$_6$H$_5$	Ni(CO)$_4$	S	21
p-CH$_3$C$_6$H$_4$C(CO$_2$H):CHC$_6$H$_5$, C$_6$H$_5$C(CO$_2$H):CHC$_6$H$_4$-p-CH$_3$	p-CH$_3$C$_6$H$_4$C:CC$_6$H$_5$	Ni(CO)$_4$	S	21
m-CH$_3$OC$_6$H$_4$C(CO$_2$H):CHC$_6$H$_5$, C$_6$H$_5$C(CO$_2$H):CHC$_6$H$_4$-m-OCH$_3$	m-CH$_3$OC$_6$H$_4$C:CC$_6$H$_5$	Ni(CO)$_4$	S	21
p-ClC$_6$H$_4$C(CO$_2$H):CHC$_6$H$_5$, C$_6$H$_5$C(CO$_2$H):CHC$_6$H$_4$-p-Cl	p-ClC$_6$H$_4$C:CC$_6$H$_5$	Ni(CO)$_4$	S	21
m-ClC$_6$H$_4$C(CO$_2$H):CHC$_6$H$_5$, C$_6$H$_5$C(CO$_2$H):CHC$_6$H$_4$-m-Cl	m-ClC$_6$H$_4$C:CC$_6$H$_5$	Ni(CO)$_4$	S	21
p-NO$_2$C$_6$H$_4$C(CO$_2$H):CHC$_6$H$_5$, C$_6$H$_5$C(CO$_2$H):CHC$_6$H$_4$-p-NO$_2$	p-NO$_2$C$_6$H$_4$C:CC$_6$H$_5$	Ni(CO)$_4$	S	21
o-CH$_3$OC$_6$H$_4$C(CO$_2$H):CHC$_6$H$_5$, C$_6$H$_5$C(CO$_2$H):CHC$_6$H$_4$-o-OCH$_3$	o-CH$_3$OC$_6$H$_4$C:CC$_6$H$_5$	Ni(CO)$_4$	S	21
o-CH$_3$C$_6$H$_4$C(CO$_2$H):CHC$_6$H$_5$, C$_6$H$_5$C(CO$_2$H):CHC$_6$H$_4$-o-CH$_3$	o-CH$_3$C$_6$H$_4$C:CC$_6$H$_5$	Ni(CO)$_4$	S	21
o-ClC$_6$H$_4$C(CO$_2$H):CHC$_6$H$_5$, C$_6$H$_5$C(CO$_2$H):CHC$_6$H$_4$-o-Cl	o-ClC$_6$H$_4$C:CC$_6$H$_5$	Ni(CO)$_4$	S	21
[structure: 3-(phenylmethylene)indolin-2-one]	o-NO$_2$C$_6$H$_4$C:CC$_6$H$_5$	Ni(CO)$_4$	S	21, 105

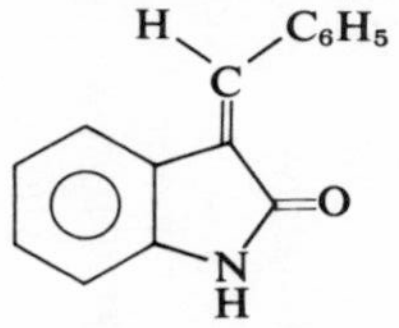

^{a} S = stoichiometric reaction; C = catalytic reaction; SC = semicatalytic reaction.

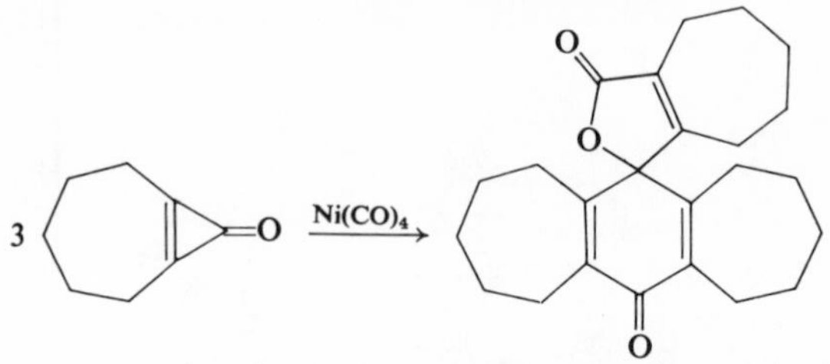

a compound mistakenly identified as a tris(cyclopropenone)nickel complex is obtained (17, 20, 244). Based on these results it has been suggested (10) that the reaction of cyclopropenones with aqueous acidic nickel tetracarbonyl involves the same nickelacyclobutenone intermediate postulated to be involved in the carbonylation of alkynes (see Section II, E). In agreement with

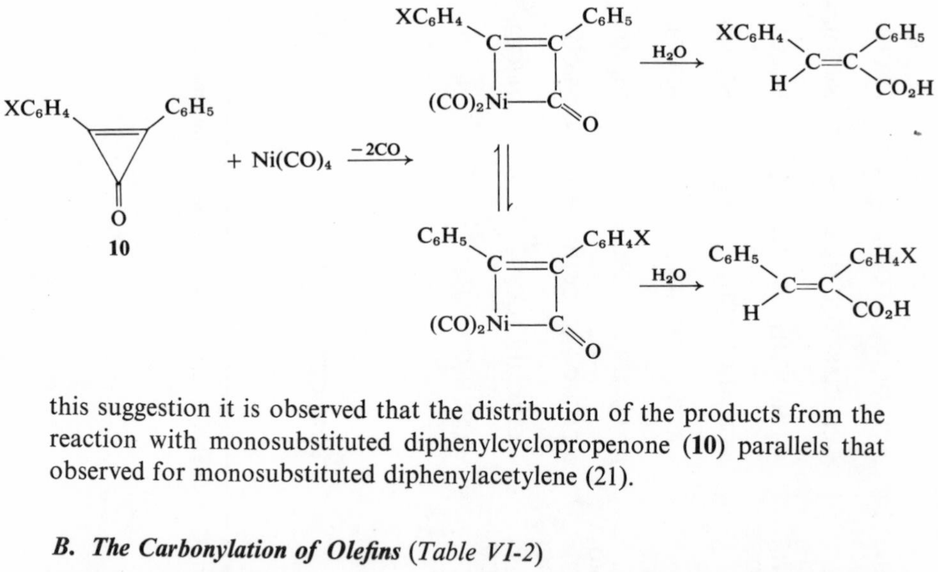

this suggestion it is observed that the distribution of the products from the reaction with monosubstituted diphenylcyclopropenone (10) parallels that observed for monosubstituted diphenylacetylene (21).

B. The Carbonylation of Olefins *(Table VI-2)*

The carbonylation of olefins, in the presence of nickel, was patented as early as 1934 (P110). However, the results of a systematic study were not widely available until 1953 (49).

Olefins are, in general, more difficult to carbonylate than alkynes and both stoichiometric and catalytic reactions require elevated temperatures and pressure. The product in the presence of water is a carboxylic acid or the corresponding ester, thioester (49, P40), amide (49, P40, P102), or anhydride (26, 49, P9, P28, P31, P104) when carried out in the presence of an alcohol, thiol, amine, or carboxylic acid.

Catalysts for the carbonylation of olefins have been described based on nickel tetracarbonyl alone (26, 37, 45, 47, 49, P28, P30, P101, P104, P109, P162) or combined with various additives, e.g., NiX_2, I_2 (49, P9, P17, P26, P27, P35, P57, P105, P106), or triphenylphosphine (37); nickel salts (12, 22, 23, 26, 49, 84, 102, 114, 249, P9, P10, P28, P52, P94, P98, P102) that may also be modified by the addition of triphenylphosphine (36) or additives, e.g., HI, CuI (49, 112, P26, P27, P55, P107); and metallic nickel with or without additives (31, 42, 49, P105, P110). Catalysts or additives containing iodine, e.g., NiI_2, HI, or CuI, are particularly effective, whereas the addition of boric acid is reported to suppress the formation of insoluble polymers (P100). Typical conditions for the catalytic carbonylation of ethylene to give propionic acid are ca. 250° and 200 atm pressure.

$$CH_2{=}CH_2 + CO + H_2O \xrightarrow{\text{[Ni]}} CH_3CH_2CO_2H$$

Nickel tetracarbonyl can also be used in a stoichiometric carbonylation reaction (16, 25, 30, 49, 73, 74, P40). Under mild conditions, the stoichiometric reaction is confined to 1,5-dienes (73) or to systems containing double bonds activated by ring strain (16, 25, 30) or by ultraviolet irradiation (74).

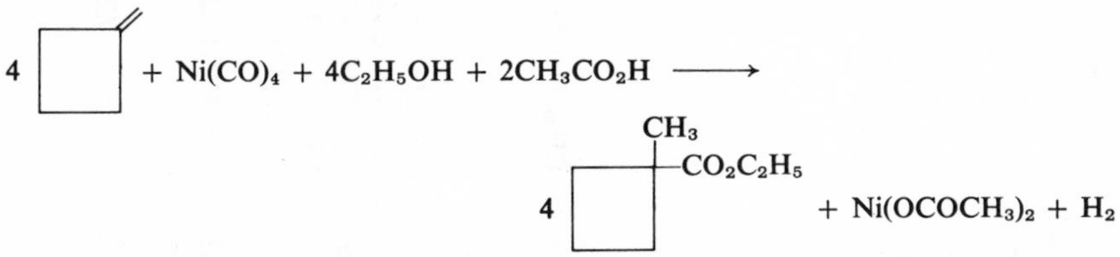

$$4\ \square + Ni(CO)_4 + 4C_2H_5OH + 2CH_3CO_2H \longrightarrow$$

$$4\ \underset{}{\square}\overset{CH_3}{\underset{CO_2C_2H_5}{}} + Ni(OCOCH_3)_2 + H_2$$

The stereochemistry of the reaction has been investigated only in the case of bicycloheptene. Using deuterated solvents it has been shown that cis addition to the exo side of the molecule occurs (25).

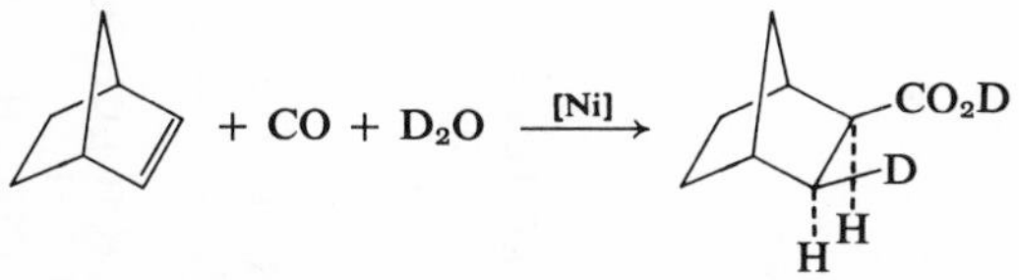

$$+ CO + D_2O \xrightarrow{\text{[Ni]}} \overset{CO_2D}{\underset{H}{\cdots}D}$$

Unsymmetrical olefins are in general converted into a mixture of the possible carbonylation products, e.g., propylene produces *n-* and *iso*-butyric acid.

$$CH_3CH{=}CH_2 + CO + H_2O \xrightarrow{\text{[Ni]}} CH_3CH_2CH_2CO_2H + (CH_3)_2CHCO_2H$$

In some cases the structure of the carbonylation products indicates that isomerization of the olefin has occurred (74, 112).

TABLE VI-2

THE CARBONYLATION OF OLEFINS

Product	Olefin	Nickel component	Type^a	Ref.
$CH_3CH_2CO_2H$	$CH_2:CH_2$	$Ni(CO)_4$	C	26, 49, 84, P9, P10, P28, P30, P31, P35, P104, P109, P162
		NiX_2	C	12, 22, 23, 26, 49, 102, 114, P9, P10, P26–P28, P52, P94, P100
		Ni	C	42, 49, P105
$CH_3CH_2CH_2CO_2H$, $(CH_3)_2CHCO_2H$	$CH_3CH:CH_2$	NiX_2–Lig	C	36, P26, P27
		$Ni(CO)_4$–Lig	C	37, 45, 47, 49, P28, P30, P31, P105
		Ni	C	42
$CH_3(CH_2)_2CH_2CO_2H$	$CH_3CH_2CH:CH_2$	$Ni(CO)_4$	S	49, P40
$(CH_3)_2CHCH_2CO_2H$, $(CH_3)_3CCO_2H$	$(CH_3)_2C:CH_2$	$Ni(CO)_4–NiI_2$	C	49
$CH_3(CH_2)_2CH_2CH_2CO_2H$, $CH_3(CH_2)_2CH(CH_3)CO_2H$	$CH_3(CH_2)_2CH:CH_2$	NiX_2	C	112
	$CH_3CH_2CH:CHCH_3$	NiX_2	C	112
$(CH_3)_2CHCH_2CH_2CO_2H$, $(CH_3)_2CHCH(CH_3)CO_2H$, $CH_3CH_2C(CH_3)HCH_2CO_2H$	$(CH_3)_2C:CHCH_3$	NiX_2	C	112
$CH_3(CH_2)_3CH_2CO_2C_2H_5$, $CH_3(CH_2)_3CH(CH_3)CO_2C_2H_5$	$CH_3(CH_2)_3CH:CH_2$	$Ni(CO)_4$	C	45
		Ni	C	31
		NiX_2	C	112
C_8 Carboxylic acids	$CH_3(CH_2)_5CH:CH_2$	$Ni(CO)_4$	S/C	49, 74, P40, P57
		NiX_2	C	249, P108

Reactant	Product	Catalyst	C/S	Reference
C9 Carboxylic acids				
$C_3H_7CH{:}CHC_3H_7$		$Ni(CO)_4$	S	74
$C_4H_9C(C_2H_5){:}CH_2$	$C_4H_9CH(C_2H_5)CH_2CO_2H$	NiX_2	C	49
$CH_3(CH_2)_9CH{:}CH_2$	$CH_3(CH_2)_{10}CH_2CO_2H$, $CH_3(CH_2)_9CH(CH_3)CO_2H$	Ni	C	49, P104
C17 Carboxylic acids				
$CH_3(CH_2)_{13}CH{:}CH_2$		NiX_2	C	49, P55
$CH_3(CH_2)_{15}CH{:}CH_2$	$CH_3(CH_2)_{15}CH(CH_3)CO_2H$	NiX_2	C	49, P40
$CH_2{:}CHCH_2OH$	$CH_3CH{:}CHCO_2H$	$Ni(CO)_4$	C	49; see also 115
$CH_2{:}CHCH_2CH_2OH$	[δ-valerolactone]	$Ni(CO)_4$	C	49
$CH_2{:}CHCO_2H$	$HO_2CCH_2CH_2CO_2H$	NiX_2	C	98
$CH_2{:}CH(CH_2)_8CO_2H$	$CH_3CH(CO_2H)(CH_2)_8CO_2H$, $HO_2CCH_2CH_2(CH_2)_8CO_2H$	$Ni(CO)_4$	C	49, P101
C18 Dicarboxylic acids				
$CH_3(CH_2)_7CH(CH_2)_7CO_2H$	$CH_2{:}CH(CH_2)_2CH{:}CH_2$	$Ni(CO)_4{-}NiI_2$	C	49, P17, P101, P106
C_8H_8 Dimer	$C_{16}H_{16}$ Dicarboxylic acid	$Ni(CO)_4$	C	109
$CH_2{:}CH(CH_2)_2CH{:}CH_2$	$CH_2{:}CH(CH_2)_2CH(CH_3)CO_2H$	$Ni(CO)_4$	S	49, 73, P82
$CH_2{:}CH(CH_2)_2CH{:}CH_2$	[methyl-cyclohexanone]	$Ni(CO)_4$	S	73
C8 Carboxylic acid				
$CH_2{:}CH(CH_2)_4CH{:}CH_2$	[dimethyl-cyclopentanone]	$Ni(CO)_4$	S	73
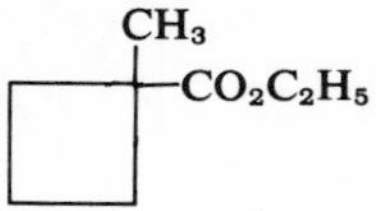 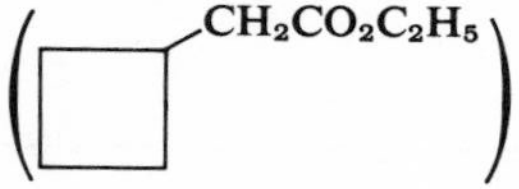 [methylenecyclobutane]	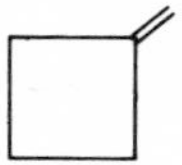 [cyclobutane-CH_3, $CO_2C_2H_5$] / [cyclobutane-$CH_2CO_2C_2H_5$]	$Ni(CO)_4$	S	30

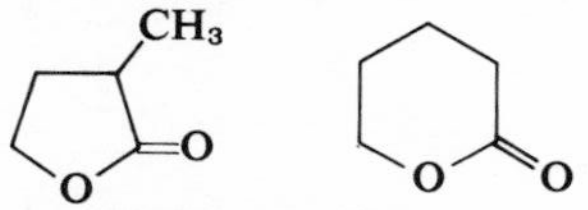

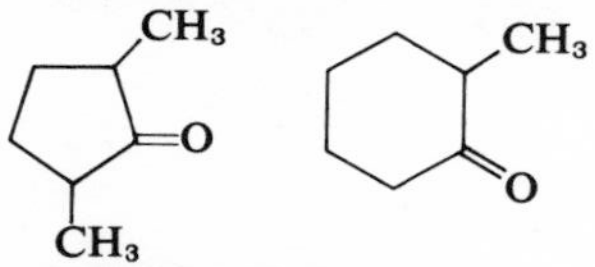

(continued)

TABLE VI-2 (*continued*)

Product	Olefin	Nickel component	Type^a	Ref.
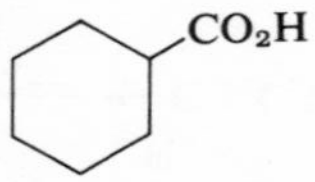	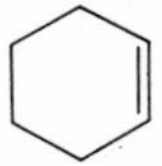	Ni(CO)$_4$	C	31, 49, 112, P82
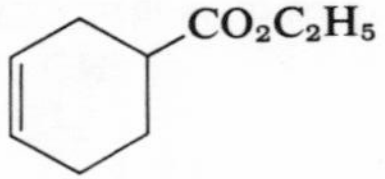	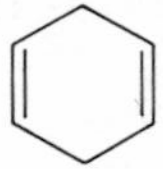	Ni-KI	C	P105
		Ni(CO)$_4$	S	25
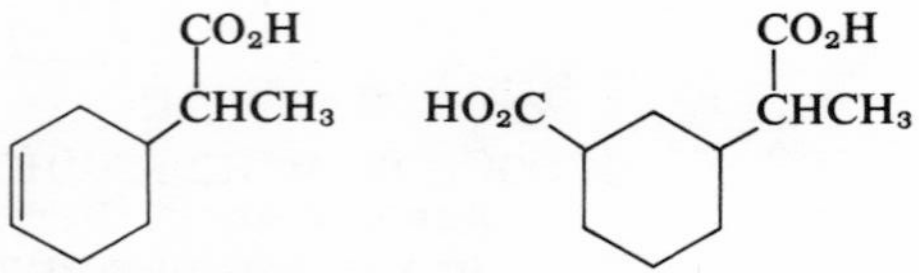	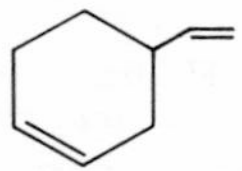	Ni(CO)$_4$	C	49, P82, P105
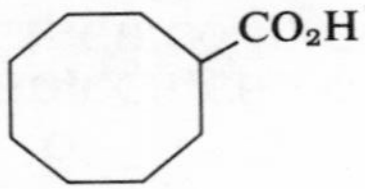	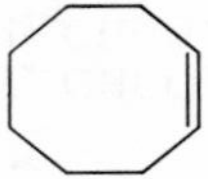	Ni(CO)$_4$	C	109
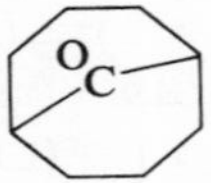	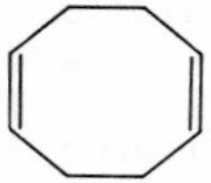	Ni(CO)$_4$	S	73; see also P107
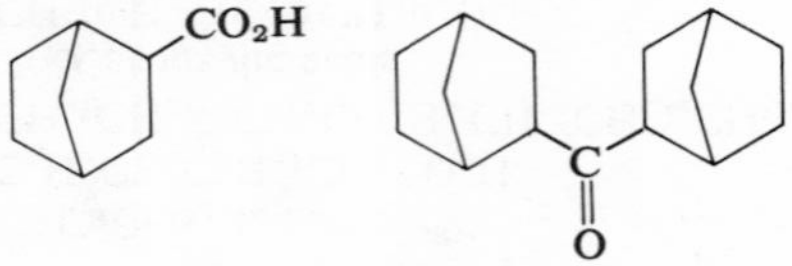	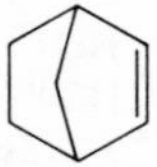	Ni(CO)$_4$	S	25

(continued)

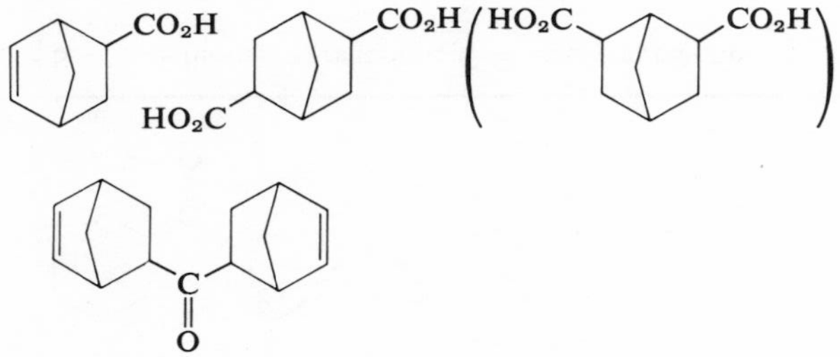 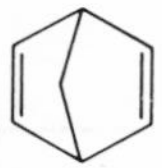 $Ni(CO)_4$	S	16, 25
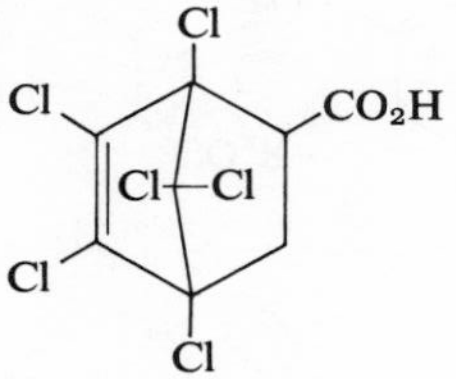 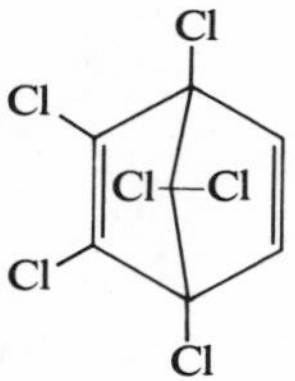 $Ni(CO)_4$	S	25
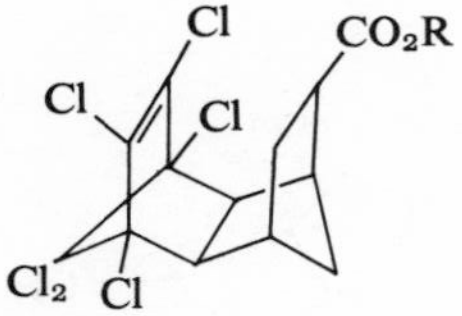 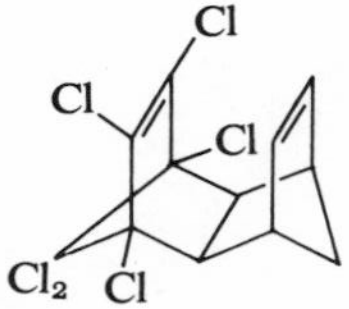 $Ni(CO)_4$	S	25
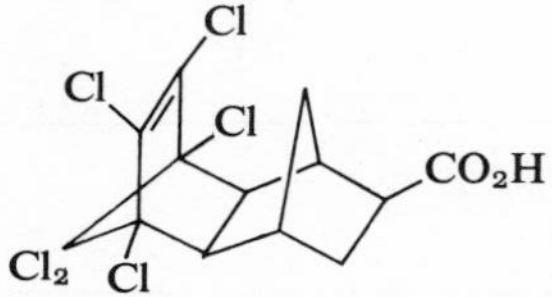 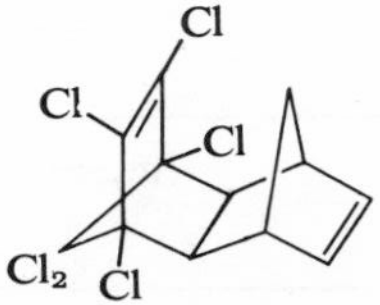 $Ni(CO)_4$	S	25

TABLE VI-2 (*continued*)

Product	Olefin	Nickel component	Type^a	Ref.
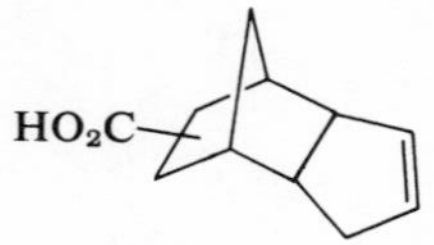	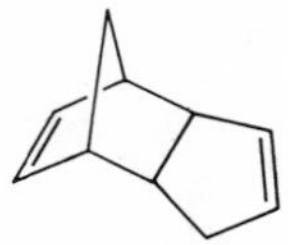	Ni(CO)$_4$	S	25
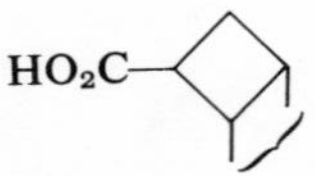	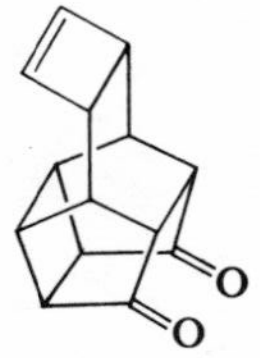	Ni(CO)$_4$	S	25
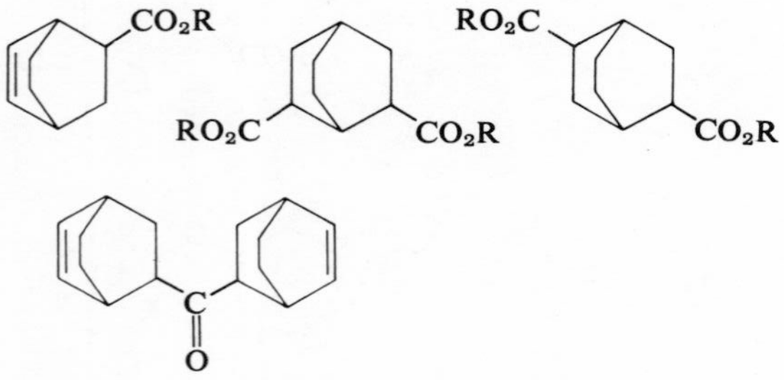	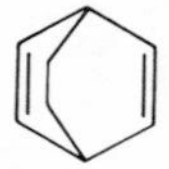	Ni(CO)$_4$	S	25

^{a} S = stoichiometric reaction; C = catalytic reaction.

A few examples of the conversion of olefins into aldehydes and ketones have been reported. These may be the result of a nickel-catalyzed "oxo reaction," although evidence is lacking. In some cases the aldehyde (49) or ketone (49, P103) is present only as a minor component; however, the use of a complex cyanide of nickel promotes the formation of ketones and from the reaction with ethylene a mixture of polyketones and polyketocarboxylic acids has been obtained, e.g., $(C_2H_5)_2CO$, $CH_3(CH_2COCH_2)_nCH_3$ (n = 2,3,4), and $CH_3CH_2COCH_2CH_2CO_2H$ (P9, P52). Considerable conversion to a ketone is also observed in the reactions involving strained ring systems (16, 25), e.g., **11** is formed in the reaction with bicycloheptene, while bicyclo-3,3,1-nonanone-9 (**12**) is the principal reaction product from the stoichiometric carbonylation of 1,5-cyclooctadiene (73). A "true" oxo reaction has been

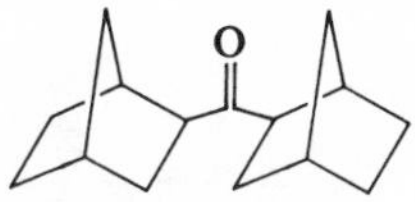

11    **12**

observed in the reaction of 1-hexene, hydrogen, and carbon monoxide to give heptanal, which is catalyzed by tetrakisligand nickel complexes (P51; see also P29).

$$CH_3(CH_2)_3CH{=}CH_2 + CO + H_2 \xrightarrow{[Ni]} CH_3(CH_2)_5CHO$$

RELATED REACTIONS

Diazoalkanes react with nickel tetracarbonyl to give a ketene which, it is suggested, results from the carbonylation of an intermediate nickel carbene complex (52).

$$(C_6H_5)_2CN_2 + Ni(CO)_4 \xrightarrow{-N_2/CO} [(C_6H_5)_2CNi(CO)_3] \xrightarrow{Ni(CO)_2} (C_6H_5)_2C{=}C{=}O$$

C. The Carbonylation of Alcohols, Ethers, Esters, and Aldehydes (*Tables VI-3 and VI-4*)

The carbonylation of an alcohol to the next higher carboxylic acid is catalyzed by nickel. The catalyst may be based on nickel tetracarbonyl, nickel halides, or metallic nickel in the presence of a halogen or halide spender (e.g., CuI). Systems involving iodine or iodide have the greatest activity. Relatively high temperatures (ca. 250–300°) and high pressures (ca. 200 atm) are needed.

TABLE VI-3

THE CATALYTIC CARBONYLATION OF ALCOHOLS

Product	Alcohol	Nickel component	Ref.
CH_3CO_2H	CH_3OH	NiX_2	14, 22, 23, 50, 84, 101, P1, P7, P8, P26, P27, P46, P112
$CH_3CH_2CO_2H$	CH_3CH_2OH	Ni–CuI	6, 7, 22, 23, 49, 50, 101, P26, P27
$CH_3CH_2CH_2CO_2H$, $(CH_3)_2CHCO_2H$	$CH_3CH_2CH_2OH$	$Ni(CO)_4$–NiI_2	50, 114
$(CH_3)_2CHCO_2H$	$(CH_3)_2CHOH$	Ni–$NiCl_2$–NaI	P26, P27
$CH_3CH_2CH(CH_3)CO_2H$	$CH_3(CH_2)_2CH_2OH$	$Ni(CO)_4$–$NiCl_2$	4, P26, P27, P46, P112
	$CH_3CH_2CH(CH_3)OH$	$Ni(CO)_4$–$NiCl_2$	4
$(CH_3)_2CHCH_2CO_2H$, $(CH_3)_3CCO_2H$	$(CH_3)_2CHCH_2OH$	$Ni(CO)_4$–NiI_2	50, 114
$CH_3(CH_2)_2CH(CH_3)CO_2H$	$CH_3(CH_2)_3CH_2OH$	$Ni(CO)_4$–$NiCl_2$	4
$CH_3CH_2C(CH_3)_2CO_2H$	$CH_3CH_2C(CH_3)_2OH$	$Ni(CO)_4$–$NiCl_2$	4
C_6 Carboxylic acid	$(CH_3)_3CCH_2OH$	$Ni(CO)_4$–$NiCl_2$	4
$CH_3(CH_2)_3CH(CH_3)CO_2H$	$CH_3(CH_2)_4CH_2OH$	$Ni(CO)_4$–$NiCl_2$	4
	$CH_3(CH_2)_3CH(CH_3)OH$	$Ni(CO)_4$–$NiCl_2$	4
$(CH_3CH_2)_2C(CH_3)CO_2H$	$(CH_3CH_2)_2CHCH_2OH$	$Ni(CO)_4$–$NiCl_2$	4
$CH_3(CH_2)_4CH(CH_3)CO_2H$	$CH_3(CH_2)_5CH_2OH$	$Ni(CO)_4$–$NiCl_2$	4
$CH_3(CH_2)_5CH(CH_3)CO_2H$	$CH_3(CH_2)_6CH_2OH$	$Ni(CO)_4$–$NiCl_2$	4
	$CH_3(CH_2)_5CH(CH_3)OH$	$Ni(CO)_4$–$NiCl_2$	4

Substrate	Product	Catalyst	Ref.
 cyclohexyl–$CH_2C(CH_3)CO_2H$	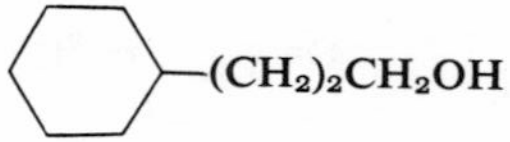 cyclohexyl–$(CH_2)_2CH_2OH$	$Ni(CO)_4$–$NiCl_2$	4
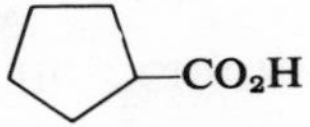 cyclopentyl–CO_2H	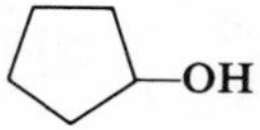 cyclopentyl–OH	$Ni(CO)_4$–$NiCl_2$	4
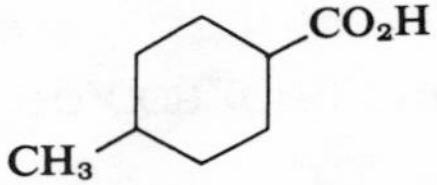 4-methylcyclohexyl–CO_2H	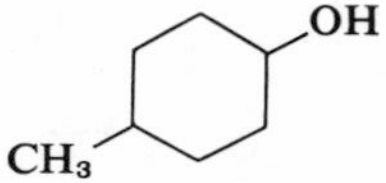 4-methylcyclohexyl–OH	$Ni(CO)_4$–$NiCl_2$	4
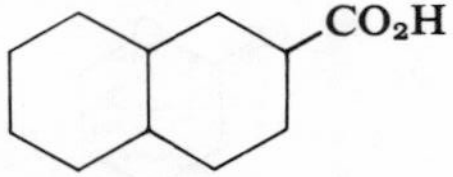 decahydronaphthyl–CO_2H	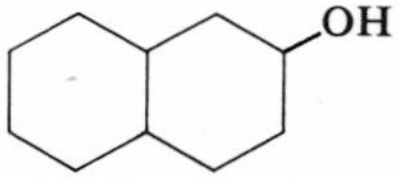 decahydronaphthyl–OH	$Ni(CO)_4$–$NiCl_2$	4
$HO_2CCH_2CH_2CO_2H$	$HOCH_2CH_2OH$	$Ni(CO)_4$–I_2	50
$HO_2CCH_2(CH_2)_2CH_2CO_2H$	$HOCH_2(CH_2)_2CH_2OH$	$Ni(CO)_4$–I_2	50, P111, P112
$HO_2CCH_2(CH_2)_3CH_2CO_2H$, $CH_3(CH_2)_2CH_2CH(CH_3)CO_2H$	$HOCH_2(CH_2)_3CH_2OH$	$Ni(CO)_4$–I_2	4, 50
$HO_2CCH_2(CH_2)_4CH_2CO_2H$, $CH_3(CH_2)_3CH_2CH(CH_3)CO_2H$	$HOCH_2(CH_2)_4CH_2OH$	$Ni(CO)_4$–I_2	4, 50
$HO_2CCH_2(CH_2)_8CH_2CO_2H$	$HOCH_2(CH_2)_8CH_2OH$	$Ni(CO)_4$–I_2	50
$HO_2CCH_2(CH_2)_{10}CH_2CO_2H$	$HOCH_2(CH_2)_{10}CH_2OH$	$Ni(CO)_4$–I_2	50
$HO_2CCH_2(CH_2)_{12}CH_2CO_2H$	$HOCH_2(CH_2)_{12}CH_2OH$	$Ni(CO)_4$–I_2	50

TABLE VI-4

THE CARBONYLATION OF ETHERS, ESTERS, AND ALDEHYDES

Product	Reactant	Nickel component	Ref.
CH_3CO_2H	CH_3OCH_3	NiI_2	50, P112, P114
$CH_3CH_2CO_2CH_2CH_3$	$CH_3CH_2OCH_2CH_3$	NiI_2	6, 13, 22, 23, P26, P27
iso-$C_3H_7CO_2$-iso-C_3H_7	iso-C_3H_7O-o-iso-C_3H_7	NiI_2	22, 23
$CH_3(CH_2)_2CH_2CO_2H$, $HO_2CCH_2(CH_2)_2CH_2CO_2H$, 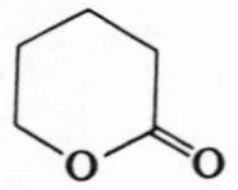	 tetrahydrofuran ring with O	$Ni(CO)_4$–I_2	6, 22, 23, 51, P26, P27, P32
$HO_2CCH_2(CH_2)_2CH(CH_3)CO_2H$	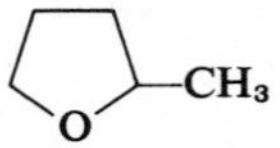 2-methyltetrahydrofuran ring	$NiBr_2$–I_2	51
$HO_2CCH(CH_3)CH_2CH_2CH(CH_3)CO_2H$	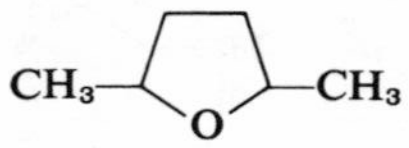 2,5-dimethyltetrahydrofuran ring	$Ni(CO)_4$–I_2	51
$HO_2CCH_2CH_2CH(CH_3)CO_2H$	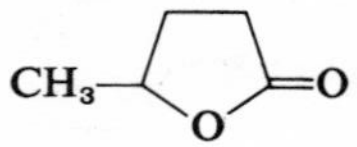 5-methyl lactone ring	$Ni(CO)_4$–I_2	51

Product	Catalyst	Reference
$HO_2CCH_2(CH_2)_3CH_2CO_2H$ (2-tetrahydrofurylmethanol structure)	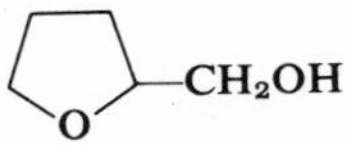 $Ni(CO)_4–I_2$	51
$HO_2CCH_2(CH_2)_2CH_2CO_2H$, $CH_3(CH_2)_2CH_2CO_2H$ (δ-valerolactone structure)	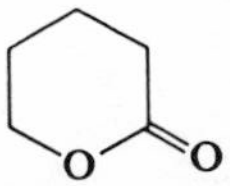 $Ni(CO)_4–I_2$	51
$HO_2CCH_2CH_2CH_2CO_2H$ (1,4-dioxane structure)	—	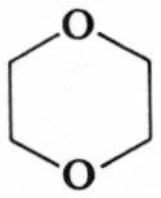 51
$HO_2CCH_2(CH_2)_2CH_2CO_2H$ (tetrahydrothiophene structure)	 $Ni(CO)_4–NiCl_2–I_2$	P34
$CH_3COOCOCH_3$	$[(C_6H_3)_3PC_4H_9]_2NiBr_4$	P112
$CH_3CH_2CO_2H$ ($CH_3COOCH_2CH_3$)	$Ni–NiI_2–NaI$	P26, P27
$HOCH_2CO_2H$, CH_3CO_2H (HCHO)	NiI_2	22, 23, P113
$CH_3CH(OH)CO_2H$ (CH_3CHO)	NiI_2	22, 23

An unusual aspect of the reaction is that straight-chain alcohols are reported to give principally branched-chain acids (4). In contrast, isomeriza-

$$CH_3(CH_2)_5CH_2CH_2OH$$
or $\xrightarrow{CO/[Ni]}$ $CH_3(CH_2)_5CH(CH_3)CO_2H$
$$CH_3(CH_2)_5CH(CH_3)OH$$

tion of the carbon skeleton is not observed in the carbonylation of straight-chain diols.

The behavior of ethers is similar to that of alcohols: cyclic ethers react more readily than alicyclic. An investigation of the conversion of THF into adipic acid indicates that δ-valerolactone is first formed (51). Hydrogenolysis

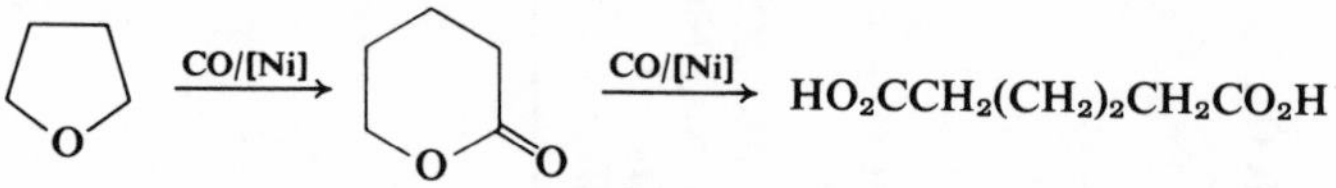

to the saturated monocarboxylic acid is observed as a side reaction.

The nickel-catalyzed carbonylation of esters has been mentioned in the patent literature (P26, P27, P112) but has not, apparently, been investigated in detail.

An interesting reaction is the formation of glycolic acid from the carbonylation of formaldehyde using a nickel iodide catalyst (22, 23).

$$HCHO + CO + H_2O \longrightarrow HOCH_2CO_2H$$

Other authors have obtained acetic acid from this reaction (P113).

D. The Catalytic Carbonylation of Amines and Related Compounds (Table VI-5)

Primary, secondary, and tertiary aliphatic amines may be carbonylated in the presence of a nickel catalyst to give the corresponding N-substituted formamide accompanied, in some cases, by a derivative of urea. Tertiary

$$(C_2H_5)_2NH + CO \longrightarrow (C_2H_5)_2NCHO + (C_2H_5)_2NCON(C_2H_5)_2$$

aromatic amines (in contrast to tertiary aliphatic amines) react to give a substituted amide (48).

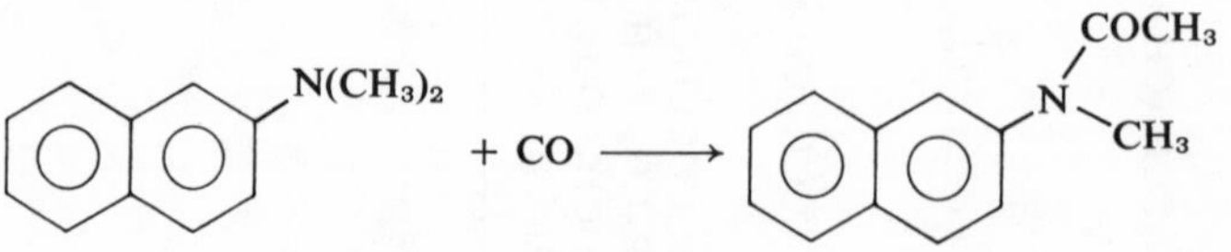

Related to these reactions is that of nitrobenzene with carbon monoxide and nickel tetracarbonyl in acetic acid. At 310° reduction to acetanilide is observed and has been suggested to involve the intermediate formation of phenyl isocyanate (35). The preparation of polyisocyanate by reaction of an

$$C_6H_5NO_2 \xrightarrow[-2CO_2]{3CO} C_6H_5NCO \xrightarrow[-CO_2]{CH_3CO_2H} C_6H_5NHCOCH_3$$

aromatic polyamine with nickel tetracarbonyl has been patented (P4).

The carbonylation of azobenzene and the Schiff base, N-benzylidene-aniline, has been studied. Although dicobalt octacarbonyl is the preferred carbonylation reagent, and nickel tetracarbonyl has been reported to be inactive (116), under suitable conditions reaction does occur to give the lactam 13 [the original formulation of 13 as an indazole derivative has been revised (117)] and 2-phenylphthalimidine (14), respectively (P33, P36).

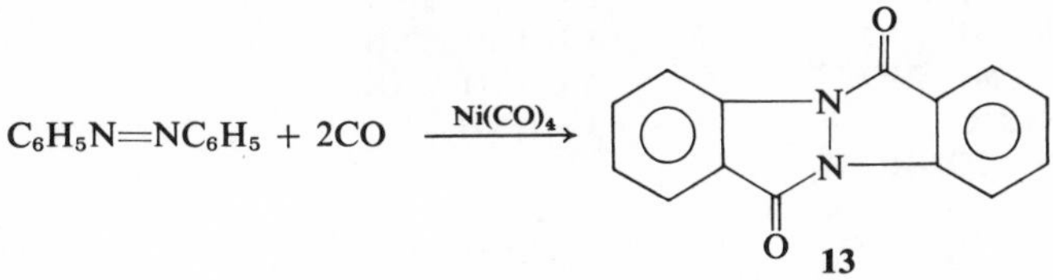

$$C_6H_5N{=}NC_6H_5 + 2CO \xrightarrow{Ni(CO)_4}$$

13

$$C_6H_5CH{=}NC_6H_5 + CO \xrightarrow{Ni(CO)_4}$$

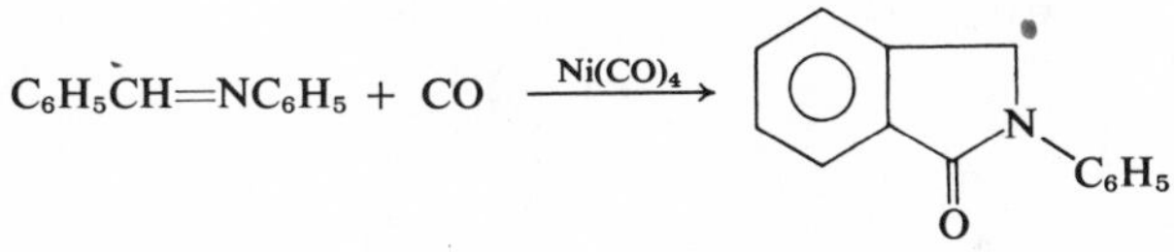

14

E. Mechanistic Considerations

Speculations by various authors on the mechanism of the carbonylation reaction have led to the rather unsatisfactory situation that several mechanisms have been formulated without there being direct evidence for any.

The original suggestion that a cyclopropenone or cyclopropanone intermediate is involved (48, 98) has been shown to be unlikely (10, 17, 99). Alternatively, it might be supposed that the initial reaction is that of the acid with the alkyne or olefin to give a substituted vinyl or alkyl compound that then reacts further with nickel tetracarbonyl as described in Section III, B. This is probably an oversimplification, however: the observed cis addition indicates that the olefin or alkyne is initially bonded to the nickel atom, and it is improbable that relatively weak acids are able to protonate an alkyne or olefin (2, 3).

TABLE VI-5
THE CATALYTIC CARBONYLATION OF AMINES[a]

Product	Amine	Ref.
$(CH_3)_2NCHO$	$(CH_3)_2NH$	1
	$(CH_3)_3N$	P18
$(C_2H_5)_2NCHO$, $(C_2H_5)_2NCON(C_2H_5)_2$	$(C_2H_5)_2NH$	1, 6, 9
$(C_2H_5)_2NCHO$	$(C_2H_5)_3N$	6, 9
$iso\text{-}C_3H_7NHCHO$	$iso\text{-}C_3H_7NH_2$	6
$(iso\text{-}C_3H_7)_2NCHO$	$(iso\text{-}C_3H_7)_2NH$	49
C_4H_9NHCHO, $C_4H_9NHCONHC_4H_9$	$C_4H_9NH_2$	6, 9, 48, 49, P18
$(C_4H_9)_2NCHO$	$(C_4H_9)_2NH$	48, 49, P18
$(C_4H_9)_2NCHO$	$(C_4H_9)_3N$	48, P18
$(iso\text{-}C_4H_9)_2NCHO$	$(iso\text{-}C_4H_9)_2NH$	9
$CH_3(CH_2)_4NHCHO$	$CH_3(CH_2)_4NH_2$	6, 9
$CH_3(CH_2)_5NHCHO$	$CH_3(CH_2)_5NH_2$	6, 9
C_6H_5NHCHO, $C_6H_5NHCONHC_6H_5$	$C_6H_5NH_2$	1, 6, 48, P18
$(C_6H_5)_2NCHO$	$(C_6H_5)_2NH$	6
$C_6H_5N(C_2H_5)COC_2H_5$	$C_6H_5N(C_2H_5)_2$	6, 48

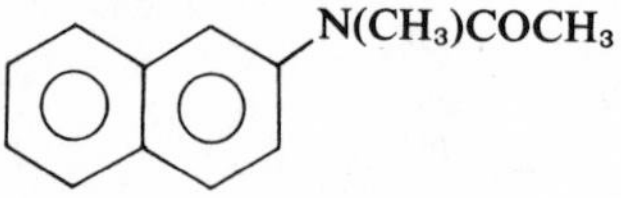

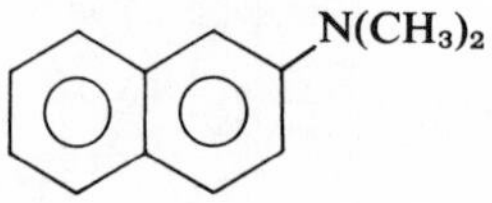

naphthalene-$N(CH_3)COCH_3$	naphthalene-$N(CH_3)_2$	48
 pyrrolidine-CHO	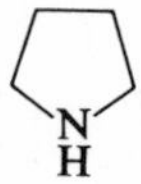 pyrrolidine-NH	48, 49, P18
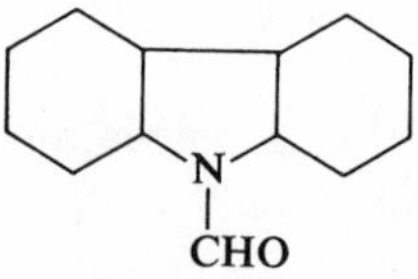 octahydrocarbazole-CHO	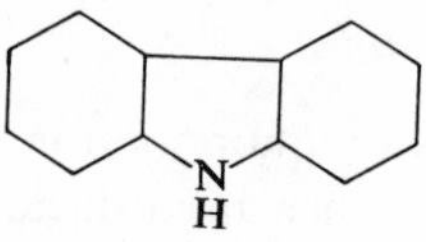 octahydrocarbazole-NH	48, 49, P18
 piperidine-CHO	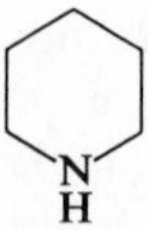 piperidine-NH	6, 8, 48, 49, P18
	piperidine-C_3H_7	P18

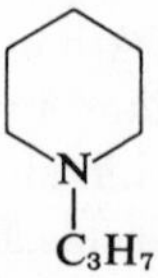

(continued)

TABLE VI-5 (*continued*)

Product	Amine	Ref.
		5, 6
		5, 6
		6
		P18

[a] Catalyst: NiI_2.

By analogy with reactions involving $HCo(CO)_4$ it has been suggested that a nickel carbonyl hydride (**15**) is the active catalyst (31, 39, 76). Although tetrakisligand nickel complexes may be readily protonated, nickel tetracarbonyl decomposes in the presence of acids under milder conditions than those normally present in the carbonylation reaction (see Volume I, p. 150).

An alternative mechanism involves initial complexation of the alkyne to the nickel atom (21, 32, 53) followed by formation of a nickelacyclobutenone intermediate (16) that reacts further with acid (10, 21). The decrease in the

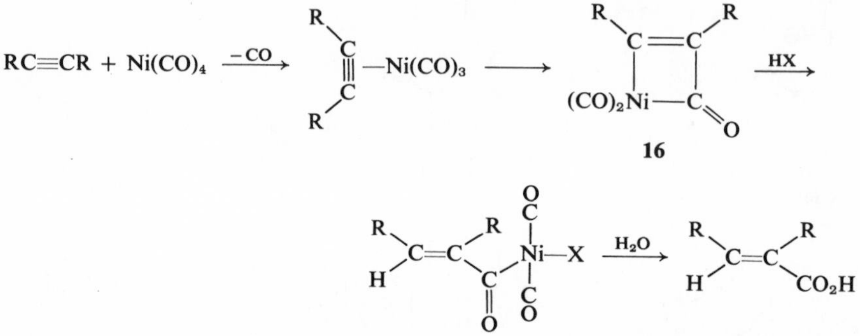

$$RC{\equiv}CR + Ni(CO)_4 \xrightarrow{-CO} \text{(complex)} - Ni(CO)_3 \longrightarrow \text{(16)} \xrightarrow{HX}$$

16

$$\xrightarrow{H_2O} \text{(product, } CO_2H)$$

length of the induction period on the introduction of pyridine (80) or the removal of liberated CO (32), as well as the effect of irradiation (74, 94, 95), are consistent with the assumption that nickel tricarbonyl is the active species.

Speculation on the mechanism of the carbonylation of alcohols and ethers is limited to the suggestions that initial conversion into an alkyl halide (50, 51, 95) or olefin (4) occurs.

A recent suggestion is that the active catalyst for these carbonylation reactions is an anionic tricarbonyl nickel halide, e.g., $[(OC)_3NiI]^-$ (71, 75).

III. The Carbonylation of Organic Halides and Related Compounds

Although Reppe investigated the nickel-catalyzed carbonylation of organic halides as part of an attempt to elucidate the mechanism of the carbonylation of alcohols and ethers (51), the full potential of this reaction was not realized until some years later. Under suitable conditions, it is possible to carbonylate most classes of organic halide, although the majority of the publications are concerned with the carbonylation of allyl halides.

The synthetic utility of this reaction has been greatly increased with the discovery that other unsaturated compounds (e.g., alkynes, olefins) may be incorporated, in addition to CO. This enables relatively complex organic molecules to be produced catalytically.

Much of the work in this area has been carried out by a research team led by G. P. Chiusoli and, because most of the detailed information is to be found in Italian patents and journals (the former being practically unavailable), it is fortunate that a number of review articles exist (104, 113, 136, 137, 142, 144–146, 153, 183, 185).

A. *The Carbonylation of Allyl Halides and Related Reactions* (*Tables VI-6 to VI-9*)

The carbonylation of allyl halides has generally been carried out using nickel tetracarbonyl as the catalyst. Other active systems can be prepared *in situ* by reducing nickel salts in the presence of CO: reducing agents reported include CO itself (140, P129, P130), iron powder (140, 151, P125), and a manganese–iron alloy (143, 183, P121). Raney nickel has also been used (140, 151, 152, P125). The activity of the catalyst can be increased by adding thiourea (75, 140, 143, 151, 152, P121, P125) and it is also claimed that the complex π-allylNiX(thiourea) is active (P118). In addition the use of the anionic nickel tricarbonyl halides, $[(CO)_3NiX]^-$, has been mentioned (75, P132). The life of the catalyst is limited by a number of side reactions (e.g., dimerization or substitutive hydrogenation of the allyl halide) in which an inactive nickel halide is produced.

In the discussion that follows we shall first consider the carbonylation reaction and then the extension to the insertion of alkynes and carbon monoxide. In general we have confined ourselves to the reaction involving unsubstituted allyl halides and acetylene: information on those involving substituted allyl compounds and substituted alkynes can be taken from Tables VI-7 and VI-8.

1. THE REACTION OF ALLYL HALIDES WITH CARBON MONOXIDE

Allyl halides react with nickel tetracarbonyl at ambient temperature and pressure either by the coupling of two allyl fragments or by substitutive hydrogenation (see Chapter V, Sections II-A and II-E). The reaction takes

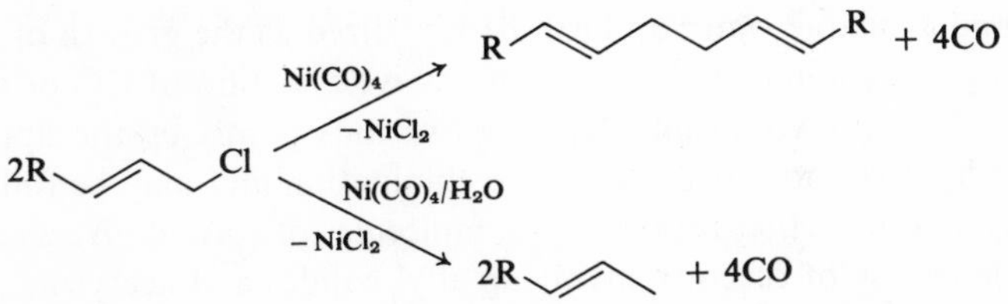

a different course under a slight pressure of carbon monoxide (2–3 atm); carbonylation occurs to give the unsaturated carboxylic acid in aqueous solvents, or its ester in alcohol (Table VI-6). At higher CO pressure (ca. 400 atm) all reaction ceases, presumably because of the increased stability of

$$\text{CH}_2\text{:CHCH}_2\text{Cl} + \text{CO} + \text{H}_2\text{O} \xrightarrow[-\text{HCl}]{\text{Ni(CO)}_4} \text{CH}_2\text{:CHCH}_2\text{CO}_2\text{H}$$

nickel tetracarbonyl at high pressure. In inert solvents the primary product has been shown to be the acyl halide (76, 134).

Alkyl-substituted allyl halides react similarly. Allyl halides having electron-withdrawing substituents (e.g., $CH_3OCOCH:CHCH_2Br$) normally react with nickel tetracarbonyl to give products resulting from coupling or substitutive hydrogenation. However, carbonylation is also observed if the reaction is carried out in ketonic solvents (126, 127).

TABLE VI-6
THE CARBONYLATION OF ALLYL HALIDES

Product	Allyl halide	Ref.
$CH_2\text{:}CHCH_2CO_2H$	$CH_2\text{:}CHCH_2Cl$	75, 76, 134, 136, 139, 147, 155, 156, 160, 166, 170, P116; see also P163
$CH_3CH\text{:}CHCH_2CO_2H$	$CH_3CH\text{:}CHCH_2Cl$	134, 136, 147
$CH_2\text{:}C(CH_3)CH_2CO_2H$	$CH_2\text{:}C(CH_3)CH_2Cl$	75
$CH_3O_2CCH\text{:}CHCH_2CO_2H$	$CH_3O_2CCH\text{:}CHCH_2Br$	126, 127
$CNCH_2CH\text{:}CHCH_2CO_2H$	$CNCH_2CH\text{:}CHCH_2Cl$	134, 136, 147, P124
$C_6H_5CH\text{:}CHCH_2CO_2CH_3$	$C_6H_5CH\text{:}CHCH_2Cl$	75

2. THE REACTION OF ALLYL HALIDES WITH CARBON MONOXIDE AND ALKYNES OR OLEFINS

The synthetic utility of the carbonylation reaction may be extended by reacting the allyl halide with a mixture of acetylene and carbon monoxide. The course of the reaction is particularly sensitive to reactant concentration, solvent, and available water and can be visualized as the growth of a nickel-bonded organic chain caused by the successive insertion of CO or acetylene molecules. This growth is interrupted at various points by the uptake of a proton or hydroxyl ion or by reductive elimination and may be followed by rearrangement or further reaction with molecules of solvent. We shall briefly outline the course of reaction involving allyl halides and acetylene and then mention the effect on the reaction of the introduction of substituents. For the sake of clarity we have indicated some of the possible organonickel intermediates involved.

The first step is the insertion of an acetylene molecule followed by a CO molecule to give either *cis*-2,5-hexadienoyl chloride (**18**) or *cis*-2,5-hexadienoic acid (19). This is a general reaction that may be used to prepare substituted hexadienoic acids (or esters) in moderate yield (Table VI-7).

Alternatively, the intermediate 17 in this reaction can cyclize to give the cyclopentenone or cyclohexenone derivatives 20 and 21 which in turn can incorporate a second molecule of CO to give the carboxylic acids 22 and 23.

The initial product leading to 22 has been shown to be the isomeric form of the acyl halide (24), which is hydrolyzed to 22 (125, 128, 186).

TABLE VI-7

THE SYNTHESIS OF *cis*-2,5-HEXADIENOIC ACID ESTERS

$$RCH{=}CHCH_2X + HC{\equiv}CH + CO \xrightarrow[R'OH]{Ni(CO)_4} RCH{=}CHCH_2CH{=}CHCO_2R' + HX$$

2,5-Hexadienoic acid ester	Allyl halide	Alkyne	Ref.
$CH_2{:}CHCH_2CH{:}CHCO_2R'$	$CH_2{:}CHCH_2Cl$	$HC{:}CH$	75, 76, 127, 133, 135–137, 140, 143, 148, 151, 163, 164, 169, 170, P118, P123, P125–P130, P132, P154
$CH_3CH{:}CHCH_2CH{:}CHCO_2R'$	$CH_3CH{:}CHCH_2Cl$	$HC{:}CH$	75, 133, 135–137, 143, 148, P125, P126
$CH_3CH{:}CHCH_2CH{:}CHCO_2R'$	$CH_2{:}CHCHClCH_3$	$HC{:}CH$	135, 136, 151
$CH_2{:}C(CH_3)CH_2CH{:}CHCO_2R'$	$CH_2{:}C(CH_3)CH_2Cl$	$HC{:}CH$	75, 135–137, 143, 148, 151, 159, P125
$CH_3O_2CCH{:}CHCH_2CH{:}CHCO_2R'$	$CH_3O_2CCH{:}CHCH_2Br$	$HC{:}CH$	126, 127, 137
$CH_3O_2C(CH{:}CHCH_2)_2CH{:}CHCO_2R'$	$CH_3O_2C(CH{:}CHCH_2)_2Cl$	$HC{:}CH$	137, 174, P115, P131
$CH_3O_2CCH_2CH{:}CHCH_2CH{:}CHCO_2R'$	$CH_3O_2CCH_2CH{:}CHCH_2Cl$	$HC{:}CH$	135–137, P126
$CH_3O_2C(CH{:}CHCH_2)_2{-}CH_2CH{:}CHCH_2CH{:}CHCO_2R'$	$CH_3O_2C(CH{:}CHCH_2)_2{-}CH_2CH{:}CHCH_2Cl$	$HC{:}CH$	137
$CH_3O_2CCH{:}C(CH_3)CH_2CH{:}CHCO_2R'$	$CH_3O_2CCH{:}C(CH_3)CH_2Cl$	$HC{:}CH$	137
$NCCH_2CH{:}CHCH_2CH{:}CHCO_2R'$	$NCCH_2CH{:}CHCH_2Cl$	$HC{:}CH$	135–137, P126

NCCH₂CH:CHCH₂CH:CHCO₂R'	NCCH₂CHClCH:CH₂	HC:CH	136, P126
ClCH:CHCH₂CH:CHCO₂R'	ClCH:CHCH₂Cl	HC:CH	149, 150
CH₂:C(Cl)CH₂CH:CHCO₂R'	CH₂:C(Cl)CH₂Cl	HC:CH	149
CH₃CH:CHCH(CH₃)CH:CHCO₂R'	CH₃CH:CHCH(CH₃)Cl	HC:CH	135–137
(CH₃)₂C:CHCH₂CH:CHCO₂R'	(CH₃)₂C:CHCH₂Cl	HC:CH	135–137
CH₃(CH₂)₂CH:CHCH₂CH:CHCO₂R'	CH₃(CH₂)₂CH:CHCH₂Cl	HC:CH	137
CH₃(CH₂)₃CH:CHCH₂CH:CHCO₂R'	CH₃(CH₂)₃CH:CHCH₂OCH₃	HC:CH	241
CH₃(CH₂)₄CH:CHCH₂CH:CHCO₂R'	CH₃(CH₂)₄CH:CHCH₂Cl	HC:CH	137
CH₃(CH₂)₆CH:CHCH₂CH:CHCO₂R'	CH₃(CH₂)₆CH:CHCH₂Cl	HC:CH	137
CH₃(CH₂)₁₄CH:CHCH₂CH:CHCO₂R'	CH₃(CH₂)₁₄CH:CHCH₂Cl	HC:CH	137
(CH₃)₃CCH₂CH:CHCH₂CH:CHCO₂R'	(CH₃)₃CCH₂CH:CHCH₂Cl	HC:CH	136, 137, 151, P125
(cyclopentenyl)—CH:CHCO₂R'	(cyclopentenyl)—Cl	HC:CH	135–137
(cyclohexenyl)—CH:CHCO₂R'	(cyclohexenyl)—Cl	HC:CH	133, 136, 137
(R'O₂CCH:CHCH₂CH:CHCH₂)₂	(ClCH₂CH:CHCH₂)₂	HC:CH	135, 136, P126
C₆H₅CH:CHCH₂CH:CHCO₂R'	C₆H₅CH:CHCH₂Cl	HC:CH	132, 133, 135–137

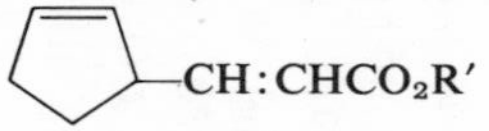

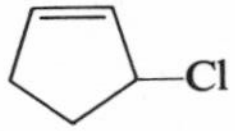

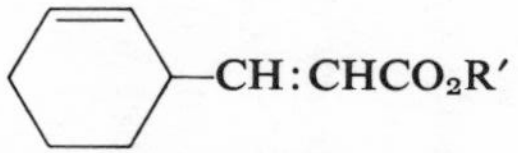

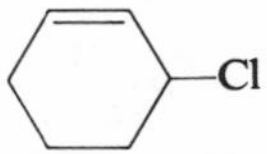

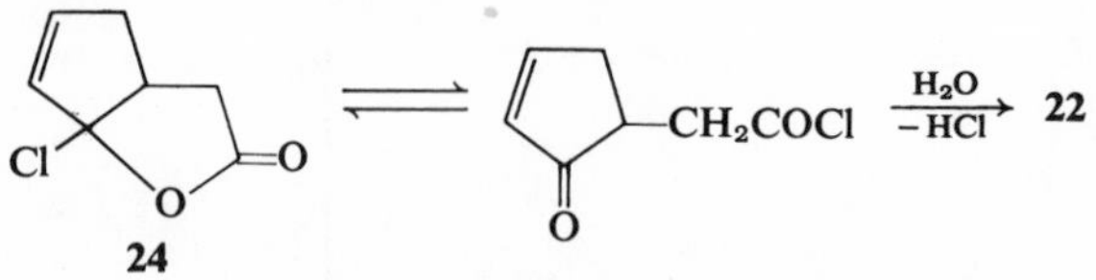

The further reaction with acetylene and CO involves the cyclopentenone derivative **25**. Insertion of a second acetylene molecule produces the intermediate **26**, which can react either by protonation to **27**, by insertion of a third CO molecule and cyclization to give the γ-lactone **28**, or by insertion of a third acetylene molecule followed by a third CO molecule and cyclization to give the ϵ-lactone **29** as in Scheme VI-1 (page 329).

A number of side reactions are also observed. These include the formation of the diketones **30** and **31** (presumably from the reaction of **25** and **26** with unreacted allyl halide) (128, 186), and the Reformatsky-type of reaction

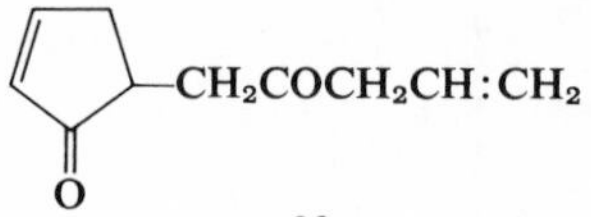

between the γ-lactone **28** and ketonic solvents to give **32** and **33** (125–127),

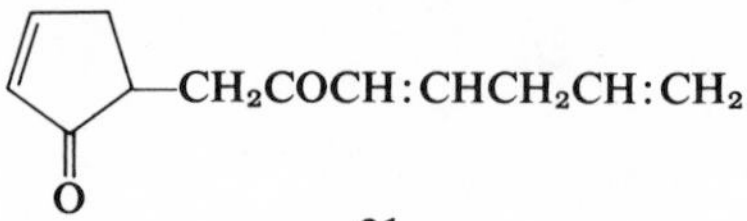

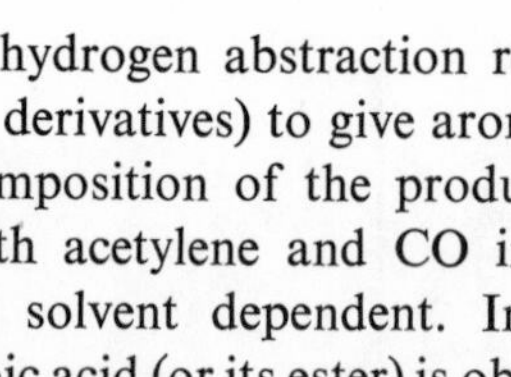

and also hydrogen abstraction reactions (presumably involving the cyclohexenone derivatives) to give aromatic compounds.

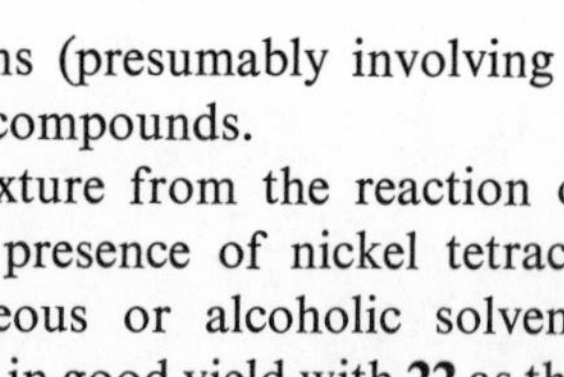

The composition of the product mixture from the reaction of an allyl halide with acetylene and CO in the presence of nickel tetracarbonyl is extremely solvent dependent. In aqueous or alcoholic solvents *cis*-2,5-hexadienoic acid (or its ester) is obtained in good yield with **22** as the principal byproduct. In other solvents, mainly cyclic products are formed: in wet acetone (ca. 0.5% water) **22** and **28** are the main products, with lesser amounts of phenol and the acyl halide **24**; in dry acetone ($< 0.1\%$ water) the product is a mixture of **32** and **33** ($R = R' = CH_3$); in dry ether a mixture of the chloride **24** and **31** with lesser amounts of **30** is obtained.

Substituted allyl halides also react with CO and acetylene. The product depends on the position and nature of the substituent, the acetylene molecule

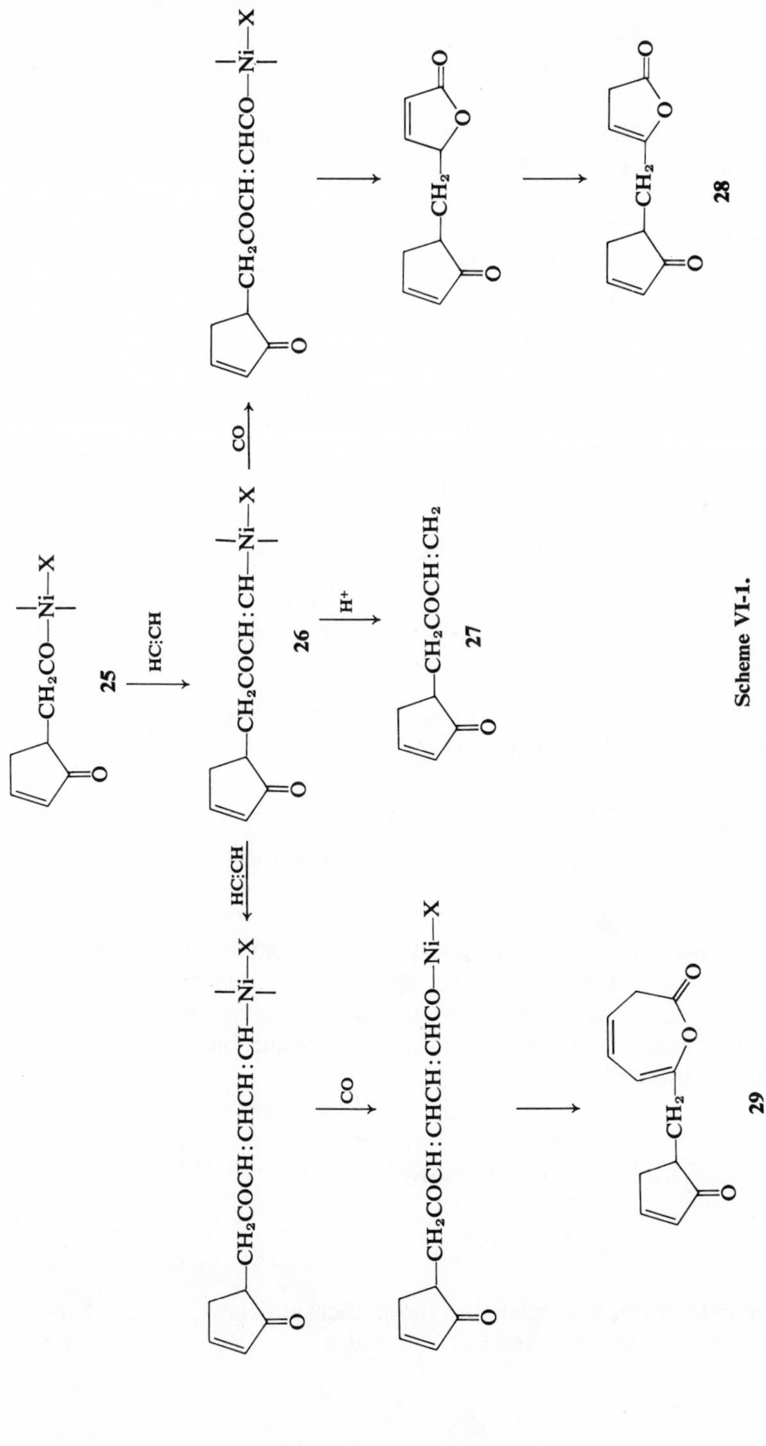

Scheme VI-1.

inserting preferentially into the least substituted carbon atom of the allyl group. Crotyl chloride and allyl chloride react analogously, the former giving mainly cyclopentenone derivatives. Methallyl chloride however, reacts to give mainly six-membered ring products (see Table VI-8). Similar reactions involving allyl halides having electron-withdrawing substituents (e.g., $CH_3OCOCH:CHCH_2Br$) are observed only in ketonic solvents (126, 127).

The reaction with CO and acetylene is not confined to allyl halides and the carbonylation of allyl esters and alcohols at elevated pressure and temperature has been reported (140, 163, 164, P154). In addition it has been demonstrated that it is advantageous to generate the allyl halide *in situ* by reacting an alcohol, ether, or ester with hydrogen halide (133, 137, 148, 163, 174, 182, 241, P115, P128, P132) and this technique has been used to synthesize dimethyl-2,5,8-decatriene-1,10-dioate (**34**) by reacting *cis*-2-butene-1,4-diol with acetylene, CO, and HBr (generated *in situ* using a weak acid–inorganic halide combination, e.g., $NaBr-H_3PO_4$) (174). A related reaction is that of the

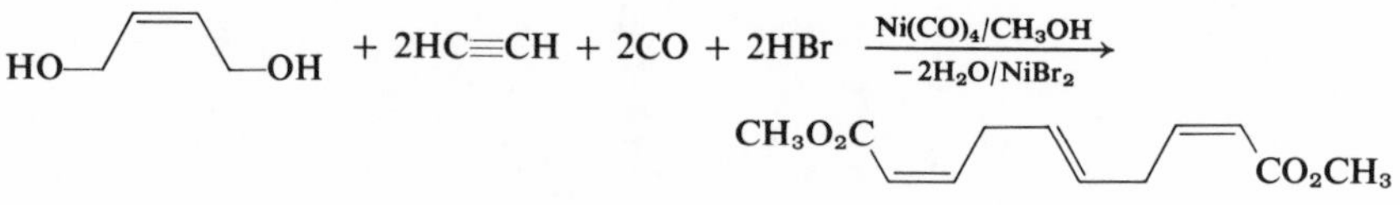

$$HO\diagup\diagdown OH + 2HC\equiv CH + 2CO + 2HBr \xrightarrow[-2H_2O/NiBr_2]{Ni(CO)_4/CH_3OH}$$

$$CH_3O_2C\diagdown\diagup\diagdown\diagup\diagdown CO_2CH_3$$

34

epoxide of butadiene with acetylene and nickel tetracarbonyl, which produces a 7-hydroxyheptadienoate (P157).

$$CH_2{=}CHCH\overset{O}{\diagup\diagdown}CH_2 + HC\equiv CH + CO \xrightarrow{CH_3OH}$$

$$HOCH_2CH{=}CHCH_2CH{=}CHCO_2CH_3$$

The insertion reaction has been reported for a variety of monosubstituted alkynes (Table VI-8). The allyl group adds preferentially to the unsubstituted carbon atom of the alkyne. For example, the reaction of allyl chloride, phenylacetylene, and CO in wet acetone produces a mixture of cyclic products consisting mainly of **35** and **36**. Substituted hexadienoic acid derivatives have not been observed.

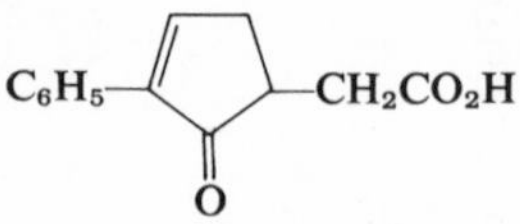

35 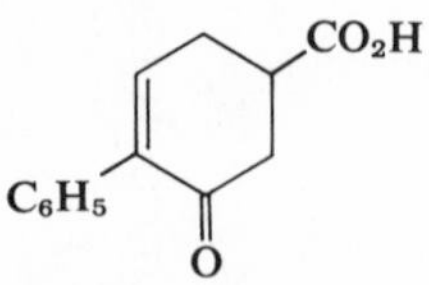   **36**

An unusual reaction, related to those discussed above, is that of the vinyl-lactone **37** with acetylene and CO from which the methyl ester of 2-*cis*-5-*trans*-

nonadiene-1,9-dioic acid has been isolated. An intermediate π-allylnickel system (38) is believed to be involved (146, 187).

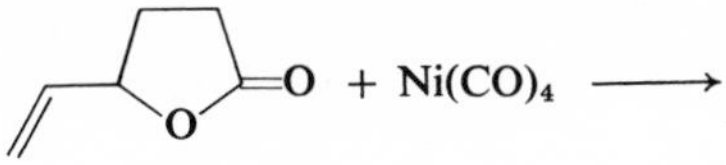

37

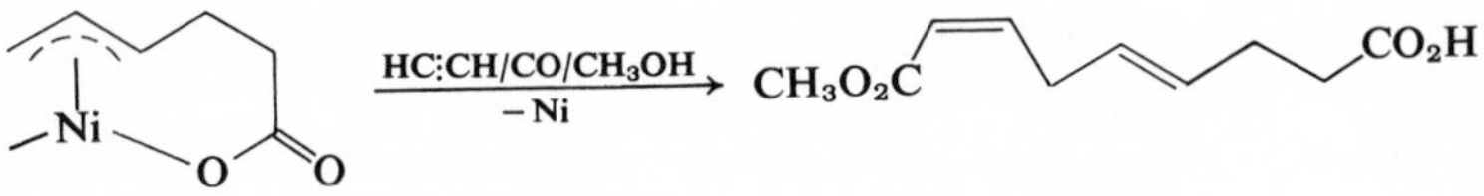

38

The insertion reactions are not confined to alkynes and recently a number of examples have been reported involving olefins (Table VI-9).

Under pressure ethylene and CO react with allyl chloride to give hexenoic acid in modest yield (184, 248). The insertion of higher olefins is accompanied

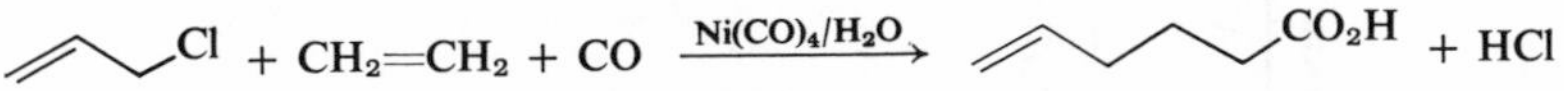

by cyclization to give cyclopentanone systems, e.g., **39**, derived from the reaction of crotyl chloride with 1-hexene, and **40**, derived from the reaction of allyl chloride with 1,5-hexadiene. 1,5-Cyclooctadiene reacts with crotyl

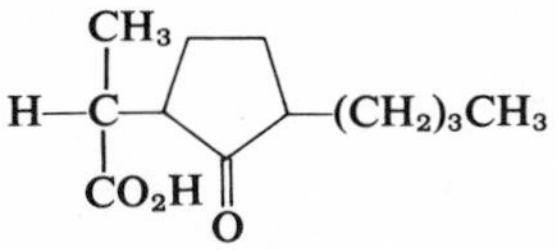

39

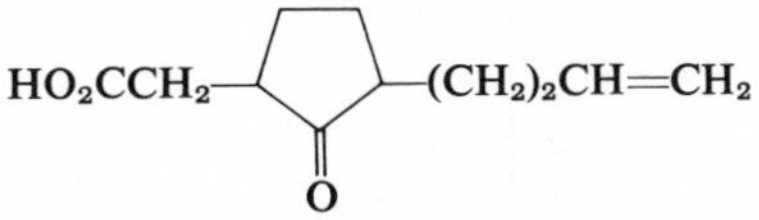

40

chloride and CO to give a mixture of **41** and **42**. It has also been reported that

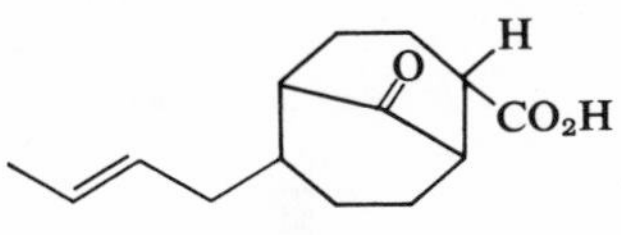

41

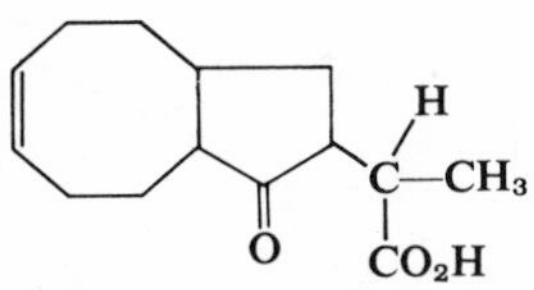

42

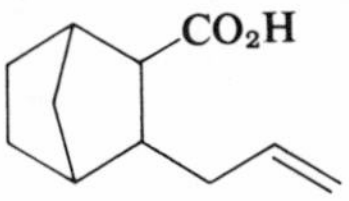

43

π-allylnickel halide complexes react with bicycloheptene and CO to give the acid **43** (75, see also 250).

The vinyl acetic acid produced by carbonylation of the allyl halides can itself take part in an insertion reaction with further allyl halide: 3-butenyl succinic acid (44) or the anhydride 45 results. The yield is increased by the addition of MgO to absorb the liberated acid (155, 156). Substituted allyl

TABLE VI-8

The Reaction of Allyl Halides with Alkynes and Carbon Monoxide

Product	Allyl halide	Alkyne	Ref.
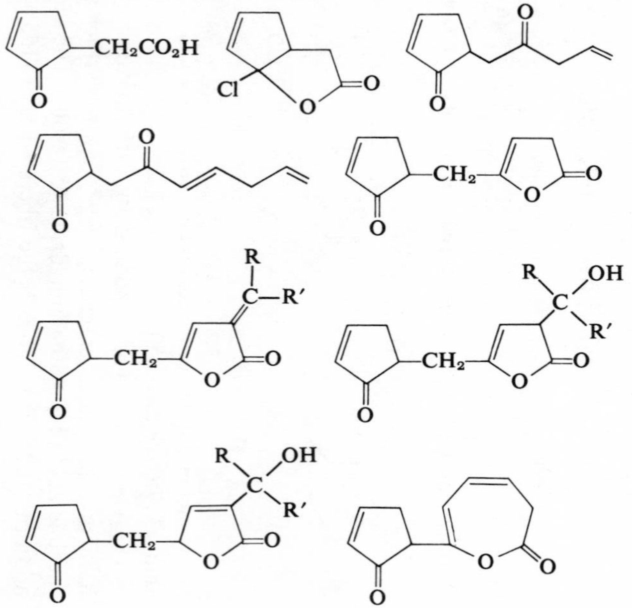	$CH_2:CHCH_2Cl$	$HC:CH$	75, 125–128, 135–137, 143, 167, 186, P 117, P119–P122

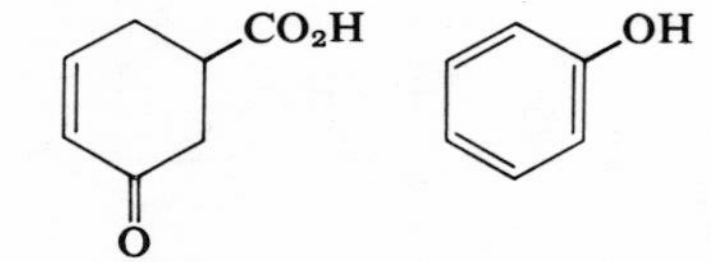

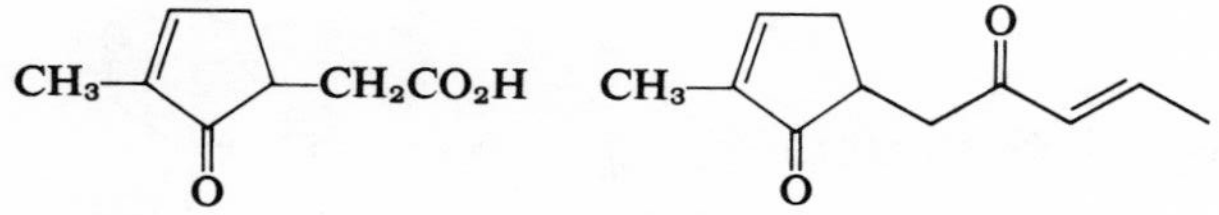

CH$_3$C:CH $\qquad$ 131, 137, 154

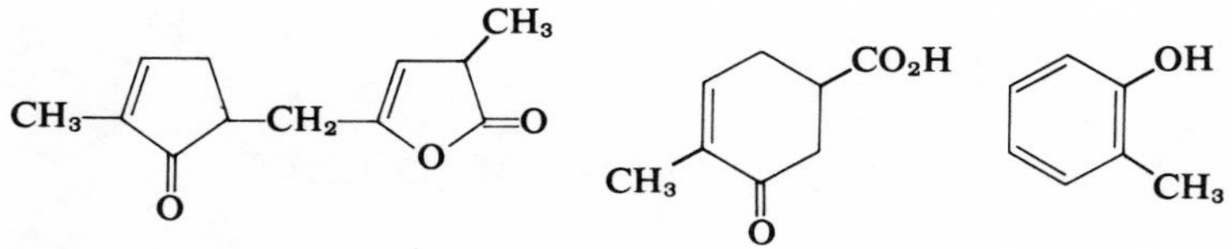

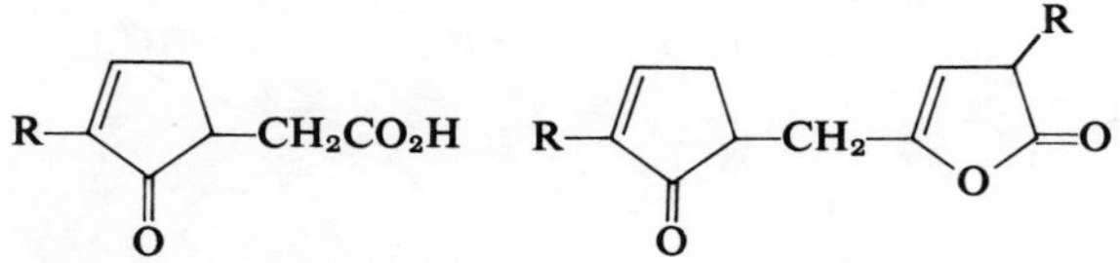

CH$_2$:CHCH$_2$Cl

CH$_3$(CH$_2$)$_n$C:CH (n = 3, 4, 5) $\qquad$ 131

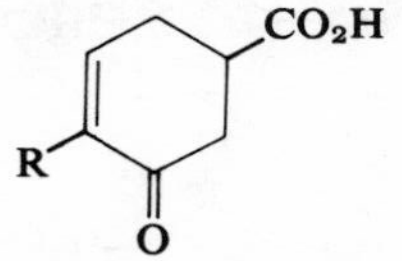

CH$_2$:CHCH$_2$Cl

R = CH$_3$(CH$_2$)$_n$ — (n = 3, 4, 5)

(continued)

333

TABLE VI-8 (*continued*)

Product	Allyl halide	Alkyne	Ref.
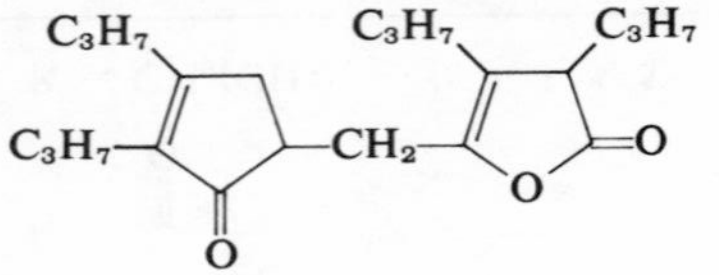	$CH_2:CHCH_2Cl$	$C_3H_7C:CC_3H_7$	131
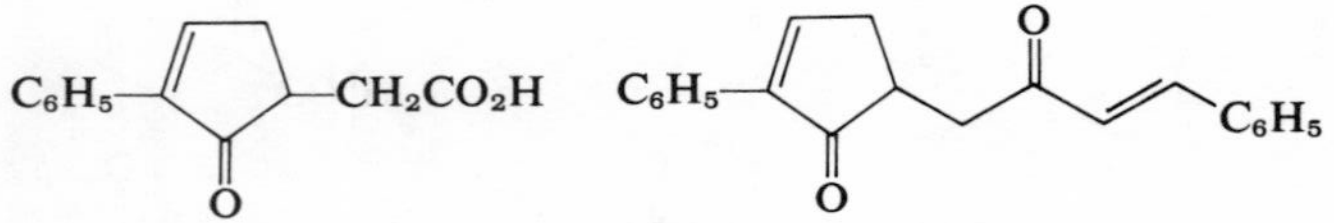	$CH_2:CHCH_2Cl$	$C_6H_5C:CH$	130, 131, 137, 138
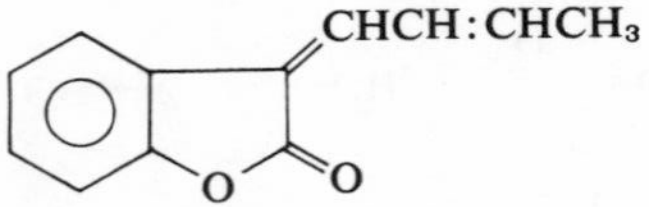	$CH_2:CHCH_2Cl$	$o\text{-}HOC_6H_4C:CH$	185
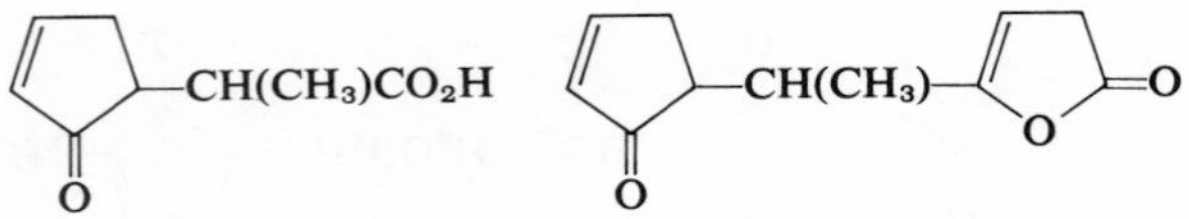	$CH_3CH:CHCH_2Cl$	$HC:CH$	127, 137, 186, P121

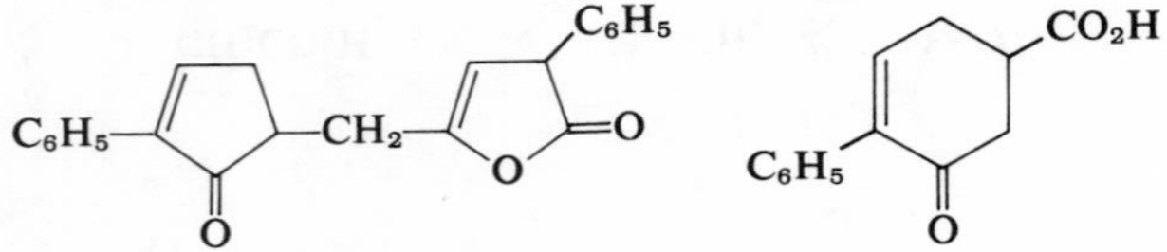

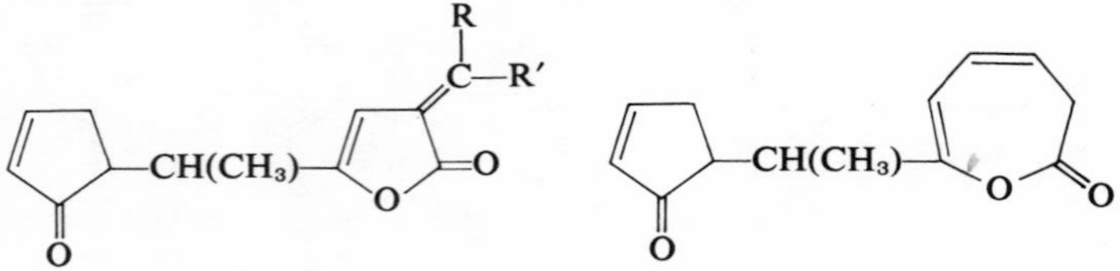

334

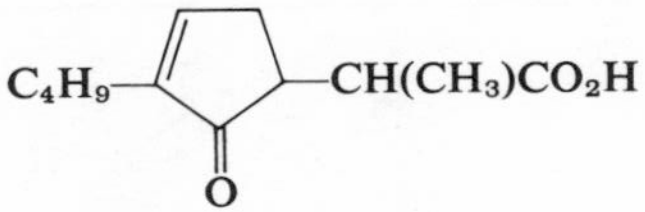 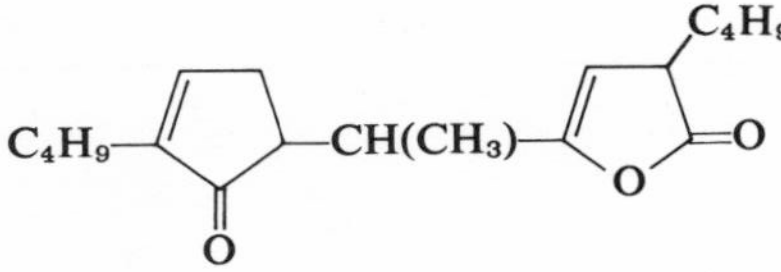 $CH_3CH:CHCH_2Cl$	$C_4H_9C:CH$	138
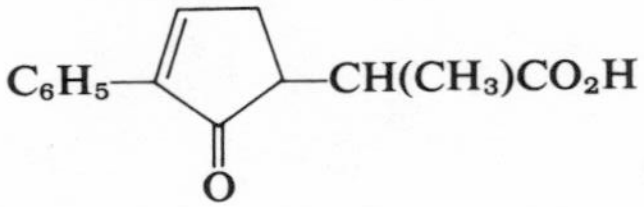 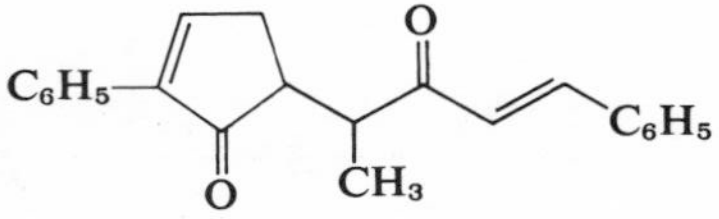 $CH_3CH:CHCH_2Cl$	$C_6H_5C:CH$	130, 137, 138
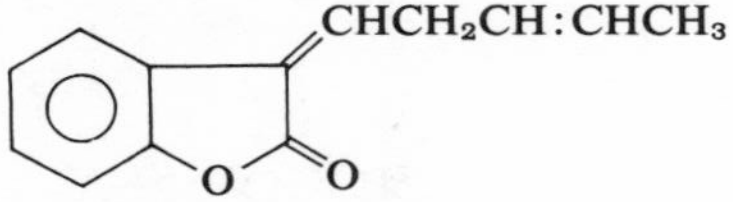 $CH_3CH:CHCH_2Cl$	$o\text{-}HOC_6H_4C:CH$	185
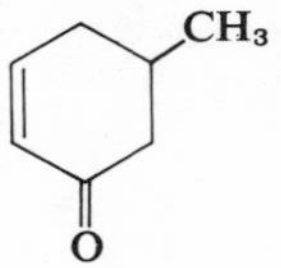 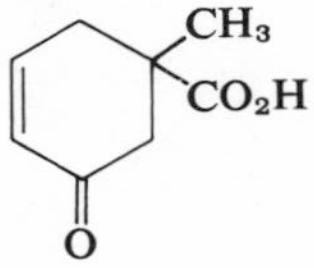 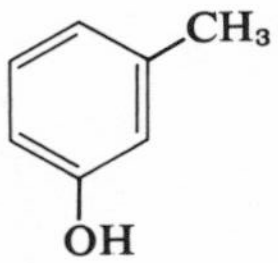 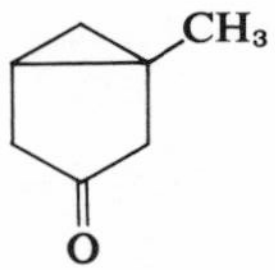 $CH_2:C(CH_3)CH_2Cl$	$HC:CH$	127, 137, 146, 185
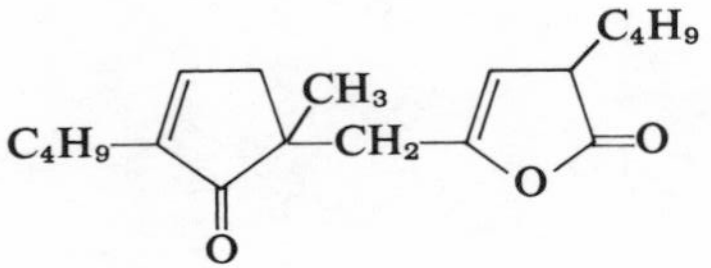 $CH_2:C(CH_3)CH_2Cl$	$C_4H_9C:CH$	138

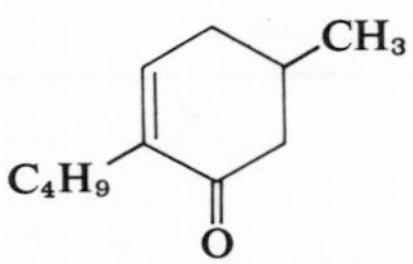

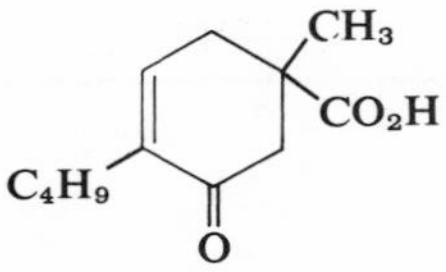

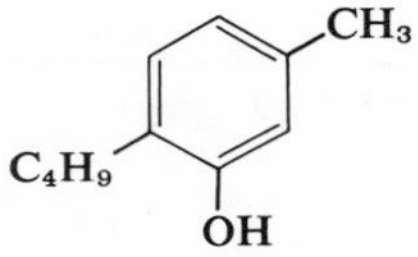

(continued)

TABLE VI-8 (*continued*)

Product	Allyl halide	Alkyne	Ref.
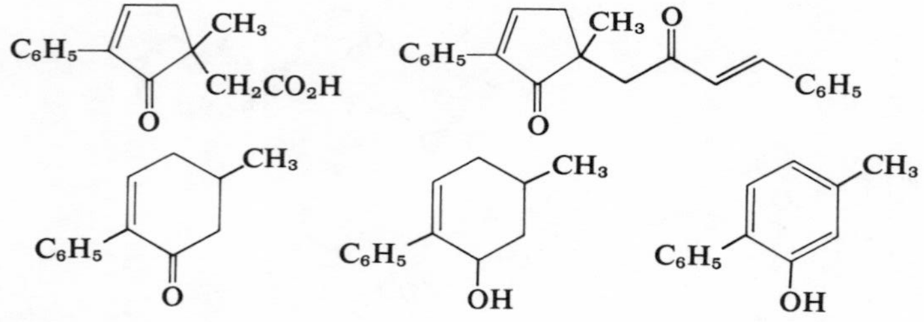	$CH_2:C(CH_3)CH_2Cl$	$C_6H_5C:CH$	130, 137, 138
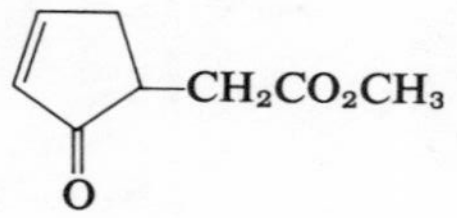	$CH_3O_2CCH:CHCH_2Br$	$HC:CH$	126
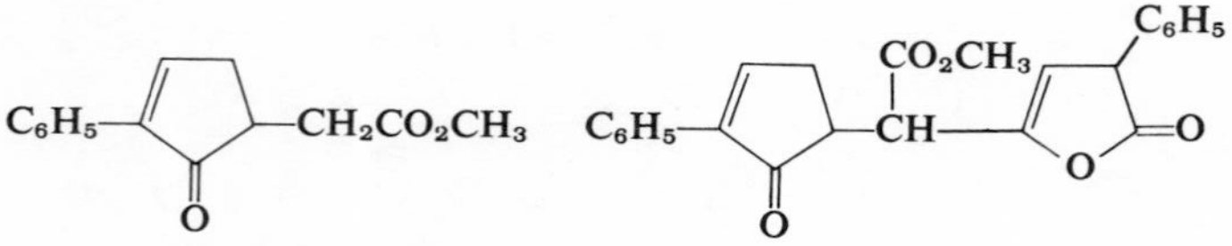	$CH_3O_2CH:CHCH_2Br$	$C_6H_5C:CH$	137, 138
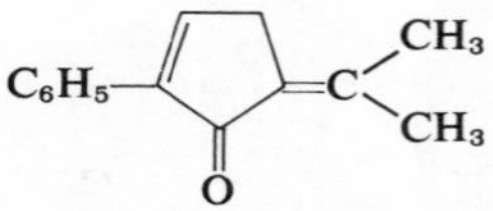	$(CH_3)_2C:CHCH_2Cl$	$C_6H_5C:CH$	185

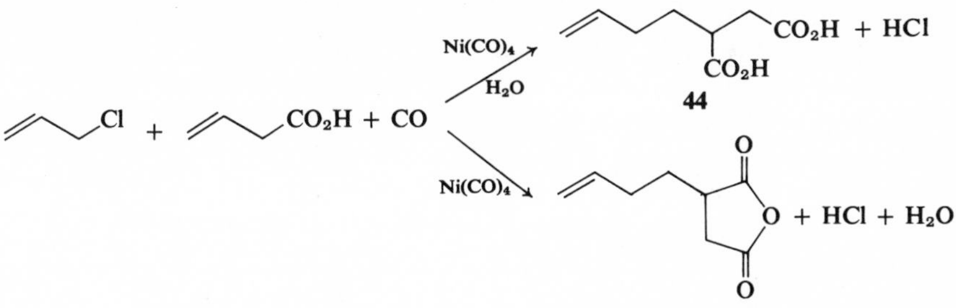

halides react similarly (Table VI-9). An indication of the possible reaction course is given by the observation that allyl vinylacetate also reacts with nickel tetracarbonyl to give **44**: presumably an allylnickel–vinylacetate intermediate is first formed and this then reacts further with insertion of the vinyl group into the allylnickel moiety.

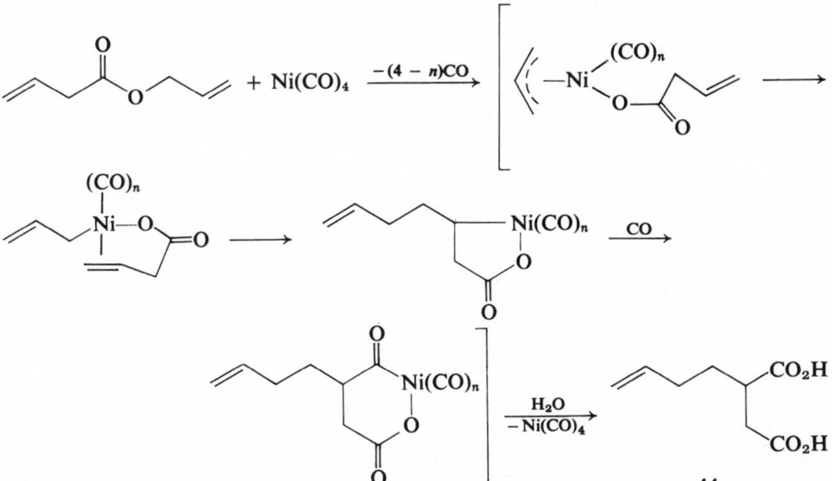

3. MECHANISTIC CONSIDERATIONS

In the preceding section we have already indicated mechanistic pathways for the various insertion products. The evidence for these is based partly on the chemistry of π-allylnickel complexes and partly on the further reaction of some of the products, e.g., acyl halides, with nickel tetracarbonyl and alkynes. The second source is discussed in Section III-B.

It can be assumed that π-allylnickel complexes are formed when an allyl halide is reacted in the presence of nickel tetracarbonyl. Indeed, this is one of the standard methods for preparing π-allylnickel halides. These complexes

TABLE VI-9

THE REACTION OF ALLYL HALIDES WITH OLEFINS AND CARBON MONOXIDE

Product	Allyl halide	Olefin	Ref.
$CH_2:CHCH_2CH_2CH_2CO_2H$	$CH_2:CHCH_2Cl$	$CH_2:CH_2$	183, 184, 248
$CH_3CH:CHCH_2CH_2CH_2CO_2H$	$CH_3CH:CHCH_2Cl$	$CH_2:CH_2$	183, 184, 248
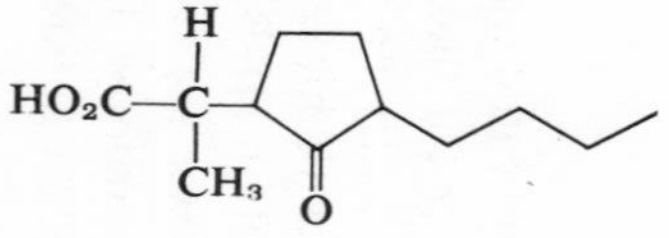	$CH_3CH:CHCH_2Cl$	$CH_2:CH(CH_2)_3CH_3$	183, 184, 248
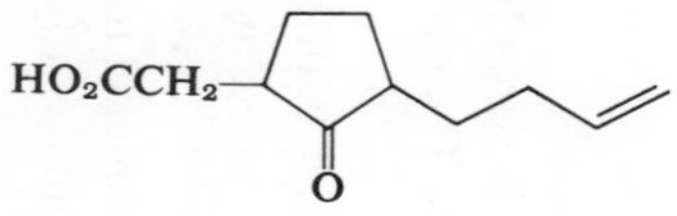 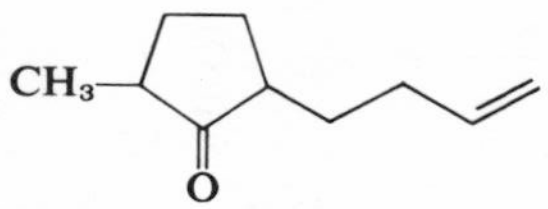	$CH_2:CHCH_2Cl$	$(CH_2:CHCH_2)_2$	183, 184, 248
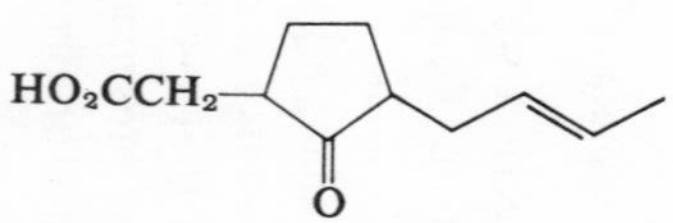 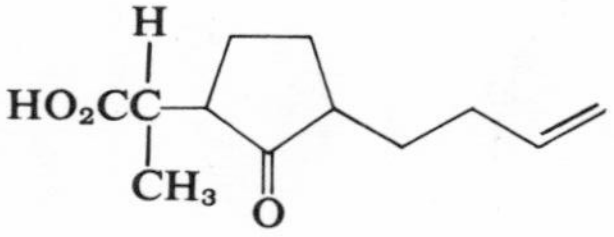 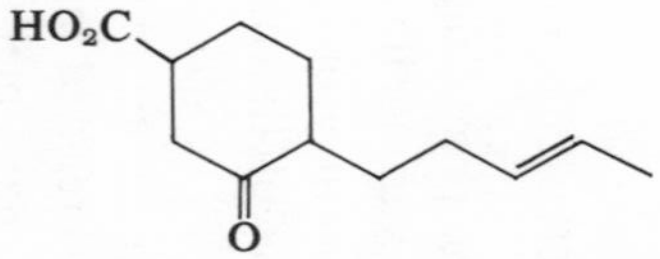	$CH_3CH:CHCH_2Cl$	$(CH_2:CHCH_2)_2$	183, 184, 248
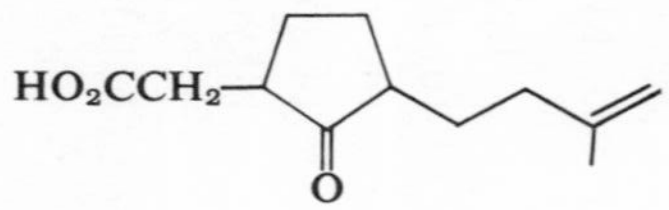 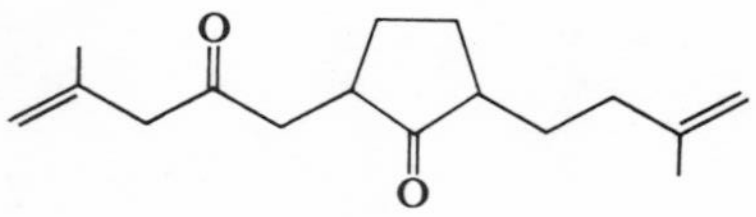	$CH_2:C(CH_3)CH_2Cl$	$(CH_2:CHCH_2)_2$	248

338

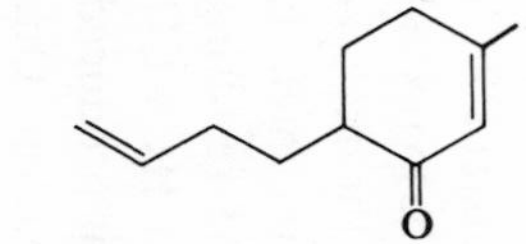

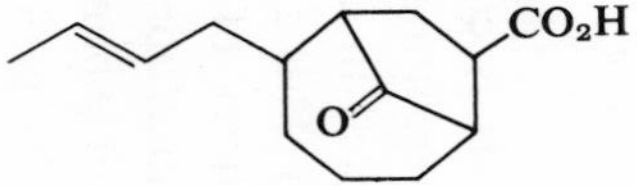 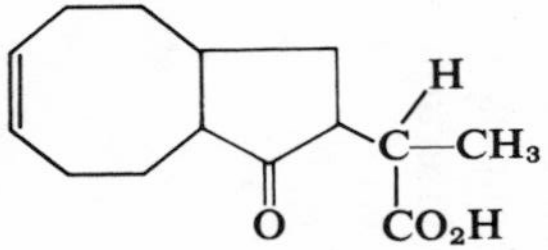 $CH_3CH:CHCH_2Cl$	1,5-COD	183, 250
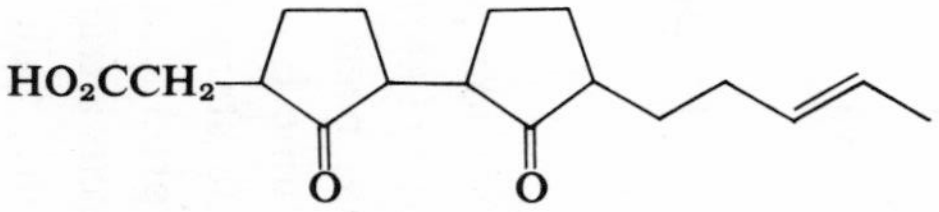 $CH_3CH:CHCH_2Cl$	1,5,9-$C_{12}H_{20}$	183, 250
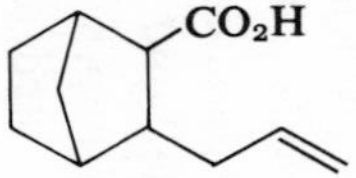 $(\pi-C_3H_5NiX)_2$	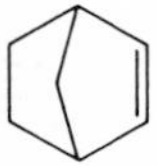	183, see also 250
$CH_2:CHCH_2Cl$	$CH_2:CHCH_2CO_2H$	155, 156, 183
$CH_3CH:CHCH_2Br$	$CH_2:CHCH_2CO_2Na$	155, 156
$CH_2:C(CH_3)CH_2Cl$	$CH_2:C(CH_3)CH_2CO_2H$	155, 156
$CH_3CH:CHCH_2Br$	$CH_3CH:CHCH_2CO_2H$	155, 156, 183

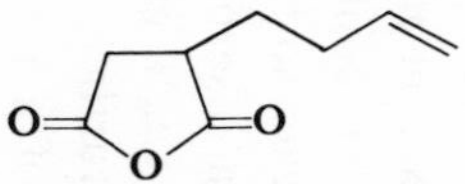

$CH_2:CHCH_2CH_2CH(CO_2H)CH_2CO_2H$

$CH_3CH:CHCH_2CH_2CH(CO_2H)CH_2CO_2H$
$CH_2:C((CH_3)CH_2CH_2CH_2CCH_3(CO_2H)CH_2CO_2H$
$CH_3CH:CHCH_2CH(CH_3)CH(CO_2H)CH_2CO_2H,$
$CH_3CH:CHCH_2CH_2CH(CO_2H)CH(CH_3)CO_2H$

have been shown to take part in stoichiometric insertion reactions with CO and acyl halides, acids, or esters have been isolated, depending on the reaction media (76, 139, 160, 166, 170, 171, 180, 189).

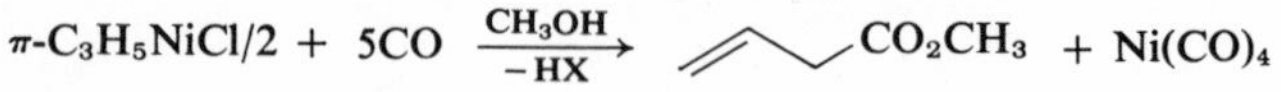

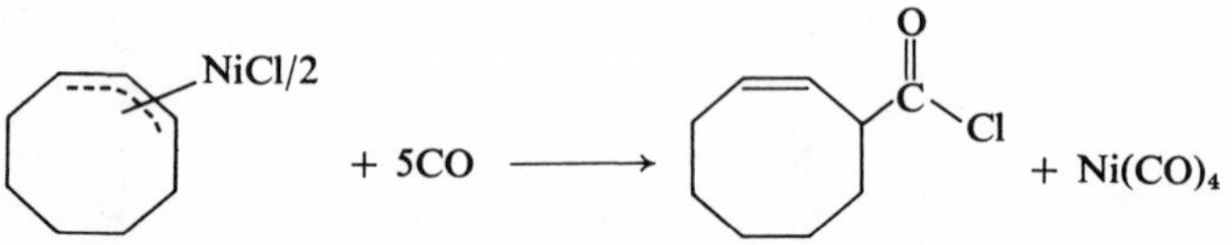

π-C₃H₅NiCl/2 + 5CO
CH₃OH
−HX
CO₂CH₃
+ Ni(CO)₄

NiCl/2
+ 5CO
O
C
Cl
+ Ni(CO)₄

The triphenylphosphine derivative **46** reacts with CO and acetylene to give *cis*-2,5-hexadienoic acid ester in 47% yield (169, 170; see also 76).

O
C
Ni—P(C₆H₅)₃ + HC≡CH + 3CO
CH₃OH
−HX
Cl
CO₂CH₃
+ (C₆H₅)₃PNi(CO)₃

46

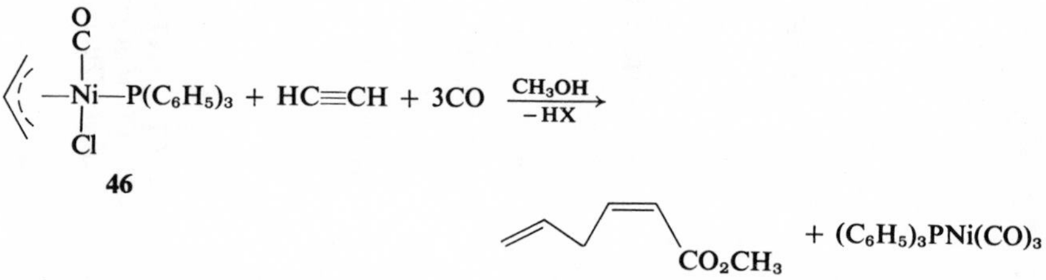

The incorporation of a substituted allyl halide as a terminally substituted trans double bond indicates that the substituent occupies a syn position in the intermediate π-allylnickel complex and that the insertion occurs into the least substituted side of the allyl group.

Cl
+ Ni(CO)₄
CO
Ni
Cl
CO
Ni—Cl
O
CH₃OH
−HCl
OCH₃
O

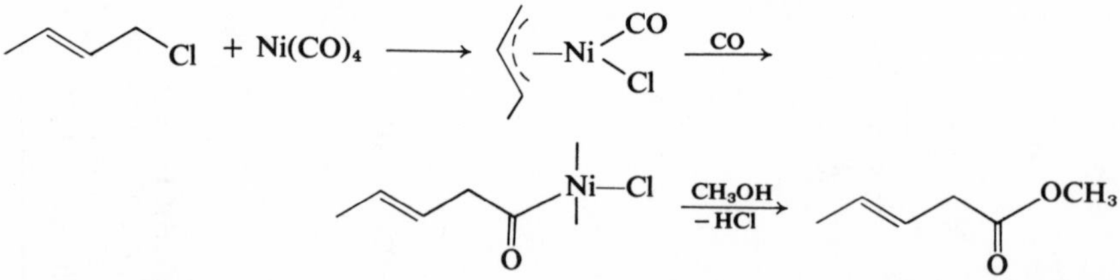

The stereochemistry of the alkyne insertion products is consistent with initial complexation of the alkyne to the nickel atom where it may be expected to adopt that cis configuration which has been observed in all the nickel–alkyne complexes so far studied (see Volume I, p. 305). This arrangement is maintained on insertion and dictates the stereochemistry of the resulting double bond, i.e., a cis double bond is produced on insertion of HC≡CH followed by a further organic group, whereas a trans double bond is produced on insertion of a monosubstituted alkyne, followed by protonation. One exception should be mentioned: unexpectedly, the diketone **48** (presumably formed by reaction of allyl chloride with **47**) contains a trans double bond (128, 186).

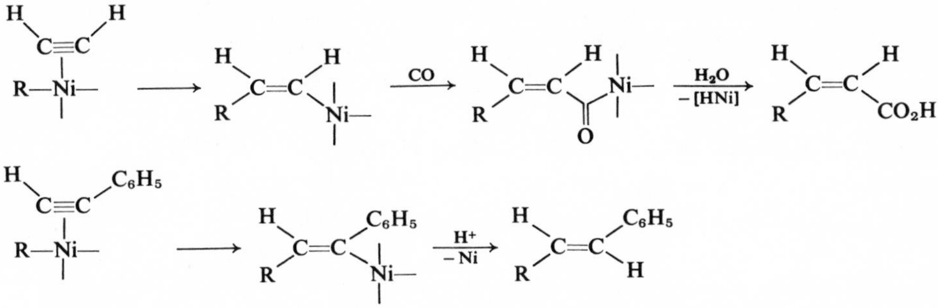

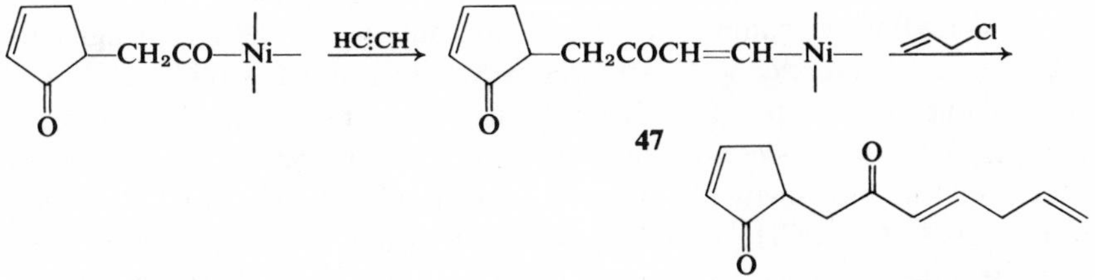

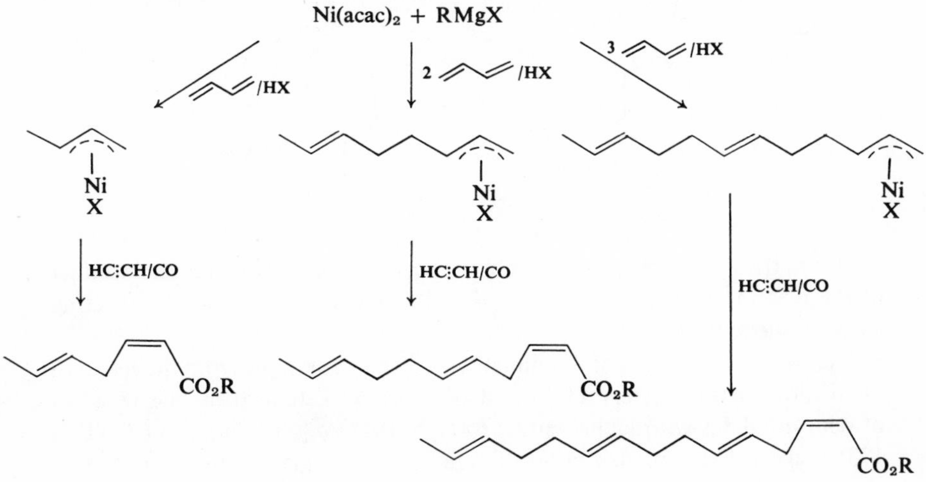

The stoichiometric reactions described above have recently been extended to include the insertion of CO and acetylene into the various π-allylnickel systems formed by treating ligand-free zerovalent nickel complexes (prepared by reacting nickel salts with Grignard reagents) with butadiene: a mixture of the expected dienoic, trienoic, and tetraenoic acids is formed (247). The growing chain in the polymerization of butadiene catalyzed by π-allylnickel bromide has also been subject to reaction with CO and acetylene and traces of trienoic and tetraenoic acids have been isolated (246).

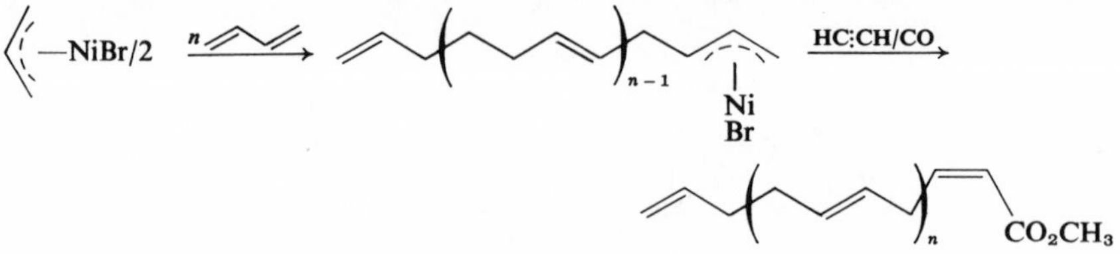

4. THE CARBONYLATION OF BIS(π-ALLYL)NICKEL COMPLEXES

It is appropriate to include at this point the stoichiometric carbonylation of bis(π-allyl)nickel complexes. In general coupling of the allyl fragments occurs (see Chapter V, Section II). However, in a number of cases, this is accompanied by insertion of a CO molecule to give a substituted ketone. It is not clear what factors control the course of reaction and insertion has been observed for bis(π-methallyl)nickel (189), $[\pi\text{-}(CH_3)_2CCHCH_2]_2Ni$ (196), $[\pi\text{-}CH_2C(CO_2C_2H_5)CH_2]_2Ni$ (197), bis(π-cyclooctenyl)nickel (157, 180, 189), and the $\pi\text{-}C_{12}H_{18}$nickel species **49** (123) but not for $(\pi\text{-}C_3H_5)_2Ni$, bis(π-crotyl)nickel, or $\pi\text{-}C_8H_{12}NiP(C_6H_5)_3$. Presumably an intermediate π-allyl-nickel-acyl species is first formed and this is then reductively eliminated. In

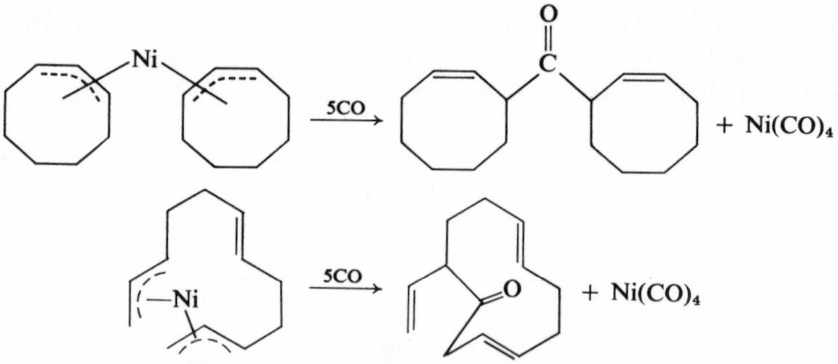

addition, traces of a precursor of muscone have been identified in the product of the reaction of CO with the bis(π-allyl)nickel species formed by reacting **49** with allene (122).

The isolation of traces of 3,6-dimethylenecycloheptanone from the reaction of dichloromethylethylene (158) and of dibenzyl ketone from the reaction of benzyl halides with nickel tetracarbonyl (181) or $[Ni_2(CN)_6]^{4-}$–CO (172, 173) suggest that here also bis(π-allyl)nickel species may be involved.

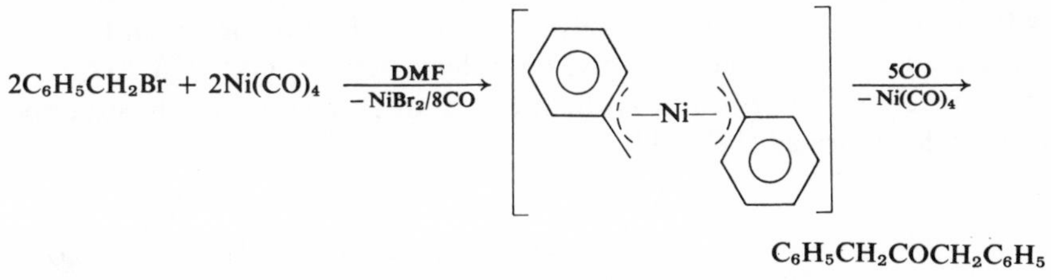

$$2C_6H_5CH_2Br + 2Ni(CO)_4 \xrightarrow[-NiBr_2/8CO]{DMF} \left[\text{Ni complex} \right] \xrightarrow[-Ni(CO)_4]{5CO}$$

$$C_6H_5CH_2COCH_2C_6H_5$$

B. The Carbonylation of Alkyl and Aryl Halides and Related Reactions

1. THE CARBONYLATION OF ALKYL AND VINYL HALIDES (Tables VI-10 to VI-14)

The relevance of the stepwise insertion sequence described in the preceding section to account for the products formed in the reaction of allyl halides with acetylene and CO is underlined by the isolation of related products (i.e., vinyl ketones, γ-lactones, and ϵ-lactones) from the reactions involving acyl halides (125, 128, 137, 144, 167, 186, P117, P119, P120) or halovinyl ketones (129).

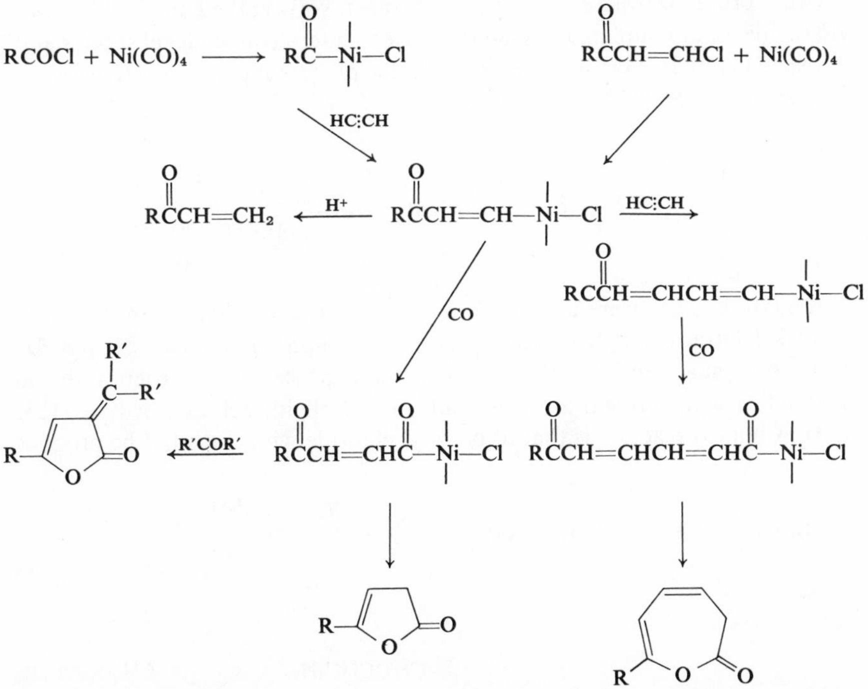

Particularly interesting in this context is the reaction involving *cis*-2,5-hexadienoyl chloride [the primary product of the reaction of allyl chloride with acetylene and CO in inert solvents (76)]; cyclization occurs in the presence of nickel tetracarbonyl to give (after hydrolysis) **22** (125, 128, 137, 186). If the reaction is carried out in the presence of acetylene then the expected lactone **28** is formed (125, 137).

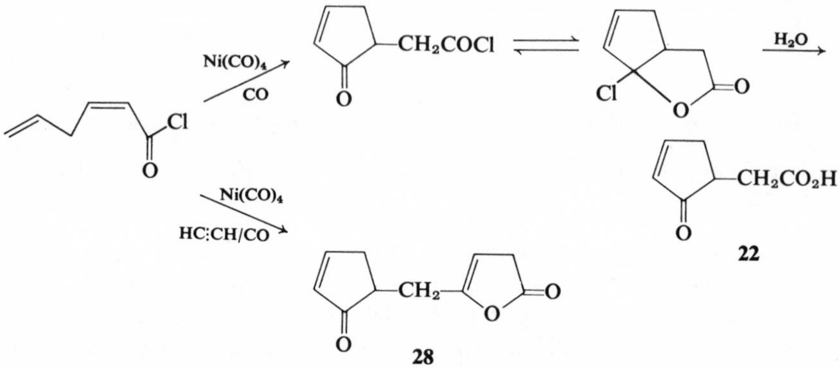

Other products isolated from the reactions involving the halovinyl ketones are the result of coupling between the various intermediate nickel species and unreacted halovinyl ketone. For example, the precursor of the γ-lactone reacts further to give **50** (129).

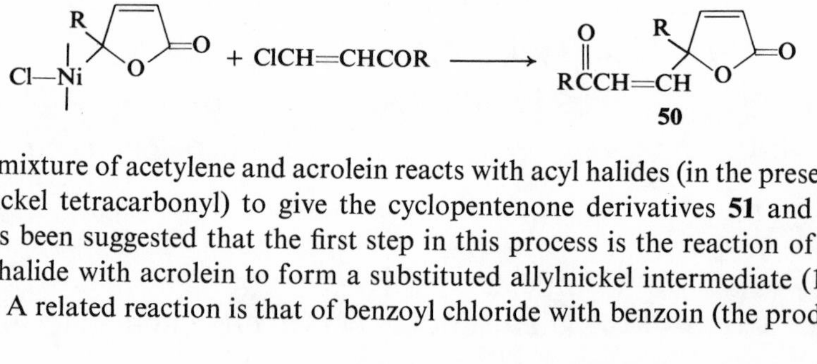

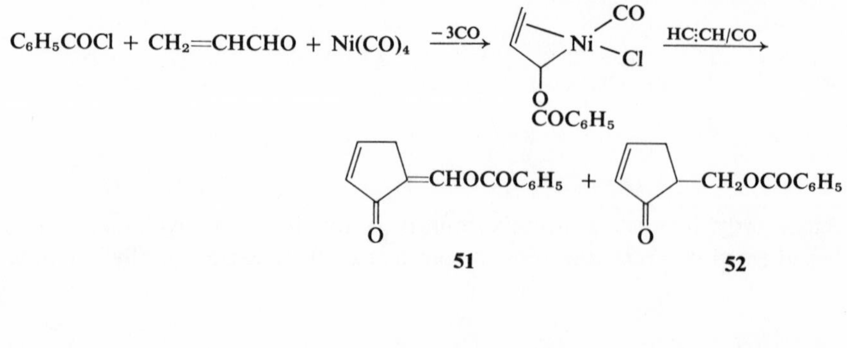

A mixture of acetylene and acrolein reacts with acyl halides (in the presence of nickel tetracarbonyl) to give the cyclopentenone derivatives **51** and **52**. It has been suggested that the first step in this process is the reaction of the acyl halide with acrolein to form a substituted allylnickel intermediate (125, 144). A related reaction is that of benzoyl chloride with benzoin (the product of the coupling of two benzoyl groups) to give the dienediol ester **53** in THF, or the benzoate of benzoin **54** in moist acetone (125, 186, 198, 199). It is suggested that the benzil coordinates to the nickel atom and reacts further with benzoyl chloride to form a nickel–allyl intermediate. Perfluoroacyl halides react similarly (200).

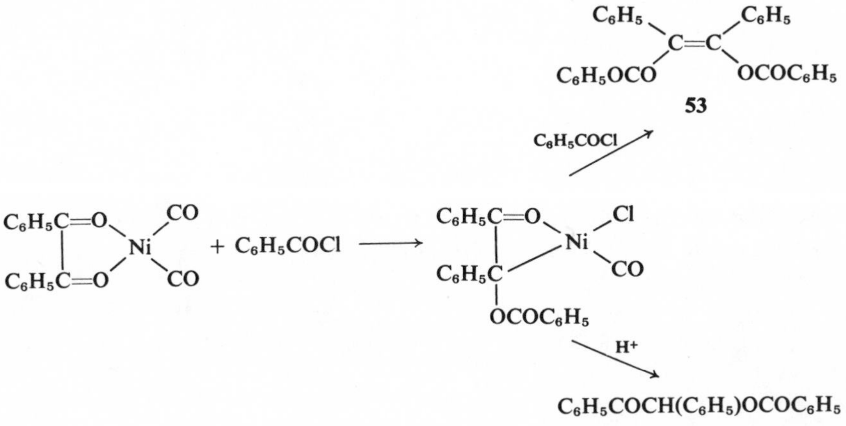

It should also be mentioned that vinyl acetyl chloride, $CH_2:CHCH_2COCl$ [the primary product of the carbonylation of allyl chloride (76)], is decarbonylated in the presence of nickel tetracarbonyl to give diallyl (141). Other acyl halides are reported to react with nickel tetracarbonyl in esters or ethers to give anhydrides (167).

The carbonylation of vinyl halides is not limited to the halovinyl ketones described above but is apparently a general reaction: iodides and bromides reacting more readily than chlorides (129, 183). The activity can be increased by adding potassium methoxide or butoxide to the nickel tetracarbonyl (202, P155)—this is probably associated with the generation of a nickel–carbene species, viz., $(CO)_3Ni-C(OK)O\text{-}tert\text{-}C_4H_9$ (201)—see Section III-C.

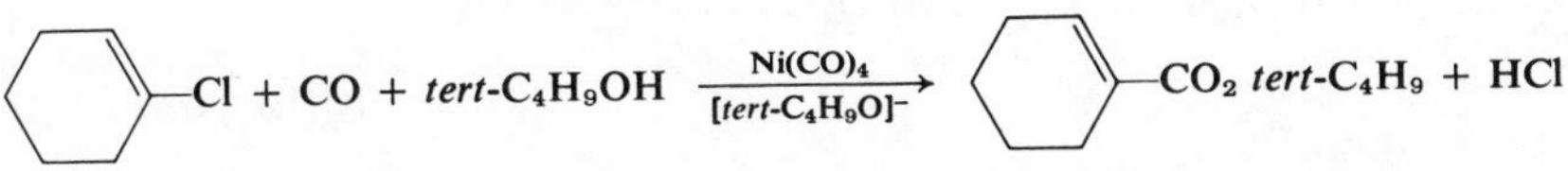

The reactions involving vinyl halides have been extended in individual cases to the insertion of acrylonitrile (203) and alkynes (204). In the first case the ketone **55** results and in the second a γ-lactone, e.g., **56**.

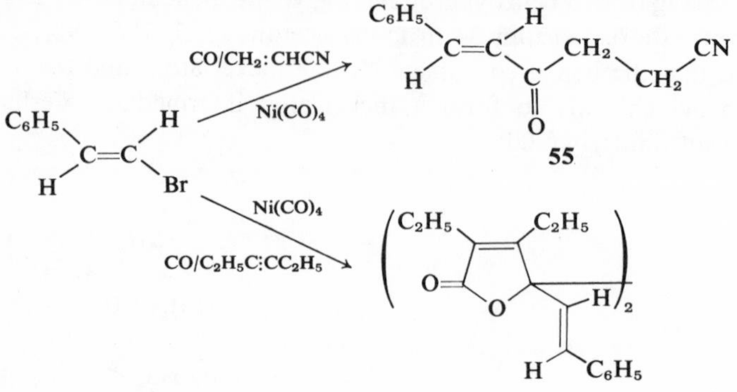

55

56

Aminocarbonylation has been observed, in a few cases, if the reaction is carried out in the presence of pyrrolidine (202).

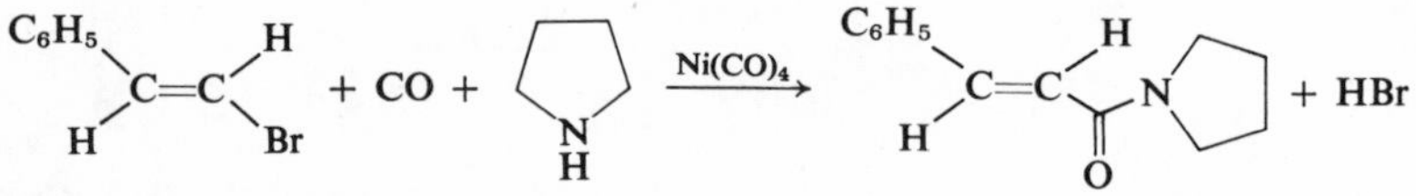

Diiodotetraphenylbutadiene reacts readily with nickel tetracarbonyl to give tetraphenylcyclopentadienone (242).

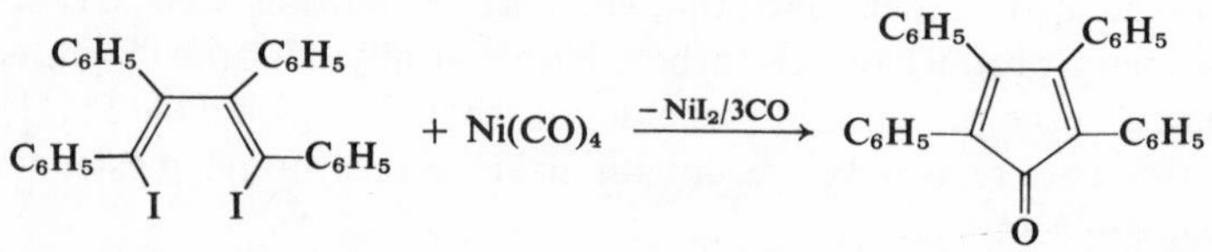

Alkyl halides cannot, in general, be directly carbonylated in the presence of nickel tetracarbonyl [traces of dibutyl ketone have been detected in the reaction of butyl iodide with $[Ni_2(CN)_6]^{4-}$ and CO (172, 173)]. However, efficient carbonylation of alkyl iodides (202) and ω-iodoalkynes (205) has been observed using the $Ni(CO)_4$-$[O\text{-}tert\text{-}C_4H_9]^-$ reagent. Under more forcing

$$C_4H_9C{\equiv}C(CH_2)_3I + CO + tert\text{-}C_4H_9OH \xrightarrow[-HI]{Ni(CO)_4/tert\text{-}C_4H_9O^-}$$

$$C_4H_9C{\equiv}C(CH_2)_3CO_2tert\text{-}C_4H_9$$

conditions (ca. 300°) it is possible to carbonylate alkyl halides stoichiometrically (51) or catalytically (112, P136, P156).

Benzyl halides are more easily carbonylated and successful reactions have been reported involving nickel tetracarbonyl (172, 173, 181) and $Ni(CO)_4$-I^-

(71, 183). The latter reagent is more efficient and it has been suggested that $[INi(CO)_3]^-$ is the active species. The possibility that π-benzylnickel intermediates are involved is mentioned on page 342. In addition, the reaction

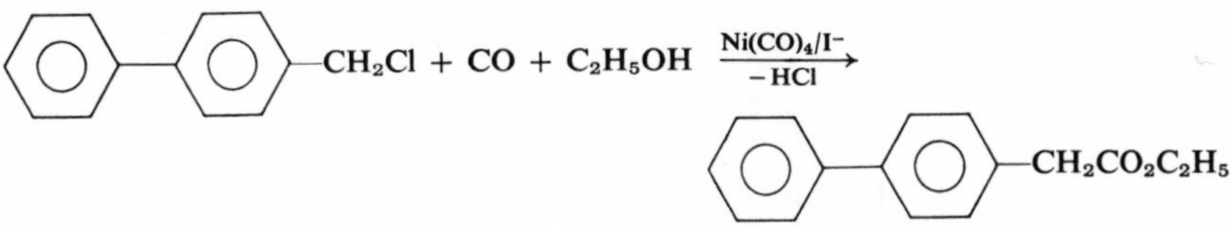

between benzyl bromide and nickel tetracarbonyl in the presence of 3-hexyne has been studied: a dimeric γ-but-2-enolactone is formed (204).

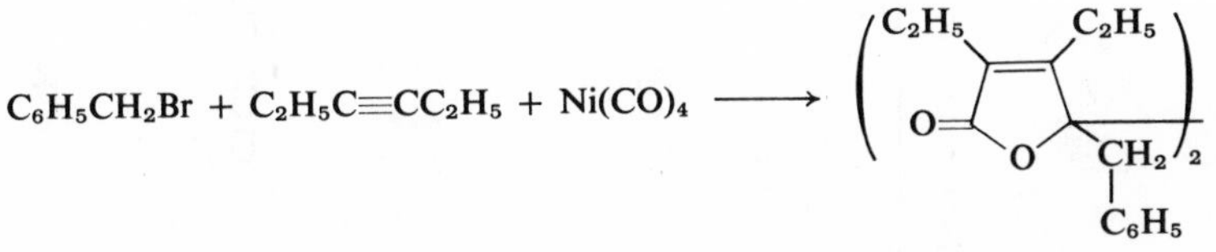

α-Chloroalkynes react with nickel tetracarbonyl in acidic media to give modest yields of allenic acids (83, 136, 206–209, P133). The reaction involving

$$HC{\equiv}C\underset{CH_3}{\overset{H}{C}}{-}Cl + CO + H_2O \xrightarrow{Ni(CO)_4} HO_2CCH{=}C{=}CHCH_3 + HCl$$

propargyl chloride produces itaconic acid, in addition to butadienoic acid. Bromo- or iodoalkynes react to give the ketoacid **57** and maleic anhydride derivatives. The mechanism of these reactions has not been investigated but

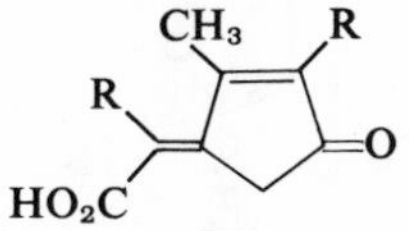

57

because carbonylation occurs at that acetylenic carbon atom farthest removed from the chloride substituent, direct attack by the nickel tetracarbonyl at the halide center seems to be unlikely. A concerted mechanism has been suggested (210).

1-Haloalkynes are reduced by nickel tetracarbonyl (80).

TABLE VI-10

The Reaction of Acyl Halides with Carbon Monoxide and Alkynes

Product	Acyl halide	Alkyne	Ref.
$C_6H_5COCH:CH_2$	C_6H_5COCl	$HC:CH-HCl$	186, P120
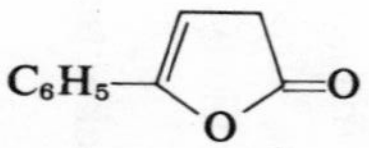 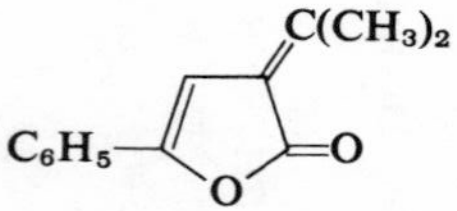 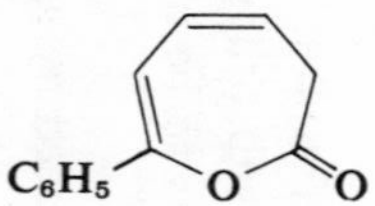	C_6H_5COCl	$HC:CH$	125, 144, 167, 186, 251, P117, P119
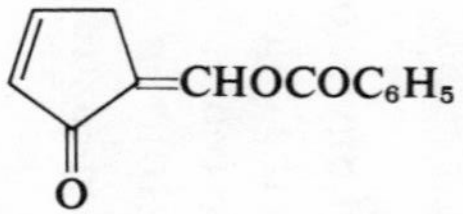 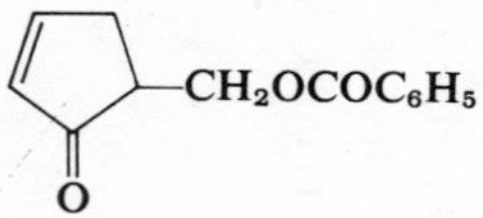	C_6H_5COCl	$HC:CH-CH_2:CHCHO$	125, 144, 186
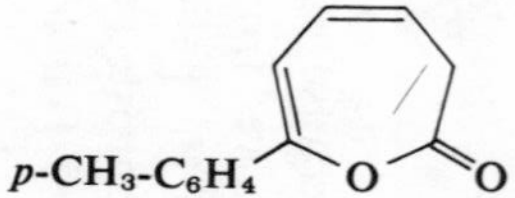	$p\text{-}CH_3C_6H_4COCl$	$HC:CH$	167

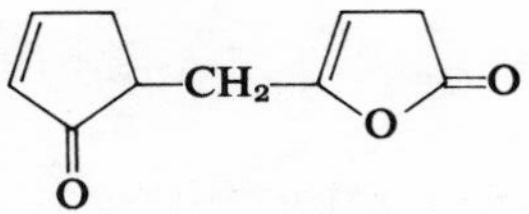 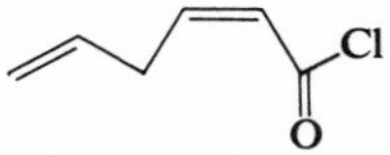 HC:CH		125, 128, 186
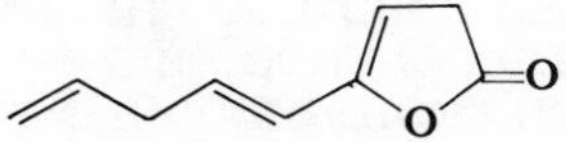 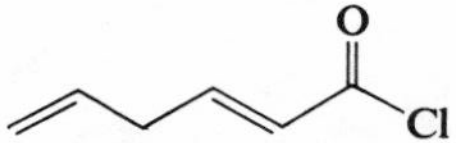 HC:CH		128, 186
HC:CH	C$_4$H$_9$COCl	P117
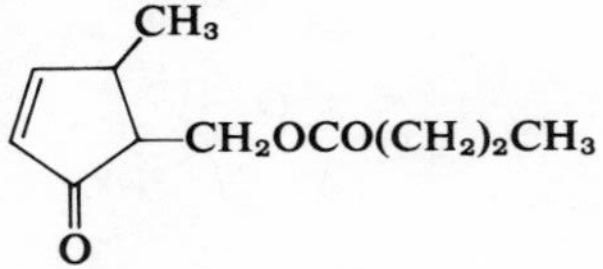 HC:CH–CH$_3$CH:CHCHO	CH$_3$(CH$_2$)$_2$COCl	186
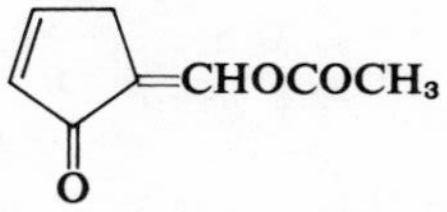 HC:CH–CH$_2$:CHCHO	CH$_3$COCl	186

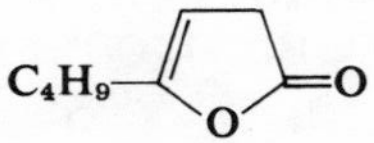

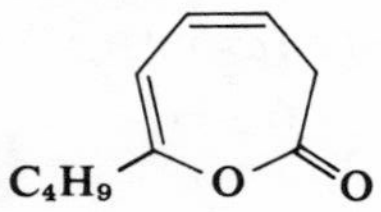

CH$_2$

C$_4$H$_9$

C$_4$H$_9$

CH$_3$ CH$_2$OCO(CH$_2$)$_2$CH$_3$

CHOCOCH$_3$

TABLE VI-11

THE CARBONYLATION OF VINYL HALIDES

Product	Vinyl halide	Ref.
$C_6H_5COCH:CH_2$, $C_6H_5COCH:CH(CH_2)_2COC_6H_5$, $C_6H_5COCH_2CH_2CO_2CH_3$, 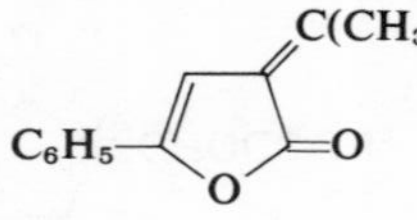 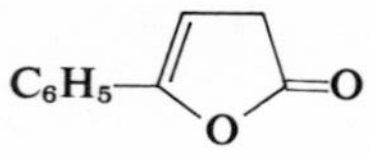 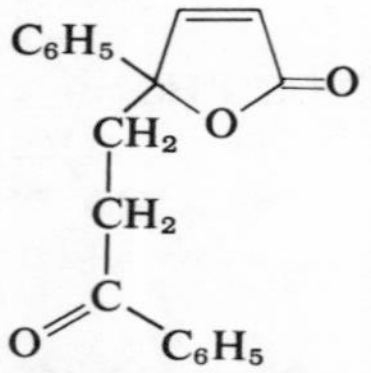 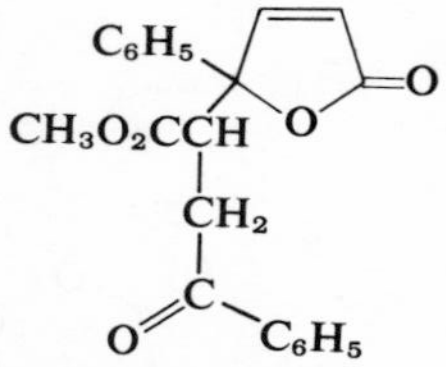 ...	$C_6H_5COCH:CHCl$	129
$CH_3COCH:CHCO_2CH_3$, $CH_3COCH_2CH_2CO_2CH_3$, $CH_3COCH:CHCH_2COCH_3$	$CH_3COCH:CHCl$	129
$iso\text{-}C_3H_7COCH:CHCO_2CH_3$, $iso\text{-}C_3H_7COCH_2CH_2CO_2CH_3$, $iso\text{-}C_3H_7COCH_2C(CO_2CH_3):CHCH_2COC_3H_7$	$iso\text{-}C_3H_7COCH:CHCl$	129
$tert\text{-}C_4H_9COCH:CHCO_2CH_3$, $tert\text{-}C_4H_9COCH_2CH_2CO_2CH_3$, $tert\text{-}C_4H_9COCH_2C(CO_2CH_3):CHCH_2CO\text{-}tert\text{-}C_4H_9$, 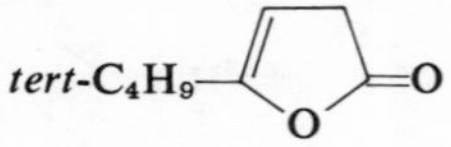	$tert\text{-}C_4H_9COCH:CHCl$	129

Product	Halide	References
trans-C₆H₅CH:CHCO₂CH₃	*trans*-C₆H₅CH:CHBr	129, 183
	trans-C₆H₅CH:CHCl	202
trans-C₆H₅CH:CHCO₂CH₃, C₆H₅CH:CHCHO	*trans*-C₆H₅CH:CHBr	140, 173
(cyclohexenyl)–CO₂*tert*-C₄H₉	(cyclohexenyl)–Cl	202
tert-C₄H₉–(cyclohexenyl)–CO₂*tert*-C₄H₉	*tert*-C₄H₉–(cyclohexenyl)–Br	202
(C₆H₅)₂C:CHCO₂CH₃	(C₆H₅)₂C:CHBr	202
cis-C₂H₅OCH:CHCO₂CH₃	*cis*-C₂H₅OCH:CHBr	202
CH₂:C((C₆H₅)CO₂CH₃	CH₂:CBrC₆H₅	129, 183
CH₃:CHCO₂CH₃	CH₂:CHBr	129, 183, P155
CH₃O₂CCH:CHCO₂CH₃	ICH:CHI	129, 183
CH₃O₂CCH:CHCO₂CH₃	CH₃O₂CCH:CHBr	129, 183

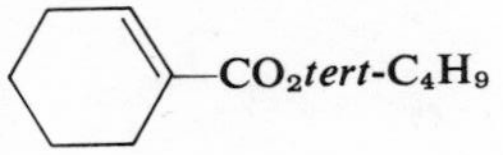

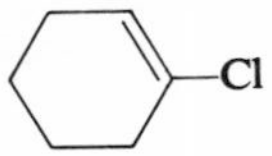

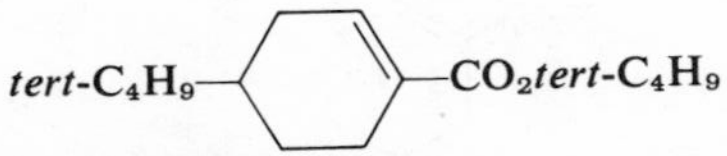

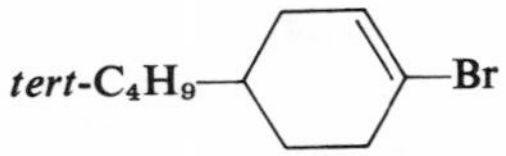

TABLE VI-12

THE REACTION OF VINYL HALIDES WITH CARBON MONOXIDE AND ALKYNES OR OLEFINS

Product	Vinyl halide	Alkyne or olefin	Ref.
$C_6H_5COCH:CHCH:CHCO_2CH_3$, $C_6H_5COCH_2CH_2(OCH_3)_2$ 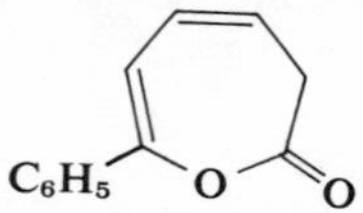	$C_6H_5COCH:CHCl$	$HC:CH$	129
$CH_3COCH:CHCH:CHCO_2CH_3$ $trans\text{-}C_6H_5CH:CHCOCH_2CH_2CN$	$CH_3COCH:CHCl$ $trans\text{-}C_6H_5CH:CHBr$	$HC:CH$ $CH_2:CHCN$	129, 183 203
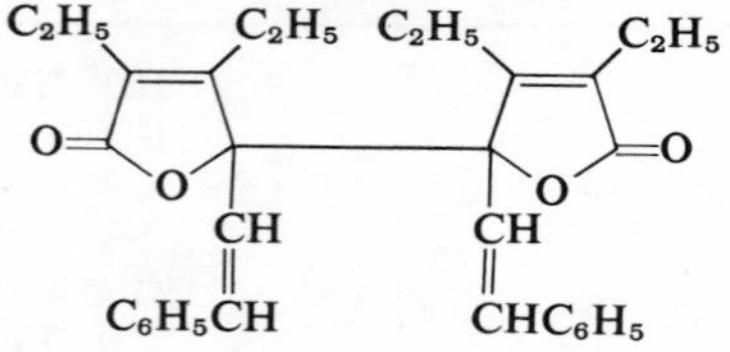	$trans\text{-}C_6H_5CH:CHBr$	$C_2H_5C:CC_2H_5$	204
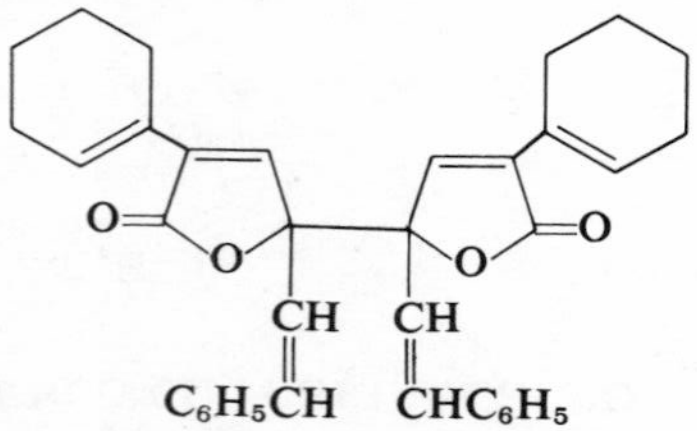	$trans\text{-}C_6H_5CH:CHBr$	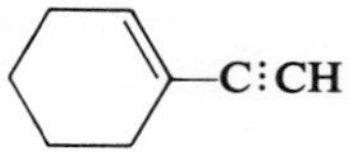	204

TABLE VI-13
THE CARBONYLATION OF ALKYL AND BENZYL HALIDES

Product	Alkyl halide	Ref.
CH_3CO_2H, $CH_3CO_2CH_3$	CH_3Cl	136
$C_4H_9COC_4H_9$	C_4H_9I	172, 173; see also 151
$CH_3(CH_2)_4CO_2H$, $CH_3(CH_2)_2CH(CH_3)CO_2H$	$CH_3(CH_2)_4Cl$	112
$CH_3(CH_2)_6CO_2tert\text{-}C_4H_9$	$CH_3(CH_2)_6I$	202
$C_8F_{17}CH_2CH_2CO_2H$	$C_8F_{17}CH_2CH_2I$	P156
$HO_2C(CH_2)_4CO_2H$	$I(CH_2)_4CO_2H$	51
	$I(CH_2)_4I$	51
$tert\text{-}C_4H_9O_2C(CH_2)_6CO_2tert\text{-}C_4H_9$	$I(CH_2)_6I$	202
$(p\text{-}ClC_6H_4)_2C\!:\!CHCO_2H$, $(p\text{-}ClC_6H_4)_2CHCH(Cl)CO_2H$	$(p\text{-}ClC_6H_4)_2CHCCl_3$	188
$C_6H_5CH_2COCH_2C_6H_5$, $C_6H_5CH_2CO_2C_2H_5$	$C_6H_5CH_2Br$	172, 173, 181
$C_6H_5CH_2CO_2H$	$C_6H_5CH_2Cl$	71, 183
$p\text{-}CH_3C_6H_4CH_2COCH_2C_6H_4\text{-}p\text{-}CH_3$	$p\text{-}CH_3C_6H_4CH_2Br$	172, 173
$p\text{-}C_6H_5C_6H_4CH_2CO_2C_2H_5$	$p\text{-}C_6H_5C_6H_4CH_2Cl$	71, 183

2. THE CARBONYLATION OF ARYL HALIDES (Tables VI-15 and VI-16)

The carbonylation of aryl halides, catalyzed by nickel, has received considerable attention and is a useful method for preparing aromatic carboxylic acids or their esters. The reactivity of the aryl halides decreases in the order

$$ArX + CO + H_2O \xrightarrow{[Ni]} ArCO_2H + HX$$

I > Br > Cl > F, which enables, for example, p-chlorobenzoic acid to be synthesized directly from p-chlorobromobenzene. Under more forcing conditions the dicarboxylic acid, e.g., terephthalic acid, results.

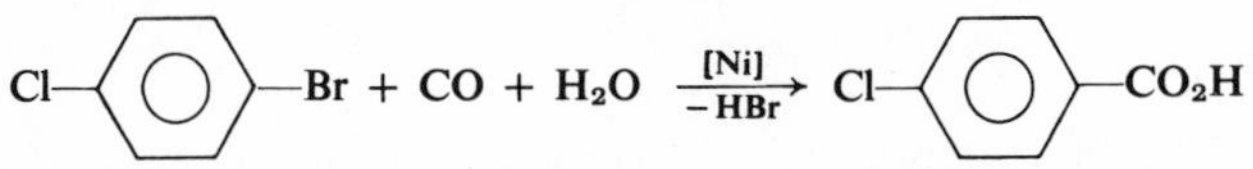

$$Cl\text{—}\langle\bigcirc\rangle\text{—}Br + CO + H_2O \xrightarrow[-HBr]{[Ni]} Cl\text{—}\langle\bigcirc\rangle\text{—}CO_2H$$

The nickel may be present as a stoichiometric reagent in the form of nickel tetracarbonyl or as a catalyst if the reaction is carried out in an atmosphere of carbon monoxide. The stoichiometric reaction under mild conditions is confined to the carbonylation of aryl iodides [or diazonium compounds (69, 211, 222)]. Under more vigorous reaction conditions (e.g., 250°) aryl chlorides and bromides also take part in a stoichiometric carbonylation reaction (51, P134). The addition of sodium methoxide (202) or potassium acetate greatly improves the efficiency of the reaction.

TABLE VI-14
THE CARBONYLATION OF HALOALKYNES

Product	Haloalkyne	Ref.
$CH_2:C:CHCO_2H$, $CH_2:C(CO_2H)CH_2CO_2H$	$HC:CCH_2Cl$	136, 206, 207, P133; see also 208
$CH_3CH:C:CHCO_2H$	$HC:CCH(CH_3)Cl$	83
$C_3H_7CH:C:CHCO_2H$	$HC:CCH(C_3H_7)Cl$	83
$C_6H_5CH:C:CHCO_2H$	$HC:CCH(C_6H_5)Cl$	83
$(CH_3)_2C:C:CCO_2H$, 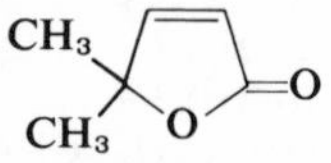	$HC:CC(CH_3)_2Cl$	70, 83
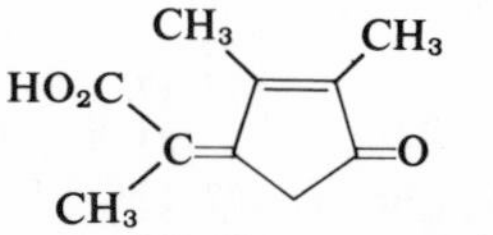 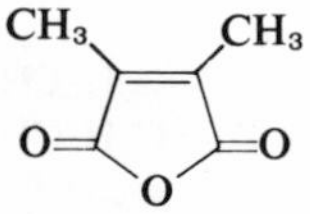	$CH_3C:CCH_2I$	207
$CH_2:C:C(C_4H_9)CO_2H$	$C_4H_9C:CH_2Cl$	83
$CH_2:C:C(C_4H_9)CO_2H$, 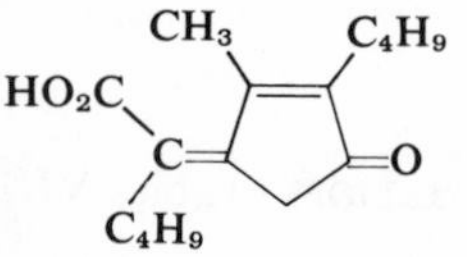	$C_4H_9C:CCH_2Br$	207
$(CH_3)_2C:C:C(C_4H_9)CO_2H$	$C_4H_9C:CC(CH_3)_2Cl$	83
$C_4H_9C:C(CH_2)_3CO_2tert\text{-}C_4H_9$	$C_4H_9C:C(CH_2)_3I$	205
$C_4H_9C:C(CH_2)_4CO_2tert\text{-}C_4H_9$, 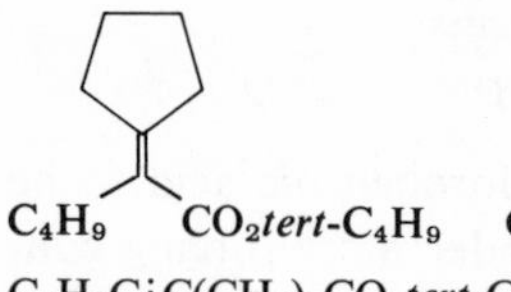 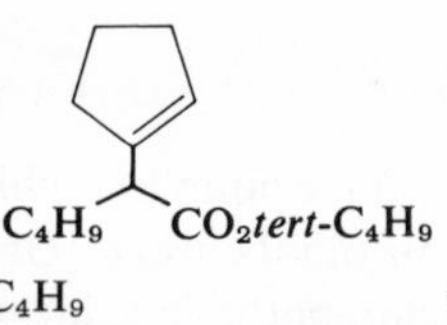	$C_4H_9C:C(CH_2)_4I$	205
$C_4H_9C:C(CH_2)_5CO_2tert\text{-}C_4H_9$	$C_4H_9C:C(CH_2)_5I$	205

Catalytic carbonylation has been described using catalysts based on $Ni(CO)_4$ (214, P134–P140), $[(C_6H_5)_3P]_2Ni(CO)_2$ (P141, P142), NiX_2 (215, 216, P137, P143), $[(C_6H_5)_3P]_2NiBr_2$ (215), or metallic nickel (216, 217, P137, P141, P144–P146). The activity and efficiency of these systems is increased by the addition of a base, which may either be the salt of an organic acid, e.g., potassium acetate (218–221), sodium benzoate (220, 221, P153), or sodium formate (213, P141, P144), or an inorganic base, e.g., calcium or sodium hydroxide (223), ammonium hydroxide (P135), or sodium methoxide (202, P147). The catalyst derived from $Ni(CO)_4$ in polar solvents (DMF) with added

TABLE VI-15
THE CARBONYLATION OF ARYL HALIDES

Product	Aryl halide	Ref.
$C_6H_5CO_2H$	C_6H_5Cl	215, 216, P135, P143
	C_6H_5Br	218, 220, 221, 223, P134, P141, P153
	C_6H_5I	137, 202, 216, 217
$o\text{-}CH_3C_6H_4CO_2H$	$o\text{-}CH_3C_6H_4Br$	216, 220, 223, 224, P148
$m\text{-}CH_3C_6H_4CO_2H$	$m\text{-}CH_3C_6H_4Br$	220
$p\text{-}CH_3C_6H_4CO_2H$	$p\text{-}CH_3C_6H_4Br$	220, 223
	$p\text{-}CH_3C_6H_4N_2BF_4$	69
3,4-Dimethyl benzoic acid	4-Br, o-xylene	220
2,4-Dimethyl benzoic acid	4-Br, m-xylene	220
2,5-Dimethyl benzoic acid	2-Br, p-xylene	220
2,4,5-Trimethyl benzoic acid	5-Br, pseudocumene	220
2,4,6-Trimethyl benzoic acid	Bromomesitylene	220
$o\text{-}CH_3OC_6H_4CO_2H$	$o\text{-}CH_3OC_6H_4Br$	223
$m\text{-}CH_3OC_6H_4CO_2H$	$m\text{-}CH_3OC_6H_4Br$	223
$p\text{-}CH_3OC_6H_4CO_2H$	$p\text{-}CH_3OC_6H_4Br$	223, P141
$p\text{-}C_2H_5OC_6H_4CO_2C_2H_5$	$p\text{-}C_2H_5OC_6H_4N_2BF_4$	69
$p\text{-}CH_3OC_6H_4CO_2H$	$p\text{-}CH_3OC_6H_4N_2BF_4$	69
$p\text{-}NCC_6H_4CO_2H$	$p\text{-}NCC_6H_4Br$	223
$o\text{-}ClC_6H_4CO_2H$	$o\text{-}ClC_6H_4Br$	223
$m\text{-}ClC_6H_4CO_2H$	$m\text{-}ClC_6H_4Br$	223
$p\text{-}ClC_6H_4CO_2H$	$p\text{-}ClC_6H_4Br$	223
	$p\text{-}ClC_6H_4N_2BF_4$	69
$p\text{-}HO_2CC_6H_4CO_2H$	$p\text{-}BrC_6H_4Br$	241–217, 221, P138, P141, P142, P144, P146, P147 P152, P153
	$p\text{-}HO_2CC_6H_4N_2BF_4$	69
1-Naphthoic acid	1-Chloronaphthalene	271, 219, 221, 223, P137
	1-Naphthyl N_2BF_4	69
2-Naphthoic acid	2-Chloronaphthalene	216, 223

calcium hydroxide is particularly active and may be used to carbonylate aryl chlorides and bromides almost quantitatively at ca. 100° under 1 atm of carbon monoxide (223). The base neutralizes the liberated hydrogen halide, thereby suppressing undesirable side reactions.

In several cases anomalous behavior has been observed. The reaction of bromo- or chlorobenzene with CO and Ni(CO)$_4$ at 275–325° in the presence of sodium carbonate or phosphate is reported to give a mixture of benzoic anhydride, phthalic anhydride, and benzene. The benzoic anhydride has been shown to be the probable precursor of the phthalic anhydride. This reaction

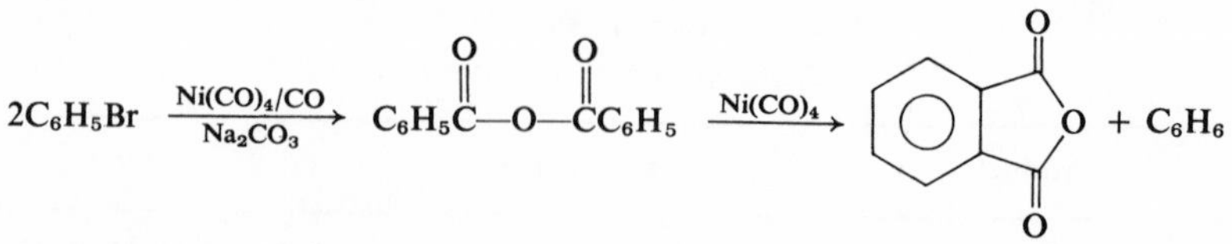

$$2C_6H_5Br \xrightarrow[Na_2CO_3]{Ni(CO)_4/CO} C_6H_5\overset{O}{\overset{\|}{C}}-O-\overset{O}{\overset{\|}{C}}C_6H_5 \xrightarrow{Ni(CO)_4} \text{(phthalic anhydride)} + C_6H_6$$

is general and has been used to prepare methyl, phenyl, and naphthalic anhydrides as well as *N*-phenylphthalimide from *N,N*-dibenzoyl aniline (224, P148, P149). Pentafluorophenyl halides also react anomalously with nickel

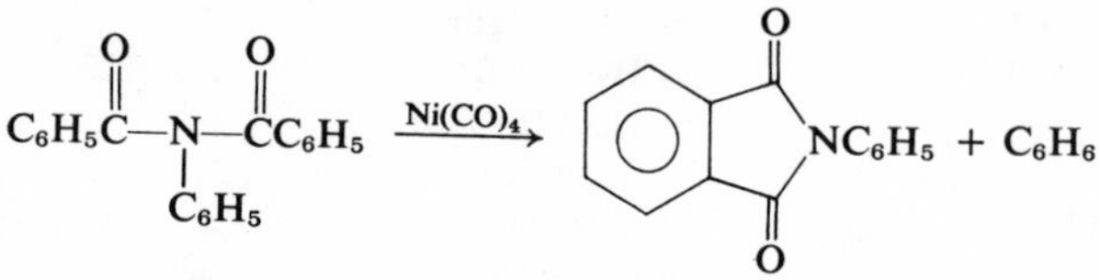

$$C_6H_5\overset{O}{\overset{\|}{C}}-\underset{\underset{C_6H_5}{|}}{N}-\overset{O}{\overset{\|}{C}}C_6H_5 \xrightarrow{Ni(CO)_4} \text{(N-phenylphthalimide)} + C_6H_6$$

tetracarbonyl; the principal reaction products are decafluorobenzophenone and decafluorobiphenyl and a radical mechanism has been discussed (225). Anomalous behavior is also observed in the reaction of an aryl iodide with nickel tetracarbonyl when THF is used as the solvent: instead of the expected carboxylic acid, a mixture of an aril (**58**) and an enediol diester (**59**) is obtained. The enediol diester is believed to be the product of the reaction of the aril and an aroyl iodide with nickel tetracarbonyl (see Section III-B1) (198, 199).

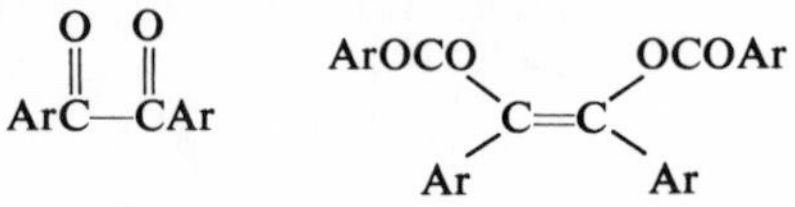

$$\underset{\textbf{58}}{Ar\overset{O}{\overset{\|}{C}}-\overset{O}{\overset{\|}{C}}Ar} \qquad\qquad \underset{\textbf{59}}{\overset{ArOCO \qquad OCOAr}{\underset{Ar \qquad\quad Ar}{C=C}}}$$

A number of related carbonylation reactions are conveniently mentioned in this section. These include methods for synthesizing unsymmetrical ketones by reacting an aryl iodide and an aryl mercuric halide with nickel tetracarbonyl (226) or symmetrical ketones by carbonylation of an aryl mercuric halide (226), and the formation of benzoyl fluoride by carbonylation of chlorobenzene in the presence of sodium fluoride (P150). Benzoyl fluoride is

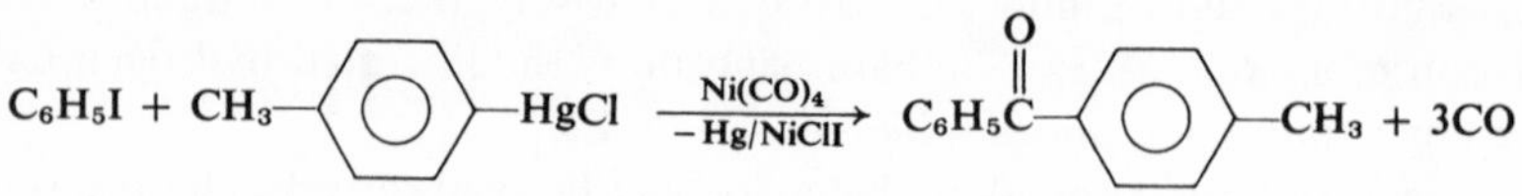

$$C_6H_5I + CH_3-\text{(C}_6\text{H}_4)-HgCl \xrightarrow[-Hg/NiCl_2]{Ni(CO)_4} C_6H_5\overset{O}{\overset{\|}{C}}-\text{(C}_6\text{H}_4)-CH_3 + 3CO$$

$$C_6H_5Cl + CO + NaF \xrightarrow[-NaCl]{Ni(CO)_4} C_6H_5COF$$

also the product of the reaction of nickel tetracarbonyl with benzene diazonium fluoroborate (P145).

The carbonylation reaction has been extended to the synthesis of aromatic amides and nitriles by allowing the reaction to proceed in the presence of an amine or amide (P151). Presumably the amide is the product of a thermal reaction with the intermediate aroyl halide and the nitrile the result of dehydration of the amide. It should also be mentioned that benzene was

$$C_6H_5Cl + CO \xrightarrow{[Ni]} C_6H_5COCl \xrightarrow{HCONH_2} C_6H_5CONH_2 + CO + HCl$$

reported at an early date to react with nickel tetracarbonyl in the presence of aluminum trichloride to give benzaldehyde (33) while aromatic carboxylic acids can also be prepared by the carboxylation of an aromatic hydrocarbon in the presence of nickel tetracarbonyl and CO (P69).

$$Ar-H + RCO_2H \xrightarrow{Ni(CO)_4/CO} ArCO_2H + RH$$

The carbonylation of iodobenzene can be extended, in much the same way as has already been described for the allyl halides, by carrying out the reaction in the presence of carbon monoxide and an olefin or alkyne. The result is a series of insertion steps leading to linear ketones and cyclic lactones (Table VI-16). The reaction is frequently accompanied by hydrogenation of the product. The reaction of iodobenzene with a mixture of CO and acetylene catalyzed by nickel tetracarbonyl needs more forcing conditions than the analogous carbonylation reaction (i.e., 120–140°, 30 atm) and can be used as a method for synthesizing benzyl propionates in high yield (141). The olefins

$$C_6H_5I + 2CO + HC{\equiv}CH \xrightarrow[\text{[Ni]}]{ROH/HX} C_6H_5\overset{O}{\overset{\|}{C}}CH_2CH_2CO_2R + NiIX$$

styrene, acrylonitrile, and ethyl acrylate react similarly. The products are of two types: the first is the result of the insertion of a CO molecule followed by an olefin molecule and the second is the result of the insertion of a second CO molecule. Because the reaction is carried out under anhydrous conditions, an acid or ester is not formed in this last case; instead cyclization to a lactone occurs (227). N-Benzylidene alkylamine also takes part in an insertion reaction.

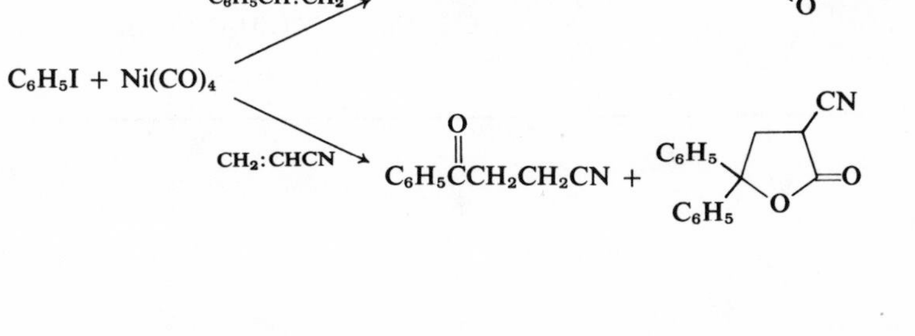

TABLE VI-16
THE REACTION OF ARYL HALIDES WITH CARBON MONOXIDE AND
ALKYNES OR OLEFINS

Product	Aryl halide	Alkyne or olefin	Ref.
$C_6H_5COCH_2CH_2C_6H_5$	C_6H_5I	$CH_2:CHC_6H_5$	227
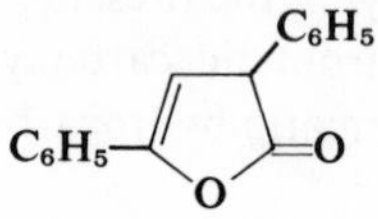			
$C_6H_5COCH_2CH_2CN$	C_6H_5I	$CH_2:CHCN$	227
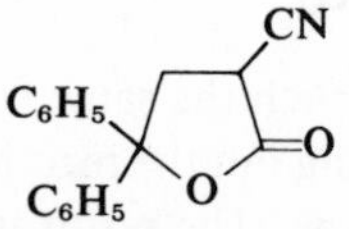			
$C_6H_5COCH_2CH_2CO_2C_2H_5$	C_6H_5I	$CH_2:CHCO_2C_2H_5$	227
$(C_6H_5CON(CH_3)CHC_6H_5)_{\frac{1}{2}}$	C_6H_5I	$C_6H_5CH:NCH_3$	228
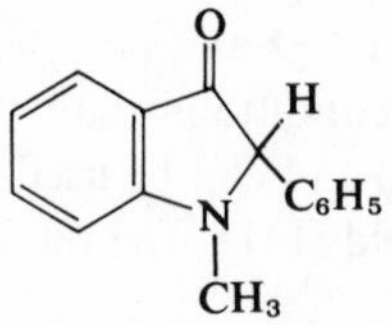			
$C_6H_5CONCH_3CHC_6H_5COC_6H_5$			
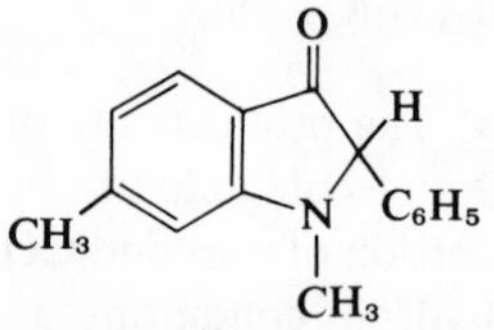	p-$CH_3C_6H_4I$	$C_6H_5CH:NCH_3$	228
$C_6H_5CON(C_2H_5)CHC_6H_5COC_6H_5$	C_6H_5I	$C_6H_5CH:NC_2H_5$	228
$C_6H_5COOCH:CHC_6H_5$	C_6H_5HgCl		226
$C_6H_5COCH_2CH_2CO_2R$	C_6H_5I	$HC:CH$	141
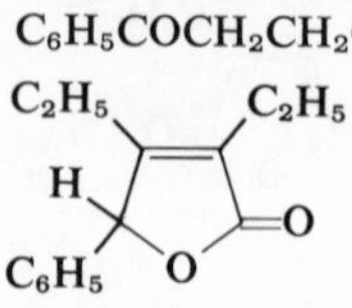	C_6H_5HgCl	$C_2H_5C:CC_2H_5$	226

The product depends on the solvent used: in DMF a mixture of the symmetrical ketone **60** and the indoline derivative **61** is obtained, whereas in benzene the principal reaction product is an *N*-alkyl-*N*-(α-phenylphenacyl)-benzamide **62** (228). A reaction related to those described above is that of

60 **61** **62**

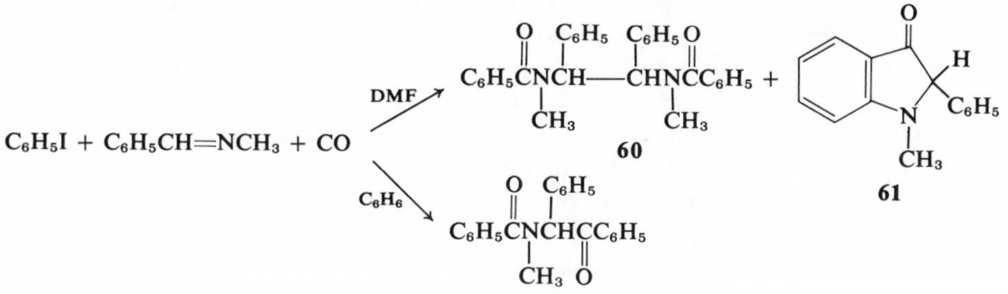

phenyl mercuric chloride and nickel tetracarbonyl with styrene oxide or 3-hexyne in DMF. Although simple carbonylation is the main reaction, insertion of the oxide or alkyne, giving **63** and **64**, is also observed (226).

63 **64**

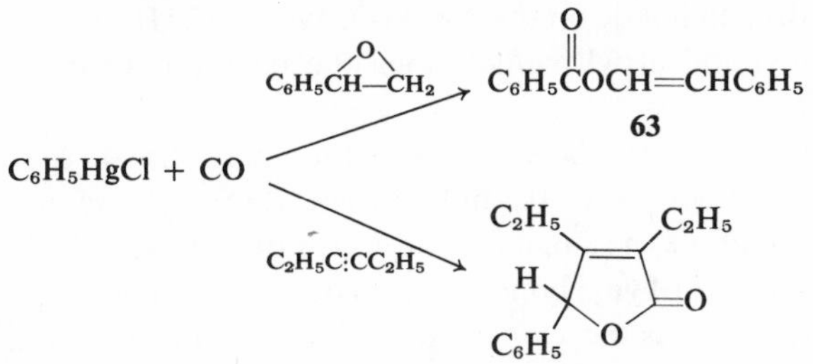

3. Mechanistic Considerations

The mechanism of the carbonylation of alkyl and aryl halides remains mainly speculative. It is normally assumed that a sequence of reactions familiar in organometallic chemistry is involved: oxidative addition of the organic halide to a nickel carbonyl species, insertion of a CO molecule or migration of the organic group, and finally reductive elimination to give the carboxylic acid either via the aroyl chloride or in a concerted reaction.

$$RCOX \xrightarrow{H_2O} RCO_2H + HX$$

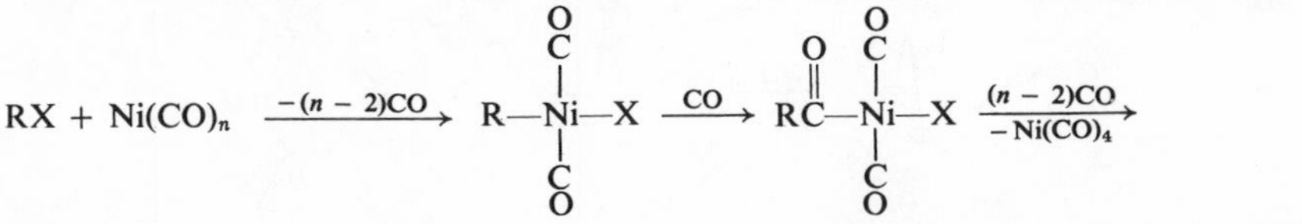

The kinetics of the carbonylation of aryl halides have been studied for the reactions catalyzed by nickel tetracarbonyl in the presence of $Ca(OH)_2$ (223) or the presence of potassium acetate (219, 220). The rate of the reaction in the first study is found to decrease in the order I > Br > Cl > F, which correlates qualitatively with the carbon–halogen bond energies and suggests that oxidative addition is the rate-determining step. A reasonable linear correlation between the relative rates of carbonylation of substituted aryl halides (e.g., RC_6H_4Br) and the Hammett σ constant for the substituent is interpreted as evidence for the nucleophilic character of the nickel species taking part in the oxidative addition step. The possibility that this species is a polynuclear nickel carbonyl anion has been considered (the suggested anion $[Ni_3(CO)_8]^{2-}$ should be treated with caution, however, because recent work has suggested that a reformulation of this species may be necessary—see Volume I, p. 20). The base, $Ca(OH)_2$, is assigned the double function of generating the active catalyst by reacting with nickel tetracarbonyl and of neutralizing the acid produced.

Additional evidence supporting the general mechanism is to be found in the formation of the hydrocarbon ArH in the absence of base—the result of attack of hydrogen halide on the Ar–Ni–X system (223)—and in the isolation of various acyl and aroyl halides when the reaction is carried out in inert media (P150).

Various examples can be cited from the chemistry of the organonickel complexes as precedents for the more extended insertion sequences observed in the presence of alkynes, olefins, and CO. For example, the insertion of an alkyne into a nickel alkynyl bond has been observed in the reaction of benzyne with **65** (193), whereas bicycloheptene inserts into the π-allylnickel group of

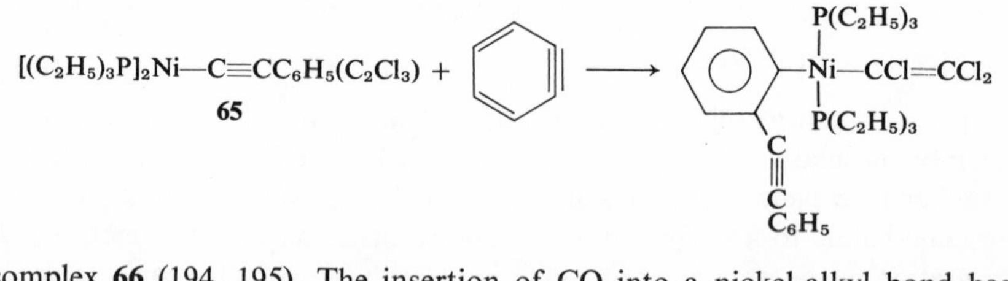

[(C2H5)3P]2Ni—C≡CC6H5(C2Cl3) +
65
P(C2H5)3
Ni—CCl=CCl2
P(C2H5)3
C
C
C6H5

complex **66** (194, 195). The insertion of CO into a nickel-alkyl bond has

Cl
Ni
66
2CH3CO2Na
− 2NaCl
CH3
H2C
CH3
C
C
O
O—Ni
CH2
2

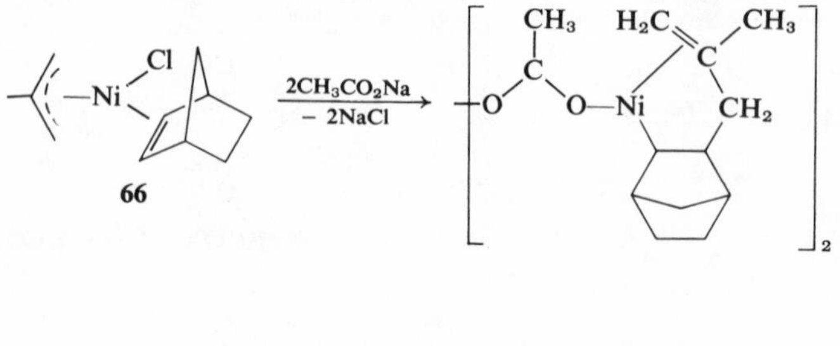

precedence in the reactions of the σ-cyclooctenylnickel systems with CO (190–192) and in the reaction of bis(trimethylphosphine) nickel chloro(methyl)

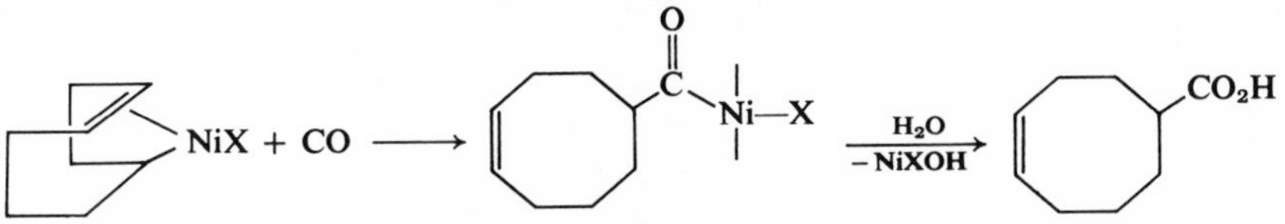

with CO (66).

$$[(CH_3)_3P]_2Ni(CH_3)Cl + CO \longrightarrow [(CH_3)_3P]_2Ni(COCH_3)Cl$$

This last reaction proceeds readily at room temperature.

C. Carbonylation Reactions Involving Organolithium Reagents *(Table VI-17)*

A number of stoichiometric carbonylation reactions have been reported that have used a reagent prepared by reacting organolithium compounds with nickel tetracarbonyl (a recent review is to be found in ref. 177). The intermediates involved in this reaction have customarily been formulated as ionic acyl–nickel systems but a formulation as a carbene species (e.g., **67**) is equally acceptable, particularly as this class of compound may be prepared by reacting transition metal carbonyl complexes with organolithium reagents (see Volume I, p. 37, and ref. 229). These intermediates are unstable and react

$$RLi + Ni(CO)_4 \longrightarrow (OC)_3Ni-C\begin{smallmatrix} OLi \\ \\ R \end{smallmatrix}$$

67

further to give ketonic derivatives. The course of reaction is markedly sensitive to the reaction conditions and in particular to the temperature and the solvent medium. The main product obtained by hydrolysis of the aryllithium intermediate at $-70°$ is an acyloin, whereas an α-diketone is formed preferentially if the reaction is carried out at $50°$. Carboxylic acids and mono-

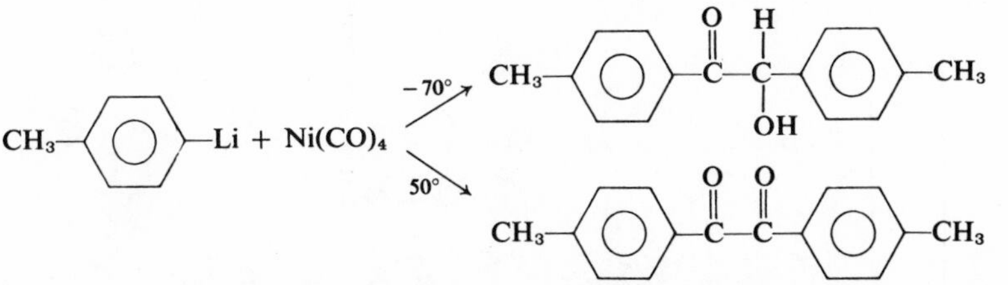

ketones are also produced in varying amounts (176, 230). Lithium phenylacetylide is reported to react with nickel tetracarbonyl to give a mixture of

TABLE VI-17

THE REACTION OF ORGANOLITHIUM REAGENTS WITH NICKEL TETRACARBONYL

Product	Reactant	Ref.
$CH_3COC(C_6H_5)HCH_2COCH_3$	$C_6H_5CH:CHCOCH_3-CH_3Li$	232
$C_4H_9COC(C_6H_5)HCH_2COCH_3$	$C_6H_5CH:CHCOCH_3-C_4H_9Li$	232
$C_6H_5COC(C_6H_5)HCH_2COCH_3$	$C_6H_5CH:CHCOCH_3-C_6H_5Li$	232
(structure) CH_3CO — cyclohexanone	(structure) O — CH_3Li cyclohexenone	232
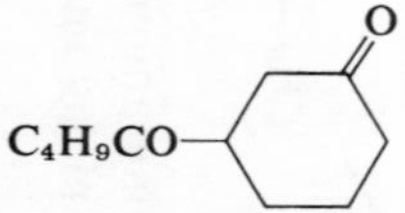 (structure) C_4H_9CO — cyclohexanone	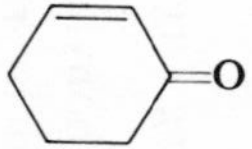 (structure) O — C_4H_9Li cyclohexenone	232
$C_4H_9COCH_2CH_2COCH_3$	$CH_2:CHCOCH_3-C_4H_9Li$	232
$CH_3COCH(CH_3)CH_2CO_2CH_3$	$trans\text{-}CH_3CH:CHCO_2CH_3-CH_3Li$	232
$C_4H_9COCH(CH_3)CH_2CO_2CH_3$	$trans\text{-}CH_3CH:CHCO_2CH_3-C_4H_9Li$	232
$CH_3COCH(C_6H_5)CH_2CO_2CH_3$	$C_6H_5CH:CHCO_2CH_3-CH_3Li$	232
$C_4H_9COCH(C_6H_5)CH_2CO_2CH_3$	$C_6H_5CH:CHCO_2CH_3-C_4H_9Li$	232
$CH_3COCH(CH_3)CH(CH_3)COCH_3$	$CH_3CH:C(CH_3)COCH_3-CH_3Li$	232
$C_4H_9COCH(CH_3)CH(CH_3)COCH_3$	$CH_3CH:C(CH_3)COCH_3-C_4H_9Li$	232
$CH_3COC(CH_3)_2CH_2COCH_3$	$(CH_3)_2C:CHCOCH_3-CH_3Li$	232
$C_4H_9COC(CH_3)_2CH_2COCH_3$	$(CH_3)_2C:CHCOCH_3-C_4H_9Li$	232
$p\text{-}CH_3C_6H_4COC(CH_3)HCH_2COC_6H_4\text{-}p\text{-}CH_3,$ 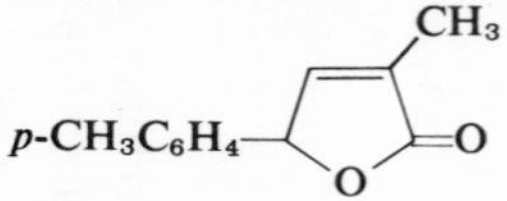	$CH_3C:CH-LiC_6H_4\text{-}p\text{-}CH_3$	233
(structure with CH_3 and $p\text{-}CH_3C_6H_4$ furanone)		

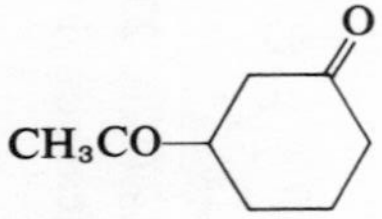

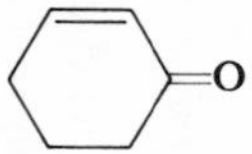

Reactant	Product	Ref.
$C_6H_5C:CH-LiC_6H_4-p-CH_3$	$p-CH_3C_6H_4COC(C_6H_5)HCH_2COC_6H_4-p-CH_3$, (furanone with C_6H_5, $p-CH_3C_6H_4$)	233
$HC:CH-LiC_6H_4-p-CH_3$	$p-CH_3C_6H_4COCH_2CH_2COC_6H_4-p-CH_3$	233
$C_6H_5C:CH-LiC_6H_5$	$C_6H_5COCH(C_6H_5)CH_2COC_6H_5$	233
$CH_3C::CH-LiC_6H_5$	$C_6H_5COCH(CH_3)CH_2COC_6H_5$	233
$HC::CH-LiC_6H_5$	$C_6H_5COCH_2CH_2COC_6H_5$	233
$HC:CH-LiC_6H_4-p-OCH_3$	$p-CH_3OC_6H_4COCH_2CH_2COC_6H_4-p-OCH_3$	233
$HC:CH-LiC_4H_9$	$C_4H_9COCH_2CH_2COC_4H_9$	233
$C_6H_5C:CH-LiN(CH_3)_2$	$(CH_3)_2NCOCH(C_6H_5)CON(CH_3)_2$	234
$trans-C_6H_5CH:CHBr-LiN(CH_3)_2$	$trans-C_6H_5CH:CHCON(CH_3)_2$	201
$C_6H_5I-LiN(CH_3)_2$	$C_6H_5CON(CH_3)_2$	201
$C_6H_5CH_2Br-LiN(CH_3)_2$	$C_6H_5CH_2CON(CH_3)_2,\ C_6H_5CH_2N(CH_3)_2$	201
$CH_2:CHCH_2Br-LiN(CH_3)_2$	$CH_3CH:CHCON(CH_3)_2$	201
(cyclohexenyl)$-Br-LiN(CH_3)_2$	(cyclohexenyl)$-CON(CH_3)_2$	201
$CH_3COCl-LiN(CH_3)_2$	$CH_3CON(CH_3)_2$	201
$C_4H_9COCl-LiN(CH_3)_2$	$C_4H_9CON(CH_3)_2$	201
$C_6H_5COCl-LiN(CH_3)_2$	$C_6H_5CON(CH_3)_2$	201
$C_6H_5COC_6H_5-LiN(CH_3)_2$	$C_6H_5COH(C_6H_5)CON(CH_3)_2$	201
$C_6H_5CHO-LiN(CH_3)_2$	$C_6H_5CON(CH_3)_2$	201
$C_6H_5COCl-p-CH_3C_6H_4Li$	$p-CH_3C_6H_4C(OCOC_6H_5):C(OCOC_6H_5)C_6H_4-p-CH_3$	176
$C_6H_5CH_2Cl-p-CH_3C_6H_4Li$	$p-CH_3C_6H_4COC(CH_2C_6H_5)OHC_6H_4-p-CH_3$	176
$C_6H_5CHCH_2-LiC_6H_5-CO$ (epoxide)	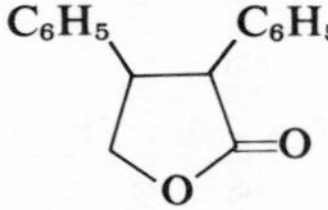 (furanone with C_6H_5, C_6H_5), $C_6H_5COC(OH)HC_6H_5,\ C_6H_5CO_2H$	235
$C_6H_5CHCH_2-LiC_6H_4-p-CH_3-CO$ (epoxide)	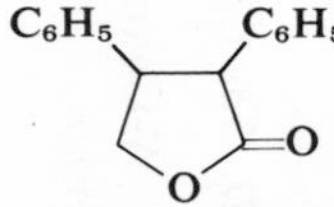 (furanone with C_6H_5, C_6H_5), $(p-CH_3C_6H_4)_2CO,$ $C_6H_5CH_2COC_6H_4-p-CH_3$	235

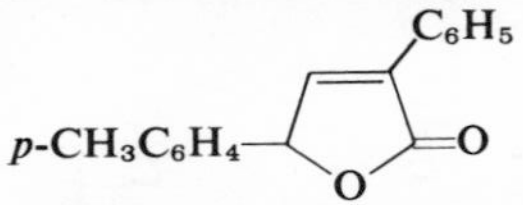

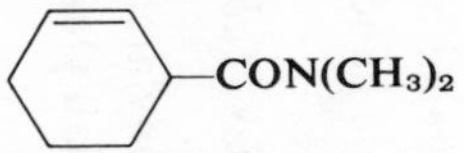

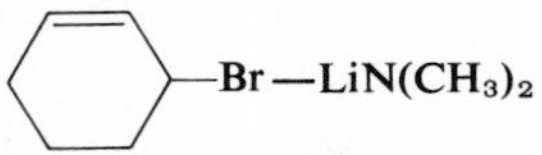

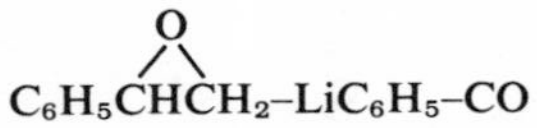

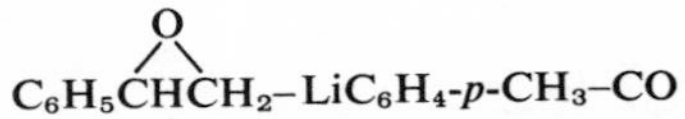

diphenylbutadiyne and methylphenylpropiolate (231). Related behavior has been observed with Grignard reagents (121, 161, 162, 168, 176): for example, phenylmagnesium bromide reacts to give benzoin or a mixture of benzophenone, benzil, benzoin, and benzoic acid depending on the reaction conditions.

A number of interesting synthetic reactions have been developed using the nickel tetracarbonyl–organolithium reagent (Table VI-17): 1,4-dicarbonyl compounds may be prepared by either insertion of an α,β-unsaturated carbonyl compound (232)

$$(OC)_3Ni—C(OLi)CH_3 + C_6H_5CH=CHCOCH_3 \longrightarrow CH_3\overset{\overset{O}{\|}}{C}-\overset{\overset{H}{|}}{\underset{\underset{C_6H_5}{|}}{C}}-CH_2\overset{\overset{O}{\|}}{C}CH_3$$

or insertion of an alkyne (233, 234)

$$(OC)_3Ni—C(OLi)C_6H_5 + C_6H_5C≡CH \longrightarrow C_6H_5\overset{\overset{O}{\|}}{C}-\overset{\overset{H}{|}}{\underset{\underset{C_6H_5}{|}}{C}}-CH_2\overset{\overset{O}{\|}}{C}C_6H_5$$

whereas the reaction with organic halides produces ketones in high yield (176, 201). The reaction with epoxides has also been studied (235); a mixture

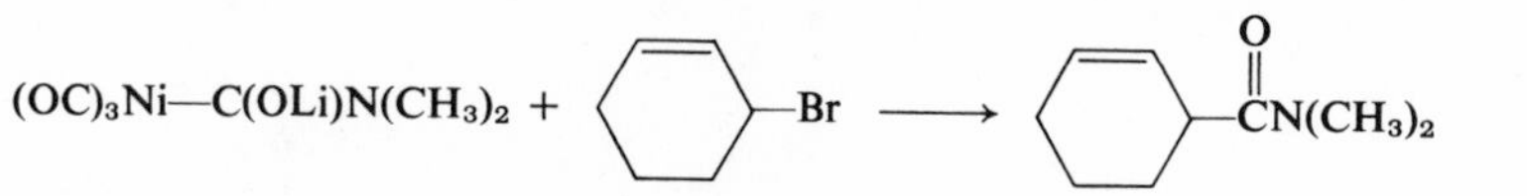

$$(OC)_3Ni—C(OLi)N(CH_3)_2 + \text{⬡}—Br \longrightarrow \text{⬡}—\overset{\overset{O}{\|}}{C}N(CH_3)_2$$

of lactones, ketones, and acids results.

It is possible that the carbonylation reactions observed in basic media also involve nickel-carbene intermediates. Although in most cases concrete evidence does not exist, infrared spectral data have been presented which indicate that $(OC)_3Ni–C(OK)O\text{-}tert\text{-}C_4H_9$ is formed on treating nickel tetracarbonyl with potassium *tert*-butoxide (201). Candidates for such a mechanism include the carbonylation of organic halides in the presence of alkoxides, salts of carboxylic acids, inorganic bases, or halide ions (71, 183). However, in the

$$CH_3(CH_2)_6I + CO + tert\text{-}C_4H_9OH \xrightarrow{Ni(CO)_4/tert\text{-}C_4H_9O^-} CH_3(CH_2)_6CO_2H + HI$$

absence of conclusive evidence these reactions have been included in the section devoted to the particular organic halide.

D. The Carbonylation of N-Chloroamines (Table VI-18)

N-Chloroamines may also be carbonylated by nickel tetracarbonyl. Dichloroalkylamines react to give *N,N'*-dialkylurea in high yield (236), whereas

TABLE VI-18
THE CARBONYLATION OF N-CHLOROAMINES

Product	N-Chloroamine	Ref.
$tert\text{-}C_4H_9NHCONHtert\text{-}C_4H_9$	$tert\text{-}C_4H_9NCl_2$	236
 [cyclohexyl]—NHCONH—[cyclohexyl]	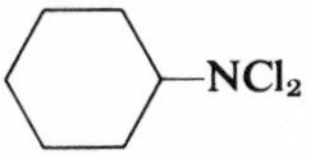 [cyclohexyl]—NCl$_2$	236
$(CH_3)_2NCON(CH_3)_2$	$(CH_3)_2NCl$	236
$(C_2H_5)_2NCON(C_2H_5)_2$	$(C_2H_5)_2NCl$	237
$C_2H_5NHCONHC_2H_5$	C_2H_5NHCl	237
$C_6H_5CH_2NHCONHCH_2C_6H_5$	$C_6H_5CH_2NHCl$	237
$C_6H_5CONHCONHCOC_6H_5$	$C_6H_5CONHCl$	237

monochloroamines react to give tetraalkylurea in modest yields (237). N-Chlorobenzamide reacts to give N,N'-dibenzoylurea.

$$2RNCl_2 + CO + H_2O \xrightarrow[-2NiCl_2]{2Ni(CO)_4} RNHCONHR$$

$$2RR'NCl + CO \xrightarrow[-NiCl_2]{Ni(CO)_4} RR'N\overset{\overset{\displaystyle O}{\|}}{-}C-NRR'$$

IV. Decarbonylation

The use of nickel or nickel complexes as reagents for the decarbonylation of organic compounds has not been investigated systematically. However, the strength of the Ni–CO bond in substituted nickel–carbonyl complexes suggests that this may well be a rewarding area.

The only direct reaction to have been reported is that between methanol, bis(cyclooctadiene)nickel, and triphenylphosphine in the presence of a conjugated diene: the methanol is decarbonylated and the hydrogen liberated hydrogenates the conjugated diene (118). Decarbonylation has also been

$$CH_3OH + (COD)_2Ni + 3P(C_6H_5)_3 \longrightarrow [(C_6H_5)_3P]_3NiCO + 2COD + [2H_2]$$

observed in the reaction of vinylacetyl chloride with nickel tetracarbonyl from which 1,5-hexadiene may be isolated (141), whereas urea reacts with metallic nickel at 60° to give hydrazine and nickel tetracarbonyl (P160, P161). Other

$$4NH_2CONH_2 + Ni \longrightarrow Ni(CO)_4 + 4NH_2NH_2$$

examples include the decarbonylation of oxalyl chloride by nickel chloride to give carbon monoxide (240) and the decarbonylation of benzoyl chloride by nickel on charcoal to give chlorobenzene (P99).

A reaction that is perhaps related to these is the formation of an olefin, e.g., **68**, on treating an anhydride or thioanhydride with bistriphenylphosphine nickel dicarbonyl (119, 120). Olefins are also the product of the reaction of

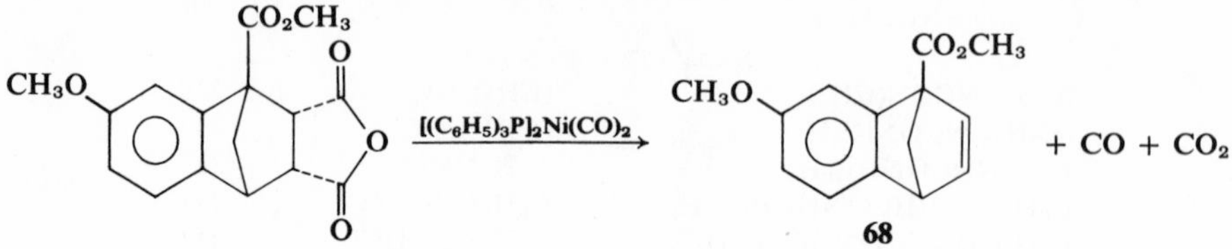

$$[(C_6H_5)_3P]_2Ni(CO)_2 \longrightarrow \quad + CO + CO_2$$

68

the thionocarbonates of vicinal diols with bis(cyclooctadiene)nickel (245).

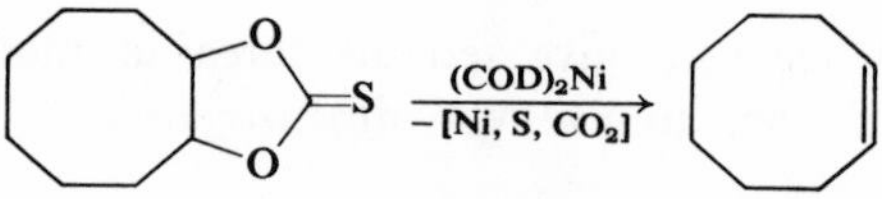

$$\xrightarrow[- [Ni, S, CO_2]]{(COD)_2Ni}$$

V. Isonitrile Insertion

Isonitriles are isoelectronic with carbon monoxide and therefore can be expected to undergo analogous insertion reactions. Such reactions have been observed with nickel-alkyl or -aryl complexes, e.g., the formation of the nickel-imino complex **69** (212, 238; see Volume I, pp. 213 and 216). In contrast

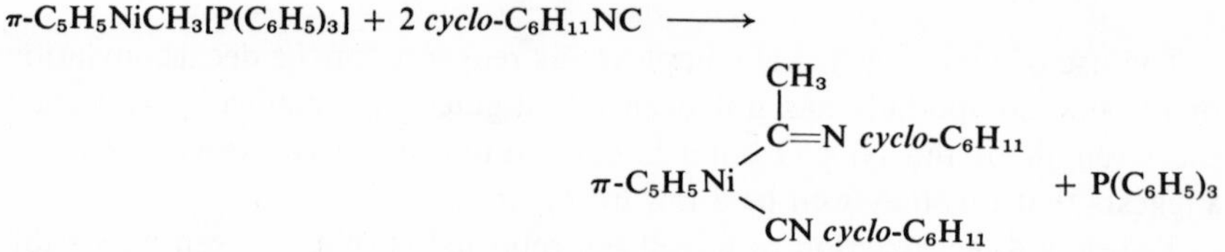

$$\pi\text{-}C_5H_5NiCH_3[P(C_6H_5)_3] + 2\ cyclo\text{-}C_6H_{11}NC \longrightarrow$$

$$\pi\text{-}C_5H_5Ni \begin{array}{c} CH_3 \\ | \\ C=N\ cyclo\text{-}C_6H_{11} \\ \\ CN\ cyclo\text{-}C_6H_{11} \end{array} + P(C_6H_5)_3$$

69

to carbon monoxide, the isonitrile group readily undergoes multiple insertion.

Analogous behavior to that of carbon monoxide has been observed in the reaction of an isonitrile with the bis(π-allyl)C_{12}-nickel complex **70**; insertion

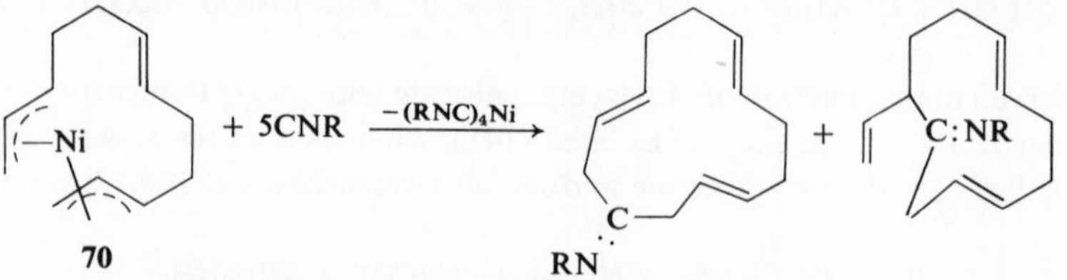

$$+ 5CNR \xrightarrow{-(RNC)_4Ni} \quad + \quad$$

70 $\qquad$ RN $\qquad$

103. L. I. Smith and H. H. Hoehn, *J. Amer. Chem. Soc.* **63**, 1180 (1941).
104. Y. T. Eidus, A. L. Lapidus, K. J. Puzitskii, and B. K. Nefedov, *Russ. Chem. Rev.* **42**, 199 (1973).
105. C. W. Bird, *J. Organometal. Chem.* **47**, 281 (1973).
106. J. W. Copenhaver and M. H. Bigelow, "Acetylene and Carbon Monoxide Chemistry," Van Nostrand-Reinhold, Princeton, New Jersey, 1949.
107. F. L. Resen, *Oil Gas J.* **51**, 92 (1953).
108. S. Otsuka, A. Nakamura, and T. Yoshida, *J. Amer. Chem. Soc.* **91**, 7196 (1969).
109. W. Reppe, O. Schlichting, K. Klager, and T. Toepel, *Justus Liebigs Ann. Chem.* **560** 1 (1948).
110. H. Kröper, *in* "Houben-Weyl (Methoden der Org. Chem.)," (E. Müller, ed.) Vol. IV, Part 2, p. 359 (1955).
111. H. Kröper, *in* "Ullmann's Encyklopädie der technischen Chemie," Vol. 5, p. 121. Verlag Chemie (1954).
112. D. R. Levering and A. L. Glasebrook, *J. Org. Chem.* **23**, 1836 (1958).
113. G. P. Chiusoli, *Accounts Chem. Res.* **6**, 422 (1973).
114. S. Sourirajan, *Advan. Catal.* **9**, 618 (1957).
115. J. Falbe, H.-J. Schulze-Steinen, and F. Korte, *Chem. Ber.* **98**, 886 (1965).
116. S. Horiie and S. Murahashi, *Bull. Chem. Soc. Jap.* **33**, 88 (1960); S. Murahashi and S. Horiie, *J. Amer. Chem. Soc.* **78**, 4816 (1956).
117. W. L. Mosby, *Chem. Ind.* (*London*) p. 17 (1957).
118. Y. Inoue, M. Hidai, and Y. Uchida, *Chem. Lett.* p. 1119 (1972).
119. B. M. Trost and F. Chen, *Tetrahedron Lett.* p. 2603 (1971).
120. P.-T. Ho, S. Oida, and K, Wiesner, *Chem. Commun.* p. 883 (1972).
121. W. L. Gilliland and A. A. Blanchard, *J. Amer. Chem. Soc.* **48**, 410 (1926).
122. R. Baker, B. N. Blackett, and R. C. Cookson, *Chem. Commun.* p. 802 (1972).
123. B. Bogdanović, P. Heimbach, M. Kröner, G. Wilke, E. G. Hoffmann, and J. Brandt, *Justus Liebigs Ann. Chem.* **727**, 143 (1969).
124. H. Breil and G. Wilke, *Angew. Chem.* **82**, 355 (1970).
125. L. Cassar and G. P. Chiusoli, *Tetrahedron Lett.* p. 2805 (1966).
126. L. Cassar and G. P. Chiusoli, *Tetrahedron Lett.* p. 3295 (1965).
127. L. Cassar and G. P. Chiusoli, *Chim. Ind.* (*Milan*) **48**, 323 (1966).
128. L. Cassar, G. P. Chiusoli, and M. Foa, *Tetrahedron Lett.* p. 285 (1967).
129. L. Cassar and M. Foa, *Chim. Ind.* (*Milan*) **51**, 673 (1969).
130. G. P. Chiusoli, G. Bottaccio, and C. Venturello, *Tetrahedron Lett.* p. 2875 (1965).
131. G. P. Chiusoli and G. Bottaccio, *Chim. Ind.* (*Milan*) **47**, 165 (1965).
132. G. P. Chiusoli, *Atti Accad. Naz. Lincei, Cl. Sci. Fis., Mat. Natur., Rend.* **26**, 790 (1959).
133. G. P. Chiusoli, *Chim. Ind.* (*Milan*) **41**, 762 (1959).
134. G. P. Chiusoli, *Chim. Ind.* (*Milan*) **41**, 503 (1959).
135. G. P. Chiusoli, *Chim. Ind.* (*Milan*) **41**, 506 (1959).
136. G. P. Chiusoli, *Angew. Chem.* **72**, 74 (1960).
137. G. P. Chiusoli and L. Cassar, *Angew. Chem.* **79**, 177 (1967).
138. G. P. Chiusoli, G. Bottaccio, and C. Venturello, *Chim. Ind.* (*Milan*) **48**, 107 (1966).
139. G. P. Chiusoli and S. Merzoni, *Z. Naturforsch.* **17b**, 850 (1962).
140. G. P. Chiusoli, S. Merzoni, and G. Mondelli, *Chim. Ind.* (*Milan*) **46**, 743 (1964).
141. G. P. Chiusoli, S. Merzoni, and G. Mondelli, *Tetrahedron Lett.* p. 2777 (1964).
142. G. P. Chiusoli, *Riv. Combust.* **15**, 647 (1961).
143. G. P. Chiusoli, M. Dubini, M. Ferraris, F. Guerrieri, S. Merzoni, and G. Mondelli, *J. Chem. Soc., C* p. 2889 (1968).

144. G. P. Chiusoli, *Corsi Semin. Chim.* No. 10, p. 77 (1968).
145. G. P. Chiusoli, *Chim. Ind.* (*Milan*) **43**, 638 (1961).
146. G. P. Chiusoli, *Aspects Homogen. Catal.* **1**, 77 (1970).
147. G. P. Chiusoli, *Gazz. Chim. Ital.* **89**, 1332 (1959).
148. G. P. Chiusoli and S. Merzoni, *Chim. Ind.* (*Milan*) **45**, 6 (1963).
149. G. P. Chiusoli, G. Bottaccio, and A. Cameroni, *Chim. Ind.* (*Milan*) **44**, 131 (1962).
150. G. P. Chiusoli and G. Bottaccio, *Chim. Ind.* (*Milan*) **43**, 1022 (1961).
151. G. P. Chiusoli and S. Merzoni, *Chim. Ind.* (*Milan*) **43**, 259 (1961).
152. G. P. Chiusoli, *Chim. Ind.* (*Milan*) **43**, 365 (1961).
153. G. P. Chiusoli, *Petrol. Mater. Prima Ind. Chim. Mod.* (*Relaz. Commun. G. Chim. Milan*) p. 248 (1961).
154. G. P. Chiusoli and G. Bottaccio, *Chim. Ind.* (*Milan*) **44**, 1129 (1962).
155. G. P. Chiusoli and S. Merzoni, *Chem. Commun.* p. 522 (1971).
156. G. P. Chiusoli, G. Cometti, and S. Merzoni, *Organometal. Chem. Syn.* **1**, 439 (1972).
157. E. J. Corey, M. F. Semmelhack, and L. S. Hegedus, *J. Amer. Chem. Soc.* **90**, 2416 (1968).
158. E. J. Corey and M. F. Semmelhack, *Tetrahedron Lett.* p. 6237 (1966).
159. E. J. Corey and M. Jautelat, *J. Amer. Chem. Soc.* **89**, 3912 (1967).
160. W. T. Dent, R. Long, and G. H. Whitfield, *J. Chem. Soc. London* p. 1588 (1964).
161. N. Zelinsky, *J. Russ. Phys.-Chem. Soc.* **36**, 339 (1904).
162. H. O. Jones, *Chem. News* **90**, 144 (1904).
163. M. Dubini, G. P. Chiusoli, and F. Montino, *Tetrahedron Lett.* p. 1591 (1963).
164. M. Dubini, G. P. Chiusoli, and F. Montino, *Chim. Ind.* (*Milan*) **45**, 1237 (1963).
165. P. Hong, K. Sonogashira, and N. Hagihara, *Tetrahedron Lett.* p. 1633 (1970).
166. E. O. Fischer and G. Bürger, *Z. Naturforsch.* **17b**, 484 (1962).
167. M. Foa, L. Cassar, and M. T. Venturi, *Tetrahedron Lett.* p. 1357 (1968).
168. F. L. Benton, M. C. Voss, and P. A. McCusker, *J. Amer. Chem. Soc.* **67**, 82 (1945).
169. F. Guerrieri and G. P. Chiusoli, *Chem. Commun.* p. 781 (1967).
170. F. Guerrieri and G. P. Chiusoli, *J. Organometal. Chem.* **15**, 209 (1968).
171. R. F. Heck, J. C. W. Chien, and D. S. Breslow, *Chem. Ind.* (*London*) p. 986 (1961).
172. I. Hashimoto, M. Ryang, and S. Tsutsumi, *Tetrahedron Lett.* p. 3291 (1969).
173. I. Hashimoto, N. Tsuruta, M. Ryang, and S. Tsutsumi, *J. Org. Chem.* **35**, 3748 (1970).
174. J. B. Mettalia and E. H. Specht, *J. Org. Chem.* **32**, 3941 (1967).
175. Y. Yamamoto, T. Takizawa, and N. Hagihara, *Nipon Kagaku Zasshi* **87**, 1355 (1966); *Chem. Abstr.* **67**, 32995 (1967).
176. M. Ryang, S. K. Myeong, Y. Sawa, and S. Tsutsumi, *J. Organometal. Chem.* **5**, 305 (1966).
177. M. Ryang and S. Tsutsumi, *Synthesis* p. 55 (1971).
178. R. D. Johnston, F. Basolo, and R. G. Pearson, *Inorg. Chem.* **10**, 247 (1971).
179. Y. Suzuki and T. Takizawa, *Chem. Commun.* p. 837 (1972).
180. G. Wilke *et al.*, *Angew. Chem.* **78**, 157 (1966).
181. E. Yoshisato and S. Tsutsumi, *J. Org. Chem.* **33**, 869 (1968).
182. G. P. Chiusoli, *Chim. Ind.* (*Milan*) **41**, 513 (1959).
183. L. Cassar, G. P. Chiusoli, and F. Guerrieri, *Synthesis* p. 509 (1973).
184. G. P. Chiusoli and G. Cometti, *Chem. Commun.* p. 1015 (1972).
185. G. P. Chiusoli, *Plenary Lect. Int. Congr. Pure Appl. Chem. 23rd, 1971* Vol. 6, p. 169 (1971).
186. L. Cassar, G. P. Chiusoli, and M. Foa, *Chim. Ind.* (*Milan*) **50**, 515 (1968).

187. G. P. Chiusoli, M. Ferraris, and S. Merzoni, *Proc. Int. Conf. Coord. Chem. 12th,* (1969) quoted in ref. 146 p. 97.
188. T. Kunieda, T. Tamura, and T. Takizawa, *Chem. Commun.* p. 885 (1972).
189. W. Keim, Ph.D. Thesis, Technische Hochschule Aachen, 1968.
190. K. Fischer, K. Jonas, P. Misbach, R. Stabba, and G. Wilke, *Angew. Chem.* **85,** 1002 (1973).
191. P. Misbach, Ph.D. Thesis, University of Bochum, 1969.
192. K. W. Barnett, *J. Organometal. Chem.* **21,** 477 (1970).
193. R. G. Miller, D. R. Fahey, and D. P. Kuhlman, *J. Amer. Chem. Soc.* **90,** 6248 (1968).
194. M. C. Gallazzi, T. L. Hanlon, G. Vitulli, and L. Porri, *J. Organometal. Chem.* **33,** C45 (1971).
195. C. Tieghi and M. Zocchi, *J. Organometal. Chem.* **57,** C90 (1973).
196. E. J. Corey, L. S. Hegedus, and M. F. Semmelhack, *J. Amer. Chem. Soc.* **90,** 2417 (1968).
197. M. F. Semmelhack, *Org. React.* **19,** 117 (1972) (see p. 125).
198. N. L. Bauld, *Tetrahedron Lett.* p. 1841 (1963).
199. N. L. Bauld, *J. Amer. Chem. Soc.* **84,** 4345 (1962).
200. J. J. Drysdale and D. D. Coffman, *J. Amer. Chem. Soc.* **82,** 5111 (1960).
201. S. Fukuoka, M. Ryang, and S. Tsutsumi, *J. Org. Chem.* **36,** 2721 (1971).
202. E. J. Corey and L. S. Hegedus, *J. Amer. Chem. Soc.* **91,** 1233 (1969).
203. I. Hashimoto, M. Ryang, and S. Tsutsumi, *Tetrahedron Lett.* p. 4567 (1970).
204. M. Ryang, Y. Sawa, S. N. Somasundaram, S. Murai, and S. Tsutsumi, *J. Organometal. Chem.* **46,** 375 (1972).
205. J. K. Crandall and W. J. Michaely, *J. Organometal. Chem.* **51,** 375 (1973).
206. G. P. Chiusoli, *Chim. Ind. (Milan)* **41,** 513 (1959).
207. P. J. Ashworth, G. H. Whitham, and M. C. Whiting, *J. Chem. Soc., London* p. 4633 (1957).
208. R. W. Rosenthal and L. H. Schwartzman, *J. Org. Chem.* **24,** 836 (1959).
209. R. A. Raphael, "Acetylene Compounds in Organic Synthesis," p. 138. Academic Press, New York, 1955.
210. C. W. Bird, "Transition Metal Intermediates in Organic Synthesis," p. 182. Academic Press, New York, 1967.
211. G. N. Schrauzer, *Chem. Ber.* **94,** 1891 (1961).
212. Y. Yamamoto, H. Yamazaki, and N. Hagihara, *J. Organometal. Chem.* **18,** 189 (1969).
213. H. Kröper, F. Wirth, and O. Huchler, *Angew. Chem.* **72,** 867 (1960).
214. V. I. Romanovskii and A. A. Artemev. *Zh. Vses. Khim. Obshchest.* **5,** 476 (1960); *Chem. Abstr.* **55,** 1506 (1961).
215. S. E. Yakushkina and N. V. Kislyakova, *Izv. Akad. Nauk SSSR, Otd. Khim. Nauk.* p. 1119 (1958); *Chem. Abstr.* **53,** 3134 (1959).
216. K. Yamamoto and K. Sato, *Bull. Chem. Soc. Jap.* **27,** 389 (1954).
217. S. Matsuda, S. Kikkawa, and K. Tsuchino, *Yuki Gosei Kagaku Kyokai Shi* **19,** 61 (1961); *Chem. Abstr.* **55,** 5408 (1961).
218. M. Nakayama and T. Mizoroki, *Bull. Chem. Soc. Jap.* **44,** 508 (1971).
219. M. Nakayama and T. Mizoroki, *Bull. Chem. Soc. Jap.* **43,** 569 (1970).
220. M. Nakayama and T. Mizoroki, *Bull. Chem. Soc. Jap.* **42,** 1124 (1969).
221. T. Mizoroki and M. Nakayama, *Bull. Chem. Soc. Jap.* **40,** 2203 (1967).
222. S. Yaroslavsky, *Chem. Ind. (London)* p. 765 (1965).

223. L. Cassar and M. Foa, *J. Organometal. Chem.* **51**, 381 (1973).
224. W. W. Prichard, *J. Amer. Chem. Soc.* **78**, 6137 (1956).
225. W. F. Beckert and J. U. Lowe, *J. Org. Chem.* **32**, 1215 (1967).
226. Y. Hirota, M. Ryang, and S. Tsutsumi, *Tetrahedron Lett.* p. 1531 (1971).
227. E. Yoshisato, M. Ryang, and S. Tsutsumi, *J. Org. Chem.* **34**, 1500 (1969).
228. M. Ryang, Y. Toyoda, S. Murai, N. Sonoda, and S. Tsutsumi, *J. Org. Chem.* **38**, 62 (1973).
229. R. J. Angelici, *Accounts Chem. Res.* **5**, 335 (1972).
230. K. M. Song, Y. Sawa, M. Ryang, and S. Tsutsumi, *Bull. Chem. Soc. Jap.* **38**, 330 (1965).
231. I. Rhee, M. Ryang, and S. Tsutsumi, *Tetrahedron Lett.* p. 4593 (1969).
232. E. J. Corey and L. S. Hegedus, *J. Amer. Chem. Soc.* **91**, 4926 (1969).
233. Y. Sawa, I. Hashimoto, M. Ryang, and S. Tsutsumi, *J. Org. Chem.* **33**, 2159 (1968).
234. S. Fukuoka, M. Ryang, and S. Tsutsumi, *J. Org. Chem.* **33**, 2973 (1968).
235. S. Fukuoka, M. Ryang, and S. Tsutsumi, *J. Org. Chem.* **35**, 3184 (1970).
236. H. Bock and K. L. Kompa, *Angew. Chem.* **78**, 114 (1966).
237. S. Fukuoka, M. Ryang, and S. Tsutsumi, *Tetrahedron Lett.* p. 2553 (1970).
238. Y. Yamamoto and H. Yamazaki, *Coord. Chem. Rev.* **8**, 225 (1972).
239. T. Kashiwagi, M. Hidai, Y. Uchida, and A. Misono, *J. Polym. Sci., Part B* **8**, 173 (1970).
240. H. O. Jones and H. S. Tasker, *J. Chem. Soc., London* p. 1904 (1909).
241. V. P. Nechiporenko, I. A. Korneeva, A. D. Treboganov, A. B. Prokkorov, G. I. Myagkova, and N. A. Preobrazhenskii, *Uch. Zap. Mosk. Inst. Tonkoi Khim. Tekhnol.* **1**, 76 (1970); *Chem. Abstr.* **76**, 13706 (1972).
242. C. E. Berkoff, R. C. Cookson, J. Hudec, D. W. Jones, and R. O. Williams, *J. Chem. Soc., London* p. 194 (1965).
243. W. Reppe, *Justus Liebigs Ann. Chem.* **582**, 116 (1953).
244. C. W. Bird and E. M. Briggs, *J. Organometal. Chem.* **69**, 311 (1974).
245. M. F. Semmelhack and R. D. Stauffer, *Tetrahedron Lett.* p. 2667 (1973).
246. V. P. Nechiporenko, A. D. Treboganov. V. P. Chernova, G. I. Myagkova, and N. A. Preobrazhenskii, *J. Org. Chem. USSR* **9**, 236 (1973).
247. V. P. Nechiporenko, A. A. Glazkov, G. I. Myagkova, and R. P. Evstigneeva, *J. Org. Chem. USSR* **9**, 241 (1973).
248. G. P. Chiusoli, G. Cometti, and V. Bellotti, *Gazz. Chim. Ital.* **103**, 569 (1973).

Patents

P1. D. F. Friederich (B.A.S.F.), Ger. Patent 933,148 (1955); *Chem. Abstr.* **52**, 19952 (1958).
P2. R. D. Anderson and E. M. Smolin (American Cyanamid Co.), U.S. Patent 3,025,319 (1962); *Chem. Abstr.* **57**, 11027 (1962).
P3. E. M. Smolin and R. D. Anderson (American Cyanamid Co.), U.S. Patent 3,025,322 (1962); *Chem. Abstr.* **57**, 11025 (1962).
P4. F. E. Drummond (Basic Research Corp.), U.S. Patent 3,070,618 (1962); *Chem. Abstr.* **59**, 9886 (1963).
P5. B.A.S.F., Brit. Patent 802, 544 (1958); *Chem. Abstr.* **53**, 7016 (1959).
P6. W. Reppe and A. Magin (B.A.S.F.), Brit. Patent 943,721 (1963); *Chem. Abstr.* **60**, 5340 (1969).

to give an imine occurs (124). Alkynes and *tert*-butylisonitrile react together, in the presence of a nickel catalyst, to give derivatives of pyrrole, e.g., **71** (72).

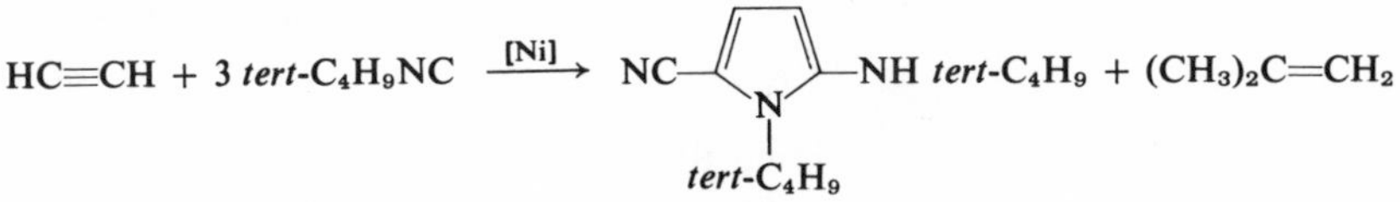

$$HC \equiv CH + 3 \; tert\text{-}C_4H_9NC \xrightarrow{[Ni]} NC \text{—} \langle \text{pyrrole} \rangle \text{—} NH \; tert\text{-}C_4H_9 + (CH_3)_2C = CH_2$$

71

A related stoichiometric reaction is that of tetrakisarylisonitrile nickel complexes with diphenylacetylene: the diiminocyclobutene **72** is formed—possibly through the intermediacy of a tris(alkyne)nickel bisisonitrile complex (179).

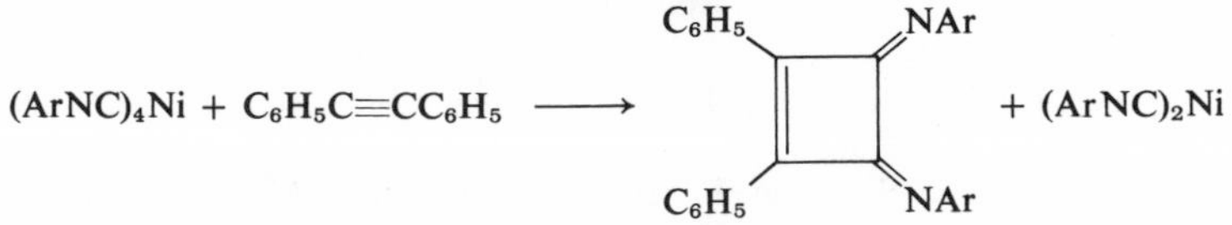

$$(ArNC)_4Ni + C_6H_5C \equiv CC_6H_5 \longrightarrow \text{[cyclobutene]} + (ArNC)_2Ni$$

72

The nickel-catalyzed reaction of *tert*-butylisonitrile with aqueous methanol leads to a mixture of products in which **73** predominates (43). If the reaction

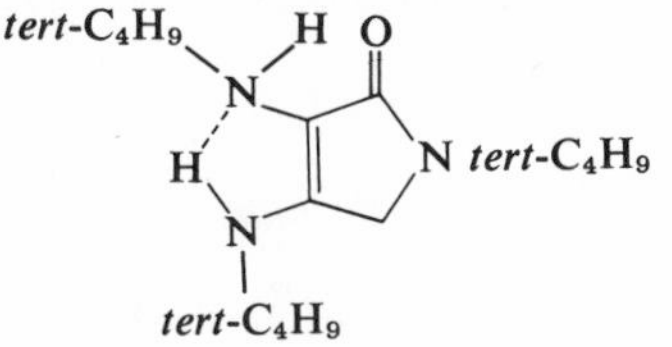

73

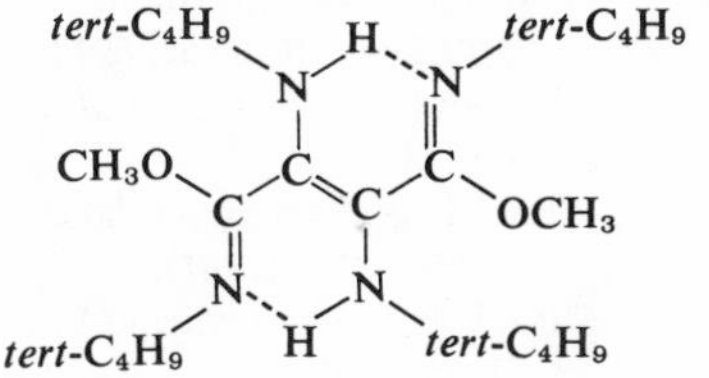

74

is carried out in anhydrous methanol, **74** is formed.

Hydrolysis of the compound **75** (formed by reacting methyl iodide or benzyl bromide with tetrakis-*tert*-butylisonitrile nickel) produces a mixture of products that have been identified as the aminonitrile **76** and the amino acid derivative **77** (44). Compound **75** is reported to catalytically polymerize

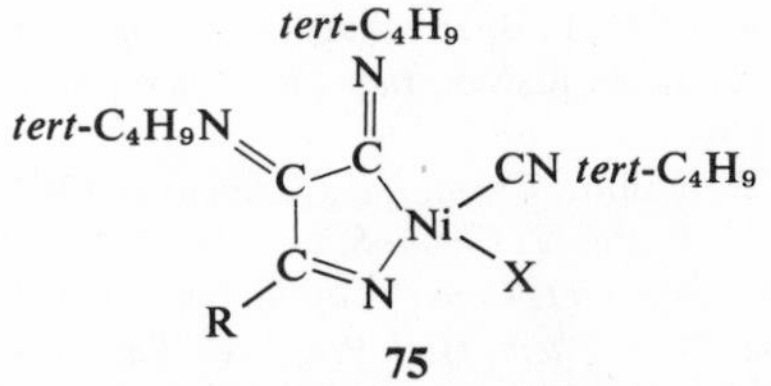

75

$$R = CH_3, \; X = I; \; R = C_6H_5CH_2, \; X = Br$$

isonitriles into a polymer containing $RN{=}C$ as repeating unit (108). Nickel tetracarbonyl, tetrakistrifluorophosphine nickel, and nickelocene also act as catalysts for the polymerization of isonitriles (175, 178). Isocyanates are catalytically polymerized and trimerized by tetrakisligand nickel complexes (239; see also 165, P4).

References

1. Y. Y. Aliev, I. B. Romanova, and L. K. Freidlin, *Uzb. Khim. Zh.* p. 54 (1961); *Chem. Abstr.* **57**, 8413 (1962).
2. M. Almasi, L. Szabo, S. Farkas, F. Kacso, O. Vegh, and I. Muresan, *Acad. Repub. Pop. Rom. Stud. Cercet. Chim.* **8**, 509 (1960).
3. M. Almasi and L. Szabo, *Acad. Repub. Pop. Rom. Stud. Cercet. Chim.* **8**, 519 (1960).
4. H. Adkins and R. W. Rosenthal, *J. Amer. Chem. Soc.* **72**, 4550 (1950).
5. Y. Y. Aliev, I. B. Romanova, and L. K. Freidlin, *Uzb. Khim. Zh.* **6**, 58 (1962); *Chem. Abstr.* **59**, 3882 (1963).
6. Y. Y. Aliev and I. B. Romanova, *Neftekhim. Mater. Sredneaziat. Soveshch., 1962* p. 204 (1963); *Chem. Abstr.* **61**, 6913 (1964).
7. Y. Y. Aliev and Y. I. Isakov, *Issled. Miner. Rast. Syriya Uzb.* p. 95 (1962); *Chem. Abstr.* **59**, 3768 (1963).
8. Y. Y. Aliev, I. B. Romanova, and L. K. Freidlin, *Uzb. Khim. Zh.* p. 72 (1960); *Chem. Abstr.* **55**, 10444 (1961).
9. Y. Y. Aliev, I. B. Romanova and L. K. Freidlin, *Uzb. Khim. Zh.* **6**, 67 (1962); *Chem. Abstr.* **58**, 5492 (1963).
10. G. Ayrey, C. W. Bird, E. M. Briggs, and A. F. Harmer, *Organometal. Chem. Syn.* **1**, 187 (1970/1971).
11. M. H. Bigelow, *Chem. Eng. News* **25**, 1038 (1947).
12. S. K. Bhattacharyya and S. N. Nag, *J. Appl. Chem.* **12**, 182 (1962).
13. S. K. Bhattacharyya and S. K. Palit, *J. Appl. Chem.* **12**, 174 (1962).
14. S. K. Bhattacharyya and S. Sourirajan, *J. Sci. Ind. Res. Sect.* **II-B** p. 123 (1952).
15. S. K. Bhattacharyya and A. K. Sen, *J. Appl. Chem.* **13**, 498 (1963).
16. C. W. Bird, R. C. Cookson, and J. Hudec, *Chem. Ind. (London)* p. 20 (1960).
17. C. W. Bird and J. Hudec, *Chem. Ind. (London)* p. 570 (1959).
18. S. K. Bhattacharyya and D. P. Bhattacharyya, *J. Appl. Chem.* **16**, 18 (1966).
19. S. K. Bhattacharyya and A. K. Sen, *Ind. Eng. Chem. Process Des. Develop.* **3**, 169 (1964).
20. C. W. Bird and E. M. Hollins, *Chem. Ind. (London)* p. 1362 (1964).
21. C. W. Bird and E. M. Briggs, *J. Chem. Soc., C* p. 1265 (1967).
22. S. K. Bhattacharyya, *Proc. Int. Congr. Catal., 2nd, 1960* Vol. 2 p. 2401 (1961).
23. S. K. Bhattacharyya, *Phys. Chem. High Pressures, Pap. Symp. 1962* p. 202 (1963).
24. M. H. Bigelow, *Chem. Corps J.* **1**, 6 (1947).

223. L. Cassar and M. Foa, *J. Organometal. Chem.* **51**, 381 (1973).
224. W. W. Prichard, *J. Amer. Chem. Soc.* **78**, 6137 (1956).
225. W. F. Beckert and J. U. Lowe, *J. Org. Chem.* **32**, 1215 (1967).
226. Y. Hirota, M. Ryang, and S. Tsutsumi, *Tetrahedron Lett.* p. 1531 (1971).
227. E. Yoshisato, M. Ryang, and S. Tsutsumi, *J. Org. Chem.* **34**, 1500 (1969).
228. M. Ryang, Y. Toyoda, S. Murai, N. Sonoda, and S. Tsutsumi, *J. Org. Chem.* **38**, 62 (1973).
229. R. J. Angelici, *Accounts Chem. Res.* **5**, 335 (1972).
230. K. M. Song, Y. Sawa, M. Ryang, and S. Tsutsumi, *Bull. Chem. Soc. Jap.* **38**, 330 (1965).
231. I. Rhee, M. Ryang, and S. Tsutsumi, *Tetrahedron Lett.* p. 4593 (1969).
232. E. J. Corey and L. S. Hegedus, *J. Amer. Chem. Soc.* **91**, 4926 (1969).
233. Y. Sawa, I. Hashimoto, M. Ryang, and S. Tsutsumi, *J. Org. Chem.* **33**, 2159 (1968).
234. S. Fukuoka, M. Ryang, and S. Tsutsumi, *J. Org. Chem.* **33**, 2973 (1968).
235. S. Fukuoka, M. Ryang, and S. Tsutsumi, *J. Org. Chem.* **35**, 3184 (1970).
236. H. Bock and K. L. Kompa, *Angew. Chem.* **78**, 114 (1966).
237. S. Fukuoka, M. Ryang, and S. Tsutsumi, *Tetrahedron Lett.* p. 2553 (1970).
238. Y. Yamamoto and H. Yamazaki, *Coord. Chem. Rev.* **8**, 225 (1972).
239. T. Kashiwagi, M. Hidai, Y. Uchida, and A. Misono, *J. Polym. Sci., Part B* **8**, 173 (1970).
240. H. O. Jones and H. S. Tasker, *J. Chem. Soc., London* p. 1904 (1909).
241. V. P. Nechiporenko, I. A. Korneeva, A. D. Treboganov, A. B. Prokkorov, G. I. Myagkova, and N. A. Preobrazhenskii, *Uch. Zap. Mosk. Inst. Tonkoi Khim. Tekhnol.* **1**, 76 (1970); *Chem. Abstr.* **76**, 13706 (1972).
242. C. E. Berkoff, R. C. Cookson, J. Hudec, D. W. Jones, and R. O. Williams, *J. Chem. Soc., London* p. 194 (1965).
243. W. Reppe, *Justus Liebigs Ann. Chem.* **582**, 116 (1953).
244. C. W. Bird and E. M. Briggs, *J. Organometal. Chem.* **69**, 311 (1974).
245. M. F. Semmelhack and R. D. Stauffer, *Tetrahedron Lett.* p. 2667 (1973).
246. V. P. Nechiporenko, A. D. Treboganov. V. P. Chernova, G. I. Myagkova, and N. A. Preobrazhenskii, *J. Org. Chem. USSR* **9**, 236 (1973).
247. V. P. Nechiporenko, A. A. Glazkov, G. I. Myagkova, and R. P. Evstigneeva, *J. Org. Chem. USSR* **9**, 241 (1973).
248. G. P. Chiusoli, G. Cometti, and V. Bellotti, *Gazz. Chim. Ital.* **103**, 569 (1973).

Patents

P1. D. F. Friederich (B.A.S.F.), Ger. Patent 933,148 (1955); *Chem. Abstr.* **52**, 19952 (1958).
P2. R. D. Anderson and E. M. Smolin (American Cyanamid Co.), U.S. Patent 3,025,319 (1962); *Chem. Abstr.* **57**, 11027 (1962).
P3. E. M. Smolin and R. D. Anderson (American Cyanamid Co.), U.S. Patent 3,025,322 (1962); *Chem. Abstr.* **57**, 11025 (1962).
P4. F. E. Drummond (Basic Research Corp.), U.S. Patent 3,070,618 (1962); *Chem. Abstr.* **59**, 9886 (1963).
P5. B.A.S.F., Brit. Patent 802, 544 (1958); *Chem. Abstr.* **53**, 7016 (1959).
P6. W. Reppe and A. Magin (B.A.S.F.), Brit. Patent 943,721 (1963); *Chem. Abstr.* **60**, 5340 (1969).

187. G. P. Chiusoli, M. Ferraris, and S. Merzoni, *Proc. Int. Conf. Coord. Chem. 12th*, (1969) quoted in ref. 146 p. 97.
188. T. Kunieda, T. Tamura, and T. Takizawa, *Chem. Commun.* p. 885 (1972).
189. W. Keim, Ph.D. Thesis, Technische Hochschule Aachen, 1968.
190. K. Fischer, K. Jonas, P. Misbach, R. Stabba, and G. Wilke, *Angew. Chem.* **85**, 1002 (1973).
191. P. Misbach, Ph.D. Thesis, University of Bochum, 1969.
192. K. W. Barnett, *J. Organometal. Chem.* **21**, 477 (1970).
193. R. G. Miller, D. R. Fahey, and D. P. Kuhlman, *J. Amer. Chem. Soc.* **90**, 6248 (1968).
194. M. C. Gallazzi, T. L. Hanlon, G. Vitulli, and L. Porri, *J. Organometal. Chem.* **33**, C45 (1971).
195. C. Tieghi and M. Zocchi, *J. Organometal. Chem.* **57**, C90 (1973).
196. E. J. Corey, L. S. Hegedus, and M. F. Semmelhack, *J. Amer. Chem. Soc.* **90**, 2417 (1968).
197. M. F. Semmelhack, *Org. React.* **19**, 117 (1972) (see p. 125).
198. N. L. Bauld, *Tetrahedron Lett.* p. 1841 (1963).
199. N. L. Bauld, *J. Amer. Chem. Soc.* **84**, 4345 (1962).
200. J. J. Drysdale and D. D. Coffman, *J. Amer. Chem. Soc.* **82**, 5111 (1960).
201. S. Fukuoka, M. Ryang, and S. Tsutsumi, *J. Org. Chem.* **36**, 2721 (1971).
202. E. J. Corey and L. S. Hegedus, *J. Amer. Chem. Soc.* **91**, 1233 (1969).
203. I. Hashimoto, M. Ryang, and S. Tsutsumi, *Tetrahedron Lett.* p. 4567 (1970).
204. M. Ryang, Y. Sawa, S. N. Somasundaram, S. Murai, and S. Tsutsumi, *J. Organometal. Chem.* **46**, 375 (1972).
205. J. K. Crandall and W. J. Michaely, *J. Organometal. Chem.* **51**, 375 (1973).
206. G. P. Chiusoli, *Chim. Ind. (Milan)* **41**, 513 (1959).
207. P. J. Ashworth, G. H. Whitham, and M. C. Whiting, *J. Chem. Soc., London* p. 4633 (1957).
208. R. W. Rosenthal and L. H. Schwartzman, *J. Org. Chem.* **24**, 836 (1959).
209. R. A. Raphael, "Acetylene Compounds in Organic Synthesis," p. 138. Academic Press, New York, 1955.
210. C. W. Bird, "Transition Metal Intermediates in Organic Synthesis," p. 182. Academic Press, New York, 1967.
211. G. N. Schrauzer, *Chem. Ber.* **94**, 1891 (1961).
212. Y. Yamamoto, H. Yamazaki, and N. Hagihara, *J. Organometal. Chem.* **18**, 189 (1969).
213. H. Kröper, F. Wirth, and O. Huchler, *Angew. Chem.* **72**, 867 (1960).
214. V. I. Romanovskii and A. A. Artemev. *Zh. Vses. Khim. Obshchest.* **5**, 476 (1960); *Chem. Abstr.* **55**, 1506 (1961).
215. S. E. Yakushkina and N. V. Kislyakova, *Izv. Akad. Nauk SSSR, Otd. Khim. Nauk.* p. 1119 (1958); *Chem. Abstr.* **53**, 3134 (1959).
216. K. Yamamoto and K. Sato, *Bull. Chem. Soc. Jap.* **27**, 389 (1954).
217. S. Matsuda, S. Kikkawa, and K. Tsuchino, *Yuki Gosei Kagaku Kyokai Shi* **19**, 61 (1961); *Chem. Abstr.* **55**, 5408 (1961).
218. M. Nakayama and T. Mizoroki, *Bull. Chem. Soc. Jap.* **44**, 508 (1971).
219. M. Nakayama and T. Mizoroki, *Bull. Chem. Soc. Jap.* **43**, 569 (1970).
220. M. Nakayama and T. Mizoroki, *Bull. Chem. Soc. Jap.* **42**, 1124 (1969).
221. T. Mizoroki and M. Nakayama, *Bull. Chem. Soc. Jap.* **40**, 2203 (1967).
222. S. Yaroslavsky, *Chem. Ind. (London)* p. 765 (1965).

144. G. P. Chiusoli, *Corsi Semin. Chim.* No. 10, p. 77 (1968).
145. G. P. Chiusoli, *Chim. Ind.* (*Milan*) 43, 638 (1961).
146. G. P. Chiusoli, *Aspects Homogen. Catal.* 1, 77 (1970).
147. G. P. Chiusoli, *Gazz. Chim. Ital.* 89, 1332 (1959).
148. G. P. Chiusoli and S. Merzoni, *Chim. Ind.* (*Milan*) 45, 6 (1963).
149. G. P. Chiusoli, G. Bottaccio, and A. Cameroni, *Chim. Ind.* (*Milan*) 44, 131 (1962).
150. G. P. Chiusoli and G. Bottaccio, *Chim. Ind.* (*Milan*) 43, 1022 (1961).
151. G. P. Chiusoli and S. Merzoni, *Chim. Ind.* (*Milan*) 43, 259 (1961).
152. G. P. Chiusoli, *Chim. Ind.* (*Milan*) 43, 365 (1961).
153. G. P. Chiusoli, *Petrol. Mater. Prima Ind. Chim. Mod.* (*Relaz. Commun. G. Chim. Milan*) p. 248 (1961).
154. G. P. Chiusoli and G. Bottaccio, *Chim. Ind.* (*Milan*) 44, 1129 (1962).
155. G. P. Chiusoli and S. Merzoni, *Chem. Commun.* p. 522 (1971).
156. G. P. Chiusoli, G. Cometti, and S. Merzoni, *Organometal. Chem. Syn.* 1, 439 (1972).
157. E. J. Corey, M. F. Semmelhack, and L. S. Hegedus, *J. Amer. Chem. Soc.* 90, 2416 (1968).
158. E. J. Corey and M. F. Semmelhack, *Tetrahedron Lett.* p. 6237 (1966).
159. E. J. Corey and M. Jautelat, *J. Amer. Chem. Soc.* 89, 3912 (1967).
160. W. T. Dent, R. Long, and G. H. Whitfield, *J. Chem. Soc. London* p. 1588 (1964).
161. N. Zelinsky, *J. Russ. Phys.-Chem. Soc.* 36, 339 (1904).
162. H. O. Jones, *Chem. News* 90, 144 (1904).
163. M. Dubini, G. P. Chiusoli, and F. Montino, *Tetrahedron Lett.* p. 1591 (1963).
164. M. Dubini, G. P. Chiusoli, and F. Montino, *Chim. Ind.* (*Milan*) 45, 1237 (1963).
165. P. Hong, K. Sonogashira, and N. Hagihara, *Tetrahedron Lett.* p. 1633 (1970).
166. E. O. Fischer and G. Bürger, *Z. Naturforsch.* 17b, 484 (1962).
167. M. Foa, L. Cassar, and M. T. Venturi, *Tetrahedron Lett.* p. 1357 (1968).
168. F. L. Benton, M. C. Voss, and P. A. McCusker, *J. Amer. Chem. Soc.* 67, 82 (1945).
169. F. Guerrieri and G. P. Chiusoli, *Chem. Commun.* p. 781 (1967).
170. F. Guerrieri and G. P. Chiusoli, *J. Organometal. Chem.* 15, 209 (1968).
171. R. F. Heck, J. C. W. Chien, and D. S. Breslow, *Chem. Ind.* (*London*) p. 986 (1961).
172. I. Hashimoto, M. Ryang, and S. Tsutsumi, *Tetrahedron Lett.* p. 3291 (1969).
173. I. Hashimoto, N. Tsuruta, M. Ryang, and S. Tsutsumi, *J. Org. Chem.* 35, 3748 (1970).
174. J. B. Mettalia and E. H. Specht, *J. Org. Chem.* 32, 3941 (1967).
175. Y. Yamamoto, T. Takizawa, and N. Hagihara, *Nipon Kagaku Zasshi* 87, 1355 (1966); *Chem. Abstr.* 67, 32995 (1967).
176. M. Ryang, S. K. Myeong, Y. Sawa, and S. Tsutsumi, *J. Organometal. Chem.* 5, 305 (1966).
177. M. Ryang and S. Tsutsumi, *Synthesis* p. 55 (1971).
178. R. D. Johnston, F. Basolo, and R. G. Pearson, *Inorg. Chem.* 10, 247 (1971).
179. Y. Suzuki and T. Takizawa, *Chem. Commun.* p. 837 (1972).
180. G. Wilke *et al.*, *Angew. Chem.* 78, 157 (1966).
181. E. Yoshisato and S. Tsutsumi, *J. Org. Chem.* 33, 869 (1968).
182. G. P. Chiusoli, *Chim. Ind.* (*Milan*) 41, 513 (1959).
183. L. Cassar, G. P. Chiusoli, and F. Guerrieri, *Synthesis* p. 509 (1973).
184. G. P. Chiusoli and G. Cometti, *Chem. Commun.* p. 1015 (1972).
185. G. P. Chiusoli, *Plenary Lect. Int. Congr. Pure Appl. Chem. 23rd, 1971* Vol. 6, p. 169 (1971).
186. L. Cassar, G. P. Chiusoli, and M. Foa, *Chim. Ind.* (*Milan*) 50, 515 (1968).

103. L. I. Smith and H. H. Hoehn, *J. Amer. Chem. Soc.* **63**, 1180 (1941).
104. Y. T. Eidus, A. L. Lapidus, K. J. Puzitskii, and B. K. Nefedov, *Russ. Chem. Rev.* **42**, 199 (1973).
105. C. W. Bird, *J. Organometal. Chem.* **47**, 281 (1973).
106. J. W. Copenhaver and M. H. Bigelow, "Acetylene and Carbon Monoxide Chemistry," Van Nostrand-Reinhold, Princeton, New Jersey, 1949.
107. F. L. Resen, *Oil Gas J.* **51**, 92 (1953).
108. S. Otsuka, A. Nakamura, and T. Yoshida, *J. Amer. Chem. Soc.* **91**, 7196 (1969).
109. W. Reppe, O. Schlichting, K. Klager, and T. Toepel, *Justus Liebigs Ann. Chem.* **560** 1 (1948).
110. H. Kröper, *in* "Houben-Weyl (Methoden der Org. Chem.)," (E. Müller, ed.) Vol. IV, Part 2, p. 359 (1955).
111. H. Kröper, *in* "Ullmann's Encyklopädie der technischen Chemie," Vol. 5, p. 121. Verlag Chemie (1954).
112. D. R. Levering and A. L. Glasebrook, *J. Org. Chem.* **23**, 1836 (1958).
113. G. P. Chiusoli, *Accounts Chem. Res.* **6**, 422 (1973).
114. S. Sourirajan, *Advan. Catal.* **9**, 618 (1957).
115. J. Falbe, H.-J. Schulze-Steinen, and F. Korte, *Chem. Ber.* **98**, 886 (1965).
116. S. Horiie and S. Murahashi, *Bull. Chem. Soc. Jap.* **33**, 88 (1960); S. Murahashi and S. Horiie, *J. Amer. Chem. Soc.* **78**, 4816 (1956).
117. W. L. Mosby, *Chem. Ind. (London)* p. 17 (1957).
118. Y. Inoue, M. Hidai, and Y. Uchida, *Chem. Lett.* p. 1119 (1972).
119. B. M. Trost and F. Chen, *Tetrahedron Lett.* p. 2603 (1971).
120. P.-T. Ho, S. Oida, and K, Wiesner, *Chem. Commun.* p. 883 (1972).
121. W. L. Gilliland and A. A. Blanchard, *J. Amer. Chem. Soc.* **48**, 410 (1926).
122. R. Baker, B. N. Blackett, and R. C. Cookson, *Chem. Commun.* p. 802 (1972).
123. B. Bogdanović, P. Heimbach, M. Kröner, G. Wilke, E. G. Hoffmann, and J. Brandt, *Justus Liebigs Ann. Chem.* **727**, 143 (1969).
124. H. Breil and G. Wilke, *Angew. Chem.* **82**, 355 (1970).
125. L. Cassar and G. P. Chiusoli, *Tetrahedron Lett.* p. 2805 (1966).
126. L. Cassar and G. P. Chiusoli, *Tetrahedron Lett.* p. 3295 (1965).
127. L. Cassar and G. P. Chiusoli, *Chim. Ind. (Milan)* **48**, 323 (1966).
128. L. Cassar, G. P. Chiusoli, and M. Foa, *Tetrahedron Lett.* p. 285 (1967).
129. L. Cassar and M. Foa, *Chim. Ind. (Milan)* **51**, 673 (1969).
130. G. P. Chiusoli, G. Bottaccio, and C. Venturello, *Tetrahedron Lett.* p. 2875 (1965).
131. G. P. Chiusoli and G. Bottaccio, *Chim. Ind. (Milan)* **47**, 165 (1965).
132. G. P. Chiusoli, *Atti Accad. Naz. Lincei, Cl. Sci. Fis., Mat. Natur., Rend.* **26**, 790 (1959).
133. G. P. Chiusoli, *Chim. Ind. (Milan)* **41**, 762 (1959).
134. G. P. Chiusoli, *Chim. Ind. (Milan)* **41**, 503 (1959).
135. G. P. Chiusoli, *Chim. Ind. (Milan)* **41**, 506 (1959).
136. G. P. Chiusoli, *Angew. Chem.* **72**, 74 (1960).
137. G. P. Chiusoli and L. Cassar, *Angew. Chem.* **79**, 177 (1967).
138. G. P. Chiusoli, G. Bottaccio, and C. Venturello, *Chim. Ind. (Milan)* **48**, 107 (1966).
139. G. P. Chiusoli and S. Merzoni, *Z. Naturforsch.* **17b**, 850 (1962).
140. G. P. Chiusoli, S. Merzoni, and G. Mondelli, *Chim. Ind. (Milan)* **46**, 743 (1964).
141. G. P. Chiusoli, S. Merzoni, and G. Mondelli, *Tetrahedron Lett.* p. 2777 (1964).
142. G. P. Chiusoli, *Riv. Combust.* **15**, 647 (1961).
143. G. P. Chiusoli, M. Dubini, M. Ferraris, F. Guerrieri, S. Merzoni, and G. Mondelli, *J. Chem. Soc., C* p. 2889 (1968).

P7. B.A.S.F., Brit. Patent 775, 689 (1957); *Chem. Abstr.* **52**, 15568 (1958).

P8. W. Reppe and N. von Kutepow (B.A.S.F.), Brit. Patent 699,556 (1953); *Chem. Abstr.* **49**, 2484 (1955).

P9. B.A.S.F., Brit. Patent 714,659 (1954); *Chem. Abstr.* **50**, 6500 (1956).

P10. W. Reppe, N. von Kutepow, and W. Koelsch (B.A.S.F.), U.S. Patent 2,768,968 (1956); *Chem. Abstr.* **51**, 7405 (1957).

P11. W. Reppe, H. Friederich, E. Henkel, and H. Lautenschlager (B.A.S.F.), U.S. Patent 2,806,040 (1957); *Chem. Abstr.* **52**, 2891 (1958).

P12. W. Reppe, H. Friedrich, H. Lautenschlager, and H. Laib (B.A.S.F.), U.S. Patent 2,809,976 (1957); *Chem. Abstr.* **52**, 3854 (1958).

P13. W. Reppe and A. Magin (B.A.S.F.), U.S. Patent 3,396,191 (1968); *Chem. Abstr.* **69**, 76662 (1968).

P14. W. Reppe and R. Stadler (B.A.S.F.), U.S. Patent 3,023,237 (1962); *Chem. Abstr.* **58**, 5518 (1963).

P15. H. Lautenschlager, H. Friedrich, E. Henkel, N. von Kutepow, W. Himmele, and P. Raff (B.A.S.F.), U.S. Patent 2,845,451 (1958); *Chem. Abstr.* **53**, 1150 (1959).

P16. H. Lautenschlager and H. Friedrich (B.A.S.F.), Ger. Patent 1,058,048 (1959); *Chem. Abstr.* **55**, 8294 (1961).

P17. W. Himmele, K. Bittler, P. Hornberger, J. Gnad, and D. Fischer (B.A.S.F.), Ger. Offen. 2,018,160 (1971); *Chem. Abstr.* **76**, 24708 (1972).

P18. H. Krzikalla and E. Woldan (B.A.S.F.), Ger. Patent 863,800 (1953); *Chem. Z.* p. 2674 (1953).

P19. W. Reppe and A. Simon (B.A.S.F.), Ger. Patent 857,635 (1952); *Chem. Abstr.* **50**, 1915 (1956).

P20. R. Stadler, F. Becke, and H. Pirzer (B.A.S.F.) Ger. Patent 1,152,409 (1963); *Chem. Abstr.* **60**, 422 (1964).

P21. W. Reppe, H. H. Friederich, H. Lautenschlager, and H. Laib (B.A.S.F.), Ger. Patent 944,654 (1956); *Chem. Abstr.* **53**, 9064 (1959).

P22. W. Reppe and W. Schweckendiek (B.A.S.F.), Ger. Patent 805, 641 (1951); *Chem. Abstr.* **47**, 602 (1953).

P23. S. Kunichika and Y. Sakakibara (Chiyoda Chem. Eng. and Constr. Co., Ltd.), Brit. Patent 1,110,565 (1968); *Chem. Abstr.* **69**, 51643 (1968).

P24. O. Albrecht and A. Maeda (Ciba Ltd.), U.S. Patent 2,653,969 (1953); *Chem. Abstr.* **49**, 4707 (1955).

P25. K. Ohashi and S. Suzuki (East Asia Syn. Chem. Ind. Co.), Jap. Patent 1581('53) (1953); *Chem. Abstr.* **48**, 12169 (1954).

P26. H. J. Hagemeyer (Eastman Kodak Co.), U.S. Patent 2,739,169 (1956); *Chem. Abstr.* **50**, 16835 (1956).

P27. H. J. Hagemeyer (Eastman Kodak Co.), U.S. Patent 2,593,440 (1952); *Chem. Abstr.* **47**, 601 (1953).

P28. W. F. Gresham and R. E. Brooks (E.I. du Pont), Brit. Patent 631,001 (1949); *Chem. Abstr.* **44**, 4493 (1950).

P29. W.F. Gresham and R. E. Brooks (E. I. du Pont), U.S. Patent 2,497,303 (1950); *Chem. Abstr.* **44**, 4492 (1950).

P30. W. F. Gresham and R. E. Brooks (E. I. du Pont), U.S. Patent 2,448,368 (1948); *Chem. Abstr.* **43**, 669 (1949).

P31. W. F. Gresham and R. E. Brooks (E. I. du Pont), U.S. Patent 2,549,453 (1951); *Chem. Abstr.* **45**, 8551 (1951).

P32. W. F. Gresham (E. I. du Pont), U.S. Patent 2,432,474 (1947); *Chem. Abstr.* **42**, 1961 (1948).

P33. W. W. Prichard (E. I. du Pont), U.S. Patent 2,769,003 (1956); *Chem. Abstr.* **51**, 7412 (1957).

P34. R. A. Hines (E. I. du Pont), U.S. Patent 2,809,991 (1957); *Chem. Abstr.* **52**, 3856 (1957).

P35. A. T. Larson (E.I. du Pont), U.S. Patent 2,448,375 (1948); *Chem. Abstr.* **43**, 670 (1949).

P36. W. W. Prichard (E. I. du Pont), U.S. Patent 2,841,591 (1958); *Chem. Abstr.* **52**, 20197 (1958).

P37. W. A. Raczynski (Hercules Powder Co.), U.S. Patent 2,738,368 (1956); *Chem. Abstr.* **50**, 15577 (1956).

P38. M. L. Noble (I.C.I.), Brit Patent 713,325 (1954); *Chem. Abstr.* **50**, 6500 (1956).

P39. I. G. Farbenind, A.G., Fr. Patent 930,368 (1948); *Chem. Abstr.* **43**, 5412 (1949).

P40. W. Reppe and H. Kröper (I. G. Farbenind, A.G.), Ger. Patent 765,969 (1953); *Chem. Abstr.* **51**, 13904 (1957).

P41. S. Kunichika and Y. Sakakibara (Idemitsu Kosan Co. Ltd.), Ger. Offen. 1,952,976 (1970); *Chem. Abstr.* **73**, 26061 (1970).

P42. T. Kiryu and Y. Shimokawa (Chiyoda Chem. Eng. Constr. Co.), Jap Patent 73 42,624 (1973); *Chem. Abstr.* **81**, 64285 (1974).

P43. E. R. H. Jones and M. C. Whiting, Brit. Patent 640,489 (1950); *Chem. Abstr.* **45**, 642 (1951).

P44. M. Tanaka *et al.* (Mitsui Chem. Ind. Co.), Jap. Patent 4918('53) (1953); *Chem. Abstr.* **49**, 6992 (1955).

P45. M. Tanaka *et al.* (Mitsui Chem. Ind. Co.), Jap. Patent 4919('53) (1953); *Chem. Abstr.* **49**, 6992 (1955).

P46. K. Yamamoto *et al.* (Mitsui Chem. Ind. Co.), Jap. Patent 8268('54) (1954); *Chem. Abstr.* **50**, 13987 (1956).

P47. J. Happel and H. Blanck (National Lead Co.), U.S. Patent 3,420,753 (1969); *Chem. Abstr.* **70**, 67650 (1969).

P48. J. Happel, J. H. Blanck, and Y. Sakakibara (National Lead Co.), U.S. Patent 3,496,221 (1970); *Chem. Abstr.* **72**, 31256 (1970).

P49. National Lead Co., Brit. Patent 887,433 (1962); *Chem. Abstr.* **57**, 11027 (1962).

P50. National Lead Co., Brit. Patent 997,923 (1965); Conc. Belg. Patent 639,260 (1964); *Chem. Abstr.* **62**, 9019 (1965).

P51. J. Berthoux, J. P. Martinaud, and R. Poilblanc (Progil S.A.), Ger. Offen. 2,039,938 (1971); *Chem. Abstr.* **74**, 99461 (1971).

P52. W. Reppe and A. Magin, U.S. Patent 2,577,208 (1951); *Chem. Abstr.* **46**, 6143 (1952).

P53. H. T. Neher, E. A. Specht and A. Neuman (Rohn & Haas Co.), U.S. Patent 2,582,911 (1952); *Chem. Abstr.* **46**, 11231 (1952).

P54. E. H. Specht, A. Neuman, and H. T. Neher (Rohm & Haas Co.), U.S. Patent 2,613,222 (1952); *Chem. Abstr.* **47**, 11226 (1953).

P55. P. M. Bakker (Shell Int. Res. Maatsch.), Ger. Offen. 2,057,521 (1971); *Chem. Abstr.* **75**, 88125 (1971).

P56. E. Konto, K. Takao, and E. Uchida (Toa Gosei Chem. Ind. Co.), Jap. Patent 70 38,129 (1970); *Chem. Abstr.* **75**, 98679 (1971).

P57. R. W. Rosenthal (Texas Co.), U.S. Patent 2,652,413 (1953); *Chem. Abstr.* **48**, 5209 (1954).

P58. J. T. Dunn (Union Carbide Corp.), Brit. Patents 879,307, 879,308, and 879,346 (1961); *Chem. Abstr.* **58**, 12427 (1963).

P59. J. T. Dunn and W. R. Proops (Union Carbide Corp.), Brit. Patent 850,509 (1960); *Chem. Abstr.* **55**, 8294 (1961).

P60. J. T. Dunn (Union Carbide Corp.), Brit. Patent 879,306 (1961); *Chem. Abstr.* **59**, 1494 (1963).

P61. J. T. Dunn (Union Carbide Corp.), Brit. Patents 879,009 and 879,010 (1959); *Chem. Abstr.* **54**, 1495 (1963).

P62. J. T. Dunn (Union Carbide Corp.), U.S. Patent 3,013,067 (1961); *Chem. Abstr.* **56**, 9973 (1962).

P63. H. H. Mathews (Air Reduction Co., Inc.), U.S. Patent 2,903,479 (1959); *Chem. Abstr.* **54**, 2173 (1960).

P64. H. Lautenschlager and H. Friederich (B.A.S.F.), U.S. Patent 2,886,591 (1959); *Chem. Abstr.* **53**, 19884 (1959).

P65. H. T. Neher, E. H. Specht, and E. J. Kelley (Rohm & Haas Co.), U.S. Patent 2,888,480 (1959); *Chem. Abstr.* **53**, 21670 (1959).

P66. I. Dakle, B. Arsizio, and L. Corsi (Montecatini Soc. Gen.), U.S. Patent 2,881,205 (1959); *Chem. Abstr.* **53**, 15982 (1959).

P67. A. Neuman, H. T. Neher, and E. H. Specht (Rohm & Haas Co.), U.S. Patent 2,778,848 (1957); *Chem. Abstr.* **51**, 8129 (1957).

P68. E. H. Specht, A. Neuman, and H. T. Neher (Rohm & Haas Co.), U.S. Patent 2,773,063 (1956); *Chem. Abstr.* **51**, 8778 (1957).

P69. W. W. Prichard (E. I. du Pont), U.S. Patent 2,729,673 (1956); *Chem. Abstr.* **51**, 484 (1957).

P70. N. T. Gehshan and E. H. Specht (Rohm & Haas Co.), U.S. Patent 2,990,403 (1961); *Chem. Abstr.* **56**, 9933 (1962).

P71. W. Reppe (B.A.S.F.), Ger. Patent 855,110 (1952); *Chem. Abstr.* **50**, 10131 (1956).

P72. W. Reppe (B.A.S.F.), Ger. Patent 854,948 (1952); *Chem. Abstr.* **50**, 10132 (1956).

P73. B.A.S.F., Brit. Patent 779,277 (1957); *Chem. Abstr.* **52**, 1206 (1958).

P74. K. Yamamoto *et al.* (Mitsui Chem. Ind. Co.), Jap. Patent 612('55) (1955); *Chem. Abstr.* **51**, 1248 (1957).

P75. W. Reppe and R. Stadler (B.A.S.F.), Ger. Patent 942,809 (1956); *Chem. Abstr.* **50**, 16833 (1956).

P76. H. Lautenschlager and H. H. Friederich (B.A.S.F.), Ger. Patent 1,046,030 (1958); *Chem. Abstr.* **55**, 2485 (1961).

P77. J. T. Dunn (Union Carbide Co.), U.S. Patent 3,019, 256 (1962); *Chem. Abstr.* **56**, 12747 (1962).

P78. W. Reppe and A. Magin (B.A.S.F.), Ger. Patent 1,215,139 (1966); *Chem. Abstr.* **65**, 8766 (1966).

P79. Dow Chemical Co., Brit. Patent 805,259 (1958); *Chem. Abstr.* **53**, 9064 (1959).

P80. J. T. Dunn (Union Carbide Co.), U.S. Patent 2,992, 270 (1961); *Chem. Abstr.* **56**, 329 (1962).

P81. J. T. Dunn and W. R. Proops (Union Carbide Corp.), U.S. Patent 2,966,510 (1960); *Chem. Abstr.* **55**, 22134 (1961).

P82. W. Reppe and H. Kröper (B.A.S.F.), Ger. Patent 863,194 (1953); *Chem. Abstr.* **48**, 1425 (1954).

P83. K. Yamamoto *et al.* (Mitsui Chem. Ind. Co.), Jap. Patent 1031('52) (1952); *Chem. Abstr.* **48**, 1426 (1954).

P84. W. Reppe, F. Reicheneder, G. Stengel, and A. Zieger (B.A.S.F.), U.S. Patent 2,925,436 (1960); *Chem. Abstr.* **54**, 17270 (1960).

P85. E. M. Smolin and B. J. Luberoff (American Cyanamid Co.), U.S. Patent 2,882,299 (1959); Chem. Abstr. 53, 15983 (1959).

P86. B. J. Luberoff (American Cyanamid Co.), U.S. Patent 2,882,297 (1959); Chem. Abstr. 53, 15982 (1959).

P87. G. A. Elliott and S. A. Furbush (Allied Chemical Corp.), U.S. Patent 3,002,016 (1961); Chem. Abstr. 56, 9971 (1962).

P88. B.A.S.F., Brit. Patent 824,520 (1959); Chem. Abstr. 54, 7563 (1960).

P89. K. Yamamoto and M. Oku (Mitsui Chem. Ind. Co.), Jap. Patent 5968('51) (1951); Chem. Abstr. 47, 9997 (1953).

P90. W. Reppe, W. Schweckendiek, and H. Friederich (B.A.S.F.), U.S. Patent 2,738,364 (1956); Chem. Abstr. 50, 15578 (1956).

P91. C. Ujiie and T. Tsurutu (New Nippon Nitrogeneous Fertilizers Co.), Jap. Patent 1274('54) (1954); Chem. Abstr. 49, 11687 (1955).

P92. K. Yamamoto and K. Sato (Mitsui Chem. Ind. Co.), Jap. Patent 3763('56) (1956); Chem. Abstr. 51, 14788 (1957).

P93. K. Yamamoto and K. Sato (Mitsui Chem. Ind. Co.), Jap. Patent 3224('52) (1952); Chem. Abstr.. 48, 2763 (1954).

P94. W. Reppe, W. Schweckendiek, and K. Kroeper (B.A.S.F.), U.S. Patent 2,658,075 (1953); Chem. Abstr. 49, 11688 (1961).

P95. O.Hecht, E. Gassenmeier, and W. Reppe (B.A.S.F.), Ger. Patent 859, 611 (1952); Chem. Z. p. 5256 (1953).

P96. O. Hecht, E. Gassenmeier, and W. Reppe (B.A.S.F.), Ger. Patent 851,339 (1952); Chem. Z. p. 3475 (1953).

P97. O. Hecht, E. Gassenmeier, and W. Reppe (B.A.S.F.), Ger. Patent 857,634 (1952); Chem. Z. p. 2519 (1953).

P98. W. Reppe (B.A.S.F.), Ger. Patent 888,099 (1953); Chem. Z. p. 9194 (1955).

P99. E. B. McCall and P. J. S. Bain (Monsanto Chem. Ltd.), Brit. Patent 957, 957 (1964); Chem. Abstr. 61, 5563 (1964).

P100. J. W. H. McCoy and N. Swanson (Dow Chemical Co.), U.S. Patent 3,151,155 (1964); Chem. Abstr. 62, 448 (1965).

P101. W. Reppe and H. Kröper (B.A.S.F.), Ger. Patent 861,243 (1952); Chem. Z. p. 2493 (1954).

P102. W. Reppe and H. Kröper (B.A.S.F.), Ger. Patent 868,149 (1953); Chem. Z. p. 5255 (1953).

P103. W. Reppe and H. Kröper (B.A.S.F.), Ger. Patent 860,350 (1952); Chem. Z. p. 1587 (1954).

P104. W. Reppe and H. Kröper (B.A.S.F.), Ger. Patent 848,355 (1952); Chem. Z. p. 2218 (1954).

P105. W. Reppe and H. Kröper (B.A.S.F.), Ger. Patent 862,748 (1953); Chem. Abstr. 48, 10059 (1954).

P106. W. Reppe, N. von Kutepow, and H. Detzer (B.A.S.F.), Ger. Patent 1,006,849 (1957); Chem. Abstr. 53, 14945 (1959).

P107. H. Detzer, H. Metzger, and H. Urbach (B.A.S.F.), Belg. Patent 613,730 (1962); Chem. Abstr. 58, 455 (1963).

P108. W. Reppe and H. Kröper (B.A.S.F.), Ger. Patent 879,987 (1953); Chem. Abstr. 52, 11899 (1958).

P109. H. Kröper, N. von Kutepow, O. Huchler, W. Kölsch, and W. Himmele (B.A.S.F.), Ger. Patent 920,244 (1954); Chem. Abstr. 52, 13780 (1958).

P110. J. Schalch (Van Schaack Bros. Chem. Works), U.S. Patent 1,973,662 (1934); Chem. Abstr. 28, 6723 (1934).

P111. F. Codignola and M. Piacenza, Ital. Patent 431,407 (1948); *Chem. Abstr.* **44**, 1134 (1950).

P112. W. Reppe, H. Friederich, N. von Kutepow, and W. Morsch (B.A.S.F.), U.S. Patent 2,729,651 (1956); *Chem. Abstr.* **50**, 13081 (1956); *Chem. Z.* p. 6620 (1955).

P113. J. Kato, R. Iwanaga, and H. Wakamatsu (Ajinomoto Co. Inc.), Ger. Patent 1,135,884 (1962); *Chem. Abstr.* **58**, 4430 (1963).

P114. W. Reppe, H. Kröper, and N. von Kutepow (B.A.S.F.), Ger. Patent 879,988 (1953); *Chem. Z.* p. 5650 (1955).

P115. G. P. Chiusoli and G. Bottaccio (Montecatini Soc. Gen.), Brit. Patent 967,299 (1964); *Chem. Abstr.* **61**, 577 (1964).

P116. F. Montino (Montecatini Edison S.p.A.), Ger. Offen. 1,936,725 (1970); *Chem. Abstr.* 73, 77847 (1970).

P117. Montecatini Edison S.p.A., Ital. Patent 827,704 (1967); *Chem. Abstr.* 74, 141001 (1971).

P118. Montecatini Edison S.p.A., Ital. Patent 846,011 (1969); *Chem. Abstr.* 75, 129944 (1971).

P119. Montecatini Soc. Gen., Ital. Patent 789,552 (1957); *Chem. Abstr.* **72**, 21336 (1970).

P120. Montecatini-Edison S.p.A., Ital. Patent 815,179 (1968); *Chem. Abstr.* **71**, 80951 (1969).

P121. Montecatini-Edison S.p.A., Ital. Patent 792,602 (1967); *Chem. Abstr.* **71**, 3047 (1969).

P122. Montecatini Soc. Gen., Ital. Patent 788,039 (1967); *Chem. Abstr.* **73**, 76736 (1970).

P123. M. Dubini and F. Montino (Montecatini-Edison S.p.A.), Ger. Offen. 1,916,533 (1969); *Chem. Abstr.* **72**, 42834 (1970).

P124. G. P. Chiusoli (Montecatini Soc. Gen.), U.S. Patent 3,146,257 (1964); *Chem. Abstr.* **62**, 9015 (1965).

P125. G. P. Chiusoli and S. Merzoni (Montecatini Soc. Gen.), U.S. Patent 3,032,583 (1962); *Chem. Abstr.* **57**, 11030 (1962).

P126. Montecatini Soc. Gen., Brit. Patent 888,162 (1962); *Chem. Abstr.* **61**, 8195 (1964).

P127. G. P. Chiusoli, S. Merzoni, and G. Cometti (Montecatini Soc. Gen.) Ital. Patent 719,830 (1966); *Chem. Abstr.* **69**, 51638 (1968).

P128. G. P. Chiusoli and S. Merzoni (Montecatini Soc. Gen.), U.S. Patent 3,203,978 (1965); see *Chem. Abstr.* **59**, 11268 (1963).

P129. G. P. Chiusoli and S. Merzoni (Montecatini Soc. Gen.), Fr. Patent 1,358,900 (1964); *Chem. Abstr.* **61**, 14535 (1964).

P130. G. P. Chiusoli and S. Merzoni (Montecatini Soc. Gen.), Ital. Patent 675,616 (1964); *Chem. Abstr.* **64**, 3363 (1966).

P131. J. B. Mettalia and E. H. Specht (Rohm & Haas Co.), U.S. Patent 3,476,797 (1969); *Chem. Abstr.* **72**, 12135 (1970).

P132. G. P. Chiusoli and S. Merzoni (Montecatini Soc. Gen.), Brit. Patent 1,006,008 (1965); see *Chem. Abstr.* **60**, 13146 (1964).

P133. G. P. Chiusoli (Montecatini Soc. Gen.), U.S. Patent 3,025,320 (1962); *Chem. Abstr.* **57**, 9668 (1962).

P134. W. W. Prichard and G. E. Tabet (E. I. du Pont), U.S. Patent 2,565,462 (1951); *Chem. Abstr.* **46**, 2578 (1952).

P135. H. Bliss and P. W. Southworth, U.S. Patent 2,565,461 (1951); *Chem. Abstr.* **46**, 2577 (1952).

P136. W. H. Groombridge, Brit. Patent 621,520 (1949); *Chem. Abstr.* **43**, 6650 (1949).

P137. K. Yamamoto and K. Sato (Mitsui Chem. Ind. Co.), Jap. Patent 2424 (1952); *Chem. Abstr.* **48**, 2105 (1954).

P138. H. J. Leibu (E. I. du Pont), U.S. Patent 2,773,090 (1956); *Chem. Abstr.* **51**, 7418 (1957).

P139. G. E. Tabet (E. I. du Pont), U.S. Patent 2,565,463 (1951); *Chem. Abstr.* **46**, 2578 (1952).

P140. G. E. Tabet (E. I. du Pont), U.S. Patent 2,565,464 (1951); *Chem. Abstr.* **46**, 2578 (1952).

P141. H. Kröper, F. Wirth, and O. H. Huchler (B.A.S.F.), Ger. Patent 1,074,028 (1960); *Chem. Abstr.* **55**, 12364 (1961).

P142. H. J. Leibu (E. I. du Pont), U. S. Patent 2,640,071 (1953); *Chem. Abstr.* **48**, 5214 (1954).

P143. H. Dieterle and W. Eschenbach, German Patent 537,610 (1931); *Chem. Z.* p. 1155 (1932).

P144. B.A.S.F., Brit. Patent 815,835 (1959); *Chem. Abstr.* **54**, 1449 (1960).

P145. R. G. Linville (E. I. du Pont), U.S. Patent 2,517,898 (1950); *Chem. Abstr.* **45**, 2505 (1951).

P146. H. J. Leibu (E. I. du Pont), U.S. Patent 2,734,912 (1956); *Chem. Abstr.* **50**, 10775 (1956).

P147. H. Kröper, F. Wirth, and O. Huchler (B.A.S.F.), U.S. Patent 2,914,554 (1959); *Chem. Abstr.* **54**, 3322 (1960).

P148. W. W. Prichard (E. I. du Pont), U.S. Patent 2,680,751 (1954); *Chem. Abstr.* **49**, 6308 (1955).

P149. W. W. Prichard (E. I. du Pont), U.S. Patent 2,680,750 (1954); *Chem. Abstr.* **49**, 6308 (1955).

P150. W. W. Prichard (E. I. du Pont), U.S. Patent 2,696,503 (1954); *Chem. Abstr.* **49**, 15966 (1955).

P151. G. E. Tabet (E. I. du Pont), U.S. Patent 2,691,670 (1954); *Chem. Abstr.* **49**, 14806 (1955).

P152. H. Kröper, F. Wirth, and O. H. Huchler (B.A.S.F.), Ger. Patent 1,066,574 (1959); *Chem. Abstr.* **55**, 12362 (1961).

P153. T. Mizoroki and M. Nakayama (Jap. Bur. Ind. Technol.), Jap. Patent 71 11,016 (1971); *Chem. Abstr.* **75** 48706 (1971).

P154. H. Fernholz and L. Schläfer (Farbwerke Hoechst A.G.), Ger. Patent 1,280,850 (1968); *Chem. Abstr.* **70**, 19604 (1969).

P155. M. Nakayama (Jap. Bur. Ind. Technol.), Jap. Kokai 72 29,315 (1972); *Chem. Abstr.* **78**, 59001 (1973).

P156. F. Röhrscheid (Farbwerke Hoechst A.G.), Ger. Offen. 2,140,644 (1973); *Chem. Abstr.* **78**, 124047 (1973).

P157. J. B. Mettalia and E. H. Specht (Rohm & Haas Co.), Fr. Patent 1,507,657 (1967); *Chem. Abstr.* **70**, 19597 (1969).

P158. W. Reppe and W. Schlenck (B.A.S.F.), Ger. Patent 753,618 (1940).

P159. W. Reppe (B.A.S.F.), Ger. Patent 855,845 (1952); *Chem. Z.* p. 8693 (1953)

P160. H. J. Passino (M. W. Kellogg Co.), U.S. Patent 2,675,301 (1954); *Chem. Abstr.* **48**, 9636 (1954).

P161. H. J. Passino (M. W. Kellogg Co.), U.S. Patent 2,717,201 (1955); *Chem. Abstr.* **50**, 2131 (1956).

P162. H. Hohenschutz, D. Franz, H. Buelow, and G. Dinkhauser (B.A.S.F.), Ger. Offen. 2,133,349 (1973); *Chem. Abstr.* **78**, 97133 (1973).

P163. H. Fernholz and D. Freudenberger (Farbwerke Hoechst A.G.), Ger. Offen. 2,217,534 (1973); *Chem. Abstr.* **80**, 26763 (1973).

Reviews

A selection of general review articles relevant to the material discussed in this chapter is listed below.

C. W. Bird, "Transition Metal Intermediates in Organic Synthesis." Chapters 7, 8, and 9. Academic Press, New York, 1967.

C. W. Bird, The synthesis of heterocyclic compounds via transition metal intermediates. *J. Organometal. Chem.* **47**, 281 (1973).

A. J. Chalk and J. F. Harrod, Catalysis by cobalt carbonyls. *Advan. Organometal. Chem.* **6**, 119 (1968).

Y. T. Eidus and K. V. Puzitskii, The catalytic synthesis of carboxylic acids and their esters from carbon monoxide, alkenes and alcohols. *Russ. Chem. Rev.* **33**, 438 (1964).

Y. T. Eidus, K. V. Puzitskii, A. L. Lapidus, and B. K. Nefedov, Carbonylation of mono-olefinic and mono-acetylenic hydrocarbons. *Russ. Chem. Rev.* **40**, 429 (1971).

Y. T. Eidus, A. L. Lapidus, K. V. Puzitskii, and B. K. Nefedov, Carbonylation of poly-unsaturated hydrocarbons and of saturated and unsaturated alcohols and halogeno-derivatives. *Russ. Chem. Rev.* **42**, 199 (1973).

J. Falbe, "Synthesen mit Kohlenmonoxyd." Springer-Verlag, Berlin and New York, (1967).

R. F. Heck, Organic synthesis via alkyl- and acylcobalt tetracarbonyls. *In* "Organic Synthesis via Metal Carbonyls" (I. Wender and P. Pino, eds.), Vol. I, p. 373. Wiley (Interscience), New York, 1968.

T. Mizoroki, Carbonylation reactions catalyzed by metal carbonyls. (In Japanese). *Yuki Gosei Kagaku Kyokai Shi* **28**, 696 (1970).

M. Orchin and W. Rupilius, On the mechanism of the oxo reaction. *Catal. Rev.* **6**, 85 (1972).

F. E. Paulik, Recent developments in hydroformylation catalysis. *Catl. Rev.* **6**, 49 (1972).

A. Rosenthal and I. Wender, Reactions of nitrogen compounds. *In* "Organic Syntheses via Metal Carbonyls" (I. Wender and P. Pino, eds.), Vol. I, p. 405 Wiley (Interscience), New York, 1968.

D. T. Thompson and R. Whyman, Carbonylation. *In* "Transition metals in homogeneous Catalysis" (G. N. Schrauzer, ed.), p. 149. Dekker, New York, 1971.

J. Tsuji, Organic syntheses by means of noble metal compounds. *Advan. Org. Chem.* **6**, 109 (1969).

J. Tsuji and K. Ohno, Decarbonylation reactions using transition metal compounds. *Synthesis* p. 157 (1969).

A. Wojcicki, Insertion reactions of transition metal-carbon σ-bonded compounds 1: Carbon monoxide insertion. *Advan. Organometal. Chem.* **11**, 87 (1973).

Subject Index

A

Catalyst Index

Nickel components are indexed under the name of the organic ligand, e.g., bis(cyclooctadiene)nickel and nickel tetracarbonyl are indexed as "cyclooctadiene complex" and "tetracarbonyl complex," respectively.

A

Acetylacetonate [Ni(acac)$_2$]
co-oligomerization of alkynes and allene, 121, 124
of alkynes and 1,3-dienes, 180–184
of alkynes and olefins, 114
of 1,3-dienes and olefins, 173–179, 189
of olefins, 15
hydrogenation of olefins, 67
hydrosilylation of alkynes, 106
isomerization of olefins, 58–60
oligomerization of acetylene, 95
of alkynes, 96, 98, 100, 107–111, 113
of butadiene, 134–137
of 1,3-dienes, 155
of olefins, 2, 6, 8, 10, 11, 19, 28–38
polymerization of allene, 117
of 1,3-dienes, 213, 216, 219, 221, 222, 229
of olefins, 52, 53
reaction of organic halides with organomagnesium complex, 286, 287
of organic halides with phosphites, 273
synthesis of trialkylaluminum, 5
telomerization of allene, 122
of 1,3-dienes, 190–195
Acetylene, dimerization of ethylene, 2
Acrylonitrile complex
co-oligomerization of acetylene and olefins, 114
of alkanes and olefins, 49
of olefins, 45, 46, 50

isomerization of olefins, 59
oligomerization of acetylene, 107, 108, 111, 112
of butadiene, 134, 136, 185–187
polymerization of allene, 123
of butadiene, 214
telomerization of 1,3-dienes, 188
Alkylbenzenesulfanato compound, oligomerization of olefins, 6
Alkyl complex
hydrosilylation, 69, 193
oligomerization of butadiene, 142
of olefins, 6, 28, 30, 33, 34
polymerization of acetylene, 107
of olefins, 53
Alkyl halide
oligomerization of butadiene, 161
of olefins, 7, 8
Alkynes, see also Acetylene, Phenylacetylene
isomerization of olefins, suppression, 4
oligomerization of butadiene, 142
Alkynyl complex, oligomerization of alkynes, 111
π-Allyl complex
carbonylation of allyl halides, 323
co-oligomerization of 1,3-dienes with olefins, 16, 17
of olefins, 16, 17
coupling of organic halides, 247, 249, 250, 253–255, 258–261, 263–266
isomerization of olefins, 55, 56, 58, 59
of phosphorus ylids, 66
oligomerization of alkynes, 96–99, 107–111

392